从企业级开发到云原生微服务 Spring Boot实战

汪云飞　沈永林　陈晓茜　编著

電子工業出版社
Publishing House of Electronics Industry
北京•BEIJING

内 容 简 介

本书以 Spring 5.2 和 Spring Boot 2.2 为基础，系统地讲解了在日常企业级开发和微服务开发中面临的大部分的问题，如函数式编程、Spring MVC、Spring Data、Spring Security、响应式编程、事件驱动、Spring Integration、Spring Batch、Spring Cloud、Kubernetes 与微服务等。本书内容由浅入深，适合有 Java 基础的初级程序员学习。同时，本书还介绍了较深的理论及原理知识，可供中、高级工程师提升使用。本书讲解的所有主题都附有实战案例，读者可快速将相关技术应用于工作实践中。

图书在版编目（CIP）数据

从企业级开发到云原生微服务：Spring Boot 实战 / 汪云飞，沈永林，陈晓茜编著. —北京：电子工业出版社，2020.1

ISBN 978-7-121-37792-1

Ⅰ. ①从… Ⅱ. ①汪… ②沈… ③陈… Ⅲ. ①JAVA 语言－程序设计 Ⅳ. ①TP312.8

中国版本图书馆 CIP 数据核字(2019)第 240137 号

责任编辑：安 娜
印　　刷：三河市良远印务有限公司
装　　订：三河市良远印务有限公司
出版发行：电子工业出版社
　　　　　北京市海淀区万寿路 173 信箱　　邮编：100036
开　　本：787×980　1/16　印张：31.5　字数：700 千字
版　　次：2020 年 1 月第 1 版
印　　次：2020 年 1 月第 1 次印刷
定　　价：108.00 元

凡所购买电子工业出版社图书有缺损问题，请向购买书店调换。若书店售缺，请与本社发行部联系，联系及邮购电话：（010）88254888，88258888。

质量投诉请发邮件至 zlts@phei.com.cn，盗版侵权举报请发邮件至 dbqq@phei.com.cn。

本书咨询联系方式：010-51260888-819，faq@phei.com.cn。

前　言

缘起

距离我的第一本书《Java EE 开发的颠覆者：Spring Boot 开发》的出版已经过去了四年，在这四年中，Spring Boot 从刚开始被关注到现在被广泛应用于企业级开发，我很荣幸参与了这一进程。

在这四年里，技术发生了许多的变化，微服务、云原生已经成为技术流行词，这也促使我们对技术的关注点从企业级开发逐渐向微服务、云原生应用转移。在规划本书时，本想将其作为《Java EE 开发的颠覆者：Spring Boot 开发》的第二版，但最终还是以一本新书推出，原因有二：

第一，内容完全重新编写，全部基于 Spring 5.2 和 Spring Boot 2.2 编写，只是部分章节名称和第一本书相同。

第二，添加了微服务、云原生应用方面的知识，做到让微服务、云原生应用落地。

第一本书收到了许多读者的好评，这给了我很大的信心再写一本关于最新的 Spring Boot 实战的书籍。为了能更多地照顾到应用开发的方方面面，最初我列出了 26 章，这让本书几乎无法完成。此时，我想起《人月神话》中的“第二系统效应”：在完成一个小型、优雅而成功的系统之后，人们倾向于对下一个计划有过度的期待，可能因此建造出一个巨大的、有各种特色的“怪兽系统”。

这正是我最初在写作本书时的心态，此时我将精力重新聚焦在 Spring Boot 在企业级应用和云原生应用，经过近半年的努力，终于让本书与大家见面。

特色

非常感谢各位读者对我的第一本书的厚爱，期待大家同样喜欢本书。本书特色如下：

◎ 由浅入深，适合初学者及各个级别的学习者学习。

◎ 无论是简单的技术点，还是复杂的技术点，都配有实战案例。

本书内容

第 1 章 初识 Spring Boot。以简单的方式新建 Spring Boot 应用，让初学者可以对 Spring Boot 有感性的认识。如果读者已经学习或使用过 Spring Boot，则可快速阅读或略过本章。

第 2 章 函数式编程。Spring 5.X 支持的 JDK 基线版本为 8，本书中的大量代码都涉及函数式编程的内容，响应式编程更是以函数式编程为基础，所以学好函数式编程会给后面的学习打下良好的基础。

第 3 章 Spring 5.X 基础。本章带领读者快速学习 Spring 5.2 常用的主要内容，为学习和理解 Spring Boot 打下坚实的基础。

第 4 章 深入 Spring Boot。本章首先讲解 Spring Boot 的运行原理，然后讲解 Spring Boot 2.2 的核心内容。

第 5 章 Spring Web MVC。Spring Web MVC 是工程师开发工作的核心，本章从简单应用和深层配置等各个方面对 Spring MVC 进行深入的讲解。

第 6 章 数据访问。本章讲解 Spring Data 伞形项目，Spring Data 是 Spring 生态中有魅力、能提高生产力的框架之一，它可以使用相同的编程模型对不同的数据库技术进行开发，本章包含 Spring Data JPA、Spring Data Elasticsearch 和数据缓存。

第 7 章 安全控制。Spring Security 是 Java EE 领域成熟的安全解决方案，本章学习 Spring Security 和 OAuth 2.0 的应用。

第 8 章 响应式编程。响应式编程是未来几年的技术趋势，本章从开发的各个环节完全打通响应式开发，包含 Project Reactor、Spring WebFlux、Reactive NoSQL、R2DBC 和 Reactive Spring Security。

第 9 章　事件驱动。本章讲解在事件驱动开发中的常用技术，以达到应用之间的松耦合，本章包含 JMS、RabbitMQ、Kafka、Websocket 和 RSocket。

第 10 章　系统集成与批处理。本章主要讲解系统集成框架 Spring Integration 和批处理框架 Spring Batch。

第 11 章　Spring Cloud 与微服务。本章讲解微服务和 Spring Cloud 的主要知识，还特别介绍了在 Spring Cloud 下 OAuth 2.0 的使用。

第 12 章　Kubernetes 与微服务。在微服务开发完成后，它的部署主要基于 Kubernetes 平台。本章首先讲解基于 Kubernetes、Jenkins 和 Helm 的部署，然后介绍服务网格 Istio 在微服务部署中的应用。

由于时间及作者本人水平有限，书中难免有所错漏，望各位读者及时指出书中的不当之处并与我联系：https://github.com/wiselyman/spring-boot-book-source-code。

目　录

第 1 章 初识Spring Boot

1.1 Spring Boot 概述

Spring 框架一直是 Java EE 开发的王者，但是由于其有大量的配置，因而导致学习曲线较为陡峭。Spring 在 2014 年推出了 Spring Boot，Spring Boot 提供了如下功能来简化 Spring 的开发。

（1）自动配置：Spring Boot 为绝大多数的常用开发组件提供了自动配置，如 JDBC、JPA、Kafka、Elasticsearch、Spring MVC、Spring Security、Spring Integration 及 Spring Batch，本书将对大部分自动配置的用法进行讲解。

（2）starter 项目：Spring Boot 提供了大量的 starter 项目，如 spring-boot-starter-web，它主要将相关组件的依赖信息包装分组在一起，如 spring-boot-starter-web 的依赖为 spring-boot-starter、spring-boot-starter-json、spring-boot-starter-tomcat、hibernate-validator、spring-web 和 spring-webmvc，当需要使用某个技术组件时，只需添加相关技术的 starter 即可，无须手动添加大量依赖。

（3）全局依赖版本管理：Spring Boot 提供了全局依赖版本支持，只需声明 Spring Boot 的版本号即可，无须对 Spring Boot 支持的组件技术声明版本信息，依赖组件会直接得到最佳的依赖版本。

（4）打包方式：Spring Boot 支持将整个应用打包成 jar 包形式，jar 包中内嵌了 Servlet 容器（Tomcat、Jetty 等），可独立运行。

（5）开发者工具：Spring Boot 提供了开发者工具，只要添加 spring-boot-devtools 依赖，就可以在开发过程中提供自动重启功能，减少编译等待时间。

（6）Spring Boot Actuator：Spring Boot 提供的 Actuator 为生产时对应用的监控提供了支持。

本章内容较为简单，主要是为初学者考虑的，曾经使用过 Spring Boot 的读者可直接跳过。因为本书将基于 Spring Boot 的应用讲解所有基础知识，所以在第 1 章快速引入 Spring Boot 的开发。

1.2 快速建立 Spring Boot 应用

主流的开发工具都对 Spring Boot 提供了直接支持，本节讲述最简单也是最常用的基于 Spring Initializr 建立 Spring Boot 应用的方式。Spring Boot 2.x 要求 Java 版本至少在 Java 8 及以上。

1.2.1 安装 Java

从 Oracle 官网下载适合自己操作系统的 Java SE 版本。

在安装好 Java SE 之后，在 Windows 系统下，需在环境变量中配置 JAVA_HOME 变量。

1.2.2 使用 Spring Initializr

Spring Initializr 是由 Spring 官方提供的，用来简化生成 Spring Boot 应用的网站。

（1）Spring Initializr 的首页如图 1-1 所示。

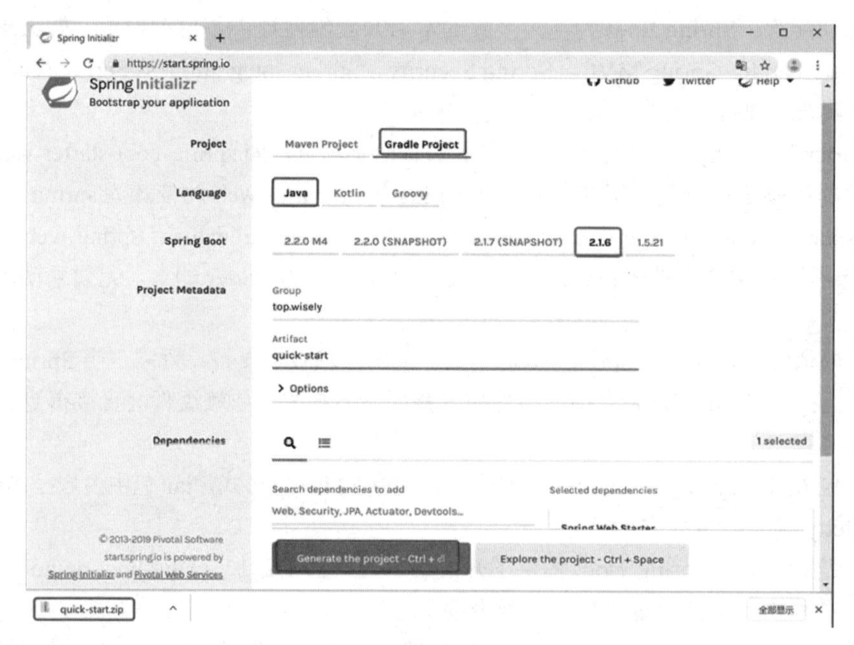

图 1-1

图 1-1 中主要选项含义如表 1-1 所示。

表 1-1

名　称	内　容	含　义
Project	Gradle Project	项目管理工具：可选 Maven Project 或 Gradle Project
Language	Java	开发语言：可选 Java、Kotlin、Groovy
Spring Boot	2.1.6	Spring Boot 版本：2.1.6，读者请选择最新正式版，本书内容基于 2.2.x
Group	top.wisely	组织域名
Artifact	quick-start	组织域名内唯一名称，即应用名称
Dependencies	Spring Web Starter	Spring Boot 依赖，可在检索框里使用关键字搜索

（2）在当前页面填写应用信息、Spring Boot 版本以及依赖信息，单击“Generate Project”按钮，会自动生成并下载应用源码。

1.2.3　第一段代码

解压缩 quick-start.zip。

（1）在 top.wisely.quickstart.QuickStartApplication 类上添加@RestController 注解，并在类上添加控制器方法。

```
package top.wisely.quickstart;

import org.springframework.boot.SpringApplication;
import org.springframework.boot.autoconfigure.SpringBootApplication;
import org.springframework.web.bind.annotation.GetMapping;
import org.springframework.web.bind.annotation.RestController;

@SpringBootApplication
@RestController
public class QuickStartApplication {

    @GetMapping("/hello-world")
    public String helloWorld() {
        return "Hello World!!!";
    }

    public static void main(String[] args) {
        SpringApplication.run(QuickStartApplication.class, args);
    }

}
```

（2）运行应用。

使用 Gradle Wrapper 运行应用，在应用目录下执行下面命令。

Windows：

```
.\gradlew.bat bootRun
```

macOS：

```
$ gradlew bootRun
```

（3）访问应用，如图 1-2 所示。

图 1-2

1.3 体验 Spring Boot

1.3.1 Spring Boot 的应用结构

下面对 1.2 节中的应用做一下解析，应用的结构如图 1-3 所示。

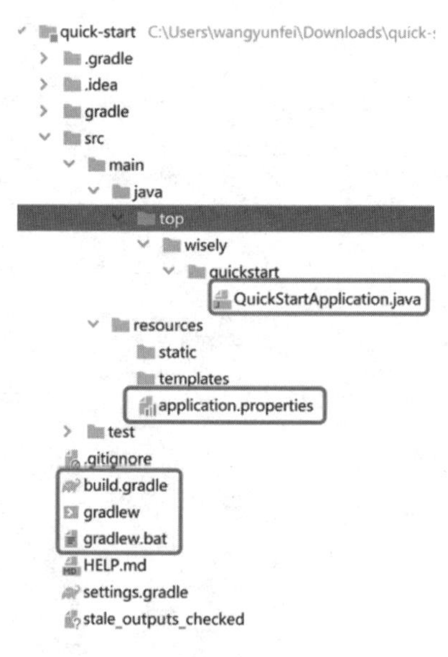

图 1-3

下面对 build.gradle、QuickStartApplication 和 application.properties 进行讲解。

1.3.2 build.gradle

```
plugins {
    id 'org.springframework.boot' version '2.1.6.RELEASE' //a
    id 'java' //b
}

apply plugin: 'io.spring.dependency-management'  //c

group = 'top.wisely'
version = '0.0.1-SNAPSHOT'
sourceCompatibility = '1.8'

repositories {//d
    mavenCentral()
}

dependencies { //e
    implementation 'org.springframework.boot:spring-boot-starter-web'
    testImplementation 'org.springframework.boot:spring-boot-starter-test'
}
```

a. Spring Boot Gradle 插件，版本为 2.1.6.RELEASE。

b. Java 插件。

c. Spring 的版本依赖管理插件。

d. jar 包下载仓库，这里使用的是 Maven 中心库。

e. 定义本应用的依赖。

1.3.3 QuickStartApplication

Spring Boot 提供了一个颠覆传统企业级应用的开发方式，在这种类似于简单的控制台程序的开发运行方式中，只需提供一个简单的入口 main 方法，即可运行应用。

```
@SpringBootApplication
@RestController
public class QuickStartApplication {

   @GetMapping("/hello-world")
   public String helloWorld(){
      return "Hello World!!!";
   }

   public static void main(String[] args) {
      SpringApplication.run(QuickStartApplication.class, args);
   }

}
```

- ◎ Spring Boot 之所以神奇，主要源于@SpringBootApplication 注解，本书将在第 4 章进行讲解，在这里只需保证代码不变即可。
- ◎ @RestController 声明了一个基于 RESTful 的 Web 控制器。
- ◎ @GetMapping("/hello-world") 声明了一个应用的端点，访问地址为"http://ip:port/hello-world"，访问得到的返回结果是字符串"Hello World!!!"。

1.3.4 application.properties

application.properties 提供了对 Spring Boot 的默认行进行定制的能力，Spring Boot 支持"properties"格式和"yml"格式，如修改 Web 容器的端口号。

```
server.port=8081
```

1.4 小结

本章带领读者快速了解了 Spring Boot。Spring Boot 提供了自动配置的能力，极大地提高了开发效率。虽然 Spring Boot 提供了许多激动人心的功能，但是建议读者依然要从最基础的 Spring 和 Spring MVC 学起，这样才能更好地掌握 Spring Boot。

第 2 章 函数式编程

2.1 了解函数式编程

顾名思义，就是用函数来编程。那什么是函数呢？函数是一种特殊的方法，它声明了如何完成任务的定义，在函数体内，它不会不修改任何共享数据，没有任何副作用。Java 8 中的函数（方法）是“值”的一种新的形式，可以将方法作为参数进行传递。而作为参数进行传递的方法主要是“Lambda 表达式”和“方法引用”。

2.2 Lambda 表达式

Lambda 表达式是一种匿名函数，在函数式编程里，它可以作为参数进行传递。

2.2.1 了解 Lambda 表达式

下面看一个简单的例子，先用“老式的”匿名类定义，再使用等同的 Lambda 表达式定义。

（1）匿名类。

```
Comparator<Person> byWeightComparator = new Comparator<Person>() {
    @Override
    public int compare(Person person1, Person person2) {
        return person1.getWeight().compareTo(person2.getWeight());
    }
  };
```

（2）Lambda 表达式。

```
Comparator<Person> byWeightComparatorUsingLambda =
        (Person person1, Person person2) ->
person1.getWeight().compareTo(person2.getWeight());
```

从上面的例子可以看出，Lambda 表达式分为三部分。

◎ 左侧：Lambda 参数列表(Person person1, Person person2)，等同于匿名类的 compare 方法的参数。

◎ 中间："->"，用来分开 Lambda 参数和 Lambda 体。

◎ 右侧：Lambda 体 person1.getWeight().compareTo(person2.getWeight())，等同于匿名类的 compare 方法的返回值。

Lambda 表达式还有以下规则。

（1）Lambda 表达式可以有 0 到多个参数，如：

◎ () -> {}。

◎ (Integer i) -> "wyf"。

◎ (Integer i, Integer j) -> { return "wyf";}。

（2）Java 编译器有类型推断（Type Inference）的能力，因而 Lambda 参数的类型可以省略，如：

```
(person1, person2) -> person1.getWeight().compareTo(person2.getWeight());
```

（3）若 Lambda 只有一个参数，则可以省略圆括号，如 a -> a+1。

（4）Lambda 体需要用花括号括起来，如果 Lambda 体内只有一句表达式，则可以省略花括号。

（5）在 Lambda 体中如果使用 return，则需要使用花括号括起来，如() -> { return "wyf";}。

2.2.2 把 Lambda 表达式作为参数

Lambda 表达式可以像参数一样传递给方法，示例如下。

```
List<Person> people = Arrays.asList(new Person("wyf", Gender.MALE, 100),
                        new Person("www", Gender.FEMALE, 80),
                        new Person("foo", Gender.FEMALE, 90));

people.sort((p1, p2) -> p1.getWeight().compareTo(p2.getWeight())); //a

people.forEach(person -> System.out.println(person.getName() + "的体重：" +
person.getWeight())); //b
```

a. Lambada 表达式作为 sort 方法的参数，是 Comparator 函数接口的实现。

b. Lambada 表达式作为 forEach 方法的参数，是 Consumer 函数接口的实现。

Person 类的定义如下。

```
public class Person {
   private String name; //名字
   private Gender gender; //性别
   private Integer weight = 0; //体重
```

```
    //无参构造
    public Person() {
       super();
    }
    //一个参数构造
    public Person(String name) {
       this.name = name;
    }
    //两个参数构造
    public Person(String name, Gender gender) {
       this.name = name;
       this.gender = gender;
    }
    //全参构造
    public Person(String name, Gender gender, Integer weight) {
       this.name = name;
       this.gender = gender;
       this.weight = weight;
    }

...
//省略 getter、setter

    @Override
    public String toString() {
       return "名字是: " + name + "性别是: " + gender + "体重是: " + weight;
    }

}
```

Gender 的定义如下。

```
public enum  Gender {
   MALE, FEMALE
}
```

2.3　函数接口

上面的例子中涉及很多函数接口，例如：

◎　Function

◎　Consumer

◎　Comparator

它们都属于函数接口，都标记了@FunctionalInterface 注解。

```
@FunctionalInterface
public interface Function<T, R>
...
```

```
@FunctionalInterface
public interface Consumer<T>
...

@FunctionalInterface
public interface Comparator<T>
...
```

任意一个只有抽象方法的接口都是函数接口（Functional Interface），这类接口都可以使用 Lambda 表达式（或方法引用）实现。函数接口只有一个抽象方法，但有很多其他方法，这些方法可归类如下。

◎ 静态方法：和接口有关的工具助手方法。使用 static 关键字实现。

◎ 默认方法：添加新的功能方法到已有的接口，在老的代码中，使用了该接口其他方法的代码不会受到影响。使用 default 关键字实现。

这也意味着从 Java 8 开始，接口内不仅可以有抽象方法，还可以有静态方法和默认方法。只要符合定义，即使没有标记@FunctionalInterface，它也是函数接口。当然，如果不符合函数接口的定义，那么即使标记了@FunctionalInterface，编译器也会报错，这就是@FunctionalInterface 的作用。

函数接口主要位于 java.util.function 包下，可分成下面几类。

◎ Predicate：有输入且只输出布尔值的函数。

◎ Function：有输入有输出的函数。

◎ Consumer：有输入无输出的函数。

◎ Supplier：无输入有输出的函数。

◎ Operator：输入和输出为相同类型的函数。

2.3.1 Predicate

Predicate（断言）的源码定义如下。

```
@FunctionalInterface
public interface Predicate<T> {
   boolean test(T t);
}
```

Lambda 表达式即为 test 方法的实现。从 test 方法的定义可以看出，test 方法可接收任意类型的参数 T，返回值为 boolean 类型，可以用下面的表达式定义。

```
Predicate<String> emptyPredicate = (String s) -> s.isEmpty()
```

根据类型推断可缩写为 Predicate<String> emptyPredicate = s -> s.isEmpty()。

使用当前 Predicate 定义，可通过 test 方法执行。

```
Predicate<String> emptyPredicate = s -> s.isEmpty();
System.out.println(emptyPredicate.test("wyf"));
```

输出的 emptyPredicate.test("wyf")返回值为 false。

1．组合 Predicate

Predicate 接口包含 negate、and 和 or 方法，可以重用已有的 Predicate，组成复杂的 Predicate。

◎ negate：已有 Predicate 的否定。

```
Predicate<String> nonEmptyPredicate = emptyPredicate.negate();
```

◎ and：相当于逻辑运算中的&&。

```
Predicate<Integer> greaterThan0 = i -> i > 0;
Predicate<Integer> lessThan100 = i -> i < 100;
Predicate<Integer> between0And100 = greaterThan0.and(lessThan100);
```

只有在 $i > 0$ 且 $i < 100$ 的情况下，test 方法的返回值才为 true。

◎ or：相当于逻辑运算中的||。

```
Predicate<Integer> greateThan0OrLessThan100 = greaterThan0.or(lessThan100);
```

当 $i > 0$ 或 $i < 100$ 时，test 方法的返回值是 true。

2．原始数据类型 Predicate

Java 会自动将包装类型拆包成原始数据类型，但这意味着性能的损失，所以当数据为原始数据类型时，Java 提供了一些特殊的 Predicate。

◎ IntPredicate：当入参是 int 类型时，

```
Predicate<Integer> greaterThan0 = i -> i > 0;
```

可修改成下面的样子。

```
IntPredicate intGreaterThan0 = i -> i > 0;
```

◎ DoublePredicate：入参为 double 类型。

◎ LongPredicate：入参为 long 类型。

3．两个参数的 Predicate

Java 还提供了表示两个入参的 Predicate，叫作 BiPredicate。

```
@FunctionalInterface
public interface BiPredicate<T, U> {
   boolean test(T t, U u);
}
```

test 方法可接收两个入参，类型分别为 T 和 U。

```
BiPredicate<String, Integer> isLongThanGivenLength = (str, len) -> str.length() > len;
```

第一个入参 T 类型为 String（str），第二个入参 U 类型为 Integer（len）。

2.3.2 Function

Function（函数）的源码定义如下。

```
@FunctionalInterface
public interface Function<T, R> {
   R apply(T t);
}
```

Lambda 表达式是 apply 方法的实现，apply 方法可接收任意类型的参数 T，返回值类型为 R，可以用下面的表达式定义。

```
Function<String, Integer> lengthFunction = str -> str.length();
```

入参 T 类型为 String（str)，返回值 R 类型为 Integer（str.length()），使用当前 Function 定义，可通过 apply 方法执行。

```
Function<String, Integer> lengthFunction = str -> str.length();
System.out.println(lengthFunction.apply("wyf"));
```

输出的 lengthFunction.apply("wyf")返回值为 3。

1. 组合 Function

Function 接口函数提供了 andThen 和 compose 方法来组合已有的 Function，组合 Function 的返回值仍为 Function。下面定义两个将被组合的 Function。

```
Function<Integer, Integer> plusFunction = x -> x + x ;
Function<Integer, Integer> multipleFunction = x -> x * x;
```

（1）andThen：新的 Function 是把组合中第一个函数的返回值作为第二个函数的输入。

```
Function<Integer, Integer> andThenFunction  = plusFunction.andThen(multipleFunction);
System.out.println(andThenFunction.apply(2));
```

执行时，plusFunction 先执行，返回值作为 multipleFunction 的入参再执行，结果为 16。

（2）compose：新的 Function 是把组合中第二个函数的返回值作为第一个函数的输入。

```
Function<Integer, Integer> composeFunction = plusFunction.compose(multipleFunction);
System.out.println(composeFunction.apply(2));
```

执行时，multipleFunction 先执行，返回值作为 plusFunction 的入参再执行，结果为 8。

2. 原始数据类型 Function

与 Predicate 一样，Function 也有原始数据类型的 Function，主要有 3 类。

第一类入参固化为函数接口，返回值类型 R，仍需在泛型中定义。

◎ IntFunction：入参为 int 类型。上面的 plusFunction 可修改为

```
IntFunction<Integer> intPlusFunction = x -> x + x 。
```

◎ LongFunction：入参为 long 类型。

◎ DoubleFunction：入参为 double 类型。

第二类是返回值固化为函数接口，入参类型 T 仍需在泛型中定义。

◎ ToIntFunction：返回值类型为 int 类型。

```
ToIntFunction<String> toIntFunction = str -> str.length();
System.out.println(toIntFunction.applyAsInt("www"));
```

◎ ToLongFunction：返回值类型为 long 类型。

◎ ToDoubleFunction：返回值类型为 double 类型。

第三类是入参和返回值都固化为函数接口。

◎ IntToLongFunction：入参为 int 类型，返回值为 long 类型。

```
IntToLongFunction intToLongFunction = i -> i;
System.out.println(intToLongFunction.applyAsLong(1));
```

◎ IntToDoubleFunction：入参为 int 类型，返回值为 double 类型。

◎ LongToIntFunction：入参为 long 类型，返回值为 int 类型。

◎ LongToDoubleFunction：入参为 long 类型，返回值为 double 类型。

◎ DoubleToIntFunction：入参为 double 类型，返回值为 int 类型。

◎ DoubleToLongFunction：入参为 double 类型，返回值为 long 类型。

3．两个入参的 Function

Java 还提供了 BiFunction，源码定义如下。

```
@FunctionalInterface
public interface BiFunction<T, U, R> {
   R apply(T t, U u);
}
```

apply 方法可接收两个参数 T 和 U，返回值为 R。可以用下面的表达式定义。

```
BiFunction<String, String, Integer> totalLengthBiFunction = (str1, str2) -> str1.length()
+ str2.length();
System.out.println(totalLengthBiFunction.apply("wyf","www"));
```

apply 方法的第一个入参 T 类型是 String（str1），第二个入参 U 类型是 String（str2），返回值是两个字符串长度之和。执行 apply 方法，输出为 6。

同样，BiFunction 也有很多原始数据类型函数。

◎ ToIntBiFunction：返回值为 int 类型。

◎ ToLongBiFunction：返回值为 long 类型。

◎ ToDoubleBiFunction：返回值为 double 类型。

2.3.3 Consumer

顾名思义，Consumer（消费者）是只消费不生产，源码定义如下。

```
@FunctionalInterface
public interface Consumer<T> {
   void accept(T t);
}
```

从 accept 方法的定义中可以看出，accept 方法可接收一个参数 T，没有返回值，可以用下面的表达式定义。

```
Consumer<String> helloConsumer = str -> System.out.println("Hello " + str + "!");
helloConsumer.accept("wyf");
```

accept 方法可接收的参数 T 类型是 String（str），没有返回值。

Consumer 有原始数据类型接口。

◎ IntConsumer：入参为 int 类型。

◎ LongConsumer：入参为 long 类型。

◎ DoubleConsumer：入参为 double 类型。

Consumer 也有表示两个参数的 BiConsumer 接口。

◎ ObjIntConsumer：第一个入参为任意类型 T，第二个入参为 int 类型。

◎ ObjLongConsumer 第一个入参为任意类型 T，第二个入参为 long 类型。

◎ ObjDoubleConsumer：第一个入参为任意类型 T，第二个入参为 double 类型。

2.3.4 Supplier

顾名思义，Supplier（提供者）是只生产不消费，源码定义如下。

```
@FunctionalInterface
public interface Supplier<T> {
   T get();
}
```

get 方法不接收参数，返回值为类型 T，可以用下面的表达式定义。

```
Supplier<Long> systemTime = () -> System.currentTimeMillis();
System.out.println(systemTime.get());
```

get 方法没有入参，返回值是 Long 类型，输出当前系统事件。

同样，Supplier 也有原始数据类型接口。

◎ IntSupplier：返回值是 int 类型。
◎ LongSupplier：返回值是 long 类型。
◎ DoubleSupplier：返回值是 double 类型。
◎ BooleanSupplier：返回值是 boolean 类型。

2.3.5 Operator

Operator（操作者）是一种特殊的 Function 接口，它的输入和返回值是同一种类型。

1. UnaryOperator

UnaryOperator 继承了 Function 接口，定义如下。

```
@FunctionalInterface
public interface UnaryOperator<T> extends Function<T, T>
```

UnaryOperator 可接收一个入参类型 T，返回值也是类型 T。

同样，UnaryOperator 也有原始数据类型的函数接口。

◎ IntUnaryOperator：入参和返回值都是 int 类型。
◎ LongUnaryOperator：入参和返回值都是 long 类型。
◎ DoubleUnaryOperator：入参和返回值都是 double 类型。

2. BinaryOperator

BinaryOperator 继承了 Function 接口，定义如下。

```
@FunctionalInterface
public interface BinaryOperator<T> extends BiFunction<T,T,T>
```

BinaryOperator 接收的两个入参类型都为 T，返回值类型也是 T。

同样，BinaryOperator 也有原始数据类型的函数接口。

◎ IntBinaryOperator：两个入参和一个返回值都是 int 类型。
◎ LongBinaryOperator：两个入参和一个返回值都是 long 类型。
◎ DoubleBinaryOperator：两个入参和一个返回值都是 double 类型。

2.3.6 Comparator

Comparator 是比较排序所用的一个函数接口，定义如下。

```
@FunctionalInterface
public interface Comparator<T> {
    int compare(T o1, T o2);
}
```

◎ 若 o1 小于 o2，则返回负数。

◎ 若 o1 等于 o2，则返回 0。

◎ 若 o1 大于 o2，则返回正数。

可以用 Lambda 表达式来定义。

```
Comparator<Person> byWeightComparatorUsingLambda =
    (person1, person2) -> person1.getWeight().compareTo(person2.getWeight());
```

Comparator 函数接口还为排序提供了 comparing 的静态方法，它可接收一个 Function 接口来获取处理数据的排序 key，上面的语句可以简写成下面的样子。

```
Comparator<Person> byWeightComparatorUsingStatic
=Comparator.comparing(Person::getWeight);
```

2.3.7 自定义函数接口

函数接口的定义主要是看入参和返回值，前面介绍了有一个入参的 Function 接口和有两个入参的 BiFunction 接口，下面自定义一个有三个入参的 TriFunction 接口。

```
@FunctionalInterface //a
public interface TriFunction<T, U, W, R> {
   R apply(T t, U u, W w); //b
}
```

a. @FunctionalInterface 标记为函数接口。

b. apply 方法可接收三个入参 T（t）、U（u）和 W（w），返回值为 R。

可以通过如下 Lambda 表达式调用。

```
TriFunction<String, String, String, Integer> lengthTriFuntion =
            (str1, str2, str3) -> str1.length() + str2.length() + str3.length();
System.out.println(lengthTriFuntion.apply("wyf", "www", "foo"));
```

第一个入参类型为 String（str1），第二个入参类型为 String（str2），第三个入参类型为 String（str3），返回值类型为 Integer。计算输出三个字符串的长度之和，输出结果为 9。

2.4 方法引用

我们可以使用已有的方法定义方法引用，并像 Lambda 表达式一样，把方法引用作为方法的参数使用。在 Java 中，方法引用使用“::”（两个冒号）表示。

2.4.1 构造器方法引用

构造器方法引用使用“类名::new”来定义。

◎ 无参数构造：构造器不接收参数，返回值为新建的 Person 对象，符合 Supplier 函数接口的定义。使用基于 Lambda 表达式的 Supplier 函数接口的实现与“方法引用”是等同的。

```
Supplier<Person> emptyConstructor = Person::new;
Supplier<Person> emptyConstructorLambda = () -> new Person();
Person person1 = emptyConstructor.get();
Person person1Lambda = emptyConstructorLambda.get();
```

◎ 一个参数构造：构造器接收一个参数，返回值为新建的 Person 对象，符合 Function 接口的定义。同样，使用基于 Lambda 表达式的 Function 接口实现与方法引用是等同的。

```
Function<String, Person> nameConstructor = Person::new;
Function<String, Person> nameConstructorLambda = name -> new Person(name);
Person person2 = nameConstructor.apply("wyf");
Person person2Lambda = nameConstructorLambda.apply("wyf");
```

◎ 两个参数构造：构造器接收两个参数，返回值为新建的 Person 对象，符合 BiFunction 接口的定义。同样，使用基于 Lambda 表达式的 BiFunction 接口实现与方法引用是等同的。

```
BiFunction<String, Gender, Person> nameAndGenderConstructor = Person::new;
BiFunction<String, Gender, Person> nameAndGenderConstructorLambda =
      (name, gender) -> new Person(name, gender);
Person person3 = nameAndGenderConstructor.apply("www", Gender.FEMALE);
Person person3Lambda = nameAndGenderConstructorLambda.apply("www", Gender.FEMALE);
```

◎ 三个参数构造：构造器接收三个参数，返回值为新建的 Person 对象，符合我们自定义的 TriFunction 接口的定义。同样，使用基于 Lambda 表达式的 TriFunction 接口实现与方法引用是等同的。

```
TriFunction<String, Gender, Integer, Person> allConstructor = Person::new;
TriFunction<String, Gender, Integer, Person> allConstructorLambda =
      (name, gender, weight) -> new Person(name, gender, weight);
Person person4 = allConstructor.apply("www", Gender.FEMALE, 110);
Person person4Lambda = allConstructorLambda.apply("www", Gender.FEMALE, 110);
```

2.4.2 静态方法引用

静态方法引用是使用方法引用的方式调用类的静态方法，格式为“类名::静态方法”。

```
IntFunction<String> intToStringFunction = Integer::toString;//a
IntFunction<String> intToStringFunLambda = i -> Integer.toString(i);//b
System.out.println(intToStringFunction.apply(123));
System.out.println(intToStringFunLambda.apply(123));
```

a. Integer::toString，toString 方法是 Integer 类型的静态方法，toString 方法的参数和返回值符合 IntFunction 接口的定义。

b. Lambda 表达式，等同于上一句。

2.4.3 实例方法引用

实例方法引用是使用方法引用的方式来调用实例对象的方法，格式为“实例对象名::实例方法”。

```
Person person = new Person("www", Gender.FEMALE, 80);
Consumer<String> walkConsumer = person::walk;
walkConsumer.accept("黄山路");
```

walk 方法是 Person 类实例对象方法。

```
public void walk(String roadName){
   System.out.println(name + "在" + roadName +"上行走");
}
```

2.4.4 引用特定类的任意对象的方法

引用特定类的任意对象的方法的格式为“类型名::实例方法”。

```
List<Person> people = Arrays.asList(new Person("wyf", Gender.MALE, 100),
        new Person("www", Gender.FEMALE, 80),
        new Person("foo", Gender.FEMALE, 90));
people.forEach(Person::sayName);
people.forEach(person -> person.sayName());
```

此处的 sayName 方法是实例对象方法，people 列表里的三个 Person 实例均可以调用这个方法。注意和上面“实例方法引用”的区别，“实例方法引用”只针对一个实例对象进行方法引用，而当前的“引用特定类的任意对象的方法”可以对 people 列表里的任意 Person 实例对象进行方法引用。sayName 方法是 Person 类的方法。

```
public void sayName(){
   System.out.println("我的名字是: " + name);
}
```

从上面 4 种方法引用与 Lambda 表达式的对比可以知道，在调用现有类的已有方法时，方法引用比 Lambda 表达式更自然，可读性更强。比如上面的例子。

```
people.sort((p1, p2) -> p1.getWeight().compareTo(p2.getWeight()));
```

可以用方法引用改写成下面的样子。

```
people.sort(Comparator.comparing(Person::getWeight));
```

2.5 Stream

前面在 “函数接口”中执行的演示都是手动调用函数接口中的函数（如 test()、apply()等）执行的，本节介绍的 Stream 就是“函数接口”的使用环境。

2.5.1　Stream 简介

Stream 是用声明式的方式来操作数据集合的一个 Java API。一般来说，可以从**数据源**（集合类、数组）获得 Stream，而 Stream 就是**数据序列**，我们可以对**数据序列**进行各种数据处理操作（过滤、转换、排序、查询等）。

很多 Stream 数据处理操作方法的返回值还是一个 Stream，此时可以对新的 Stream 再进行数据处理操作，这意味着可以把多个数据处理操作"串"成一条大的处理**管道**（pipeline）。返回值还是 Stream 的数据处理操作可以称为**中间操作**。中间操作并没有对数据做运算处理，而是对数据处理的方式做了声明。

除此之外，还有另一类操作，即在 Stream 进行一些管道处理后，把 Stream 转换成所需结果的数据操作，这些操作可以称之为**终结操作**。

在进行 Stream 开发时只需以下三步。

（1）从数据源获得 Stream。

（2）中间操作：组成处理管道。

（3）终结操作：从管道中产生处理结果。

2.5.2　获得 Stream

（1）从普通值获取：Stream.of。

```
Stream<String> singleStringStream = Stream.of("wyf");
Stream<String> stringStream = Stream.of("www", "wyf", "foo", "bar");
```

（2）空 Stream：Stream.empty。

```
Stream<String> emptyStringStream = Stream.empty();
```

（3）从空值中获取：Stream.ofNullable。

```
String str1 = "wyf";
String str2 = null;
Stream<String> str1Stream = Stream.ofNullable(str1); //等同于 Stream.of
Stream<String> str2Stream = Stream.ofNullable(str2); //等同于 Stream.empty
```

（4）从数组中获取：Arrays.stream。

```
Integer[] numArray = {1, 2, 3, 4, 5};
Stream<Integer> numStream = Arrays.stream(numArray);
```

（5）来自文件：Files 类的静态方法。

```
String filePathStr = "文件路径字符串";
Path path = Paths.get(filePathStr);
Stream<String> lineStream = Files.lines(path);//返回一个字符串行的 Stream
```

（6）来自集合类：Collection、List 和 Set。

```
Collection<String> collection = Arrays.asList("www", "wyf", "foo", "bar");
Stream<String> collectionStream = collection.stream();
```

（7）使用建造者模式（builder pattern）构建：Stream.builder。

```
Stream<String> streamBuilder =
Stream.<String>builder().add("www").add("wyf").add("foo").add("bar").build();
```

（8）来自函数：Stream.generate 和 Stream.iterate。

◎ Stream.generate 接收一个 Supplier 函数接口作为参数，且产生的 Stream 是无限的，使用时请限制数量。

```
Stream<String> generatedStream = Stream.generate(() -> "wyf").limit(10);
```

◎ Stream.iterate 接收的第一个参数是起始值，第二个参数是 UnaryOperator，产生的 Stream 是无限的，需要限制数量。

```
Stream<Integer> iteratedStream = Stream.iterate(20, n -> n+1 ).limit(5);//20,21,22,23,24
```

（9）原始数据类型的 Stream（只包含某类原始数据类型的 Stream）：IntStream、LongStream 和 DoubleStream 的静态方法。

```
IntStream intStream = IntStream.range(1,5); //开始包含，结束不包含
LongStream longStreamClosed = LongStream.rangeClosed(1,5);//开始包含，结束包含
DoubleStream doubleStream = new Random().doubles(3);//3 个随机的 double 数据
```

（10）来自字符串。

```
IntStream intStream1 = "wyf".chars();
```

（11）来自 Optional。

```
Optional<String> nameOptional = Optional.of("wyf");
Stream<String> stringOptionalStream = nameOptional.stream();
```

2.5.3 中间操作

中间操作（Intermediate Operations）不会得到最终的结果，只返回一个新的 Stream。中间操作接收函数接口作为参数，可以使用“Lambda 表达式“和“方法引用”作为实现。

1. 演示所用 Stream

从集合类获得 Stream。

```
Stream<Person> peopleStream = Arrays.asList(new Person("wyf", Gender.MALE, 100),
                                new Person("www", Gender.FEMALE, 80),
                                new Person("foo", Gender.FEMALE, 90),
                                new Person("foo", Gender.FEMALE,
                                                   90)).stream();
```

2. 过滤

（1）filter 方法。

Stream 的过滤主要是通过 Stream 的 filter 方法实现的，filter 方法接收一个 Predicate 函数接口作为参数，Predicate 接受一个参数 T，返回值为 boolean 类型。当数据运算的结果为 true 时，数据保留。下面过滤出性别为男的数据。为了演示，先引入 forEach 这个终结操作来展示 filter 方法操作的结果。

```
peopleStream.filter(person -> person.getGender().equals(Gender.MALE))
                        .forEach(person -> System.out.println(person.getName()));
```

除 Lambda 表达式可以作为函数接口的实现外，方法引用也可以作为函数接口的实现。

```
peopleStream.filter(Person::isMale)
                           .forEach(System.out::println);//只剩下性别为男的数据
```

方法引用这种形式极大地提高了代码简洁性和可读性。这里的方法引用方式属于“引用特定类的任意对象的方法”，即引用的是 Stream 中任意对象的方法。

（2）distinct 方法。

可以使用 distinct 方法过滤掉相同的数据，只留下唯一一个。当然，前提是需要覆写 Person 类的 equals 和 hashCode 方法来标识数据是相同的。

首先用 Person 类覆写 equals 和 hashCode 方法。

```
   @Override
   public boolean equals(Object obj) {
      Person person = (Person) obj;
      if(this.getGender().equals(person.getGender()) &&
              this.getName().equals(person.getName()) &&
              this.getWeight() == person.getWeight())
          return true;
      else
          return false;
   }

   @Override
   public int hashCode() {
      return Objects.hash(name, weight, gender);
   }
```

然后 Stream 直接调用 distinct 方法，返回一个新的 Stream。

```
peopleStream.distinct()
             .forEach(System.out::println); //去掉重复的数据，只剩下唯一的数据
```

（3）中间操作管道。

当然，可以将 filter 和 distinct 操作“串”成操作管道，例如：

```
peopleStream.distinct()
                  .filter(person -> person.getGender().equals(Gender.MALE))
                  .forEach(System.out::println);
```

3. 转换处理

Stream API 通过 map 和 flatMap 方法对已有数据进行转换处理。

（1）map 方法。

map 方法的参数是一个 Function 接口，接收一个参数 T，返回值类型为 R。map 方法得到的新 Stream 是包含 R 类型的数据，例如：

```
Stream<String> weightStream = peopleStream.map(person -> person.getName());
```

入参为 Person 类型的 person，返回值是字符串。新的 Stream 包含的数据类型为 String，也可使用方法引用简写成：

```
Stream<String> weightStream = peopleStream.map(Person::getName);
```

也可以链起多个 map 方法的管道，例如：

```
Stream<Integer> lengthStream = peopleStream.map(Person::getName)
                                 .map(String::length);
```

（2）flatMap 方法。

flatMap 方法用来处理 Stream 嵌套的问题，例如：

```
List<Person> people1 = Arrays.asList(new Person("wyf", Gender.MALE, 100),
                                                                       new
Person("www", Gender.FEMALE, 80));
List<Person> people2 = Arrays.asList(new Person("foo", Gender.MALE, 90),
                                                                       new
Person("bar", Gender.FEMALE, 110));
List<List<Person>> peopleList = Arrays.asList(people1, people2);

peopleList.stream()
                .flatMap(Collection::stream) //a
       .forEach(System.out::println);//b
```

a. 此时 flatMap 方法将 Stream 中包含的 List 转换成 2 个 Stream，并将两个 Stream 合并成 1 个 Stream。

b. 获取合并的 Stream，打印出来的结果是 people1 和 people2 列表里的所有 Person 实例。

若采用 map 处理，则需要处理嵌套的 Stream。

```
 peopleList.stream()
                 .map(Collection::stream)
                 .forEach(personStream -> {personStream.forEach(System.out::println);
 });
```

（3）原始数据类型方法。

Stream 中有很多针对原始数据类型的转换处理方法：mapToInt、mapToLong、mapToDouble、flatMapToInt、flatMapToLong 和 flatMapToDouble。例如：

```
IntStream intStream = peopleStream.mapToInt(Person::getWeight);
```

4．其他操作

（1）skip：忽略前 *n* 条数据。

```
peopleStream.skip(1).forEach(System.out::println);
```

（2）limit：限制只需要前 *n* 条数据。

```
peopleStream.limit(2).forEach(System.out::println);
```

（3）sorted：将数据排序。

```
peopleStream.sorted(Comparator.comparing(Person::getWeight)).forEach(System.out::pri
ntln);
```

2.5.4 终结操作

1．聚合操作

（1）count：获得 Stream 中的数据数量。

```
peopleStream.count()
```

（2）max：获得 Stream 中按照规则约定的最大值。

```
peopleStream.max(Comparator.comparing(Person::getWeight))
```

（3）min：获得 Stream 中按照规则约定的最小值。

```
peopleStream.min(Comparator.comparing(Person::getWeight))
```

2．循环

forEach：对 Stream 中的数据进行循环处理，forEach 的参数是 Consumer 函数接口，只接收参数，没有返回值。

```
peopleStream.forEach(System.out::println);
```

打印 peopleStream 中的每个 Person 对象。

3．匹配

（1）allMatch：Stream 数据是否全部匹配，返回布尔值。

```
boolean isAllMale = peopleStream.allMatch(Person::isMale);
```

（2）anyMatch：Stream 数据是否有任意匹配，返回布尔值。

```
boolean isAnyMale = peopleStream.anyMatch(Person::isMale);
```

（3）nonMatch：Stream 数据是否全部不匹配，返回布尔值。

```
boolean isNonMale = peopleStream.noneMatch(Person::isMale);
```

4．查找

（1）findAny：获得 Stream 中的任意数据。

```
Person person = peopleStream.findAny().get();
```

findAny 的返回值是 Optional，会在后面讲到，使用 Optional 的 get 方法可以获得数据。

（2）findFirst：获得 Stream 中的第一条数据。

```
Person firstPerson = peopleStream.findFirst().get();
```

5．获得数组

toArray 方法可以将 Stream 中的数据转换为 Object 数组，例如：

```
Object[] people =  peopleStream.toArray();
```

6．reduce 方法

Stream 的 reduce 方法可以进行累计的聚合操作。reduce 方法的参数主要分为两个：

（1）第一个是初始值（可选）。

（2）第二个为累计方法：按步两两计算，计算的结果作为下一步累计计算的开始值。

```
int reduced1 = Stream.of(1, 2, 3).reduce((a, b) -> a + b).get(); //a. 求和
int reduced2 = Stream.of(1, 2, 3).reduce(10, (a, b) -> a + b); //b.  带初始值的求和
int reduced3 = Stream.of(1, 2, 3).reduce((a, b) -> a * b).get(); //c. 累乘
int reduced4 =Stream.of(1, 2, 3).reduce(10, (a, b) -> a * b); //d. 带初始值的类乘
int reduced5 = Stream.of(1, 2, 3).reduce(Integer::max).get(); //e 求最大值
int reduced6 = Stream.of(1, 2, 3).reduce(10,Integer::max); //f 带初始值的求最大值
int reduced7 = Stream.of(1, 2, 3).reduce(Integer::min).get(); //g 求最小值
int reduced8 = Stream.of(1, 2, 3).reduce(10,Integer::min); //h 带初始值的求最小值
```

a. 数据有 1、2、3。第一步，计算 $1+2=3$；第二步，将第一步得到的结果 3 与 3 相加，得到 6。

b. 初始值为 10，数据有 1、2、3。第一步，$10+1=11$；第二步，$11+2=13$；第三步，$13+3=16$。

c. 数组有 1、2、3。第一步，计算 $1\times2=2$；第二步，将得到的结果 2 与 3 相乘，得到 6。

d. 初始值为 10，数据有 1、2、3。第一步，$10\times1=10$；第二步，$10\times2=20$；第三步，$20\times3=60$。

e. 数据有 1、2、3。第一步，比较 1 和 2 大小；第二步，将第一步较大值 2 与 3 相比，得到最大值为 3。

f. 初始值为 10，数据有 1、2、3。第一步，比较 10 和 1。第二步，将第一步较大值 10 与 2 相比，较大值为 10；第三步，将第二步较大值 10 与 3 相比，得到最大值为 10。

g. 数组有 1、2、3。第一步，比较 1 和 2 大小；第二步，将第一步较小值 1 与 3 相比，得到最小值为 1。

h. 初始值为 10，数据有 1、2、3。第一步，比较 10 和 1；第二步，将第一步较小值 1 与 2 相比，较小值仍为 1；第三步，将第二步较小值 1 与 3 相比，得到最小值为 1。

7. collect 方法

Stream 中的 collect 方法是终结操作中功能最丰富的一个方法，它接收一个 java.util.stream.Collector 参数，Collector 参数指定了 Stream 转换成值的方式。Java 预先定义了大量的 Collector 参数来解决常见问题。collect 方法可通过 java.util.stream.Collectors 类的静态方法构造。

Collectors 的静态方法主要分为以下几类。

（1）转换成集合类。

◎ Collectors.toList()。

```
List<String> streamToList = Stream.of("wyf", "www", "foo", "bar")
                          .collect(Collectors.toList());
```

◎ Collectors.toSet()。

```
Set<String> streamToSet = Stream.of("wyf", "wyf", "foo", "bar")
                          .collect(Collectors.toSet());
```

◎ Collectors.toCollection(集合类型)。

```
List<String> streamToCollection = Stream.of("wyf", "www", "foo", "bar")
        .collect(Collectors.toCollection(ArrayList::new));
```

（2）转换成 String。

◎ Collectors.joining()：用来给 Stream 中的字符串数据指定分隔符、前缀和后缀，然后连接成一个字符串。

```
String streamToString = Stream.of("wyf", "www", "foo", "bar")
        .collect(Collectors.joining(",", "[", "]"));
```

第一个参数以“,”为分隔符，第二个参数以“[”为前缀，第三个参数以“]”为后缀，字符串 streamToString 为“[wyf,www,foo,bar]”。另外，joining 还有一个方法是只有分隔符，没有前缀和后缀。例如：

```
String streamToStringOnlyDelimiter = Stream.of("wyf", "www", "foo", "bar")
        .collect(Collectors.joining(","));
```

字符串 streamToStringOnlyDelimiter 经转换后为“wyf,www,foo,bar”。

（3）聚合操作。

◎ Collectors.counting()：Stream 中的数据数量。

```
long count = Stream.of("wyf", "www", "foo", "bar").collect(Collectors.counting());
```

◎ Collectors.maxBy()：Stream 中的数据按照某种规则约定的最大值，下面是比较字符串长度的最大值。

```
String maxLengthString = Stream.of("wyf", "w", "fooo",
"barbar").collect(Collectors.maxBy(Comparator.comparing(String::length))).get();
```

◎ Collectors.minBy()：Stream 中的数据按照某种规则约定的最小值，下面是比较字符串长度的最小值。

```
String minLengthString = Stream.of("wyf", "ww", "fooo",
"barbar").collect(Collectors.minBy(Comparator.comparing(String::length))).get();
```

◎ Collectors.averagingInt/Long/Double：求 Stream 中的数据的平均值，下面是求字符串长度的平均值。

```
double average =  Stream.of("wyf", "ww", "fooo",
"barbar").collect(Collectors.averagingInt(String::length));
```

◎ Collectors.summingInt/Long/Double：求 Stream 中数据的和，下面是求字符串长度的和。

```
int sum = Stream.of("wyf", "ww", "fooo",
"barbar").collect(Collectors.summingInt(String::length));
```

◎ Collectors.summarizingInt/Long/Double：汇总 Stream 数据情况，包括平均值、求和、最大值、最小值和数量，所有值都包含在返回值类型 IntSummaryStatistics（LongSummaryStatistics/DoubleSummaryStatistics）中。

```
IntSummaryStatistics summary = Stream.of("wyf", "ww", "fooo",
"barbar").collect(Collectors.summarizingInt(String::length));

System.out.println(summary.getMax());
System.out.println(summary.getMin());
System.out.println(summary.getAverage());
System.out.println(summary.getSum());
System.out.println(summary.getCount());
```

（4）转换成 Map。

◎ Collectors.toMap()：toMap 方法将 Stream 转换成 Map。toMap 方法有两个参数：keyMapper（Map 实例 key 生成方式）和 valueMapper（Map 实例 value 生成方式）。这两个参数的类型都是 Function 接口。

```
Map<String, Integer> map = Stream.of("wyf", "www", "foo", "bar")
      .collect(Collectors.toMap(Function.identity(), String::length));
```

本例将字符串作为 key，Function.identity()是参数 keyMapper，它返回的是输入参数（字符串）String::length 是参数 valueMapper，它返回的是字符串长度。转换后 Map 将字符串作为 key，将字符串长度作为 value。

◎ Collectors.groupingBy()：按照特定功能处理分组，接收一个 Function 接口。

```
Map<Integer,List<String>> map = Stream.of("wyf", "www", "foobar", "barfoo")
      .collect(Collectors.groupingBy(String::length));
```

按照字符串的长度进行分组，长度相等的分到一组（List<String>），字符串长度为 key。因此上面的例子可分为两组，第一组 key 为 3，List<String>的值是：wyf、www；第二组 key 为 6，List<String>的值为：foobar 和 barfoo。

◎ Collectors.partitioningBy()：partitioningBy 是一种特殊的 groupingBy 方法，它接收 Predicate 作为参数，只将数据分成 false 和 true 两组。

```
Map<Boolean,List<String>> partitioningByMap = Stream.of("wyf", "www", "foobar", "barfoo")
      .collect(Collectors.partitioningBy(str -> str.length()== 6));
```

判断条件是字符串长度是否为 6，字符串长度不为 6 的进入 key 为 false 的分组，字符串长度为 6 的进入 key 为 true 的分组。

（5）Collectors.collectingAndThen：collect 执行后将值再进行转换。

```
int size = Stream.of("wyf", "www", "foo", "bar")
      .collect(Collectors.collectingAndThen(Collectors.toList(), List::size));
```

将 Stream 转成 List（Collectors.toList()）后，接下来是获取 List 的大小（List::size）。

为了演示，前面所有例子全部使用 Collectors 静态方法，实际上，通过静态方法导入可以让代码更简洁。例如：

```
import static java.util.stream.Collectors.collectingAndThen;
import static java.util.stream.Collectors.toList;
```

代码可以写成：

```
int size = Stream.of("wyf", "www", "foo", "bar")
      .collect(collectingAndThen(toList(),List::size));
```

2.6　Optional

Optional 类是可以解决空指针异常（NullPointException）的问题。它可以作为任意类型 T 的对象的容器，它可以在对象值不为空的时候返回值。当值为空时，可以预先做处理，而不是抛出空指针异常。

2.6.1 获得 Optional

◎ Optional.emtpty()：获得空的 Optional。

```
Optional<String> emptyOptional = Optional.empty();
```

◎ Optional.of(参数)：包含非 null 值的 Optional。

```
String str = "wyf";
Optional<String> nonNullOptional = Optional.of(str);
```

◎ Optional.ofNullable(参数)：包含 null 值的 Optional。若参数不为 null，则返回包含参数的 Optional；若参数为 null，则返回空的 Optional。

```
String str1 = "wyf";
String str2 = null;
Optional<String> nullableOptional1 = Optional.ofNullable(str1);
Optional<String> nullableOptional2 = Optional.ofNullable(str2);
System.out.println(emptyOptional.equals(nullableOptional2)); //nullableOptional2 与
emptyOptional 相同，输出 true
```

2.6.2 Optional 的用法

（1）检查值是否存在或为空：存在检查使用 isPresent；为空检查使用 isEmpty。

```
String str1 = "wyf";
String str2 = null;
Optional<String> nullableOptional1 = Optional.ofNullable(str1);
Optional<String> nullableOptional2 = Optional.ofNullable(str2);
System.out.println(nullableOptional1.isPresent()); //a
System.out.println(nullableOptional2.isEmpty());    //b
```

a. 因为 nullableOptional1 包含字符串 wyf，所以检查是否存在的结果是 true。

b. 因为 nullableOptional2 是一个空的 Optional，所以检查是否为空的结果是 true。

（2）条件运算：ifPresent，在满足数据存在的条件下，可执行自己处理语句。

```
Optional<String> nullableOptional1 = Optional.ofNullable("wyf");
Optional<String> nullableOptional2 = Optional.ofNullable(null);
nullableOptional1.ifPresent(System.out::println); //a
nullableOptional2.ifPresent(System.out::println); //b
```

a. 符合条件，输出字符串；

b. 不符合条件，没有输出。

（3）默认值：设置当 Optional 为空时的默认值；orElseGet 的参数是一个 Supplier 函数接口，它不指定默认值，而是使用函数接口实现算提供的值。

```
Optional<String> nullableOptional1 = Optional.ofNullable("wyf");
Optional<String> nullableOptional2 = Optional.ofNullable(null);
String name1 = nullableOptional1.orElse("www"); //a
```

```
String name2 = nullableOptional2.orElse("www"); //b
String name3 = nullableOptional1.orElseGet(() -> "wwwFromOrElseGet"); //c
String name4 = nullableOptional2.orElseGet(() -> "wwwFromOrElseGet"); //d
```

a. nullableOptional1 不为空，所以 name1 依然是 wyf，不需要使用 orElse 设置的默认值 www。

b. nullableOptional2 为空，所以 name2 使用的是 orElse 设置的默认值 www。

c. nullableOptional1 不为空，所以 name3 依然是 wyf，不需要使用 orElseGet 中 Lambda 表达值返回的 wwwFromOrElseGet。

d. nullableOptional2 为空，所以 name2 使用的是 orElseGet 中 Lambda 表达值返回的 wwwFromOrElseGet。

（4）获得值：只有当 Optional 不为 null 时 get 方法才能获得包含的数据。

```
Optional<String> nameOptional = Optional.of("wyf");
String name = nameOptional.get();//获得 Optional 中包含的 name 字符串
```

（5）数据过滤：可以使用 filter 方法对数据进行过滤。

```
Optional<String> nameOptional = Optional.of("wyf");
boolean isWyf = nameOptional.filter(name -> name.equals("wyf")).isPresent();
```

通过 filter 方法看 Optional 中包含的数据是否符合 name → name.equals("wyf")这个 Predicate 的实现，当前是符合的，所以运算的 isWyf 为 true。

（6）转换处理：可以通过 map 和 flatMap 方法对数据进行转换处理。

```
Optional<String> nameOptional = Optional.of("wyf");
String hello = nameOptional.map(name -> "Hello " + name).get();
```

将 Optional 中包含的数据 wyf 处理成为"Hello " + name，并获得这个值。

2.7 小结

本章系统介绍了 Java 函数式编程，函数式编程是将函数作为编程开发的“第一等公民”，即函数也是值，可以把函数当成普通的对象值进行传递。在 Java 中，函数式编程的核心是函数接口，而函数接口的两大实现分别是 Lambda 表达式和方法引用。本书后面的章节中会大量地使用 Lambda 表达式、方法引用、Stream API 和响应式编程。

第 3 章 Spring 5.X基础

3.1 IoC 容器

Spring 的核心是 IoC（Inversion of Control，控制反转）容器，它可以管理容器内的普通 Java 对象以及对象之间关系的绑定（Dependency Injection 依赖注入）。容器中被管理的对象称为 Bean。

Spring 是通过元数据和 POJO 来定义和管理 Bean 的。

◎ POJO：简单的 Java 对象。

◎ 元数据：描述如何管理 POJO 的数据。

Spring 通过读取元数据知道如何管理你的 POJO，然后按照你的要求对 POJO 进行管理（即 Bean）。在早期，Spring 的元数据主要是由 XML 实现的；现在，主要的元数据都是通过注解配置和 Java 配置实现的。

Spring 的 IoC 容器是 ApplicationContext，它拥有一个父接口 BeanFactory，用来提供管理配置任意对象的基础功能。只要新建一个 ApplicationContext 的实现，就拥有一个 Spring 的 IoC 容器。

```
ApplicationContext context = new ClassPathXmlApplicationContext("context.xml");
```

Spring Boot 可在不同的环境下自动创建正确的 IoC 容器。

◎ AnnotationConfigApplicationContext：默认创建的 IoC 容器。

◎ AnnotationConfigServletWebServerApplicationContext：在 Web 应用下创建的 IoC 容器。

◎ AnnotationConfigReactiveWebServerApplicationContext：在响应式 Web 应用下创建的 IoC 容器。

本章主要讲解 Spring 基础知识，只需建立一个简单的 Spring Boot 应用即可，它会自动创建一个 AnnotationConfigApplicationContext 的 IoC 容器。

应用信息如下。

```
Group: top.wisely
Artifact: spring-fundamentals
Dependencies: Aspects、Lombok
```

3.2　Spring Bean 的配置

3.2.1　注解配置（@Component）

当类注解为@Component、@Service、@Repository 或@Controller 时，Spring 容器会自动扫描（通过@ComponentScan 实现，Spring Boot 已经做好了配置），并将它们注册成受容器管理的 Bean。

```
@Component
public class SomeService {
  public void doSomething(){
    System.out.println("我做了一些工作");
  }
}
```

@Component、@Service、@Repository 和@Controller 在当前示例中是完全等同的。

```
@Service
public class SomeService2 {
    public void doSomething(){
        System.out.println("我也做了一些工作");
    }
}
```

上面的@Component 和@Service 都没有给 Bean 命名，Spring 容器会自动命名为类名的第一个字母的小写形式，即 someService 和 someService2。一般来说，没有必要去修改 Bean 的名称，使用默认的 Bean 名即可。当然，也可以通过@Component("SomeService")来设置 Bean 的名称。

@Service、@Repository 和@Controller 这三个注解组合了@Component 注解，它们是@Component 语义上的特例。

◎　@Component：被注解类是“组件”。

◎　@Controller：被注解类是“控制器”。

◎　@Service：被注解类是“服务”。

◎　@Repository：被注解类是“数据仓库”。

3.2.2　Java 配置（@Configuration 和@Bean）

在类上注解@Configuration（@Component 的特例，会被容器自动扫描），可使类成为配置类。如果使用@Bean 标注在类的方法上，则方法的返回值即为 Bean 的实例。假如现在有另外一个类。

```
@Getter //lombok 注解，给属性生成 get 方法
@Setter //lombok 注解，给属性生成 set 方法
public class AnotherService {
   private String person;

   public AnotherService(String person) {
      this.person = person;
   }

   public void doAnotherThing(){
      System.out.println(person + "做了另外的事情");
   }

}
```

用 Java 配置的如下。

```
@Configuration
public class JavaConfig {

   @Bean
   public AnotherService anotherService(){
      return new AnotherService("wyf");
   }

}
```

同样，没有给 Bean 命名。Spring 会将方法名 anotherService 默认成 Bean 的名称。若需要修改，则使用@Bean(name = "AnotherService")。

3.2.3 依赖注入（Dependency Injection）

1．自动注入（@Autowired）

容器已经创建了 SomeService、AnotherService 和 SomeService2 的 Bean，其他的 Bean 应如何注入使用呢？

（1）注解注入。

AnnotationInjectionService 需要使用 SomeService 和 AnotherService 的 Bean，我们只需在 AnnotationInjectionService 构造器上注解@Autowired，即可注入参数里需要的 Bean。

```
@Service
public class AnnotationInjectionService {
   private SomeService someService;

   private SomeService2 someService2;
```

```
    @Autowired
    public AnnotationInjectionService(SomeService someService,SomeService2 someService2)
{
        this.someService = someService;
        this.someService2 = someService2;
    }

    public void doMyThing(){
        someService.doSomething();
        someService2.doSomething();
    }
}
```

在构造器上注解注入是 Spring 推荐的注入方式，当然，也可以通过在属性上注解@Autowired 来注入 Bean。

```
@Service
public class AnnotationPropertyInjectionService {
    @Autowired
    private SomeService someService;

    @Autowired
    private SomeService2 someService2;

    public void doMyThing(){
        someService.doSomething();
        someService2.doSomething();
    }
}
```

还可以在 set 方法上注解@Autowired 来注入 Bean。

```
@Service
public class AnnotationSetterInjectionService {

    private SomeService someService;

    private SomeService2 someService2;

    @Autowired
    public void setSomeService(SomeService someService) {
        this.someService = someService;
    }

    @Autowired
    public void setSomeService2(SomeService2 someService2) {
        this.someService2 = someService2;
    }

    public void doMyThing(){
        someService.doSomething();
```

```
        someService2.doSomething();
    }
}
```

如果 Bean 只有一个构造器，则可以直接省略@Autowired 注解。若 Bean 有多个构造器，则需注解一个构造器用来注入，示例如下。

```
@Service
public class AnnotationOneInjectionService {
    private SomeService someService;

    public AnnotationOneInjectionService(SomeService someService) {
        this.someService = someService;
    }

    public void doMyThing(){
        someService.doSomething();
    }
}
```

（2）配置注入。

现在使用 Java 配置的方式在 Bean JavaConfigInjectService 中注入 BeanAnotherService，JavaConfigInjectService 定义如下。

```
public class JavaConfigInjectService {
    private AnotherService anotherService;

    public JavaConfigInjectService(AnotherService anotherService) {
        this.anotherService = anotherService;
    }

    public void doMyThing(){
        anotherService.doAnotherThing();
    }
}
```

前面已经将 AnotherService 通过@Bean 注解成 Bean 了，下面只需在定义 JavaConfigInjectService 的 Bean 的方法参数里注入 AnotherService 的 Bean 即可。

```
@Bean
public JavaConfigInjectService javaConfigInjectService(AnotherService anotherService){
    return new JavaConfigInjectService(anotherService);
}
```

在同一个配置类里，还可以在新建 JavaConfigInjectService 的构造里直接注入创建 SomeService2 的 Bean 的方法。

```
@Bean
public JavaConfigInjectService javaConfigInjectService(){
```

```
    return new JavaConfigInjectService(anotherService());
}
```

（3）混合注入。

注解配置的 Bean 可以直接注入给使用 Java 配置的 Bean，反之亦然。

把注解配置的 Bean 注入 Java 配置的 Bean：

```
@Service //使用注解配置的Bean
public class MixInjectionService {
    private AnotherService anotherService;//注入Java配置的Bean

    public MixInjectionService(AnotherService anotherService) {
        this.anotherService = anotherService;
    }

    public void doMyThing(){
        anotherService.doAnotherThing();
    }
}
```

把 Java 配置的 Bean 注入注解配置的 Bean。被注入的 BeanMixInjectionService2 定义如下。

```
public class MixInjectionService2 {
    private SomeService someService; //使用@Component注解配置的Bean

    public MixInjectionService2(SomeService someService) {
        this.someService = someService;
    }

    public void doMyThing(){
        someService.doSomething();
    }
}
```

在 JavaConfig 类里，可以直接在参数中注入 Bean。

```
@Bean
public MixInjectionService2 mixInjectionService2(SomeService someService){
    return new MixInjectionService2(someService);
}
```

2. @Primary

上面的例子都是通过 Bean 的名称来自动注入的。当 Bean 的名称不满足条件时，容器会根据 Bean 的类型进行自动注入。当全局只有一个类型的 Bean 时，自动注入是没有问题的，但是当全局有多个同类型的 Bean 时，会提示“required a single bean, but n were found”，此时可以通过@Primary 来注解需要优先使用的 Bean。假如有两个 Bean：

```
@Bean
public AnotherService anotherService(){
```

```
    return new AnotherService("wyf");
}

@Bean
@Primary
public AnotherService primaryAnotherService(){
         return new AnotherService("foo");
}
```

此时有两个 Bean，名称分别为 anotherService 和 primaryAnotherService。如果在注入的地方不使用这个两个名称，那么就会按照 Bean 的类型自动注入。

```
@Component
public class UsePrimaryService {
   private AnotherService service;

   public UsePrimaryService(AnotherService service) {
      this.service = service;
   }

   public void doSomething(){
      System.out.println("foo".equals(service.getPerson()));
   }
}
```

因为现在使用的 service 不符合按照名称自动注入，所以是按照类型自动注入的。因为 primaryAnotherService 注解了@Primary，所以使用 primaryAnotherService 这个 Bean。

3. @Qualifier

在上面的例子中，使用 UsePrimaryService 注入的 AnotherService 的 Bean 只能是 primaryAnotherService，这时可以使用@Qualifier 直接指定需要使用哪个 Bean。还是使用上面例子中的两个 Bean。

注入 anotherService：

```
@Component
public class UseQualifierService {
   @Autowired
   @Qualifier("anotherService") //通过@Qualifier("anotherService")指定使用
                              //anotherService
   private AnotherService service;

   public void doSomething(){
      System.out.println("wyf".equals(service.getPerson()));
   }
}
```

注入 primaryAnotherService。

```
@Component
public class UseQualifierService2 {
   private AnotherService service;

   public UseQualifierService2(@Qualifier("primaryAnotherService") AnotherService
service) {
      this.service = service;
   }

   public void doSomething(){
      System.out.println("foo".equals(service.getPerson()));
   }

}
```

3.2.4　运行检验（CommandLineRunner）

在 Spring Boot 下可以注册一个 CommandLineRunner 的 Bean，在容器启动后，这个 Bean 可用来执行一些专门的任务，如在 JavaConfig 里。

```
@Bean
CommandLineRunner configClr(AnnotationInjectionService annotationInjectionService,
                      AnnotationOneInjectionService annotationOneInjectionService,
                      AnnotationPropertyInjectionService
annotationPropertyInjectionService,
                      AnnotationSetterInjectionService
annotationSetterInjectionService,
                      JavaConfigInjectService javaConfigInjectService,
                      MixInjectionService mixInjectionService,
                      MixInjectionService2 mixInjectionService2,
                      UsePrimaryService usePrimaryService,
                      UseQualifierService useQualifierService,
                      UseQualifierService2 useQualifierService2) {
   return args -> {
      System.out.println(args);
      annotationInjectionService.doMyThing();
      annotationOneInjectionService.doMyThing();
      annotationPropertyInjectionService.doMyThing();
      annotationSetterInjectionService.doMyThing();
      javaConfigInjectService.doMyThing();
      mixInjectionService.doMyThing();
      usePrimaryService.doSomething();
      mixInjectionService2.doMyThing();
      useQualifierService.doSomething();
      useQualifierService2.doSomething();
   };
}
```

a. 通过参数注入当前的 CommandLineRunner Bean 中。

b. CommandLineRunner 是一个函数接口，输入的参数为 main 方法里接收的 args 参数。这里使用 Lambda 表达式执行每个 Bean 的 doMyThing 方法，如图 3-1 所示。

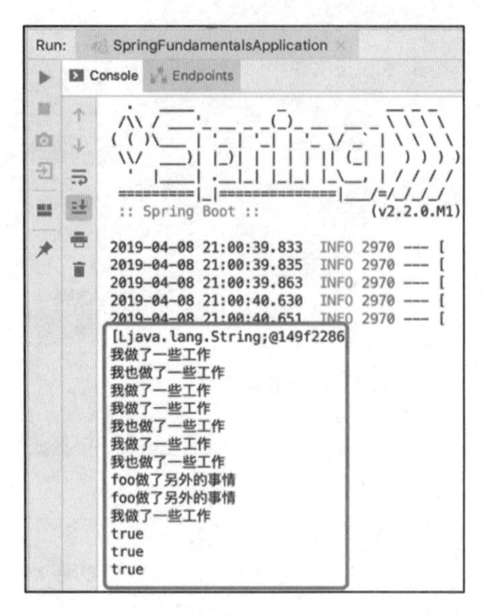

图 3-1

CommandLineRunner 有个姊妹接口叫作 ApplicationRunner，它们之间唯一的区别是 ApplicationRunner 使用 org.springframework.boot.DefaultApplicationArguments 类型的参数，示例如下。

```
@Bean
ApplicationRunner configAr(){
 return args -> System.out.println(args);
}
```

CommandLineRunner 的 args 是不定长字符串（String... args），而 ApplicationRunner 的 args 是 DefaultApplicationArguments 类型的对象。

3.2.5 Bean 的 Scope

容器中的 Bean 的 Scope 指的是 Bean 的实例在容器中创建的方式。在容器中，默认是 singleton，即整个容器中只创建一个 Bean 的实例。常用的还有 prototype，即每次请求 Bean 时都会创建一个实例。可以通过@Scope 注解来设置 Scope。

下面两种方式是相同的：

```
@Service
public class ScopeService {
}
```

```
@Service
@Scope(BeanDefinition.SCOPE_SINGLETON) //与@Scope("singleton")相同
public class ScopeService {
}
```

通过@Scope(BeanDefinition.SCOPE_PROTOTYPE)指定 Scope 为 prototype：

```
@Service
@Scope(BeanDefinition.SCOPE_PROTOTYPE) //与@Scope("prototype")相同
public class ScopeService2 {
}
```

除可以在方法上注解@Scope 外，还可以在@Bean 的类上注解@Scope，示例如下：

```
public class ScopeService3 {
}
```

在 JavaConfig 中配置的代码如下：

```
@Bean
@Scope(ConfigurableBeanFactory.SCOPE_PROTOTYPE) //与@Scope("prototype")相同
public ScopeService3 scopeService3(){
    return new ScopeService3();
}
```

这时可以在 ScopeInjectService Bean 中分别给上面三个 Bean 注入两次，由此判断相同类型的两个注入是否相等。

```
@Service
public class ScopeInjectService {
    private ScopeService scopeService;
    private ScopeService scopeService1;
    private ScopeService2 scopeService2;
    private ScopeService2 scopeService21;
    private ScopeService3 scopeService3;
    private ScopeService3 scopeService31;
     //只有一个构造器，此处可省略@Autowired
    public ScopeInjectService(ScopeService scopeService,
                        ScopeService scopeService1,
                        ScopeService2 scopeService2,
                        ScopeService2 scopeService21,
                        ScopeService3 scopeService3,
                        ScopeService3 scopeService31) {
        this.scopeService = scopeService;
        this.scopeService1 = scopeService1;
        this.scopeService2 = scopeService2;
        this.scopeService21 = scopeService21;
        this.scopeService3 = scopeService3;
        this.scopeService31 = scopeService31;
    }
```

```
    public void validateScope(){
        System.out.println(scopeService.equals(scopeService1));//容器内只有一个实例，
                                                                //所以相等
        System.out.println(scopeService2.equals(scopeService21));//请求注入创建新的
                                                                  //实例，不相等
        System.out.println(scopeService3.equals(scopeService31));//请求注入创建新的
                                                                  //实例，不相等
    }
}
```

在 JavaConfig 中配置下面的代码。

```
@Bean
CommandLineRunner scopeClr(ScopeInjectService scopeInjectService){
    return args -> {
      scopeInjectService.validateScope();
    };
}
```

执行效果如图 3-2 所示。

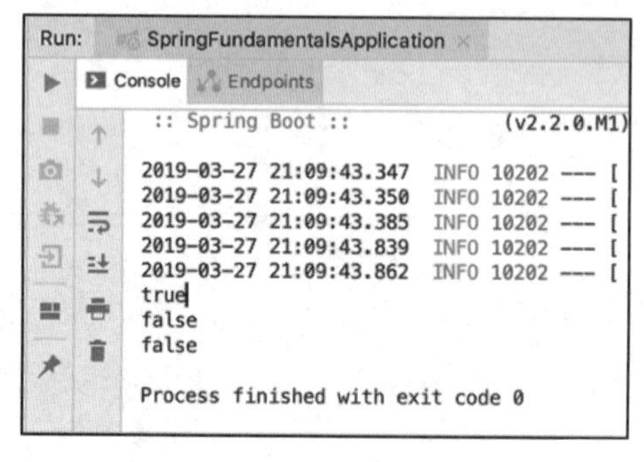

图 3-2

3.2.6 Bean 的生命周期

1. 初始化和销毁

我们可以定制 Bean 在容器中的初始化行为和销毁行为。

（1）注解配置：使用@PostConstruct 和@PreDestroy。

```
@Service
public class LifeService {

    public LifeService() {
        System.out.println("正在构造");
    }

    @PostConstruct
```

```
    public void exeAfterConstruct(){
        System.out.println("在构造完成后执行");
    }

    @PreDestroy
    public void exeBeforeDestroy(){
        System.out.println("在销毁之前执行");
    }
}
```

执行效果如图 3-3 所示。

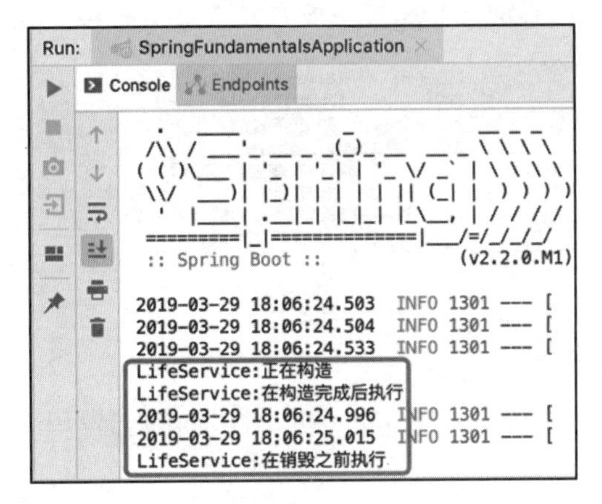

图 3-3

（2）Java 配置：使用@Bean 的 initMethod 和 destroyMethod。

Bean 的定义如下。

```
public class LifeService2 {
    public LifeService2() {
        System.out.println("LifeService2:正在构造");
    }

    public void exeAfterConstruct(){
        System.out.println("LifeService2:在构造完成后执行");
    }

    public void exeBeforeDestroy(){
        System.out.println("LifeService2:在销毁之前执行");
    }
}
```

在 JavaConfig 中配置下面的代码。

```
@Bean(initMethod = "exeAfterConstruct", destroyMethod = "exeBeforeDestroy")
public LifeService2 lifeService2(){
```

```
    return new LifeService2();
}
```

执行效果如图 3-4 所示。

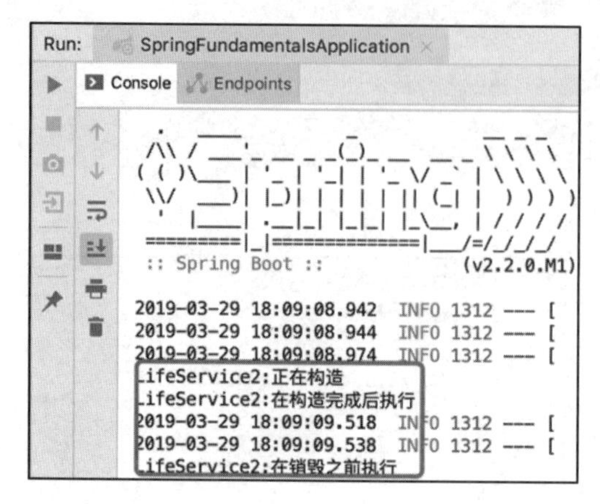

图 3-4

2．延迟初始化（@Lazy）

只要在 Bean 上注解了@Lazy，那么 Bean 在被调用时就会被初始化。它可以和@Component 类注解或@Bean 一起使用。

```
@Service
@Lazy
public class LifeService {}
```

```
@Bean(initMethod = "exeAfterConstruct", destroyMethod = "exeBeforeDestroy")
@Lazy
public LifeService2 lifeService2(){}
```

因为这两个 Bean 没有被调用过，所以没有被初始化，此时控制台没有任何输出。

3．依赖顺序（@DependsOn）

设置 Bean lifeService2 依赖于 lifeService，让 lifeService 先初始化，可以用@DependsOn 来实现。

```
@Bean(initMethod = "exeAfterConstruct", destroyMethod = "exeBeforeDestroy")
@DependsOn("lifeService")
public LifeService2 lifeService2(){
    return new LifeService2();
}
```

执行的结果是 lifeService 先于 lifeService2 初始化，如图 3-5 所示。

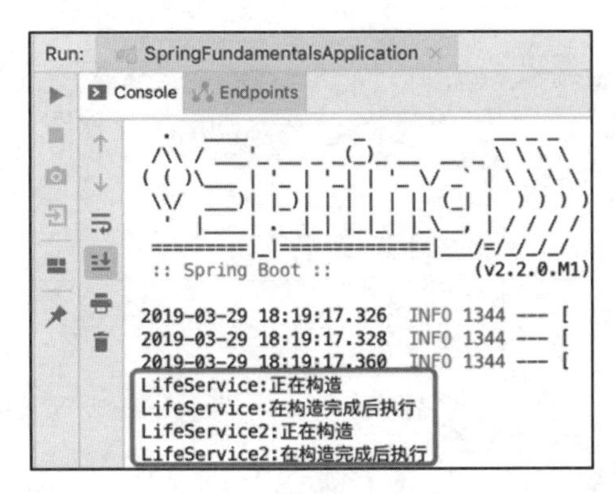

图 3-5

3.2.7　应用环境

Spring 提供了一个接口 Environment 来代表当前运行的应用环境，这个环境包含两部分。

◎　Profile：一组命名的、定义在一起的 Bean。通常为不同的应用场景（生产环境、开发环境、测试环境等）定义。

◎　Property：配置属性，可以从 properties 文件、JVM 系统属性、操作系统环境变量等外部来获得配置属性。

1．场景（@Profile）

可以通过@Profile 注解指定当前的运行场景。@Profile 可以和@Component、@Configuration、@Bean 等一起使用，当然也分别限制了@Profile 生效的 Bean 的分组。

下面使用需要显示不同操作系统的列表命令（在 Windows 下为 dir，在 Linux 下为 ls）的 Bean。

```
public class CommandService {
    private String listCommand;

    public CommandService(String listCommand) {
        this.listCommand = listCommand;
    }

    public void list(){
        System.out.println("当前系统下列表命令是：" + listCommand);
    }
}
```

在 Windows 开发环境下，场景配置如下。

```
@Configuration
@Profile("dev")
public class WindowsProfileConfig {
```

```
    @Bean
    CommandService commandService(){
        return new CommandService("dir");
    }
}
```

在 Linux 开发环境下，场景配置如下。

```
@Configuration
@Profile("production")
public class LinuxProfileConfig {
    @Bean
    CommandService commandService(){
        return new CommandService("ls");
    }
}
```

当配置好两种不同场景下的 Profile 后，我们需要在应用中配置哪个是激活的 Profile，手动配置如下。

```
AnnotationConfigApplicationContext context = new AnnotationConfigApplicationContext();
context.getEnvironment().setActiveProfiles("production");
context.scan("top.wisely");
context.refresh();
```

因为使用了 Spring Boot，所以只需在 application.properties 文件中做如下配置即可。

```
spring.profiles.active=production
```

在 JavaConfig 里，用 CommandLineRunner 分别将 Profile 配置成 production 和 dev，执行效果如图 3-6 和图 3-7 所示。

```
@Bean
CommandLineRunner profileClr(CommandService commandService){
  return args -> commandService.list();
}
```

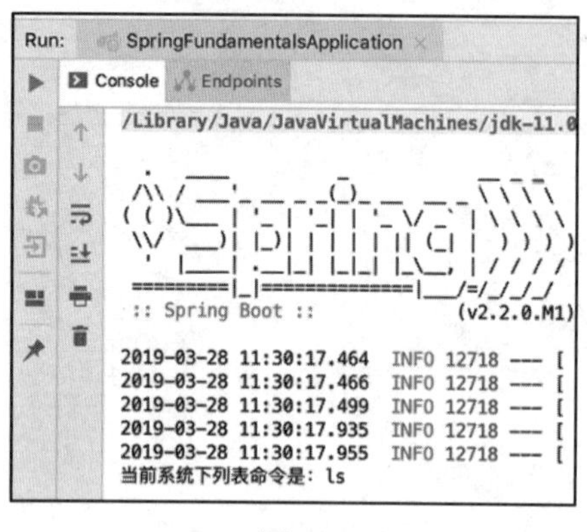

图 3-6

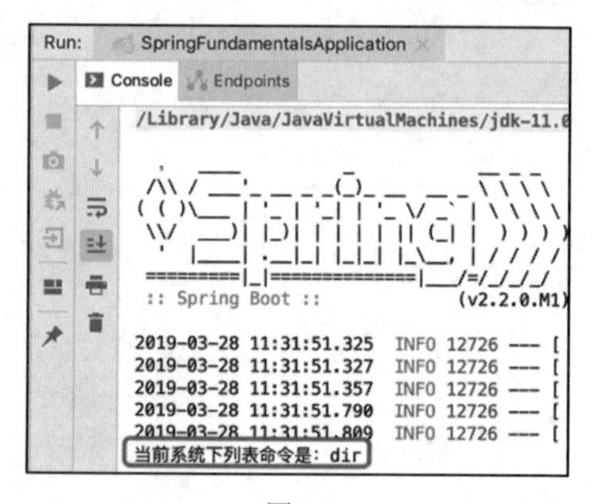

图 3-7

2. 属性配置（@PropertySource）

Spring 的 Environment 属性是由 PropertySource 组成的，我们可以通过@PropertySource 指定外部配置文件的路径。这些配置文件的属性都会以 PropertySource 的形式注册到 Environment 中，@PropertySource 支持 XML 格式和 properties 格式，不支持 Spring Boot 下的 YAML 格式。

现在添加 2 个外部配置文件。

◎ author.properties：

author.name=wyf

◎ book.properties：

book.name=spring boot in battle

在添加完成后，可以用一个配置类来接收这两个文件的配置。

```
@Configuration
@PropertySources({
        @PropertySource("classpath:author.properties"),
        @PropertySource("classpath:book.properties")
}) //a
public class ExternalConfig {

    Environment env;

    public ExternalConfig(Environment env) { //b
        this.env = env;
    }

    @Value("${book.name}") //c
    private String bookName;

    public void showEnv(){
        System.out.println("作者名字是: " + env.getProperty("author.name")); //d
        System.out.println("书籍名称是: " + bookName);
    }
}
```

a. 当有多个外部配置时，可以用@PropertySources 指定。若只有一个可用，则只使用@PropertySource("classpath:book.properties")。

b. 注入 Environment 的 Bean，因为只有一个构造器，所以可省略@Autowired。

c. 可以通过@Value 注解获得 Environment 中的属性，关于@Value 的更详细的讲解见 3.6 节。

d. 外部配置的属性都已经在 Environment 中注册过，可以直接获取。

3.2.8 条件配置（@Conditional）

通过@Conditional 我们可以定义当满足某个特定条件（Condition）时，应该做什么配置。@Conditional 同样可以和@Component、@Configuration、@Bean 一起使用，进而指定条件起作用的范围。

@Conditional 注解接收 Condition 数组作为参数，Condition 即我们的特定条件。Condition 只有一个方法 matches，当符合条件时，返回 true；当不符合条件时，返回 false。

例如，判断当前系统是否是 Windows 的条件定义：

```
public class OnWindowsCondition implements Condition { //a
   @Override
   public boolean matches(ConditionContext context, AnnotatedTypeMetadata metadata)
{ //b
      String osName = context.getEnvironment().getProperty("os.name"); //c
      if(osName.indexOf("win")>=0)
         return true;
      else
         return false;
   }
}
```

a. 条件实现 Condition 接口即可。

b. matches 的两个参数：ConditionContext 可获得容器的相关信息；AnnotatedTypeMetadata 是当前被注解的方法或类的元数据（数据的描述）信息。

c. 通过容器 context 获得运行环境 Environment 信息，从而获得操作系统信息。

配置如下。

```
@Configuration
public class SystemAutoConfig {

   @Bean
   @Conditional(OnWindowsCondition.class) // a
   public CommandService windows(){
      return new CommandService("dir");
   }
}
```

a. @Conditional 使用的是 OnWindowsCondition 条件，只有在操作系统是 Windows 的情况下，当前 Bean 才会被创建。

在 JavaConfig 中使用 CommandLineRunner 运行。

```
@Bean
CommandLineRunner windowsConditionalClr(CommandService windows){
   return args -> windows.list();
}
```

在 Windows 系统才能正常执行；在非 Windows 系统下会报错，找不到 Bean。因为不符合条件，所以没有创下这个 Bean。

3.2.9 开启配置（@Enable*和@Import）

在本书后面的内容里会出现大量以@Enable*开头的注解， @Enable*会自动对相应的功能进行自动配置，如@EnableWebMvc、@EnableCaching、@EnableScheduling、@EnableAsync、@EnableWebSocket、@EnableJpaRepositories、@EnableTransactionManagement、@EnableJpaAuditing 和@EnableAspectJAutoProxy 等。

@Enable*的开启配置的功能依赖于@Import 注解，@Import 注解支持导入如下配置：

◎ 直接导入@Configuration 配置类。

◎ 配置类选择器 ImportSelector 的实现。

◎ 动态注册器 ImportBeanDefinitionRegistrar 的实现。

◎ 混合以上三种。

下面将分别演示四种方式的实现。

1. 直接导入@Configuration 配置类

当应用注解了@Configuration 后，会被 Spring Boot 的默认组件扫描并自动注册，所以本节的注解类代码放在 io.github.wiselyman.annotations 中，配置类的代码放在 io.github.wiselyman.config 中。

定义注解：

```
@Target(ElementType.TYPE)
@Retention(RetentionPolicy.RUNTIME)
@Documented
@Import(AConfig.class) //直接导入配置类 AConfig.class
public @interface EnableA {
}
```

定义配置类：

```
@Configuration
public class AConfig {

   @Bean
   public String a(){
      return "A";
   }
}
```

在 JavaConfig 中使用@EnableA 注解，即可获得导入的配置类 AConfig 中的 Bean a。

```
@EnableA
public class JavaConfig {}
```

在 JavaConfig 中使用 CommandLineRunner 查看 Bean 的内容，执行结果如图 3-8 所示。

```
@Bean
CommandLineRunner enableAClr(String a){
   return args -> System.out.println(a);
}
```

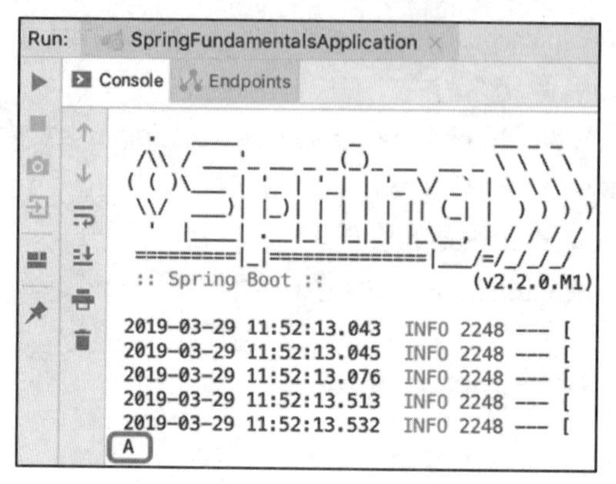

图 3-8

2．配置类选择器 ImportSelector 的实现

在这个例子中，通过注解选择生效的配置类，注解定义如下。

```
@Target(ElementType.TYPE)
@Retention(RetentionPolicy.RUNTIME)
@Documented
@Import(BSelector.class)
public @interface EnableB {

   boolean isUppercase() default true; //isUppercase 是选择条件
}
```

在 io.github.wiselyman.selector 中定义选择器。

```
public class BSelector implements ImportSelector { //a
   @Override
   public String[] selectImports(AnnotationMetadata importingClassMetadata) { //b
      AnnotationAttributes attributes =
            AnnotationAttributes.fromMap(
                  importingClassMetadata.getAnnotationAttributes
                        (EnableB.class.getName(), false));
      boolean isUppercase = attributes.getBoolean("isUppercase"); //c
      if(isUppercase == true)
         return new String[]{"io.github.wiselyman.config.BUppercaseConfig"}; //d
      else
```

```
            return new String[]{"io.github.wiselyman.config.BLowercaseConfig"}; //e
    }
}
```

a. 选择器要实现ImportSelector接口。

b. 实现接口的selectImports方法，参数AnnotationMetadata importClassMetadata是注解使用类（本例为JavaConfig）上@EnableB的元数据信息。

c. 通过@EnableB在实际使用中的元数据，获得isUppercase的值。

d. 如果isUppercase == true，则此时实际使用的是@EnableB或者@EnableB(isUppercase = true)，因而使用BUppercaseConfig提供的配置。

e. 若实际使用的是@EnableB(isUppercase = false)，则使用BLowercaseConfig提供的配置。

BUppercaseConfig的定义如下。

```
@Configuration
public class BUppercaseConfig {
    @Bean
    public String b(){
        return "B"; //返回一个大写B的Bean的配置
    }
}
```

BLowercaseConfig的定义如下。

```
@Configuration
public class BLowercaseConfig {
    @Bean
    public String b(){
        return "b"; //返回一个小写b的Bean的配置
    }
}
```

在JavaConfig中使用@EnableB，并用CommandLineRunner检验。

```
@EnableB
public class JavaConfig {}
@Bean
CommandLineRunner enbaleBClr(String b){
    return args -> System.out.println(b);
}
```

运行结果如图3-9所示。

若将isUppercase设置为false，则执行结果如图3-10所示。

```
@EnableB(isUppercase = false)
public class JavaConfig {}
```

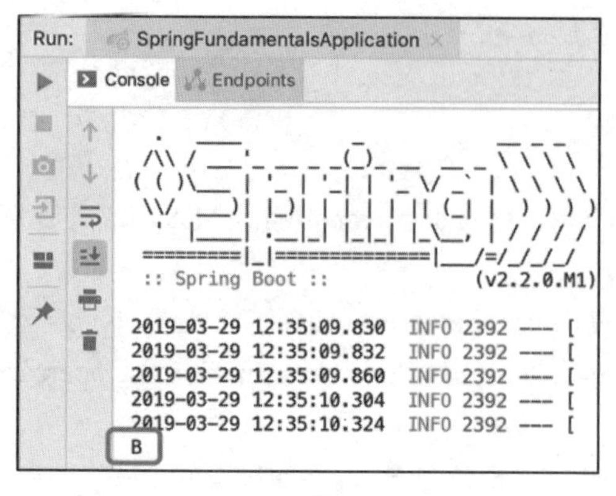

图 3-9

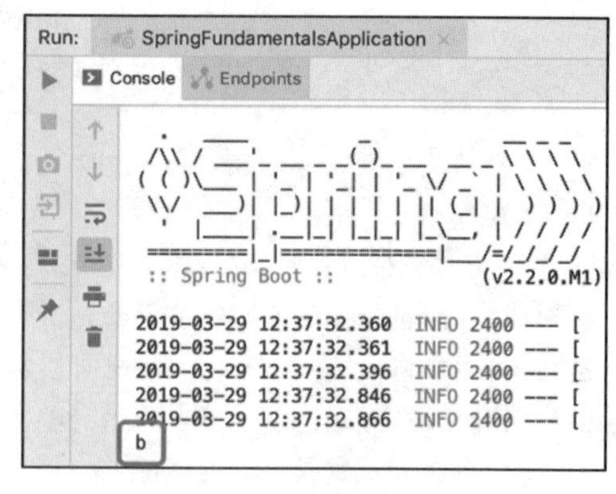

图 3-10

3. 动态注册器 ImportBeanDefinitionRegistrar 的实现

本例通过 ImportBeanDefinitionRegistrar 动态注册 Bean 到容器里。

注解定义如下：

```
@Target(ElementType.TYPE)
@Retention(RetentionPolicy.RUNTIME)
@Documented
@Import(CBeanDefinitionRegistrar.class)
public @interface EnableC {
}
```

在 io.github.wiselyman.registrar 中定义注册器。

```
public class CBeanDefinitionRegistrar implements ImportBeanDefinitionRegistrar { //a
    @Override
    public void registerBeanDefinitions(AnnotationMetadata importingClassMetadata, //b
                                    BeanDefinitionRegistry registry) { //c
        BeanDefinition bd = BeanDefinitionBuilder.genericBeanDefinition(String.class)
//d
                                        .addConstructorArgValue("C") //e
                                        .setScope(BeanDefinition.SCOPE_SINGLETON) //f
                                        .getBeanDefinition(); //g
        registry.registerBeanDefinition("c",bd); //h
    }
}
```

a. 注册器需实现 ImportBeanDefinitionRegistrar 接口。

b. 实现 registerBeanDefinitions 参数 AnnotationMetadata importClassMetadata 是注解使用类（本例为 JavaConfig）上@EnableB 的元数据信息。

c. 参数 BeanDefinitionRegistry registry 用来注册所有 Bean 的定义的接口。

d. 可以使用 BeanDefinitionBuilder 来编程实现 Bean 的定义（BeanDefinition），此句定义了一个类型为 String 的 Bean。

e. 构造 String 的值是 C。

f. 设置 Bean 的 Scope 是 singleton。

g. 获得 Bean 的定义。

h. 将 Bean 注册为名称为 c 的 Bean。

此时，在 JavaConfig 上使用@EnableC 注解，并用 CommandLineRunner 进行检验。

```
@EnableC
public class JavaConfig {}
@Bean
CommandLineRunner enableCClr(String c){
   return args -> System.out.println(c);
}
```

IntelliJ IDEA 可以检测到静态注册的 Bean，但检测不到动态注册的 Bean，因而 IDE 会标识红色，如图 3-11 中方框所示。

```
@Bean
CommandLineRunner enableCCle(String c){
    return args -> System.out.println(c);
}
```

图 3-11

但可以正常运行，运行结果如图 3-12 所示。

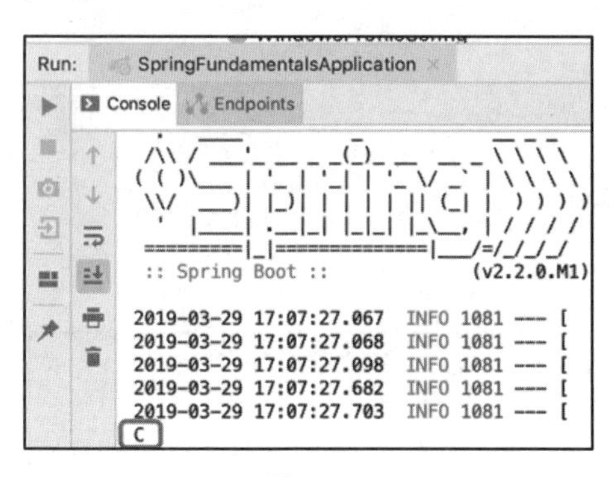

图 3-12

4. 混合使用

@Import 支持导入配置类的数组，因而我们可以混合上面三种配置，定义一个注解，使其具备上面三个功能。

```
@Target(ElementType.TYPE)
@Retention(RetentionPolicy.RUNTIME)
@Documented
@Import({AConfig.class, BForABCSelector.class, CBeanDefinitionRegistrar.class})
```

```
public @interface EnableABC {
   boolean isUppercase() default true;
}
```

因为选择器里指定了要使用的注解的类，所以需要新建一个选择器。

```
public class BForABCSelector implements ImportSelector {
   @Override
   public String[] selectImports(AnnotationMetadata importingClassMetadata) {
      AnnotationAttributes attributes =
            AnnotationAttributes.fromMap(
                  importingClassMetadata.getAnnotationAttributes
                        (EnableABC.class.getName(), false)); // 此处使用的是
                                                            // @EnableABC
      boolean isUppercase = attributes.getBoolean("isUppercase");
      if(isUppercase == true)
         return new String[]{"io.github.wiselyman.config.BUppercaseConfig"};
      else
         return new String[]{"io.github.wiselyman.config.BLowercaseConfig"};
   }
}
```

在 JavaConfig 中启用@EnableABC，并用 CommandLineRunner 进行检验。

```
@EnableABC
public class JavaConfig {}
@Bean
CommandLineRunner enableABCClr(String a, String b, String c){
   return args -> {
      System.out.println(a);
      System.out.println(b);
      System.out.println(c);
   };
}
```

校验结果如图 3-13 所示。

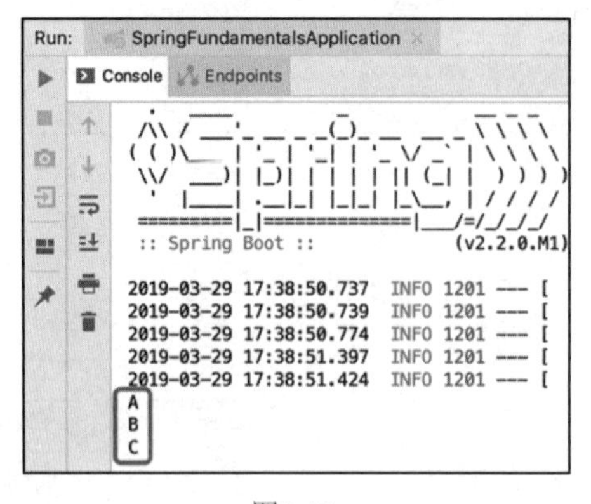

图 3-13

3.3 对 Bean 的处理（BeanPostProcessor）

可以通过实现 BeanPostProcessor 接口，在构造时对容器内所有或者部分指定 Bean 进行处理。和@PostConstruct 与@PreDestroy 不同的是，它针对的是 IoC 容器里的所有的 Bean。

```
@Component
public class GlobalPostProcessor implements BeanPostProcessor {
    @Override //初始化之前的处理
    public Object postProcessBeforeInitialization(Object bean, String beanName) throws
BeansException {
        System.out.println("----" + beanName + "----");
        System.out.println("----" + beanName.getClass() + "----");
        return bean;
    }

    @Override //初始化之后的处理
    public Object postProcessAfterInitialization(Object bean, String beanName) throws
BeansException {
        System.out.println("++++" + beanName + "++++");
        System.out.println("++++" + beanName.getClass() + "++++");
        return bean;
    }
}
```

通过覆写 postProcessBeforeInitialization 和 postProcessAfterInitialization 方法，所有的 Bean 在初始化之前都会执行 postProcessBeforeInitialization 里的处理逻辑，在初始化之后都会执行 postProcessAfterInitialization 里的处理逻辑。执行结果如图 3-14 所示。

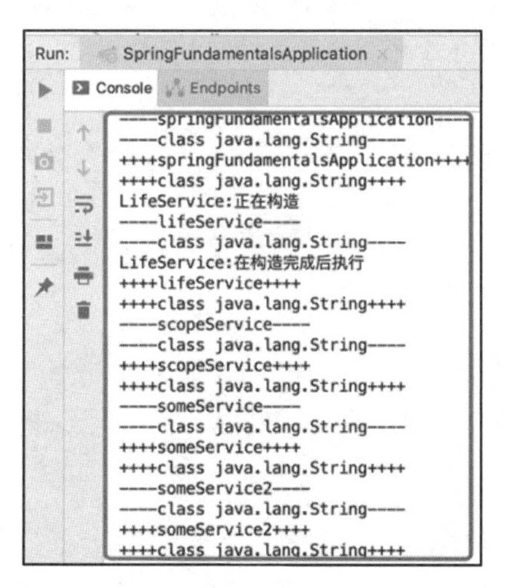

图 3-14

如果想要缩小 Processor 的处理范围，则可以通过判断 Bean 类型来实现。

```
@Override
public Object postProcessAfterInitialization(Object bean, String beanName) throws
BeansException {
    if (bean instanceof LifeService){ //用 instanceof 缩小处理范围
        System.out.println("++++" + beanName + "++++");
        System.out.println("++++" + beanName.getClass() + "++++");
    }
    return bean;
}
```

3.4 Spring Aware 容器

程序员的主要工作是编写业务逻辑代码，业务逻辑代码一般都是技术无关性的，即 Spring 代码不会侵入业务逻辑代码中。虽然我们使用了很多 Spring 的注解，但注解属于元数据（和 XML 一样），不属于代码侵入。

但有些时候却不得不让自己的代码和 Spring 框架耦合，通过实现相应的 Aware 接口，注入其对应的 Bean。

- ◎ BeanNameAware：可获得 beanName，即 Bean 的名称。
- ◎ ResourceLoaderAware：可获得 ResourceLoader，即用来加载资源的 Bean。
- ◎ BeanFactoryAware：可获得 BeanFactory，即容器的父接口，用于管理 Bean 的相关操作。
- ◎ EnvironmentAware：可获得 Environment，即当前应用的运行环境。
- ◎ MessageSourceAware：可获得 MessageSource，即用来解析文本信息的 Bean。
- ◎ ApplicationEventPublisherAware：可获得 ApplicationEventPublisher，即用来发布系统时间的 Bean。
- ◎ ApplicationContextAware：可自动注入 ApplicationContext，即容器本身。

```
@Component
public class AwareSpringService implements BeanNameAware,
                                           ResourceLoaderAware,
                                           BeanFactoryAware,
                                           EnvironmentAware,
                                           ApplicationEventPublisherAware,
                                           ApplicationContextAware{

    private String beanName;
    private ResourceLoader resourceLoader;
    private BeanFactory beanFactory;
    private Environment environment;
    private ApplicationEventPublisher publisher;
    private ApplicationContext context;
```

```
    @Override //BeanNameAware 接口的覆写方法，自动注入 name
    public void setBeanName(String name) {
        this.beanName = name;
    }

    @Override//ResourceLoaderAware 接口的覆写方法，自动注入 resourceLoader
    public void setResourceLoader(ResourceLoader resourceLoader) {
        this.resourceLoader = resourceLoader;
    }

    @Override//BeanFactoryAware 接口的覆写方法，自动注入 beanFactory
    public void setBeanFactory(BeanFactory beanFactory) throws BeansException {
        this.beanFactory = beanFactory;
    }

    @Override//EnvironmentAware 接口的覆写方法，自动注入 environment
    public void setEnvironment(Environment environment) {
        this.environment = environment;
    }

    @Override//ApplicationEventPublisherAware 接口的覆写方法，自动注入
            //applicationEventPublisher
    public void setApplicationEventPublisher(ApplicationEventPublisher
applicationEventPublisher) {
        this.publisher = applicationEventPublisher;
    }

    @Override//ApplicationContextAware 接口的覆写方法，自动注入 applicationContext
    public void setApplicationContext(ApplicationContext applicationContext) throws
BeansException {
        this.context = applicationContext;
    }

    public void doSomething() throws IOException{
        beanNameDemo();
        resourceLoaderDemo();
        beanFactoryDemo();
        environmentDemo();
        publisherDemo();
        contextDemo();
    }

    private void beanNameDemo(){
        System.out.println("注入的 beanName 为：" + beanName);
    }

    private void resourceLoaderDemo() throws IOException {
            //可以用 resourceLoader 获取外部资源
        Resource resource =
```

```java
resourceLoader.getResource("https://avatars3.githubusercontent.com/u/1981770");
        System.out.println("通过注入的 ResourceLoader 加载的文件长度为：" +
resource.contentLength());

    }

    private void beanFactoryDemo(){
        AwareSpringService service =  beanFactory.getBean(AwareSpringService.class);
        System.out.println("通过注入的 BeanFactory 获得的当前 Bean 的名称为：" +
service.beanName);
    }

    private void environmentDemo(){
        String osName = environment.getProperty("os.name");
        System.out.println("通过注入的 Environment 获得的操作系统名称为：" + osName);
    }

    private void publisherDemo(){
        System.out.println("通过注入的 ApplicationEventPublisher 发布了系统事件");
        //在 3.5 节讲解
        publisher.publishEvent(new MessageEvent("来自 AwareSpringService 的消息"));
    }

    private void contextDemo(){
        System.out.println("通过注入的 ApplicationContext 获得容器的显示名称：" +
context.getDisplayName());
    }

}
```

在 JavaConfig 中执行如下代码，执行结果如图 3-15 所示。

```java
@Bean
CommandLineRunner awareClr(AwareSpringService awareSpringService){
    return args -> awareSpringService.doSomething();
}
```

```
 :: Spring Boot ::        (v2.2.0.M1)

2019-03-29 22:01:36.491  INFO 1793 --- [  restartedMain] t.w.s.SpringFundamentalsApplication      : Starting SpringFundamentals
2019-03-29 22:01:36.492  INFO 1793 --- [  restartedMain] t.w.s.SpringFundamentalsApplication      : The following profiles are
2019-03-29 22:01:36.520  INFO 1793 --- [  restartedMain] .e.DevToolsPropertyDefaultsPostProcessor : Devtools property defaults
2019-03-29 22:01:36.956  INFO 1793 --- [  restartedMain] o.s.b.d.a.OptionalLiveReloadServer       : LiveReload server is runnin
2019-03-29 22:01:36.975  INFO 1793 --- [  restartedMain] t.w.s.SpringFundamentalsApplication      : Started SpringFundamentalsA
注入的beanName为：awareSpringService
通过注入的ResourceLoader加载的文件长度为：4729
通过注入的BeanFactory获得的当前Bean的名称为：awareSpringService
通过注入的Environment获得的操作系统名称为：Mac OS X
通过注入的ApplicationEventPublisher发布了系统事件
通过注入的ApplicationContext获得容器的显示名称：org.springframework.context.annotation.AnnotationConfigApplicationContext@7044f73d
```

图 3-15

3.5　Bean 之间的事件通信

如果 Bean 之间需要通信，比如说 BeanA 完成了处理后需要告知 BeanB，通知 BeanB 继续处理，那么我们称 BeanA 为 Publisher，称 BeanB 为 Listener。

Publisher 和 Listener 之间传递的事件数据通过继承 ApplicationEvent 来实现。

Publisher 的实现方式如下。

◎ 通过 ApplicationEventPublisherAware 注入 ApplicationEventPublisher 发布事件，前面已演示。

◎ 直接注入 ApplicationEventPublisher 发布事件。

◎ 直接注入 ApplicationContext 发布事件。

Listener 的实现方式如下。

◎ 实现 ApplicationListener 接口。

◎ 注解@EventListener 的方法接收事件。

事件数据：

```
public class MessageEvent extends ApplicationEvent {

    private String message;

    public MessageEvent(String message) {
        super(message);
       this.message = message;
    }

    public String getMessage() {
        return message;
    }

    public void setMessage(String message) {
        this.message = message;
    }
}
```

发布者：

```
@Component
public class EventPublishService {
    ApplicationEventPublisher publisher;

    public EventPublishService(ApplicationEventPublisher publisher) {
        this.publisher = publisher;
    }

    public void publish(){
```

```java
        System.out.println("EventPublishService 正在处理,处理完成后通知 EventListenerService
继续处理");
        publisher.publishEvent(new MessageEvent("EventPublishService 处理完了"));
    }
}
```

监听者：

```java
@Component
public class EventListenerService implements ApplicationListener<MessageEvent> { //1

    @Override
    public void onApplicationEvent(MessageEvent event) { //a
        System.out.println("onApplicationEvent 接收到了: " + event.getMessage());
    }

}
```

a. EventListenerService 实现了 ApplicationListener<MessageEvent>接口，泛型 MessageEvent 可缩小监听事件的范围，通过覆写 onApplicationEvent 方法来监听事件。

监听者还可以注解@EventListener 方法来监听事件。

```java
@Component
public class EventListenerService2 {

    @EventListener
    public void eventListener(MessageEvent event){
        System.out.println("@EventListener 接收到了: " + event.getMessage());
    }

}
```

推荐使用@EventListener 注解方式，耦合度更低。

除能监听自定义发布的事件外，还可以监听系统发布的事件，示例如下。

```java
@EventListener
public void contextRefreshedEventListener(ContextRefreshedEvent event){
    System.out.println("@EventListener 接收到了: " + event.getSource());
}
```

主要的系统事件如下。

- ◎ ContextRefreshedEvent：当 ApplicationContext 被初始化或刷新时发布该事件。
- ◎ ContextStartedEvent：当 ApplicationContext 开始时发布该事件。
- ◎ ContextStoppedEvent：当 ApplicationContext 停止时发布该事件。
- ◎ ContextClosedEvent：当 ApplicationContext 关闭时发布该事件。

3.6 Spring EL

Spring EL（Spring Expression Language，Spring 表达式语言）是 Spring 生态下的通用语言，在运行时使用表达式查询属性信息（使用符号$）或操作 Java 对象（使用符号#），主要用在 XML 或注解上。

本节主要使用@Value(org.springframework.beans.factory.annotation.Value)注解来演示 Spring EL 的功能，它可以获得表达式计算出来的结果。

1．数学运算

可以在表达式中进行数学运算，包括+、–、*、/、%、^、div 和 mod。

```
@Service
public class ValueService {
   @Value("#{1 + 2}") // Java 对象操作使用#
   private Integer add;
}
```

2．比较运算

可以在表达式中进行比较运算，包括<、>、==、!=、<=、>=、lt、gt、eq、ne、le 和 ge。

```
@Value("#{1 == 2}")
private boolean compare;
```

3．逻辑运算

可以在表达式中进行逻辑运算，包括 and、or、not、&&、||和 !。

```
@Value("#{1 == 2 || 1 == 1}")
private boolean compareOr;
```

4．条件运算

使用三元运算符?:可进行条件运算。

```
@Value("#{1 < 2 ? 'wyf' : 'www'}")
private String name;
```

5．正则匹配

使用 matches 可判断字符是否符合正则表达式。

```
@Value("#{'1' matches '\\d+' }")
private boolean isNumber;
```

6．调用 Bean 方法

可以直接用 Spring EL 来调用 Bean 的方法，假设被调用的 Bean 定义如下。

```
@Service
public class ForValueService {
   public String generate(String name){
      return "Hello " + name;
   }
}
```

可以通过#{bean 名.方法名(参数)}来调用。

```
@Value("#{forValueService.generate('wyf')}")
private String beanReturn;
```

7．获得 Environment 中的属性

Environment 可以从外部文件或者操作系统环境变量中获取属性信息，示例如下。

```
@Service
@PropertySource("classpath:author.properties") //a
public class ValueService {
   @Value("${author.name}") //b
   private String authorName;

   @Value("${os.name}") //c
   private String osName;

}
```

a. 通过@PropertySource 把配置文件中的属性信息加载到 Environment 中。

b. 查询属性使用符号$，格式为$(属性名)。

c. 获得操作系统环境变量信息的方式与 b 一致。

完整代码如下。

```
@PropertySource("classpath:author.properties")
public class ValueService {
   @Value("#{1 + 2}")
   private Integer add;

   @Value("#{1 == 2}")
   private boolean compare;

   @Value("#{1 == 2 || 1 == 1}")
   private boolean compareOr;

   @Value("#{1 < 2 ? 'wyf' : 'www'}")
   private String name;

   @Value("#{'1' matches '\\d+' }")
   private boolean isValidNumber;

   @Value("#{forValueService.generate('wyf')}")
```

```
    private String beanReturn;

    @Value("${author.name}")
    private String authorName;

    @Value("${os.name}")
    private String osName;

    public void doSomething(){
        System.out.println("数学运算 add 的值是: " + add);
        System.out.println("逻辑运算 compare 的值是" + compare);
        System.out.println("逻辑运算 compareOr 的值是" + compareOr);
        System.out.println("条件运算 name 的值是: " + name);
        System.out.println("正则匹配 isValidNumber 的值是: " + name);
        System.out.println("调用 Bean 的返回值 beanReturn 值是: " + beanReturn);
        System.out.println("属性查询外部配置文件 authorName 值是: " + authorName);
        System.out.println("属性查询操作系统环境变量 authorName 值是: " + osName);
    }
}
```

在 JavaConfig 中进行校验，结果如图 3-16 所示。

```
@Bean
CommandLineRunner valueClr(ValueService valueService){
    return args -> valueService.doSomething();
}
```

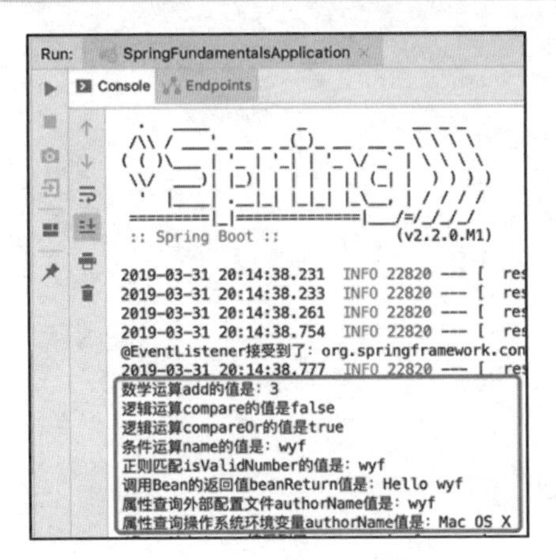

图 3-16

3.7　AOP

AOP（Aspect-Oriented Programming，面向切面编程）可以添加额外行为到现有指定条件的一批 Bean 上，但是我们并不需要修改 Bean 的代码，这样可对额外行为和 Bean 本身的行为实现关注隔离。

在学习 AOP 之前，先来熟悉下面的概念。

◎ 切面：Aspect，编写额外行为的地方。

◎ 连接点：Join Point，被拦截的方法。

◎ 切点：PointCut，通过条件匹配一批连接点。

◎ 建言：Advice，对于每个连接点需要做的行为。

◎ 目标对象：符合指定条件的 Bean。

在使用 AOP 开发之前，需要使用@EnableAspectJAutoProxy 注解来开启对 AspectJ 的支持，Spring Boot 已经自动做了配置，所以无须额外声明。

下面编写一个使用 AOP 来记录操作日志的例子。编写一个注解，作为切点拦截条件。

```
@Target(ElementType.METHOD)
@Retention(RetentionPolicy.RUNTIME)
@Documented
public @interface Logging {
    String value() default "";
}
```

目标对象：

```
@Service
public class PersonService {
    @Logging("人员新增操作")
    public void add(String name){ //每个被拦截的方法都是连接点
        System.out.println("人员新增");
    }

    @Logging("人员删除操作")
    public void remove(String name){
        System.out.println("人员删除");
    }

    @Logging("人员查询操作")
    public String query(String name){
        System.out.println("人员查询");
        return name;
    }

    @Logging("人员修改操作")
    public String modify(String name){
        System.out.println("人员修改");
        return name.toUpperCase();
    }
}
```

下面是最重要的部分，编写切面部分的代码。

```
@Aspect //a
@Component
public class LoggingAspect {

   @Pointcut("@annotation(top.wisely.springfundamentals.aop.Logging)") //b
   public void annotationPointCut(){}

   @Before("annotationPointCut()") //c
   public void beforeAnnotationPointCut(JoinPoint joinPoint){//d
      String name = (String) joinPoint.getArgs()[0]; //e
      MethodSignature methodSignature = (MethodSignature) joinPoint.getSignature();
      String action = methodSignature.getMethod().getAnnotation(Logging.class).value();
//f
      System.out.println("对" + name + "进行了"+ action);
   }

   @AfterReturning(pointcut = "annotationPointCut()", returning = "retName") //g
   public void afterReturningAnnotationPointCut(JoinPoint joinPoint, String retName){
      String name = (String) joinPoint.getArgs()[0];
      MethodSignature methodSignature = (MethodSignature) joinPoint.getSignature();
      String action = methodSignature.getMethod().getAnnotation(Logging.class).value();
      System.out.println("对" + name + "进行了"+ action + ", 返回的名字为: " + retName);
   }

}
```

a. 使用@Aspect 定义一个切面。

b. 使用@Pointcut，它会将所有注解了@Logging 注解的方法作为条件。

c. 使用@Before 建言，它使用的切点 annotationPointCut()会针对符合切点条件的 Bean 执行 beforeAnnotationPointCut()方法里的行为。

d. joinPoint 代表被拦截的方法，可以从 joinPoint 中获得方法的签名信息。

e. 通过 joinPoint 获得被拦截方法的参数。

f. 通过 joinPoint 获得被拦截方法的注解信息。

g. 使用@AfterReturning 建言，可以获得被拦截方法的返回值 retName。

在 JavaConfig 中执行下面的代码，结果如图 3-17 所示。

```
@Bean
CommandLineRunner aopCle(PersonService personService){
   return args -> {
      personService.add("wyf");
      personService.remove("wyf");
      personService.query("wyf");
      personService.modify("wyf");
   };
}
```

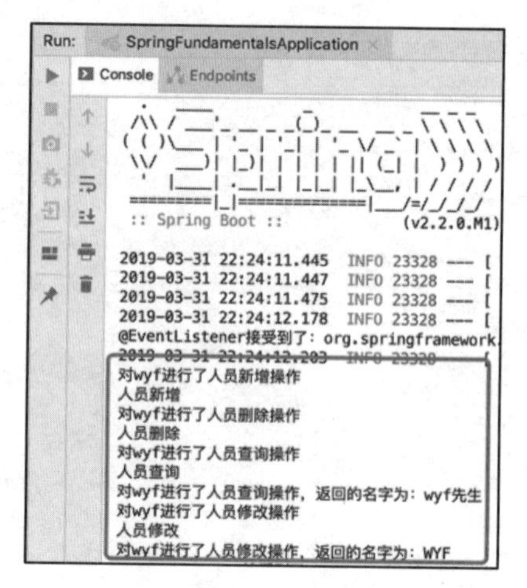

图 3-17

3.8 注解工作原理

注解本身只是元数据，即描述数据的数据，被描述的数据可以是类、方法、属性、参数或构造器等。注解本身是没有任何可执行功能的代码，但是只要标注了注解，我们就能得到想要的功能。

3.8.1 BeanPostProcessor

在 BeanPostProcessor 实现中，只要名称为*AnnotationBeanPostProcessor，就都是针对处理注解的，即对容器内标注了指定注解的 Bean，进行功能处理。

◎ AutowiredAnnotationBeanPostProcessor：让@Autowired、@Value 和@Inject 注解生效。

◎ CommonAnnotationBeanPostProcessor：让@PostConstruct 和@PreDestroy 注解生效。

◎ AsyncAnnotationBeanPostProcessor：让@Async 或@Asynchronous 注解生效。

◎ ScheduledAnnotationBeanPostProcessor：让@Scheduled 注解生效。

◎ PersistenceAnnotationBeanPostProcessor：让@PersistenceUnit 和 @PersistenceContext 注解生效。

◎ JmsListenerAnnotationBeanPostProcessor：让@JmsListener 注解生效。

它们会在构造器里指定其能处理的注解类型，并在对应的方法中进行功能处理。下面演示一个简单的例子，以帮助理解。我们定义注解@InjectLogger 向 Bean 注入 org.slf4j.Logger 来做系统日志。

首先定义要使用的注解，只能注解在类上，默认的后缀为“-Bean”。

```
@Target(ElementType.FIELD)
@Retention(RetentionPolicy.RUNTIME)
public @interface InjectLogger {
}
```

然后定义处理注解的 InjectLoggerAnnotationBeanPostProcessor，前面已介绍过，只需实现 BeanPostProcessor 接口即可。

```
@Component
public class InjectLoggerAnnotationBeanPostPorcessor implements BeanPostProcessor {
   private Class<? extends Annotation> changeAnnotationType; //a

   public InjectLoggerAnnotationBeanPostPorcessor() {
      this.changeAnnotationType = InjectLogger.class; //a
   }

   @Override
   public Object postProcessAfterInitialization(Object bean, String beanName) throws
BeansException {
      ReflectionUtils.doWithFields(bean.getClass(), field -> { //b
         ReflectionUtils.makeAccessible(field); //c
         if(field.isAnnotationPresent(changeAnnotationType)){ //d
            Logger logger = LoggerFactory.getLogger(bean.getClass());
            field.set(bean, logger); //e
         }
      });
      return bean;
   }
}
```

a. 指明当前类处理@InjectLogger 注解。

b. 通过反射机制对类的每个属性（Field）进行处理，第一个参数是 Bean 的 Class，第二个参数是入参为 Field，无返回值的函数接口的 Lambda 实现。

c. 通过反射机制让当前属性可访问。

d. 新建 Logger 的实例 logger。

e. 通过反射将 logger 值设置到 Bean 实例的当前属性（field）上。

将注解使用到其他 Bean 上。

```
@Component
public class DemoLoggerService {
   @InjectLogger
   private Logger log;

   public void doSomething(){
      log.info("通过自定义 InjectLoggerAnnotationBeanPostPorcessor 让注解@InjectLogger 注
入 Logger 对象");
   }
}
```

在 JavaConfig 中运行如下代码，结果如图 3-18 所示。

```
@Bean
CommandLineRunner changeAnnotationBeanPostProcessorClr(DemoLoggerService
demoLoggerService){
   return args -> {
      demoLoggerService.doSomething();
   };
}
```

```
t.w.s.SpringFundamentalsApplication       : Starting SpringFundamentalsApplication on wangyuneideMBP3 with PID 2087 (/Users/
t.w.s.SpringFundamentalsApplication       : The following profiles are active: dev
.e.DevToolsPropertyDefaultsPostProcessor : Devtools property defaults active! Set 'spring.devtools.add-properties' to 'fals
o.s.b.d.a.OptionalLiveReloadServer        : LiveReload server is running on port 35729
t.w.s.SpringFundamentalsApplication       : Started SpringFundamentalsApplication in 1.355 seconds (JVM running for 3.043)
t.w.s.injected.DemoLoggerService          : 通过自定义InjectLoggerAnnotationBeanPostPorcessor让注解@InjectLogger注入Logger对象
```

图 3-18

3.8.2 BeanFactoryPostProcessor

前面介绍了针对 Bean 进行处理的 BeanPostProcessor，而 BeanFactoryPostProcessor 是针对 Bean 的配置元数据（注解等）进行处理操作的，它属于 BeanFactory 的职责范畴。

◎ ConfigurationClassPostProcessor：使@PropertySource、@ComponentScan、@Component 类、@Configuration、@Bean、@Import 和@ImportResource 注解生效。

◎ EventListenerMethodProcessor：使@EventListener 注解生效。

下面通过一个自定义的注解@CustomBean 来自己自动注册 Bean，@CustomBean 的作用是配置元数据。

自定义的注解：

```
@Target(ElementType.TYPE)
@Retention(RetentionPolicy.RUNTIME)
public @interface CustomBean {
}
```

使用在 Bean 上：

```
@CustomBean
public class CustomBeanService {
   public void doSomething(){
      System.out.println("通过自定义的注解成功注册 Bean");
   }
}
```

同样，实现 BeanFactoryPostProcessor 接口：

```
@Component
public class CustomBeanDefinitionRegistryPostProcessor implements
BeanFactoryPostProcessor {
```

```
    @Override
    public void postProcessBeanFactory(ConfigurableListableBeanFactory beanFactory)
throws BeansException {
        ClassPathBeanDefinitionScanner scanner = new
ClassPathBeanDefinitionScanner((BeanDefinitionRegistry) beanFactory); //a
        scanner.addIncludeFilter(new AnnotationTypeFilter(CustomBean.class)); //b
        scanner.scan("top.wisely.springfundamentals.custom_scan"); //c

    }
}
```

a. 定义一个类路径 Bean 定义扫描器，它的入参是 BeanDefinitionRegistry 类型，而 ConfigurableListableBeanFactory 是它的子类，可强制转换使用。当然，我们可以让类直接实现 BeanDefinitionRegistryPostProcessor 接口，它的 postProcessBeanDefinitionRegistry 方法参数中直接提供了 BeanDefinitionRegistry 的对象。

b. 为扫描器添加包含注解@CustomBean 的过滤器。

c. 在包 top.wisely.springfundamentals.custom_scan 中扫描注解。

在 JavaConfig 中注入自定义的 Bean，因为是自定义的，所以 IDE 自动检测会显示红色，但可以正常执行，如图 3-19 中方框所示。

```
@Bean
CommandLineRunner customBeanDefinitionRegistryPostProcessorClr(CustomBeanService
customBeanService){
    return args -> {
        customBeanService.doSomething();
    };
}
```

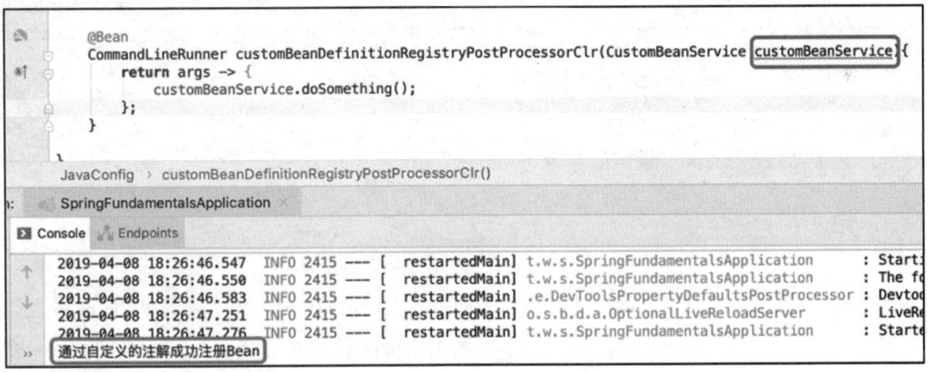

图 3-19

3.8.3 使用 AOP

还可以通过基于 AOP 来让注解具备功能，首先通过拦截标注指定注解的方法或类，然后再建言执行功能代码。

◎ AnnotationTransactionAspect：让@Transactional 注解生效。

◎ AnnotationCacheAspect：让@Cacheable 注解生效。

运作原理请参见 3.7 节。

3.8.4 组合元注解

在 Spring 中，大部分元注解均可注解到其他注解上，即用元注解（元数据）描述注解，从而使其他注解具备元注解的功能。一般来说，组合注解是元注解在新的语义下的特例。

◎ @Component 元注解：@Service、@Repository、@Controller 和@Configuration。

```
@Target({ElementType.TYPE})
@Retention(RetentionPolicy.RUNTIME)
@Documented
@Component //组合了@Component 元注解，具备了声明 Bean 的能力
public @interface Service {}
```

◎ @Import 元注解：大量的@Enable*注解。

```
@Target(ElementType.TYPE)
@Retention(RetentionPolicy.RUNTIME)
@Documented
@Import(AsyncConfigurationSelector.class) //组合了@Import 元注解，具备了导入配置的能力
public @interface EnableAsync {}
```

◎ @Conditional 元注解：@Profile 及 Spring Boot 的大量条件注解@ConditionalOn*。

```
@Target({ ElementType.TYPE, ElementType.METHOD })
@Retention(RetentionPolicy.RUNTIME)
@Documented
@Conditional(OnClassCondition.class)
public @interface ConditionalOnClass {}
```

3.9 小结

Spring 框架的核心即 IoC 容器，它负责 Bean 的管理和 Bean 之间依赖关系的注入。本章快速梳理了 Spring 框架的最新核心知识，为后面学习 Spring Boot 和 Spring MVC 打下一个良好的基础。

第 4 章 深入Spring Boot

4.1 Spring Boot 介绍

Spring Boot 提供了快速创建生产级别 Spring 应用的能力，Spring Boot 可以根据应用中包含的不同的库对其进行自动配置。在开发 Spring Boot 应用时，只需进行极少的配置即可，这让我们在研发过程中可以将主要精力集中到业务开发上，大大减少在技术相关的配置上花费的时间。

下面新建一个 Spring Boot 项目作为本章的示例。

应用信息如下。

◎ Group：top.wisely。

◎ Artifact：spring-boot-in-depth。

◎ Dependencies：Spring Web Starter。

4.2 Spring Boot 的"魔法"

Spring Boot 的"魔法"来源于自动配置，本节的内容就是探讨自动配置的"魔法"是如何实现的。

4.2.1 加载自动配置

Spring Boot 的入口类是一个简单的包含可执行 main()方法的 Java 类，示例如下。

```
@SpringBootApplication
public class SpringBootInDepthApplication {

  public static void main(String[] args) {
    SpringApplication.run(SpringBootInDepthApplication.class, args);
```

```
    }
}
```

在这个类中，有两个我们不熟悉的东西，分别是@SpringBootApplication 和 SpringApplication。@SpringBootApplication 和 SpringApplication 都位于 spring-boot-autoconfigure-2.2.x.RELEASE.jar 中。

首先看看 SpringApplication。SpringApplication 的作用是新建一个 Spring IoC 容器。在非 Web 环境中，它可以新建一个 AnnotationConfigApplicationContext；在 Web 环境中，它可以新建一个 AnnotationConfigServletWebServerApplicationContext；在响应式 Web 环境中，它可以新建一个 AnnotationConfigReactiveWebServerApplicationContext。由此看来，SpringApplication 并没有什么特别的神奇之处。

再来看看@SpringBootApplication 的定义。

```
@Target(ElementType.TYPE)
@Retention(RetentionPolicy.RUNTIME)
@Documented
@Inherited
@SpringBootConfiguration //a
@EnableAutoConfiguration //b
@ComponentScan(excludeFilters = {
     @Filter(type = FilterType.CUSTOM, classes = TypeExcludeFilter.class),
     @Filter(type = FilterType.CUSTOM, classes =
AutoConfigurationExcludeFilter.class) })//c
@ConfigurationPropertiesScan //d
public @interface SpringBootApplication {}
```

a. 组合获得@SpringBootConfiguration 注解的功能，而这个注解又组合了@Configuration 元注解，这意味着入口类是一个配置类。

b. @EnableAutoConfiguration 开启自动配置。

c. 注解@ComponentScan 获得自动扫描 Bean 的功能。

d. 自动扫描且标注了@ConfigurationProperties 的类，并将它注册成 Bean。

接着看看@EnableAutoConfiguration 的定义。

```
@Target(ElementType.TYPE)
@Retention(RetentionPolicy.RUNTIME)
@Documented
@Inherited
@AutoConfigurationPackage
@Import(AutoConfigurationImportSelector.class)
public @interface EnableAutoConfiguration {}
```

正如所预料的那样，@Import 注解导入一个自动配置选择器——AutoConfigurationImportSelector。下面聚焦 AutoConfigurationImportSelector 所实现的接口

ImportSelector 的覆写方法 selectImports，整体的调用流程如下。

◎ getAutoConfigurationEntry(autoConfigurationMetadata, annotationMetadata)：获得应导入的自动配置。

◎ getCandidateConfigurations(annotationMetadata,attributes)：获得自动配置类的名称。

◎ SpringFactoriesLoader.loadFactoryNames(getSpringFactoriesLoaderFactoryClass(), getBeanClassLoader())：从类路径下获取所有 META-INF/spring.factories（如 spring-boot-autoconfigure-2.2.x.RELEASE.jar/META-INF/spring.factories）文件，并查找类型为 getSpringFactoriesLoaderFactoryClass()的工厂，即 org.springframework.boot.autoconfigure.EnableAutoConfiguration，找到符合自动加载的配置，如图 4-1 所示。

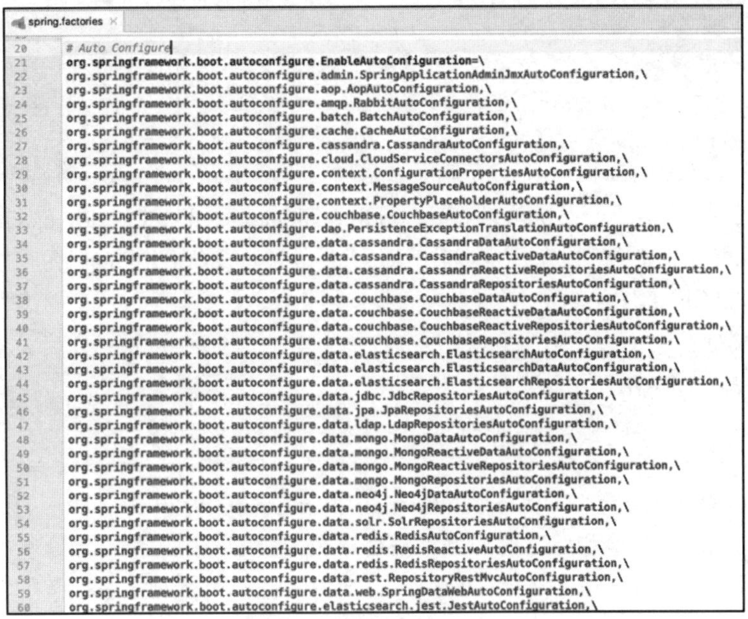

```
spring.factories
20  # Auto Configure
21  org.springframework.boot.autoconfigure.EnableAutoConfiguration=\
22  org.springframework.boot.autoconfigure.admin.SpringApplicationAdminJmxAutoConfiguration,\
23  org.springframework.boot.autoconfigure.aop.AopAutoConfiguration,\
24  org.springframework.boot.autoconfigure.amqp.RabbitAutoConfiguration,\
25  org.springframework.boot.autoconfigure.batch.BatchAutoConfiguration,\
26  org.springframework.boot.autoconfigure.cache.CacheAutoConfiguration,\
27  org.springframework.boot.autoconfigure.cassandra.CassandraAutoConfiguration,\
28  org.springframework.boot.autoconfigure.cloud.CloudServiceConnectorsAutoConfiguration,\
29  org.springframework.boot.autoconfigure.context.ConfigurationPropertiesAutoConfiguration,\
30  org.springframework.boot.autoconfigure.context.MessageSourceAutoConfiguration,\
31  org.springframework.boot.autoconfigure.context.PropertyPlaceholderAutoConfiguration,\
32  org.springframework.boot.autoconfigure.couchbase.CouchbaseAutoConfiguration,\
33  org.springframework.boot.autoconfigure.dao.PersistenceExceptionTranslationAutoConfiguration,\
34  org.springframework.boot.autoconfigure.data.cassandra.CassandraDataAutoConfiguration,\
35  org.springframework.boot.autoconfigure.data.cassandra.CassandraReactiveDataAutoConfiguration,\
36  org.springframework.boot.autoconfigure.data.cassandra.CassandraReactiveRepositoriesAutoConfiguration,\
37  org.springframework.boot.autoconfigure.data.cassandra.CassandraRepositoriesAutoConfiguration,\
38  org.springframework.boot.autoconfigure.data.couchbase.CouchbaseDataAutoConfiguration,\
39  org.springframework.boot.autoconfigure.data.couchbase.CouchbaseReactiveDataAutoConfiguration,\
40  org.springframework.boot.autoconfigure.data.couchbase.CouchbaseReactiveRepositoriesAutoConfiguration,\
41  org.springframework.boot.autoconfigure.data.couchbase.CouchbaseRepositoriesAutoConfiguration,\
42  org.springframework.boot.autoconfigure.data.elasticsearch.ElasticsearchAutoConfiguration,\
43  org.springframework.boot.autoconfigure.data.elasticsearch.ElasticsearchDataAutoConfiguration,\
44  org.springframework.boot.autoconfigure.data.elasticsearch.ElasticsearchRepositoriesAutoConfiguration,\
45  org.springframework.boot.autoconfigure.data.jdbc.JdbcRepositoriesAutoConfiguration,\
46  org.springframework.boot.autoconfigure.data.jpa.JpaRepositoriesAutoConfiguration,\
47  org.springframework.boot.autoconfigure.data.ldap.LdapRepositoriesAutoConfiguration,\
48  org.springframework.boot.autoconfigure.data.mongo.MongoDataAutoConfiguration,\
49  org.springframework.boot.autoconfigure.data.mongo.MongoReactiveDataAutoConfiguration,\
50  org.springframework.boot.autoconfigure.data.mongo.MongoReactiveRepositoriesAutoConfiguration,\
51  org.springframework.boot.autoconfigure.data.mongo.MongoRepositoriesAutoConfiguration,\
52  org.springframework.boot.autoconfigure.data.neo4j.Neo4jDataAutoConfiguration,\
53  org.springframework.boot.autoconfigure.data.neo4j.Neo4jRepositoriesAutoConfiguration,\
54  org.springframework.boot.autoconfigure.data.solr.SolrRepositoriesAutoConfiguration,\
55  org.springframework.boot.autoconfigure.data.redis.RedisAutoConfiguration,\
56  org.springframework.boot.autoconfigure.data.redis.RedisReactiveAutoConfiguration,\
57  org.springframework.boot.autoconfigure.data.redis.RedisRepositoriesAutoConfiguration,\
58  org.springframework.boot.autoconfigure.data.rest.RepositoryRestMvcAutoConfiguration,\
59  org.springframework.boot.autoconfigure.data.web.SpringDataWebAutoConfiguration,\
60  org.springframework.boot.autoconfigure.elasticsearch.jest.JestAutoConfiguration,\
```

图 4-1

至此，我们知道了 Spring Boot 是如何加载自动配置类的。下面讲解这些自动配置类是如何自动配置的。

4.2.2　实现自动配置

下面看一个最简单的自动配置类，来感受一下。

```
@Configuration
@ConditionalOnClass({ EnableAspectJAutoProxy.class, Aspect.class, Advice.class,
    AnnotatedElement.class })
```

```
@ConditionalOnProperty(prefix = "spring.aop", name = "auto", havingValue = "true",
matchIfMissing = true)
public class AopAutoConfiguration {

   @Configuration
   @EnableAspectJAutoProxy(proxyTargetClass = false)
   @ConditionalOnProperty(prefix = "spring.aop", name = "proxy-target-class",
havingValue = "false", matchIfMissing = false)
  public static class JdkDynamicAutoProxyConfiguration {

   }

   @Configuration
   @EnableAspectJAutoProxy(proxyTargetClass = true)
   @ConditionalOnProperty(prefix = "spring.aop", name = "proxy-target-class",
havingValue = "true", matchIfMissing = true)
  public static class CglibAutoProxyConfiguration {

   }

}
```

在这段代码中，我们不熟悉的注解有@ConditionalOnClass 和@ConditionalOnProperty 两个，而这两个注解用到了前面学到的两个基础知识：“条件注解@Conditional”和 “组合元注解”。以@ConditionalOnClass 为例。

```
@Target({ ElementType.TYPE, ElementType.METHOD })
@Retention(RetentionPolicy.RUNTIME)
@Documented
@Conditional(OnClassCondition.class)
public @interface ConditionalOnClass {}
```

@ConditionalOnClass 代表的是@Conditional 在 OnClassCondition 条件下的语义。通过语义可知道该注解的功能，即检测特定的一些类是否存在于类路径中。只有当这些类全部存在于类路径中时，被注解的类或方法才有效；否则，可视为被注解的类或方法不存在。

与@ConditionalOnClass 同类的条件注解如下。

◎ @ConditionalOnBean：只有当指定的一批 Bean 都已存在于容器中时才有效。

◎ @ConditionalOnMissingBean：只有当指定的一批 Bean 都不存在于容器中时才有效。

◎ @ConditionalOnClass：只有当指定的一批类都存在于当前类路径中时才有效。

◎ @ConditionalOnMissingClass：只有当指定的一批类都不存在于当前类路径中时才有效。

◎ @ConditionalOnProperty：只有当前的配置属性符合条件时才有效。

◎ @ConditionalOnResource：只有当指定的一批资源在类路径中时才有效。

◎ @ConditionalOnExpression：只有当 Spring EL 运算结果为 true 时才有效。

◎　@ConditionalOnJava：只有当 Java 版本满足要求时才有效。

◎　@ConditionalOnWebApplication：只在 Web 应用下才有效。

◎　@ConditionalOnNotWebApplication：只在非 Web 应用下有效。

◎　@ConditionalOnSingleCandidate：只有当指定的一批 Bean 在容器中已存在且只有一个候选 Bean 时才有效。

◎　@ConditionalOnCloudPlatform：只在指定的云平台下有效。

配置文件之间如果有先后依赖顺序，则既可以通过@AutoConfigureAfter 和@AutoConfigureBefore 注解指定配置顺序，也可以通过@AutoConfigureOrder 注解来指定优先级。

在 spring-boot-autoconfigure-2.2.x.RELEASE.jar 的 org.springframework.boot.autoconfigure 包中，以 AutoConfiguration 结尾的都是 Spring Boot 自带的自动配置类，如图 4-2 所示。

Spring Boot 提供的所有自动配置可参见官网。

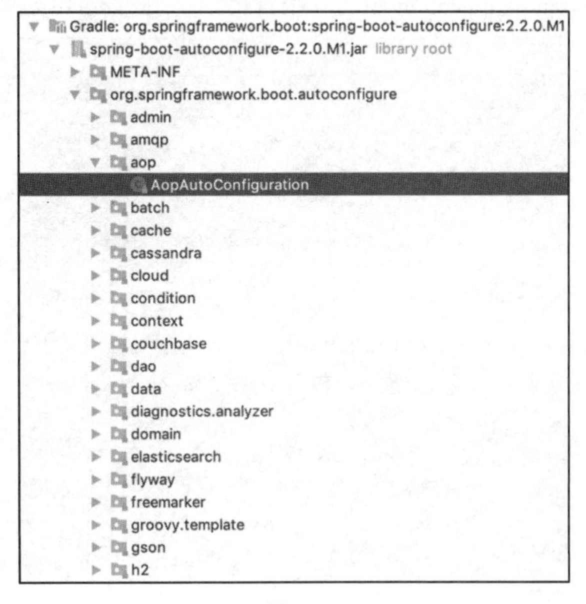

图 4-2

Spring Boot 提供的所有自动配置参见 Spring Boot Reference *Documentation* 中的 Auto-Configuration 部分。

4.3 Spring Boot 的配置

虽然 Spring Boot 已经做了绝大部分的自动配置，即只需按照默认就能满足大部分的开发需求，但有时还是需要对 Spring Boot 进行定制配置的。

4.3.1 应用配置

1. SpringApplication

入口类的 main()方法通过 SpringApplication.run(SpringBootInDepthApplication.class, args)启动应用，给 run 传入配置类 SpringBootInDepthApplication.class。配置类注解的@SpringBootApplication 所带来的自动配置在前面已经分析过了。除此之外，还可以通过 SpringApplication 类对应用启动行为进行配置，代码如下所示。

```
SpringApplication app = new SpringApplication(SpringBootInDepthApplication.class);
app.setBannerMode(Banner.Mode.OFF); //设置关闭Banner
app.addListeners(new MyListener()); //增加监听器
app.run(args);
```

除上面的演示外，SpringApplication 还支持很多和容器相关的配置，它们都是以 set 和 add 开头的方法，可以在 SpringApplication 的 API 中查找以 set 和 add 开头的方法。

监听器代码如下。

```
2 public class MyListener implements ApplicationListener<ApplicationStartingEvent> {
   @Override
   public void onApplicationEvent(ApplicationStartingEvent event) {
      System.out.println("监听到应用启动事件");
   }
}
```

2. SpringApplicationBuilder

我们也可以使用 SpringApplicationBuilder 来定制应用启动。它是一个建造者模式的类，和 Stream 运算很像，设置为中间运算，用一个终结运算来执行，示例如下。

```
new SpringApplicationBuilder()
     .bannerMode(Banner.Mode.OFF)
      .listeners(new MyListener())
     .sources(SpringBootInDepthApplication.class)
     .build(args)
     .run();
```

SpringApplication 配置与 SpringApplicationBuilder 配置是等同的，前者的方法名去掉前缀（set 和 add）即为后者的方法名，符合建造者模式的命名规则。例如，setBannerMode()变为 bannerMode()。特殊情况请参照 API 文档。

3. 通过外部配置

我们可以通过外部配置（可以是命令行、系统环境变量或 application.properties）来定制应用启动行为，例如：

```
spring.main.banner-mode=off
spring.main.lazy-initialization=true #通过开启全局延迟加载来减少启动时间，会对运行性能
# 造成影响
```

关于外部配置，会在后面着重讲解。

4. 其他默认配置

Spring Boot 除做了大量的自动配置外，还提供了一些其他默认配置。

应用监听器：在类路径下文件 META-INF/spring.factories 中的工厂名为 org.springframework.context.ApplicationListener 的所有监听器，如图 4-3 和图 4-4 所示。

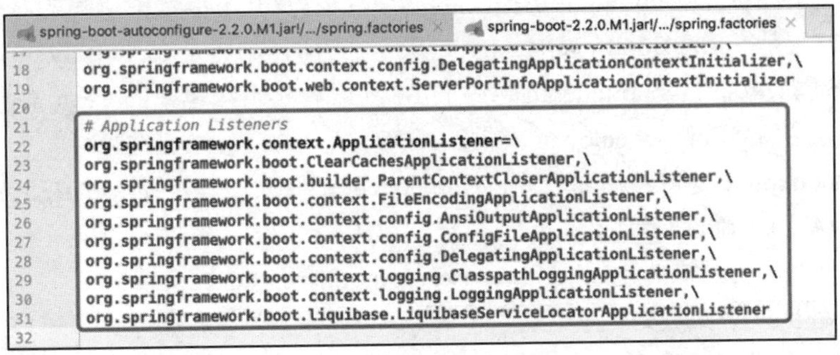

```
spring-boot-autoconfigure-2.2.0.M1.jar!/.../spring.factories    spring-boot-2.2.0.M1.jar!/.../spring.factories
18  org.springframework.boot.context.config.DelegatingApplicationContextInitializer,\
19  org.springframework.boot.web.context.ServerPortInfoApplicationContextInitializer
20
21  # Application Listeners
22  org.springframework.context.ApplicationListener=\
23  org.springframework.boot.ClearCachesApplicationListener,\
24  org.springframework.boot.builder.ParentContextCloserApplicationListener,\
25  org.springframework.boot.context.FileEncodingApplicationListener,\
26  org.springframework.boot.context.config.AnsiOutputApplicationListener,\
27  org.springframework.boot.context.config.ConfigFileApplicationListener,\
28  org.springframework.boot.context.config.DelegatingApplicationListener,\
29  org.springframework.boot.context.logging.ClasspathLoggingApplicationListener,\
30  org.springframework.boot.context.logging.LoggingApplicationListener,\
31  org.springframework.boot.liquibase.LiquibaseServiceLocatorApplicationListener
32
```

图 4-3

```
spring-boot-autoconfigure-2.2.0.M1.jar!/.../spring.factories    spring-boot-2.2.0.M1.jar!/.../spring.factories
5
6  # Application Listeners
7  org.springframework.context.ApplicationListener=\
8  org.springframework.boot.autoconfigure.BackgroundPreinitializer
9
```

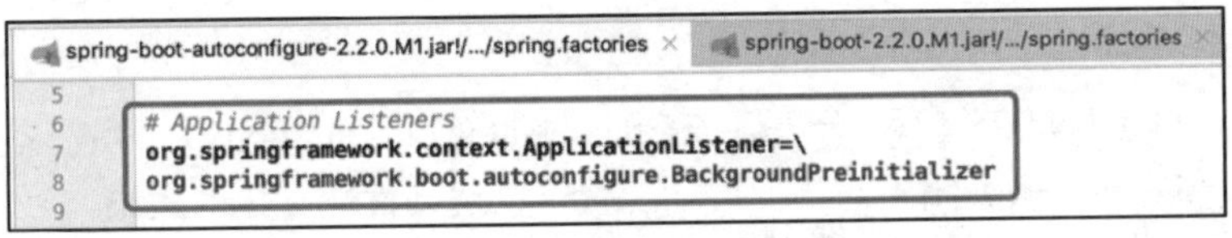

图 4-4

上面给了我们一个提示，即可以通过相同的方法来注册监听器。在当前应用中新建 resources/META-INF/spring.factories 文件，添加下面的代码：

```
org.springframework.context.ApplicationListener=top.wisely.springbootindepth.listene
r.MyListener
```

◎ 容器配置如图 4-5 所示。

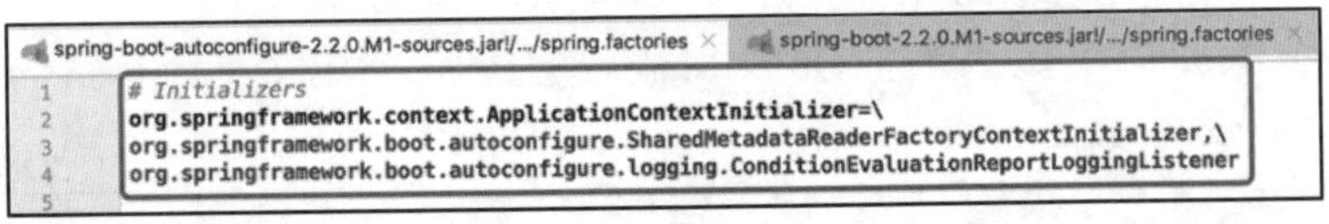

```
spring-boot-autoconfigure-2.2.0.M1-sources.jar!/.../spring.factories    spring-boot-2.2.0.M1-sources.jar!/.../spring.factories
1  # Initializers
2  org.springframework.context.ApplicationContextInitializer=\
3  org.springframework.boot.autoconfigure.SharedMetadataReaderFactoryContextInitializer,\
4  org.springframework.boot.autoconfigure.logging.ConditionEvaluationReportLoggingListener
5
```

图 4-5

◎ Environment 和应用配置如图 4-6 所示。

```
# Environment Post Processors
org.springframework.boot.env.EnvironmentPostProcessor=\
org.springframework.boot.cloud.CloudFoundryVcapEnvironmentPostProcessor,\
org.springframework.boot.env.SpringApplicationJsonEnvironmentPostProcessor,\
org.springframework.boot.env.SystemEnvironmentPropertySourceEnvironmentPostProcessor
```

图 4-6

4.3.2 修改默认配置

1．自动扫描配置

在默认情况下，Spring Boot 会自动扫描入口类所在包及其下级包中所有的 Bean（JPA 实体）。在本例中，入口类是 SpringBootInDepthApplication，它所在的包是 top.wisely.springbootindepth，这意味着在 top.wisely.springbootindepth 包中的 Bean 或者它下级包中的 Bean 都会被自动扫描。自动扫描的功能是由@ComponentScan 提供的。

@SpringBootApplication 注解组合了@ComponentScan 元注解，所以对自动扫描的配置是通过@SpringBootApplication 来完成的。

如果需要增加一个扫描的包 io.github.wiselyman，则可以这样设置：

```
@SpringBootApplication(scanBasePackages = {"io.github.wiselyman",
                        "top.wisely.springbootindepth"})
public class SpringBootInDepthApplication {}
```

2．关闭自动配置

@SpringBootApplication 组合了@EnableAutoConfiguration，所以可以通过@SpringBootApplication 来关闭不需要的自动配置，以达到性能优化的效果。

```
@SpringBootApplication(exclude = SpringApplicationAdminJmxAutoConfiguration.class)
public class SpringBootInDepthApplication {}
```

除此之外，还可以通过外部的 application.properties 文件来关闭自动配置。

```
spring.autoconfigure.exclude=org.springframework.boot.autoconfigure.admin.SpringAppl
icationAdminJmxAutoConfiguration
```

3．覆盖 Bean

我们可以用重新定义的 Bean 来覆盖已经自动配置的 Bean。

比如，Spring Boot 提供自动配置的 ObjectMapper 的 Bean，它定义在 JacksonAutoConfiguration.JacksonObjectMapperConfiguration 中。

```
@Bean
@Primary
@ConditionalOnMissingBean
public ObjectMapper jacksonObjectMapper(Jackson2ObjectMapperBuilder builder) {
```

```
  return builder.createXmlMapper(false).build();
}
```

运行下面的代码：

```
@Bean
CommandLineRunner reBeanClr(ObjectMapper jacksonObjectMapper){
  return args -> {
    AuthorProperties.Book book = new AuthorProperties.Book();
    book.setName("book3");
    book.setPrice(119);
    String content = jacksonObjectMapper.writeValueAsString(book);
    System.out.println(content);
  };
}
```

输出 JSON 格式的字符串，如图 4-7 所示。

```
2019-04-11 23:10:48.166  INFO
2019-04-11 23:10:48.333  INFO
2019-04-11 23:10:48.382  INFO
2019-04-11 23:10:48.387  INFO
{"name":"book3","price":119}
```

图 4-7

如果希望输出的结果的形式是缩进的，那么又该如何定制呢？只需用重新定义的 Bean 覆盖 jacksonObjectMapper 即可。因为原本的声明中使用了@ConditionalOnMissingBean，当重新定义 Bean 以后，意味着容器中已经存在 Bean，所以默认的配置 jacksonObjectMapper 会失效。

```
@Bean
public ObjectMapper jacksonObjectMapper(Jackson2ObjectMapperBuilder builder) {
  return builder.indentOutput(true).createXmlMapper(false).build();
}
```

运行上述代码，如图 4-8 所示输出的结果，已变为缩进形式。

```
2019-04-11 23:30:17.885
2019-04-11 23:30:18.018
2019-04-11 23:30:18.069
2019-04-11 23:30:18.074
{
  "name" : "book3",
  "price" : 119
}
```

图 4-8

4．使用 Customizer 来定制

Spring Boot 提供了大量的*Customizer 来修改或定制默认行为，我们可以继承或实现这些*Customizer。例如，想要实现前面的相同功能，只需实现 Jackson2ObjectMapperBuilderCustomizer 接口即可，代码如下。

```
@Component
public class MyJackson2ObjectMapperBuilderCustomizer implements
Jackson2ObjectMapperBuilderCustomizer {
    @Override
    public void customize(Jackson2ObjectMapperBuilder jacksonObjectMapperBuilder) {
        jacksonObjectMapperBuilder.indentOutput(true);
    }
}
```

注意，要注释掉上个例子中覆盖的 Bean 的定义，运行的结果和上面使用覆盖 Bean 定义的结果一致。通过 Intellij IDEA 可以检索 Spring Boot 提供的 Customizer，它们主要存在于 spring-boot-autoconfigure-2.2.x.RELEASE.jar 库下的 org.springframework.boot.autoconfigure 包中，如图 4-9 所示。

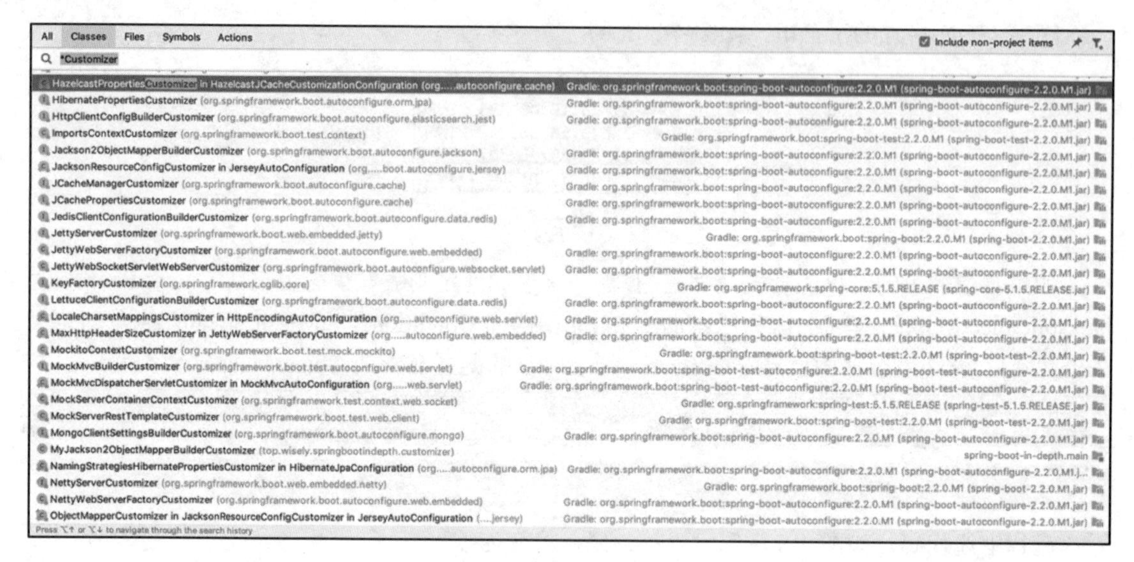

图 4-9

5. 修改依赖库

Spring Boot 可以根据类路径上不同的 jar 进行相应的自动配置。例如，默认内嵌 Servlet 容器是 Tomcat，则 Spring Boot 会为 Tomcat 做好自动配置。如果将 Servlet 容器修改成 Jetty，则 Spring Boot 也会为 Jetty 做好自动配置。

```
implementation ('org.springframework.boot:spring-boot-starter-web'){
  exclude module: 'spring-boot-starter-tomcat'
}
implementation 'org.springframework.boot:spring-boot-starter-jetty'
```

4.3.3 外部配置

Spring Boot 可以从命令行、环境变量、properties 文件以及 YAML 文件等外部获得配置，而这个能力是由 Environment 提供的。可以通过 3 种方式访问 Environment 中的属性。

◎ 使用@Value 注解，在第 3 章已演示过。

◎ 注入 Environment 的 Bean，在第 3 章已演示过。

◎ 通过@ConfigurationProperties 注解来访问，本节讲解。

为了更深入地理解外部配置的原理，首先需要了解 Environment 抽象。

1．外部配置源与 Environment

在第 3 章中曾介绍过，Environment 包含两部分内容：Profile 和 Property。Environment 的定义如下：

```
public interface Environment extends PropertyResolver {
  String[] getActiveProfiles();
  String[] getDefaultProfiles();
  boolean acceptsProfiles(Profiles profiles);
}
```

Environment 中的三个方法负责 Profile 相关的内容，而它继承的 PropertyResolver 接口负责对 Property 进行查询。

```
public interface PropertyResolver {
  boolean containsProperty(String key);
  @Nullable
  String getProperty(String key);
  String getProperty(String key, String defaultValue);
  @Nullable
  <T> T getProperty(String key, Class<T> targetType);
  <T> T getProperty(String key, Class<T> targetType, T defaultValue);
  String getRequiredProperty(String key) throws IllegalStateException;
  <T> T getRequiredProperty(String key, Class<T> targetType) throws
IllegalStateException;
  String resolvePlaceholders(String text);
  String resolveRequiredPlaceholders(String text) throws IllegalArgumentException;
}
```

在 Environment 中，每一个配置属性都是 PropertySource，多个 PropertySource 可聚集成 PropertySources。

PropertyResolver 的实现类 PropertySourcesPropertyResolver 负责对 PropertySources 进行查询操作，即 Environment 可对 PropertySources 进行查询操作。

Spring 不支持 YAML 文件作为 PropertySource，Spring Boot 使用 YamlPropertySourceLoader 来读取 YAML 文件，并获得 PropertySource。

在 Spring Boot 下，外部配置属性的加载顺序是：先列的属性配置优先级最高，先列的配置属性可覆盖后列的配置属性。

◎ 命令行参数。
◎ SPRING_APPLICATION_JSON。
◎ ServletConfig 初始化参数。
◎ ServletContext 初始化参数。
◎ JNDI（java:comp/env）。
◎ Java 系统属性（System.getProperties() ）。
◎ 操作系统变量。
◎ RandomValuePropertySource 随机值。
◎ 应用部署 jar 包外部的 application-{profile}.properties/yml。
◎ 应用部署 jar 包内部的 application-{profile}.properties/yml 。
◎ 应用部署 jar 包外部的 application.properties/yml。
◎ 应用部署 jar 包内部的 application.properties/yml。
◎ @PropertySource。
◎ SpringApplication.setDefaultProperties。

（1）命令行参数。

◎ 使用 gradle 命令行传参。

```
$ ./gradlew bootRun --args='--server.port=8888 --server.ip=192.168.1.5'
```

◎ 打包成 jar 包传参。

```
$ ./gradlew bootJar
$ java -jar build/libs/spring-boot-in-depth-0.0.1-SNAPSHOT.jar
--server.port=8888 --server.ip=192.168.1.5
```

在 IntelliJ IDEA 里设置参数，如图 4-10 和图 4-11 所示。

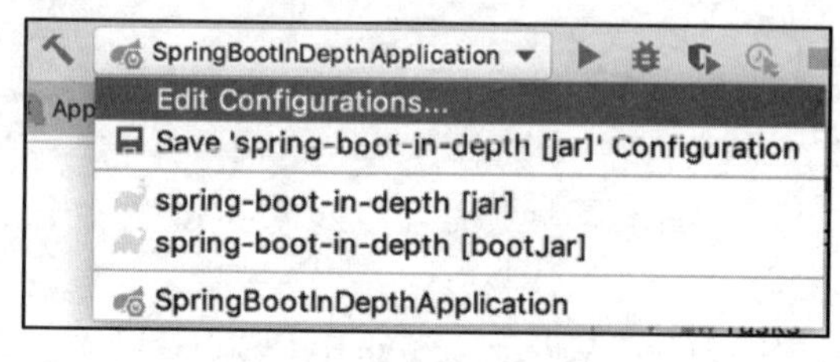

图 4-10

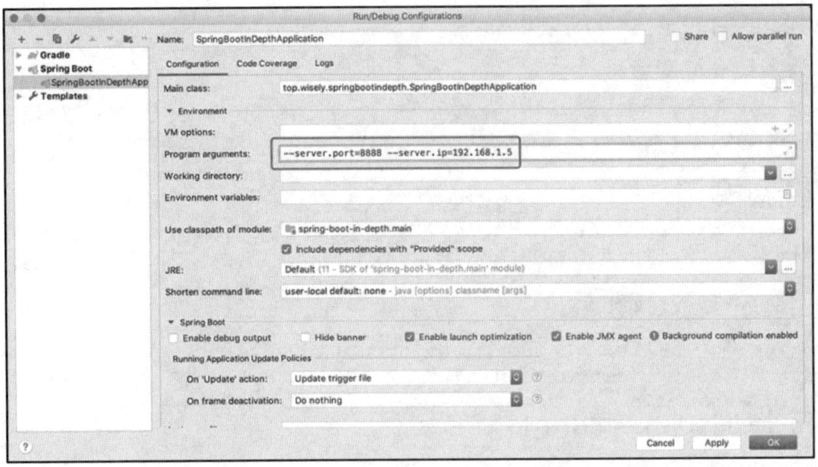
Run/Debug Configurations
Name: SpringBootInDepthApplication
Share
Allow parallel run
Gradle
Spring Boot
SpringBootInDepthApp
Templates
Configuration
Code Coverage
Logs
Main class:
top.wisely.springbootindepth.SpringBootInDepthApplication
Environment
VM options:
Program arguments:
--server.port=8888 --server.ip=192.168.1.5
Working directory:
Environment variables:
Use classpath of module:
spring-boot-in-depth.main
Include dependencies with "Provided" scope
JRE:
Default (11 - SDK of 'spring-boot-in-depth.main' module)
Shorten command line:
user-local default: none - java [options] classname [args]
Spring Boot
Enable debug output
Hide banner
Enable launch optimization
Enable JMX agent
Background compilation enabled
Running Application Update Policies
On 'Update' action:
Update trigger file
On frame deactivation:
Do nothing
Cancel
Apply
OK

图 4-11

下面进行代码校验，结果如下。

```
@Value("${server.ip}")
private String serverIp;//a

@Bean
CommandLineRunner commandLineRunner(@Value("${server.port}") String serverPort ){//b
  return args -> {
    Stream.of(args).forEach(System.out::println); //c
    System.out.println(serverPort);
    System.out.println(serverIp);
  };
}
```

a. 使用@Value 注入值到类的变量中，server.ip 不是 Spring Boot 的内置配置，可接收自用。

b. 使用@Value 注入值到方法参数中，server.port 是 Spring Boot 的内置配置，对应用配置起效，可更改当前容器的端口号。

c. 可从 args 参数中获取参数。

以上三种方式的运行结果均如图 4-12 所示。

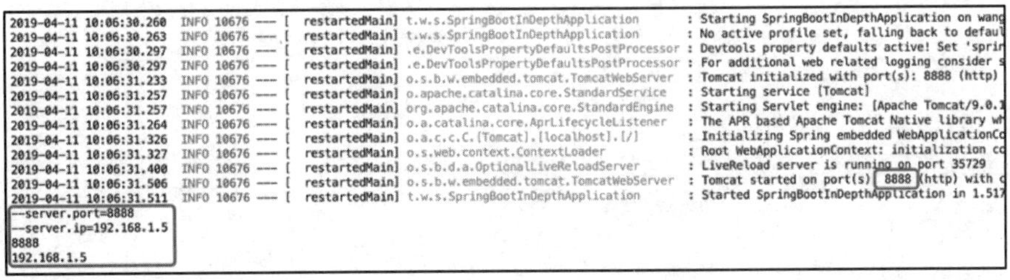
2019-04-11 10:06:30.260 INFO 10676 --- [restartedMain] t.w.s.SpringBootInDepthApplication : Starting SpringBootInDepthApplication on wang
2019-04-11 10:06:30.263 INFO 10676 --- [restartedMain] t.w.s.SpringBootInDepthApplication : No active profile set, falling back to defaul
2019-04-11 10:06:30.297 INFO 10676 --- [restartedMain] .e.DevToolsPropertyDefaultsPostProcessor : Devtools property defaults active! Set 'sprin
2019-04-11 10:06:30.297 INFO 10676 --- [restartedMain] .e.DevToolsPropertyDefaultsPostProcessor : For additional web related logging consider s
2019-04-11 10:06:31.233 INFO 10676 --- [restartedMain] o.s.b.w.embedded.tomcat.TomcatWebServer : Tomcat initialized with port(s): 8888 (http)
2019-04-11 10:06:31.257 INFO 10676 --- [restartedMain] o.apache.catalina.core.StandardService : Starting service [Tomcat]
2019-04-11 10:06:31.257 INFO 10676 --- [restartedMain] org.apache.catalina.core.StandardEngine : Starting Servlet engine: [Apache Tomcat/9.0.1
2019-04-11 10:06:31.264 INFO 10676 --- [restartedMain] o.a.catalina.core.AprLifecycleListener : The APR based Apache Tomcat Native library wh
2019-04-11 10:06:31.326 INFO 10676 --- [restartedMain] o.a.c.c.C.[Tomcat].[localhost].[/] : Initializing Spring embedded WebApplicationCo
2019-04-11 10:06:31.327 INFO 10676 --- [restartedMain] o.s.web.context.ContextLoader : Root WebApplicationContext: initialization co
2019-04-11 10:06:31.400 INFO 10676 --- [restartedMain] o.s.b.d.a.OptionalLiveReloadServer : LiveReload server is running on port 35729
2019-04-11 10:06:31.506 INFO 10676 --- [restartedMain] o.s.b.w.embedded.tomcat.TomcatWebServer : Tomcat started on port(s) 8888 (http) with c
2019-04-11 10:06:31.511 INFO 10676 --- [restartedMain] t.w.s.SpringBootInDepthApplication : Started SpringBootInDepthApplication in 1.517
--server.port=8888
--server.ip=192.168.1.5
8888
192.168.1.5

图 4-12

（2）SPRING_APPLICATION_JSON。

◎ 作为系统环境变量：

```
$ SPRING_APPLICATION_JSON='{"server":{"ip":"192.168.1.5","port":"8888"}}'
java -jar build/libs/spring-boot-in-depth-0.0.1-SNAPSHOT.jar
```

◎ 使用系统属性：

```
$ java -Dspring.application.json='{"server":{"ip":"192.168.1.5","port":
"8888"}}' -jar build/libs/spring-boot-in-depth-0.0.1-SNAPSHOT.jar
```

◎ 使用命令行参数执行：

```
$ java -jar build/libs/spring-boot-in-depth-0.0.1-SNAPSHOT.jar --spring.
application.json='{"server":{"ip":"192.168.1.5","port":"8888"}}'
```

◎ 使用 Intellij IDEA，如图 4-13 所示。

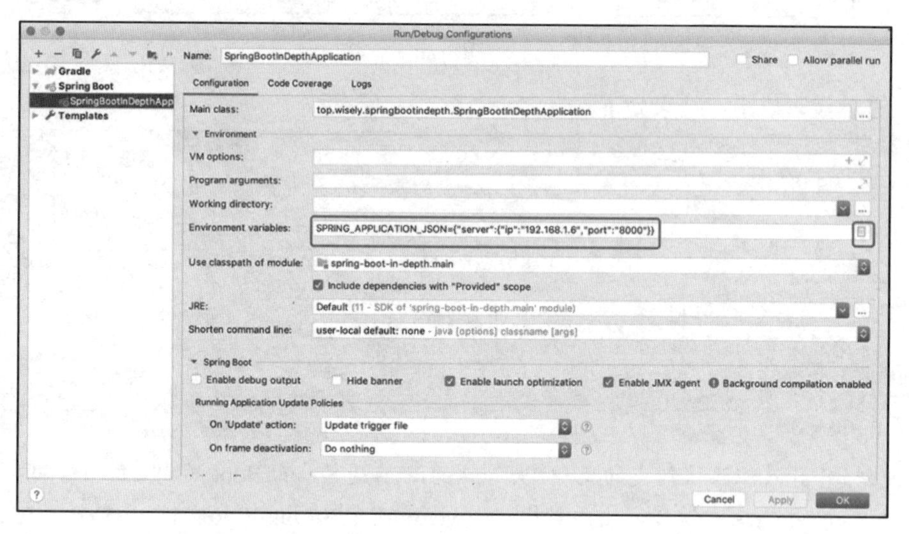

图 4-13

（3）RandomValuePropertySource 随机值。

RandomValuePropertySource 会产生随机值，示例如下：

```
$ ./gradlew bootRun --args='--server.port=${random.int[1024,10000]}
--server.ip=192.168.1.5 --some.value=${random.value}
--some.number=${random.int}'
```

使用下面代码验证：

```
@Value("${server.ip}")
private String serverIp;

@Value("${some.value}")
private String someValue;
```

```
@Value("${some.number}")
private String someNumber;

@Bean
CommandLineRunner commandLineRunner(@Value("${server.port}") String serverPort){
  return args -> {
    Stream.of(args).forEach(System.out::println);
    System.out.println(serverPort);
    System.out.println(serverIp);
    System.out.println(someValue);
    System.out.println(someNumber);
  };
}
```

启动应用，控制台显示如图 4-14 所示。

```
--server.port=${random.int[1024,10000]}
--server.ip=192.168.1.5
--some.value=${random.value}
--some.number=${random.int}
6846
192.168.1.5
4746f6dae3e8b2724ea52bc476d76d8a
376242965
```

图 4-14

2．外部文件配置

Spring Boot 会从以下位置加载外部配置文件 application.properties/yml，并读取成 PropertySources 加载到 Environment 中。

◎　入口类的当前目录的/config 子目录。

◎　入口类的当前目录。

◎　类路径下的/config 目录。

◎　类路径的根目录。

Spring Boot 提供了大量的配置属性，通过设置这些配置属性即可对当前应用进行配置。可以在 resources/application.properties 文件中进行配置。

```
spring.main.banner-mode=off
spring.main.lazy-initialization=true
spring.autoconfigure.exclude=org.springframework.boot.autoconfigure.admin.SpringAppl
icationAdminJmxAutoConfiguration
server.port=1234
```

所有这些属性设置都会被读取到 Environment，对应用的运行行为进行配置。Spring Boot 提供的所有的属性配置可参考 Spring Boot Reference *Documentation* 中“common application properties”部分的内容。

（1）YAML 文件配置。

在云计算环境中，很多配置都是基于 YAML 文件的。尽管 Spring Boot 支持基于 properties 和 YAML 文件的配置，但本书将全部基于 YAML 文件进行配置。

YAML 是 JSON 格式的“超集”，对层级配置提供了极大的便利，下面比较一下 properties 和 YAML。

application.properties

```
server.port=1234
server.address=192.168.31.199
```

application.yml

```
server:
 port: 8888
 address: 192.168.31.199
```

从上面的代码中可以看出，在层级结构下，我们无须写两个 server。层级越多的配置，配置的层次越清晰。注意，“:”后面有一个空格，这是 YAML 要求的格式。

当配置的层级性不是很强时，在 YAML 文件里仍可以按照类似于 properties 文件那样进行配置，代码如下。

```
server.port: 8888
spring.main.banner-mode: off
```

现在新建一个 application.yml 文件，后面的演示都将在这里编写。

（2）占位符。

配置文件可以从 Environment 中读取已定义的配置。

```
app:
 name: spring boot in depth
 desc: Chapter:${app.name} is hard to learn
```

用下列代码检验结果，如图 4-15 所示。

```
@Bean
CommandLineRunner placeholderClr(@Value("${app.name}") String name,
                  @Value("${app.desc}") String desc){
  return args -> {
    System.out.println(name);
    System.out.println(desc);
  };
}
```

```
spring boot in depth
Chapter:spring boot in depth is hard to learn
```

图 4-15

（3）类型安全的配置属性。

前面是通过使用@Value 获得配置属性的 Spring Boot 提供了大量的配置属性，Spring Boot 自己是如何获得并使用这些配置属性的呢？

例如，server.port=1234，当 Servlet 容器启动时，端口会被修改成 1234，这是如何做到的呢？4.2 节介绍了“加载自动配置”和“实现自动配置”。在做自动配置时，Spring Boot 开放了一个“口子”去修改默认的自动配置：一个强类型的 Bean *Properties 和外部配置文件中的内容进行映射绑定，自动配置通过使用*Properties 来配置应用的行为。例如，前面用到的 server.port 的配置是在 ServerProperties 中绑定的。

```
@ConfigurationProperties(prefix = "server", ignoreUnknownFields = true)
public class ServerProperties {
  private Integer port;
  private InetAddress address;
  ...
}
```

那么这个绑定是如何实现的呢？很显然是通过@ConfigurationProperties 来实现的。这个注解的功能是由一个 BeanPostProcessor(ConfigurationPropertiesBindingPostProcessor)提供的，前提是*Properties 是一个 Bean。我们可以通过两种方式让它成为 Bean。

在 Spring Boot 2.2 之前：

◎ 在*Properties 上注解@Component，让它成为 Bean，在配置使用的地方直接像常规 Bean 一样注入即可。

◎ ServerProperties 并没有通过注解标记成 Bean，而是在使用 ServerProperties 的地方通过@EnableConfigurationProperties({*Properties.class})来动态注册 Bean。

在 Spring Boot 2.2 及之后：

@SpringBootApplication 注解组合了@ConfigurationPropertiesScan 注解，它使用 ConfigurationPropertiesScanRegistrar 为动态扫描注册标注了在我们入口类包及其下级包中注解了@ConfigurationProperties 的类，并将其注册成 Bean。也可以通过

```
@SpringBootApplication
@ConfigurationPropertiesScan({ "com.some.other", "top.wisely" })
public class SpringBootInDepthApplication {}
```

来覆盖默认的扫描。

下面用一个简单的例子演示类型安全的配置属性的使用。

属性类：

```
@Getter
@Setter
@ConfigurationProperties(prefix = "author")
public class AuthorProperties {
```

```
    private String name = "wyf";
    private Integer age = 35;
    private String motherTongue;
    private String secondLanguage;
    private String graduatedUniversity;
    private Integer graduationYear;
    private Address address = new Address();
    private List<Book> books = new ArrayList<>();
    private Map<String, String> remarks = new HashMap<>();

    @Getter
    @Setter
    public static class Address {
        private String province;
        private String city;
    }

    @Getter
    @Setter
    public static class Book{
        private String name;
        private Integer price;
    }
}
```

在当前例子中，分别有普通属性、对象、List 和 Map，可作为有代表性的配置。

配置类：

```
@Configuration
//已自动扫描注册AuthorProperties
//无须使用@EnableConfigurationProperties({AuthorProperties.class})
public class AuthorConfiguration {

    private final AuthorProperties authorProperties;

    public AuthorConfiguration(AuthorProperties authorProperties) {
        this.authorProperties = authorProperties;
    }

    @Bean
    public String strBean(){
        String str = authorProperties.getName() + "/"
                + authorProperties.getAge() + "/"
                + authorProperties.getMotherTongue() + "/"
                + authorProperties.getSecondLanguage() + "/"
                + authorProperties.getGraduatedUniversity() + "/"
                + authorProperties.getGraduationYear() + "/"
                + authorProperties.getAddress().getProvince() + "/"
                + authorProperties.getAddress().getCity() + "/";
        System.out.println(str);
```

```
        authorProperties.getBooks().forEach(book -> {
            System.out.println(book.getName());
            System.out.println(book.getPrice());
        });
        authorProperties.getRemarks().forEach((key,value) ->{
            System.out.println(key + ":" +value);
        });
        return str;
    }
}
```

为了能够在 application.yml 中输入配置时会给出自动提示，我们需要添加 annotationProcessor 'org.springframework.boot:spring-boot-configuration-processor' 到 build.gradle 中，IntelliJ IDEA 需要开启注解处理器，如图 4-16 所示。

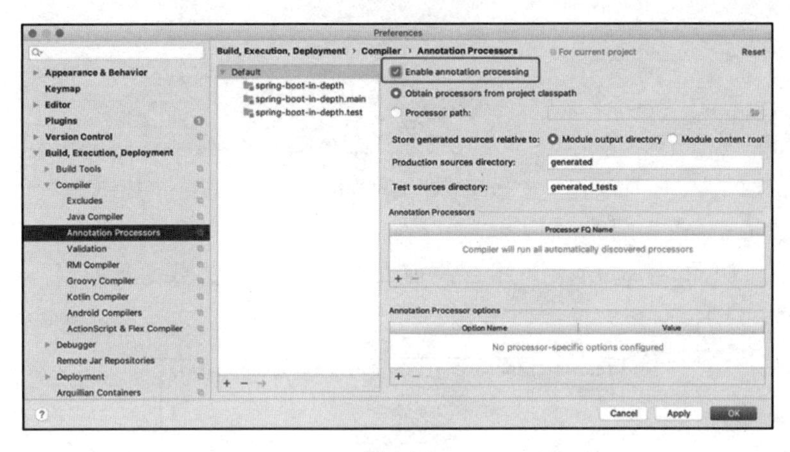

图 4-16

此时 Spring Boot 会生成 build/classes/java/main/META-INF/spring-configuration-metadata.json 文件，IDE 会根据这个文件给出自动提示，如图 4-17 所示。

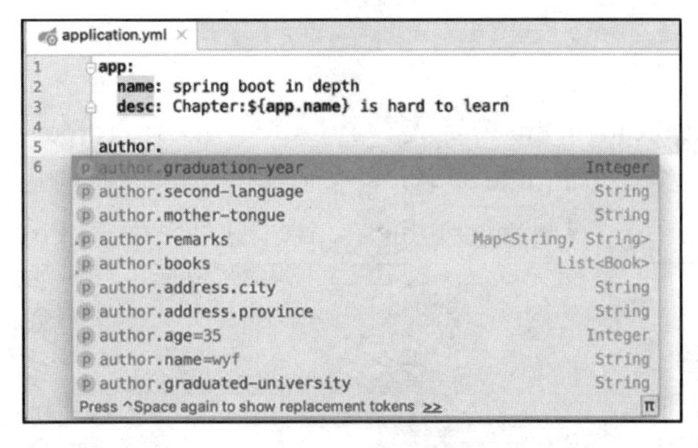

图 4-17

配置内容如下。

```
author:
 name: foo
 age: 40
 mother-tongue: Chinese # 1 烤串式
 second_language: English # 2 下划线式
 graduationYear: 2006 # 3 驼峰式
 GRADUATED_UNIVERSITY: WHUT # 4 大写字母
 address:
  province: Anhui
  city: Hefei
 books:
  - name: book1
   price: 89
  - name: book2
   price: 109
 remarks:
  hobby: reading
  some: value
```

Spring Boot 提供了一种叫作"松散绑定"的技术，属性可以是"烤串式"、"下画线式"、"驼峰式"或者是"大写字母式"。

运行的结果如图 4-18 所示（注意，前面的全局延迟加载不能打开 spring.main.lazy-initialization=false）。

```
foo/40/Chinese/English/WHUT/2006/Anhui/Hefei/
book1
89
book2
109
some:value
hobby:reading
```

图 4-18

3. Profile

（1）单文件 Profile。

我们可以在一个 application.yml 内定义多个 Profile，各个 Profile 之间用"---"隔开。我们可以通过 spring.profiles 给每个 Profile 命名，每个 Profile 还可以通过 spring.profiles.include 来包含组合其他的 Profile，最终通过 spring.profiles.active 来指定激活的 Profile，示例如下。

```
spring:
 profiles:
  active: prod #指定激活的 Profile 是 prod

---
spring.profiles: prod # Profile 的名为 prod
```

```
spring.profiles.include: # 组合prod-port和prod-lazy两个Profile
  - prod-port
  - prod-lazy

---
spring:
  profiles: prod-port # Profile的名为prod-port
server:
  port: 8888

---
spring:
  profiles: dev-port # Profile的名为dev-port
server:
  port: 8080

---
spring:
  profiles: prod-lazy # Profile的名为prod-lazy
  main:
    lazy-initialization: true
```

启动时会执行端口号为 8888 且全局延迟加载的设置。也可以通过设置多个 Profile 获得相同的结果。

```
spring:
  profiles:
    active:
      - prod-port
      - prod-lazy
```

（2）多文件 Profile。

我们可以将上面的 Profile 拆分成多个文件名称，例如，application-{profile}.yml 可以拆分成以下几种。

◎　application-prod.yml

```
spring.profiles.include:
  - prod-port
  - prod-lazy
```

◎　application-prod-port.yml

```
server:
  port: 8888
```

◎　application-prod-lazy.yml

```
spring:
  main:
    lazy-initialization: true
```

◎ application-dev-port.yml

```
server:
 port: 8080
```

同样，需要在 application.yml 中指定要激活的 Profile。

```
spring:
 profiles:
  active: prod
```

4．EnvironmentPostProcessor

在第 3 章中，对 Bean 行为的定制可以用 BeanPostProcessor，对配置元数据的定制可以用 BeanFactoryPostProcessor。Spring Boot 提供了对 Environment 和 SpringApplication 进行定制的 EnvironmentPostProcessor。

下面自定义一个 EnvironmentPostProcessor 实现类。

```
public class MyEnvironmentPostProcessor implements EnvironmentPostProcessor { // a
   @Override
   public void postProcessEnvironment(ConfigurableEnvironment environment,
SpringApplication application) { // b
       Map<String, Object> map = new HashMap<>();
       map.put("key1","value1");
       map.put("key2","value2");
       PropertySource propertySource = new MapPropertySource("map",map); //c
       environment.getPropertySources().addLast(propertySource); //d
        .setBannerMode(Banner.Mode.OFF); //e
   }
}
```

a. 实现 EnvironmentPostProcessor 接口。

b. 覆写 postProcessEnvironment，它的入参让我们可以对 Environment 和 SpringApplication 进行设置。

c. 新建一个 PropertySource，类型为 MapPropertySource，将 map 的值设置给 PropertySource。

d. 将 propertySource 添加到 Environment 中。

e. 同样，在此处可以设置 SpringApplication。

可以在本应用的 META-INF/spring.factories 中启用该处理。

```
org.springframework.boot.env.EnvironmentPostProcessor=\
top.wisely.springbootindepth.processor.MyEnvironmentPostProcessor
```

再次校验运行结果，如图 4-19 所示。

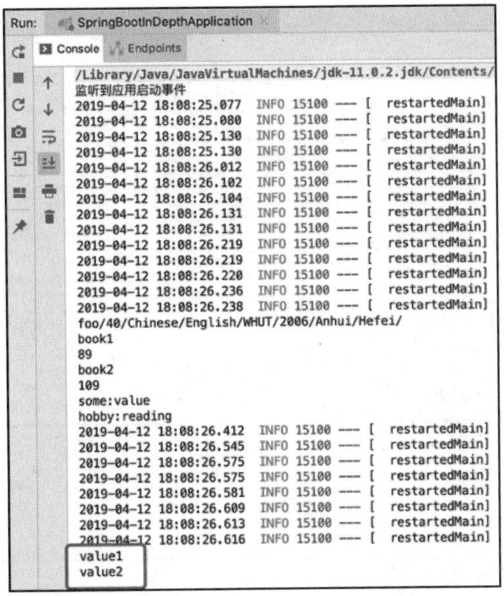

```
Run:  SpringBootInDepthApplication
Console  Endpoints
/Library/Java/JavaVirtualMachines/jdk-11.0.2.jdk/Contents/
监听到应用启动事件
2019-04-12 18:08:25.077  INFO 15100 --- [  restartedMain]
2019-04-12 18:08:25.080  INFO 15100 --- [  restartedMain]
2019-04-12 18:08:25.130  INFO 15100 --- [  restartedMain]
2019-04-12 18:08:25.130  INFO 15100 --- [  restartedMain]
2019-04-12 18:08:26.012  INFO 15100 --- [  restartedMain]
2019-04-12 18:08:26.102  INFO 15100 --- [  restartedMain]
2019-04-12 18:08:26.104  INFO 15100 --- [  restartedMain]
2019-04-12 18:08:26.131  INFO 15100 --- [  restartedMain]
2019-04-12 18:08:26.131  INFO 15100 --- [  restartedMain]
2019-04-12 18:08:26.219  INFO 15100 --- [  restartedMain]
2019-04-12 18:08:26.219  INFO 15100 --- [  restartedMain]
2019-04-12 18:08:26.220  INFO 15100 --- [  restartedMain]
2019-04-12 18:08:26.236  INFO 15100 --- [  restartedMain]
2019-04-12 18:08:26.238  INFO 15100 --- [  restartedMain]
foo/40/Chinese/English/WHUT/2006/Anhui/Hefei/
book1
89
book2
109
some:value
hobby:reading
2019-04-12 18:08:26.412  INFO 15100 --- [  restartedMain]
2019-04-12 18:08:26.545  INFO 15100 --- [  restartedMain]
2019-04-12 18:08:26.575  INFO 15100 --- [  restartedMain]
2019-04-12 18:08:26.575  INFO 15100 --- [  restartedMain]
2019-04-12 18:08:26.581  INFO 15100 --- [  restartedMain]
2019-04-12 18:08:26.609  INFO 15100 --- [  restartedMain]
2019-04-12 18:08:26.613  INFO 15100 --- [  restartedMain]
2019-04-12 18:08:26.616  INFO 15100 --- [  restartedMain]
value1
value2
```

图 4-19

```
@Bean
CommandLineRunner environmentPostProcessorClr(@Value("${key1}") String value1,
                                              @Value("${key2}") String value2){
  return args -> {
    System.out.println(value1);
    System.out.println(value2);
  };
}
```

4.4　日志和报告

4.4.1　日志

日志是进行调试和分析的重要工具。Spring Boot 使用 SLF4J 作为日志的 API，Logback、Log4j2 和 Java Util Logging 都可以作为日志提供者，Spring Boot 默认使用 Logback 作为日志提供者。

1. 日志级别配置

可以在 application.yml 文件中通过 logging.level 来配置指定包的日志级别。

```
logging.level.org.springframework.web: DEBUG
```

设置前如图 4-20 所示。

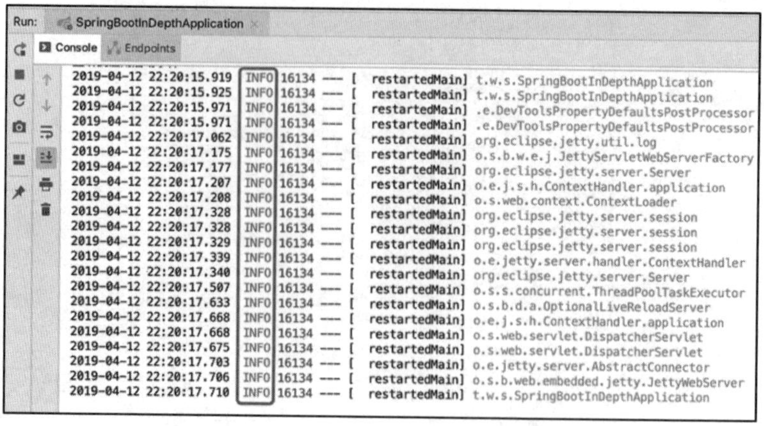

```
Run:   SpringBootInDepthApplication
Console   Endpoints
2019-04-12 22:20:15.919 INFO 16134 --- [ restartedMain] t.w.s.SpringBootInDepthApplication
2019-04-12 22:20:15.925 INFO 16134 --- [ restartedMain] t.w.s.SpringBootInDepthApplication
2019-04-12 22:20:15.971 INFO 16134 --- [ restartedMain] .e.DevToolsPropertyDefaultsPostProcessor
2019-04-12 22:20:15.971 INFO 16134 --- [ restartedMain] .e.DevToolsPropertyDefaultsPostProcessor
2019-04-12 22:20:17.062 INFO 16134 --- [ restartedMain] org.eclipse.jetty.util.log
2019-04-12 22:20:17.175 INFO 16134 --- [ restartedMain] o.s.b.w.e.j.JettyServletWebServerFactory
2019-04-12 22:20:17.177 INFO 16134 --- [ restartedMain] org.eclipse.jetty.server.Server
2019-04-12 22:20:17.207 INFO 16134 --- [ restartedMain] o.e.j.s.h.ContextHandler.application
2019-04-12 22:20:17.208 INFO 16134 --- [ restartedMain] o.s.web.context.ContextLoader
2019-04-12 22:20:17.328 INFO 16134 --- [ restartedMain] org.eclipse.jetty.server.session
2019-04-12 22:20:17.328 INFO 16134 --- [ restartedMain] org.eclipse.jetty.server.session
2019-04-12 22:20:17.329 INFO 16134 --- [ restartedMain] org.eclipse.jetty.server.session
2019-04-12 22:20:17.339 INFO 16134 --- [ restartedMain] o.e.jetty.server.handler.ContextHandler
2019-04-12 22:20:17.340 INFO 16134 --- [ restartedMain] org.eclipse.jetty.server.Server
2019-04-12 22:20:17.507 INFO 16134 --- [ restartedMain] o.s.s.concurrent.ThreadPoolTaskExecutor
2019-04-12 22:20:17.633 INFO 16134 --- [ restartedMain] o.s.b.d.a.OptionalLiveReloadServer
2019-04-12 22:20:17.668 INFO 16134 --- [ restartedMain] o.e.j.s.h.ContextHandler.application
2019-04-12 22:20:17.668 INFO 16134 --- [ restartedMain] o.s.web.servlet.DispatcherServlet
2019-04-12 22:20:17.675 INFO 16134 --- [ restartedMain] o.s.web.servlet.DispatcherServlet
2019-04-12 22:20:17.703 INFO 16134 --- [ restartedMain] o.e.jetty.server.AbstractConnector
2019-04-12 22:20:17.706 INFO 16134 --- [ restartedMain] o.s.b.web.embedded.jetty.JettyWebServer
2019-04-12 22:20:17.710 INFO 16134 --- [ restartedMain] t.w.s.SpringBootInDepthApplication
```

图 4-20

设置后如图 4-21 所示。

```
2019-04-12 22:21:44.396 DEBUG 16149 --- [ restartedMain] o.s.web.context.ContextLoader
2019-04-12 22:21:44.396  INFO 16149 --- [ restartedMain] o.s.web.context.ContextLoader
2019-04-12 22:21:44.471  INFO 16149 --- [ restartedMain] org.eclipse.jetty.server.session
2019-04-12 22:21:44.471  INFO 16149 --- [ restartedMain] org.eclipse.jetty.server.session
2019-04-12 22:21:44.472  INFO 16149 --- [ restartedMain] org.eclipse.jetty.server.session
2019-04-12 22:21:44.482  INFO 16149 --- [ restartedMain] o.e.jetty.server.handler.ContextHandler
2019-04-12 22:21:44.483  INFO 16149 --- [ restartedMain] org.eclipse.jetty.server.Server
2019-04-12 22:21:44.558 DEBUG 16149 --- [ restartedMain] o.s.w.s.handler.SimpleUrlHandlerMapping
2019-04-12 22:21:44.693  INFO 16149 --- [ restartedMain] o.s.s.concurrent.ThreadPoolTaskExecutor
2019-04-12 22:21:44.702 DEBUG 16149 --- [ restartedMain] s.w.s.m.m.a.RequestMappingHandlerAdapter
2019-04-12 22:21:44.738 DEBUG 16149 --- [ restartedMain] s.w.s.m.m.a.RequestMappingHandlerMapping
2019-04-12 22:21:44.748 DEBUG 16149 --- [ restartedMain] o.s.w.s.handler.SimpleUrlHandlerMapping
2019-04-12 22:21:44.756 DEBUG 16149 --- [ restartedMain] .m.m.a.ExceptionHandlerExceptionResolver
2019-04-12 22:21:44.854  INFO 16149 --- [ restartedMain] o.s.b.d.a.OptionalLiveReloadServer
2019-04-12 22:21:44.899  INFO 16149 --- [ restartedMain] o.e.j.s.h.ContextHandler.application
2019-04-12 22:21:44.899  INFO 16149 --- [ restartedMain] o.s.web.servlet.DispatcherServlet
2019-04-12 22:21:44.899 DEBUG 16149 --- [ restartedMain] o.s.web.servlet.DispatcherServlet
2019-04-12 22:21:44.904 DEBUG 16149 --- [ restartedMain] o.s.web.servlet.DispatcherServlet
2019-04-12 22:21:44.905  INFO 16149 --- [ restartedMain] o.s.web.servlet.DispatcherServlet
2019-04-12 22:21:44.938  INFO 16149 --- [ restartedMain] o.e.jetty.server.AbstractConnector
2019-04-12 22:21:44.943  INFO 16149 --- [ restartedMain] o.s.b.web.embedded.jetty.JettyWebServer
2019-04-12 22:21:44.948  INFO 16149 --- [ restartedMain] t.w.s.SpringBootInDepthApplication
```

图 4-21

2. 记录到文件

默认情况下，Spring Boot 只会将日志输出到控制台，如果想将日志输出到文件，则可以通过 logging.file.name 或者 logging.file.path 来定制。

```
logging.file.name: /Users/wangyunfei/log/log.log
```

```
logging.file.path: /Users/wangyunfei/log # 默认文件名为 spring.log
```

3. 切换日志提供者

我们可以很方便地将默认的日志提供者从 Logback 切换成 Log4j2，和前面从 Tomcat 切换成 Jetty 类似，但是又有区别。因为很多 Spring Boot 的 Starter 都默认以 Logback 作为日志提供者，所以建议像下面这样处理。

```
dependencies {
  implementation 'org.springframework.boot:spring-boot-starter-web'
  implementation 'org.springframework.boot:spring-boot-starter-log4j2'
}

configurations {
   all{
       exclude group: 'org.springframework.boot', module: 'spring-boot-starter-logging'
   }
}
```

4.4.2　报告

1．错误报告

Spring Boot 在 META-INF/spring.factories（位于 spring-boot-2.2.x.RELEASE.jar）中注册了失败分析器，如图 4-22 所示。

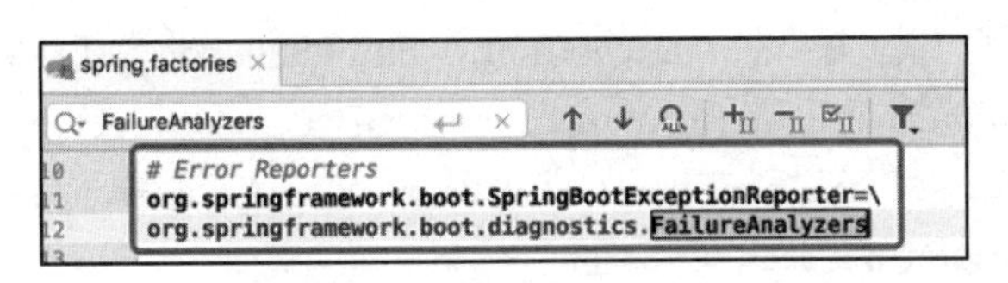

图 4-22

当应用启动因为某些原因失败时，会给出问题描述和问题的解决方案。例如，当出现端口冲突时，如图 4-23 所示。

```
org.springframework.boot.web.server.PortInUseException: Port 8888 is already in use
    at org.springframework.boot.web.embedded.jetty.JettyWebServer.start(JettyWebServer.java:153) ~[spring-boot-2.2.0.
    at org.springframework.boot.web.servlet.context.ServletWebServerApplicationContext.startWebServer(ServletWebServe
    at org.springframework.boot.web.servlet.context.ServletWebServerApplicationContext.finishRefresh(ServletWebServer
    at org.springframework.context.support.AbstractApplicationContext.refresh(AbstractApplicationContext.java:552) ~[
    at org.springframework.boot.web.servlet.context.ServletWebServerApplicationContext.refresh(ServletWebServerApplic
    at org.springframework.boot.SpringApplication.refresh(SpringApplication.java:785) [spring-boot-2.2.0.M1.jar:2.2.0
    at org.springframework.boot.SpringApplication.refreshContext(SpringApplication.java:407) [spring-boot-2.2.0.M1.ja
    at org.springframework.boot.SpringApplication.run(SpringApplication.java:322) [spring-boot-2.2.0.M1.jar:2.2.0.M1]
    at org.springframework.boot.SpringApplication.run(SpringApplication.java:1278) [spring-boot-2.2.0.M1.jar:2.2.0.M1
    at org.springframework.boot.SpringApplication.run(SpringApplication.java:1266) [spring-boot-2.2.0.M1.jar:2.2.0.M1
    at top.wisely.springbootindepth.SpringBootInDepthApplication.main(SpringBootInDepthApplication.java:41) [classes/
    at jdk.internal.reflect.NativeMethodAccessorImpl.invoke0(Native Method) ~[?:?]
    at jdk.internal.reflect.NativeMethodAccessorImpl.invoke(NativeMethodAccessorImpl.java:62) ~[?:?]
    at jdk.internal.reflect.DelegatingMethodAccessorImpl.invoke(DelegatingMethodAccessorImpl.java:43) ~[?:?] <1 inter
    at org.springframework.boot.devtools.restart.RestartLauncher.run(RestartLauncher.java:49) [spring-boot-devtools-2

2019-04-12 23:57:03.808 ERROR 16976 --- [  restartedMain] o.s.b.d.LoggingFailureAnalysisReporter   :

***************************
APPLICATION FAILED TO START
***************************

Description:

Web server failed to start. Port 8888 was already in use.

Action:

Identify and stop the process that's listening on port 8888 or configure this application to listen on another port.
```

图 4-23

2. 自动配置报告

当在 application.yml 文件中设置 debug: true 时，控制台会输出自动配置报告。报告包含以下内容。

◎ 已使用的自动配置（Positive Matches:）。

◎ 未使用的自动配置（Negative Matches）。

◎ 已关闭的自动配置（Exclusions）。

◎ 无条件执行的自动配置（Unconditional Classes）。

4.5 多线程任务和计划任务

4.5.1 Task Executor

Spring 提供了异步多线程任务的功能，只需通过@EnableAsync 开启对异步多线程任务的支持，然后使用@Async 来注解需要执行的异步多线程任务即可。

@EnableAsync 通过导入 AsyncConfigurationSelector 选择配置，在配置（ProxyAsyncConfiguration）中注册了 AsyncAnnotationBeanPostProcessor。

这个 BeanPostProcessor 在让@Async（或 EJB 3.1 的@Asynchronous）注解生效的同时，只需一个类型为 TaskExecutor 的 Bean 即可运行。

当使用 Spring 进行开发时，除使用@EnableAsync 和@Async 注解外，还需定义类型为 TaskExecutor 的 Bean。庆幸的是，Spring Boot 提供了自动配置 TaskExecutionAutoConfiguration，它自动注册了一个 Bean（名称为 applicationTaskExecutor）的 ThreadPoolTaskExecutor（TaskExecutor 的实现类）。

在 Spring Boot 中，只需使用@EnableAsync 和@Async。下面做个简单的异步多线程任务的演示。

首先，在入口类开启对异步多线程任务的支持。

```
@SpringBootApplication
@EnableAsync // SpringBootInDepthApplication 是一个配置类，可以注解在其他的配置类上
public class SpringBootInDepthApplication {}
```

定义一个包含异步多线程任务的类。

```
@Component
@Slf4j
public class AsyncTask {
   @Async //还可以注解在类上，意味着整个类的所有方法都是异步的
   public void loopPrint(Integer i){
      log.info("当前计数为：" + i);
   }
}
```

通过 CommandLineRunner 进行检验，结果如图 4-24 所示。

```
@Bean
CommandLineRunner asyncTaskClr(AsyncTask asyncTask){
  return args -> {
    for(int i = 0; i < 10 ;i++){
      asyncTask.loopPrint(i);
    }
  };
}
```

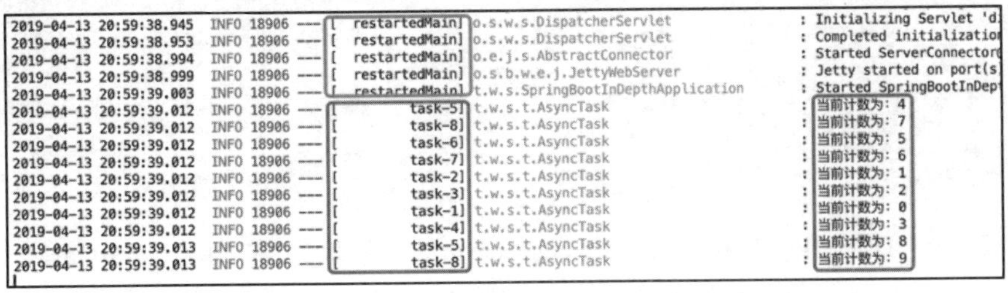

```
2019-04-13 20:59:38.945  INFO 18906 --- [  restartedMain] o.s.w.s.DispatcherServlet                : Initializing Servlet 'd
2019-04-13 20:59:38.953  INFO 18906 --- [  restartedMain] o.s.w.s.DispatcherServlet                : Completed initializatio
2019-04-13 20:59:38.994  INFO 18906 --- [  restartedMain] o.e.j.s.AbstractConnector                : Started ServerConnector
2019-04-13 20:59:38.999  INFO 18906 --- [  restartedMain] o.s.b.w.e.j.JettyWebServer               : Jetty started on port(s
2019-04-13 20:59:39.003  INFO 18906 --- [  restartedMain] t.w.s.SpringBootInDepthApplication       : Started SpringBootInDep
2019-04-13 20:59:39.012  INFO 18906 --- [         task-5] t.w.s.t.AsyncTask                        : 当前计数为: 4
2019-04-13 20:59:39.012  INFO 18906 --- [         task-8] t.w.s.t.AsyncTask                        : 当前计数为: 7
2019-04-13 20:59:39.012  INFO 18906 --- [         task-6] t.w.s.t.AsyncTask                        : 当前计数为: 5
2019-04-13 20:59:39.012  INFO 18906 --- [         task-7] t.w.s.t.AsyncTask                        : 当前计数为: 6
2019-04-13 20:59:39.012  INFO 18906 --- [         task-2] t.w.s.t.AsyncTask                        : 当前计数为: 1
2019-04-13 20:59:39.012  INFO 18906 --- [         task-3] t.w.s.t.AsyncTask                        : 当前计数为: 2
2019-04-13 20:59:39.012  INFO 18906 --- [         task-1] t.w.s.t.AsyncTask                        : 当前计数为: 0
2019-04-13 20:59:39.012  INFO 18906 --- [         task-4] t.w.s.t.AsyncTask                        : 当前计数为: 3
2019-04-13 20:59:39.013  INFO 18906 --- [         task-5] t.w.s.t.AsyncTask                        : 当前计数为: 8
2019-04-13 20:59:39.013  INFO 18906 --- [         task-8] t.w.s.t.AsyncTask                        : 当前计数为: 9
```

图 4-24

从结果可以看出，异步任务使用的不是主线程 restartedMain，而是一个线程池，默认有 8 个线程，线程名以 task-开头。执行结果是乱序的，这意味着任务是并发执行的。

还可以通过 spring.task.execution 对 ThreadPoolTaskExecutor 进行定制。

```
spring:
 task:
  execution:
   pool:
    core-size: 10 # 线程池核心线程数设置为 10，默认是 8
    max-size: 16 # 线程池最大线程数为 16
   thread-name-prefix: mytask- #定制线程名前缀
```

再次进行检验，结果如图 4-25 所示。

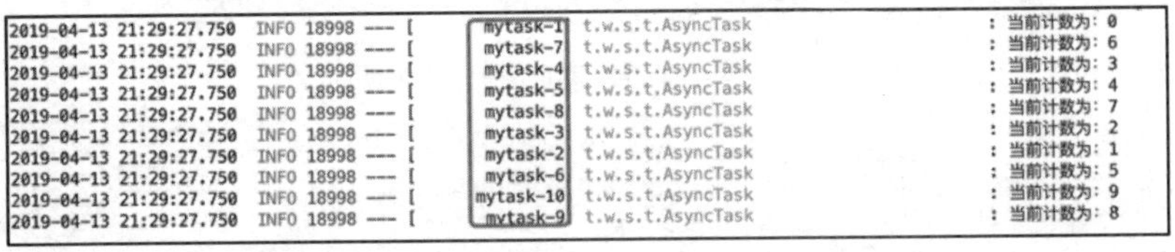

```
2019-04-13 21:29:27.750  INFO 18998 --- [      mytask-1] t.w.s.t.AsyncTask                        : 当前计数为: 0
2019-04-13 21:29:27.750  INFO 18998 --- [      mytask-7] t.w.s.t.AsyncTask                        : 当前计数为: 6
2019-04-13 21:29:27.750  INFO 18998 --- [      mytask-4] t.w.s.t.AsyncTask                        : 当前计数为: 3
2019-04-13 21:29:27.750  INFO 18998 --- [      mytask-5] t.w.s.t.AsyncTask                        : 当前计数为: 4
2019-04-13 21:29:27.750  INFO 18998 --- [      mytask-8] t.w.s.t.AsyncTask                        : 当前计数为: 7
2019-04-13 21:29:27.750  INFO 18998 --- [      mytask-3] t.w.s.t.AsyncTask                        : 当前计数为: 2
2019-04-13 21:29:27.750  INFO 18998 --- [      mytask-2] t.w.s.t.AsyncTask                        : 当前计数为: 1
2019-04-13 21:29:27.750  INFO 18998 --- [      mytask-6] t.w.s.t.AsyncTask                        : 当前计数为: 5
2019-04-13 21:29:27.750  INFO 18998 --- [     mytask-10] t.w.s.t.AsyncTask                        : 当前计数为: 9
2019-04-13 21:29:27.750  INFO 18998 --- [      mytask-9] t.w.s.t.AsyncTask                        : 当前计数为: 8
```

图 4-25

从图 4-25 可以看出，线程前缀已修改为 mytask-，线程数也变成了 10 个。

当系统中有多个不同的异步多线程任务时，可以给特定的任务指定 ThreadPoolTaskExecutor。Spring Boot 提供了 TaskExecutorBuilder 的 Bean，可以用它新建我们定制的 ThreadPoolTaskExecutor。

```
@Bean
ThreadPoolTaskExecutor customTaskExecutor(TaskExecutorBuilder taskExecutorBuilder){
  return taskExecutorBuilder
        .threadNamePrefix("customTask-")
        .corePoolSize(5)
        .build();
}

@Bean
ThreadPoolTaskExecutor myTaskExecutor(TaskExecutorBuilder taskExecutorBuilder){
  return taskExecutorBuilder
        .threadNamePrefix("myTask-")
        .corePoolSize(6)
        .build();
}
```

新建 2 个 ThreadPoolTaskExecutor 的 Bean，名称分别为 customTaskExecutor 和 myTaskExecutor。此时，Spring Boot 不会自动注册 ThreadPoolTaskExecutor 的 Bean（因为 Bean 上注解了@ConditionalOnMissingBean(Executor.class)）。在 AsyncTask 上新增一个异步方法，两个方法分别使用两个 ThreadPoolTaskExecutor。

```
@Component
@Slf4j
public class AsyncTask {
   @Async("myTaskExecutor")
   public void loopPrint(Integer i){
      log.info("当前计数为：" + i);
   }

   @Async("customTaskExecutor")
   public void loopPrint2(Integer i){
      log.info("当前计数为：" + i);
   }
}
```

修改 CommandLineRunner 并进行校验，结果如图 4-26 所示。

```
@Bean
CommandLineRunner asyncTaskClr(AsyncTask asyncTask){
  return args -> {
    for(int i = 0; i < 10 ;i++){
      asyncTask.loopPrint(i);
    }

    for(int i = 0; i < 10 ;i++){
      asyncTask.loopPrint2(i);
    }
```

```
    };
}
```

```
2019-04-13 21:46:22.595  INFO 19066 --- [   customTask-2] t.w.s.t.AsyncTask  : 当前计数为: 1
2019-04-13 21:46:22.595  INFO 19066 --- [       myTask-2] t.w.s.t.AsyncTask  : 当前计数为: 1
2019-04-13 21:46:22.595  INFO 19066 --- [   customTask-3] t.w.s.t.AsyncTask  : 当前计数为: 2
2019-04-13 21:46:22.595  INFO 19066 --- [       myTask-3] t.w.s.t.AsyncTask  : 当前计数为: 2
2019-04-13 21:46:22.596  INFO 19066 --- [   customTask-5] t.w.s.t.AsyncTask  : 当前计数为: 4
2019-04-13 21:46:22.596  INFO 19066 --- [       myTask-2] t.w.s.t.AsyncTask  : 当前计数为: 6
2019-04-13 21:46:22.596  INFO 19066 --- [   customTask-5] t.w.s.t.AsyncTask  : 当前计数为: 5
2019-04-13 21:46:22.595  INFO 19066 --- [       myTask-4] t.w.s.t.AsyncTask  : 当前计数为: 3
2019-04-13 21:46:22.596  INFO 19066 --- [   customTask-5] t.w.s.t.AsyncTask  : 当前计数为: 6
2019-04-13 21:46:22.596  INFO 19066 --- [       myTask-3] t.w.s.t.AsyncTask  : 当前计数为: 8
2019-04-13 21:46:22.596  INFO 19066 --- [       myTask-6] t.w.s.t.AsyncTask  : 当前计数为: 5
2019-04-13 21:46:22.596  INFO 19066 --- [   customTask-5] t.w.s.t.AsyncTask  : 当前计数为: 7
2019-04-13 21:46:22.595  INFO 19066 --- [   customTask-4] t.w.s.t.AsyncTask  : 当前计数为: 3
2019-04-13 21:46:22.595  INFO 19066 --- [   customTask-1] t.w.s.t.AsyncTask  : 当前计数为: 0
2019-04-13 21:46:22.595  INFO 19066 --- [       myTask-5] t.w.s.t.AsyncTask  : 当前计数为: 4
2019-04-13 21:46:22.595  INFO 19066 --- [       myTask-1] t.w.s.t.AsyncTask  : 当前计数为: 0
2019-04-13 21:46:22.596  INFO 19066 --- [   customTask-2] t.w.s.t.AsyncTask  : 当前计数为: 8
2019-04-13 21:46:22.596  INFO 19066 --- [   customTask-4] t.w.s.t.AsyncTask  : 当前计数为: 9
2019-04-13 21:46:22.596  INFO 19066 --- [       myTask-4] t.w.s.t.AsyncTask  : 当前计数为: 9
2019-04-13 21:46:22.596  INFO 19066 --- [       myTask-2] t.w.s.t.AsyncTask  : 当前计数为: 7
```

图 4-26

4.5.2 Task Scheduler

Spring 还提供了无人干预的计划任务，通过@EnableScheduling 开启对计划任务的支持，并使用@Scheduled 来注解计划执行的任务。

@EnableScheduling 会直接导入配置 SchedulingConfiguration。SchedulingConfiguration 注册了 ScheduledAnnotationBeanPostProcessor 的 Bean，这个 BeanPostProcessor 会让@Scheduled 生效，并且需要一个 TaskScheduler 的 Bean（Bean 名称为 taskScheduler）。

若使用 Spring 进行开发，则需要额外定义一个 TaskScheduler 的 Bean。同样，Spring Boot 的 TaskSchedulingAutoConfiguration 定义了名为 taskScheduler 的 Bean。TaskScheduler 的实现类是 ThreadPoolTaskScheduler，示例如下。

```
在@Bean
@ConditionalOnBean(name =
TaskManagementConfigUtils.SCHEDULED_ANNOTATION_PROCESSOR_BEAN_NAME)
@ConditionalOnMissingBean({ SchedulingConfigurer.class, TaskScheduler.class,
    ScheduledExecutorService.class })
public ThreadPoolTaskScheduler taskScheduler(TaskSchedulerBuilder builder) {
  return builder.build();
}
```

在 Spring Boot 下，只需使用@EnableScheduling 和@Scheduled 即可，无须额外的配置。

下面演示一个简单的计划任务。

在入口类（任何配置类）开启对计划任务的支持。

```
@SpringBootApplication
@EnableScheduling
public class SpringBootInDepthApplication {}
```

定义计划任务的类。

```
@Component
@Slf4j
public class ScheduledTask {

    @Scheduled(fixedRate = 5000) //a
    public void fixedRateDemo(){
        log.info("每隔 5 秒钟执行一次");
    }

    @Scheduled(fixedDelay = 10000) //b
    public void fixedDelayDemo(){
        log.info("在上次执行完成 10 秒钟之后执行");
    }

    @Scheduled(cron = "0 * * * * SAT,SUN" )// c
    public void cronDemo(){
        log.info("周六周日每分钟执行一次");
    }
}
```

a. fixedRate：每隔固定的时间执行一次，无论上次任务是否完成。

b. fixedDelay：在上次任务执行完成后，在指定时间执行新任务。

c. 使用 UNIX 的 cron 任务计划器表达式，它可接收 6 个参数。

- ◎ 秒
- ◎ 分钟
- ◎ 小时
- ◎ 日
- ◎ 月
- ◎ 星期：SUN, MON, TUE, WED, THU, FRI, SAT；之间以“,”隔开，范围可使用“-”。每周第一天为 SUN。

当然，也可以在 application.yml 中通过 spring.task.scheduling 来定制 ThreadPoolTaskScheduler，示例如下。

```
spring:
  task:
    scheduling:
      pool:
        size: 5
      thread-name-prefix: my-scheduling-
```

Spring Boot 还提供了 TaskSchedulerBuilder 的 Bean，使用它可以轻松定制 ThreadPoolTaskScheduler。和定制 ThreadPoolTaskExecutor 一样，这里不再演示，运行结果如图 4-27 所示。

```
[my-scheduling-1] t.w.s.t.ScheduledTask                    : 周六周日每分钟执行一次
[my-scheduling-3] t.w.s.t.ScheduledTask                    : 每隔5秒钟执行一次
[my-scheduling-4] t.w.s.t.ScheduledTask                    : 在上次执行完成10秒钟之后执行
[my-scheduling-2] t.w.s.t.ScheduledTask                    : 每隔5秒钟执行一次
[my-scheduling-2] t.w.s.t.ScheduledTask                    : 每隔5秒钟执行一次
[my-scheduling-5] t.w.s.t.ScheduledTask                    : 在上次执行完成10秒钟之后执行
[my-scheduling-3] t.w.s.t.ScheduledTask                    : 每隔5秒钟执行一次
[my-scheduling-3] t.w.s.t.ScheduledTask                    : 每隔5秒钟执行一次
[my-scheduling-2] t.w.s.t.ScheduledTask                    : 在上次执行完成10秒钟之后执行
[my-scheduling-4] t.w.s.t.ScheduledTask                    : 每隔5秒钟执行一次
[my-scheduling-4] t.w.s.t.ScheduledTask                    : 每隔5秒钟执行一次
[my-scheduling-5] t.w.s.t.ScheduledTask                    : 在上次执行完成10秒钟之后执行
[my-scheduling-1] t.w.s.t.ScheduledTask                    : 每隔5秒钟执行一次
[my-scheduling-1] t.w.s.t.ScheduledTask                    : 每隔5秒钟执行一次
[my-scheduling-2] t.w.s.t.ScheduledTask                    : 在上次执行完成10秒钟之后执行
[my-scheduling-5] t.w.s.t.ScheduledTask                    : 每隔5秒钟执行一次
[my-scheduling-5] t.w.s.t.ScheduledTask                    : 每隔5秒钟执行一次
[my-scheduling-3] t.w.s.t.ScheduledTask                    : 在上次执行完成10秒钟之后执行
[my-scheduling-4] t.w.s.t.ScheduledTask                    : 每隔5秒钟执行一次
[my-scheduling-1] t.w.s.t.ScheduledTask                    : 周六周日每分钟执行一次
[my-scheduling-5] t.w.s.t.ScheduledTask                    : 每隔5秒钟执行一次
[my-scheduling-2] t.w.s.t.ScheduledTask                    : 在上次执行完成10秒钟之后执行
[my-scheduling-3] t.w.s.t.ScheduledTask                    : 每隔5秒钟执行一次
[my-scheduling-3] t.w.s.t.ScheduledTask                    : 每隔5秒钟执行一次
[my-scheduling-4] t.w.s.t.ScheduledTask                    : 在上次执行完成10秒钟之后执行
[my-scheduling-5] t.w.s.t.ScheduledTask                    : 每隔5秒钟执行一次
[my-scheduling-5] t.w.s.t.ScheduledTask                    : 每隔5秒钟执行一次
[my-scheduling-2] t.w.s.t.ScheduledTask                    : 在上次执行完成10秒钟之后执行
```

图 4-27

4.6　Gradle 插件

Spring Boot 中的 Gradle 插件提供了依赖管理、运行应用、打包等功能。

一个 Spring Boot 项目主要使用三个插件：

- ◎　org.springframework.boot 插件，提供各种任务。
- ◎　java 插件。
- ◎　io.spring.dependency-management 插件，当使用此插件时，Spring Boot Gradle 插件会自动从相同版本的 spring-boot-dependencies 中导入依赖。

```
plugins {
  id 'org.springframework.boot' version '2.2.x.RELEASE'
  id 'java'
}

apply plugin: 'io.spring.dependency-management'
```

4.6.1　依赖

在 Spring Boot 中定义依赖时无须指定版本号，默认使用 Spring Boot Gradle 插件的版本号即可。

```
dependencies {
  implementation 'org.springframework.boot:spring-boot-starter-web'
  implementation 'org.springframework.boot:spring-boot-starter-data-jpa'
}
```

当然，也可以自己指定依赖的版本号：

```
ext['lombok.version'] = '1.18.4'
```

每一个 Spring Boot 的第三方依赖版本都是测试验证过的，自己覆写版本可能会导致兼容性问题。一般来说，如果没有特殊明确的需求，则无须修改版本号。

4.6.2 Spring Boot Starter

当使用 Spring Initializr 新建应用时，检索选择的依赖就是一个 Starter，例如：

◎ 若依赖是 Web，则使用的是 spring-boot-starter-web。

◎ 若依赖是 Security，则使用的是 spring-boot-starter-security。

◎ 若依赖是 JPA，则使用的是 spring-boot-starter-data-jpa。

Spring Boot Starter 提供了相关技术一站式的安装使用，如 spring-boot-starter-web 提供了所有进行 Web 开发所需要的 Spring 及第三方的依赖包。Spring Boot Starter 的 Group 是 org.springframework.boot，Artifact 命名习惯为 spring-boot-starter-*，*代表技术名称。我们可以在 IDE 里直接向 build.gradle 添加 Starter 的依赖。

```
dependencies {
    implementation 'org.springframework.boot:spring-boot-starter-web'
  implementation 'org.springframework.boot:spring-boot-starter-data-jpa'
}
```

完整官方 Starter 列表参见：Spring Boot Reference *Documentation* 中的 Starters 部分。

4.6.3 插件任务

运行任务：

```
$ ./gradlew bootRun
```

传参运行：

```
$ ./gradlew bootRun --args='--spring.profiles.active=prod'
```

打包：

```
$ ./gradlew bootJar
```

4.7 自定义 Starter

当在 Spring Boot 中使用一些技术时，需要制作一个 Starter 来简化这个技术的使用，让使用者无须做大量的胶水配置，即可一站式安装使用此技术。一般情况下，自定义 Spring Boot Starter 开发需要明确三部分内容。

◎ 需要被包装的技术库，作为依赖存在。

◎ autoconfigure 自动配置模块，对需要被包装的技术进行自动配置。

◎ starter 模块，面向使用者，它主要负责依赖：包含 autoconfigure 和必要的依赖库。

在开发时，自定义的 Starter 是由以下两部分组成的。

◎　autoconfigure 模块，命名习惯为：技术名-spring-boot-autoconfigure。

◎　starter 模块，命名习惯为：技术名-spring-boot-starter。

当然，可以把 autoconfigure 模块和 starter 模块放在一起，命名习惯为：技术名-spring-boot-starter。

4.7.1　包装技术库

下面演示如何将一个简单的功能制作成被包装的技术库。用 IntelliJ IDEA 新建一个简单的 Java 项目，如图 4-28 至图 4-30 所示。

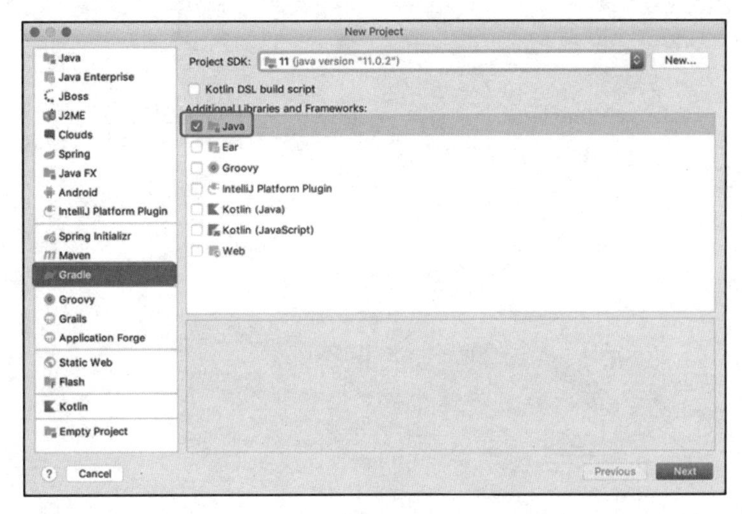

图 4-28

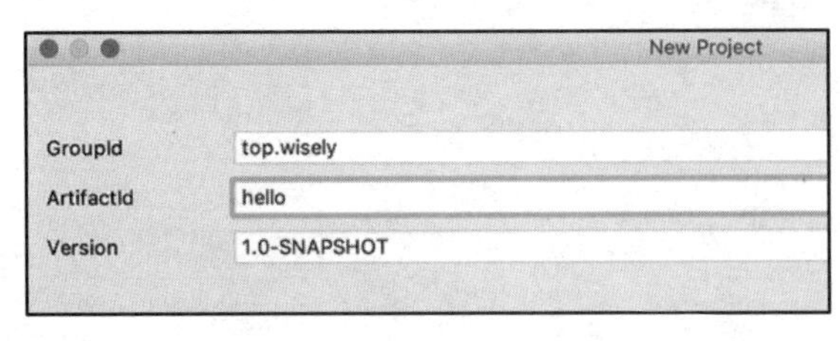

图 4-29

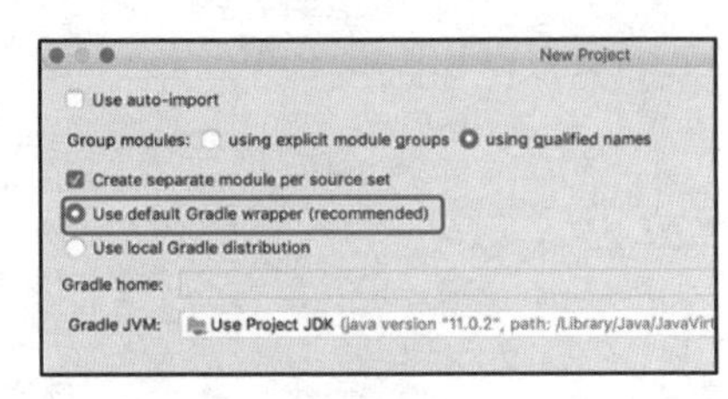

图 4-30

在应用路径 gradle/wrapper/gradle-wrapper.properties 中，修改 Gradle Wrapper 的版本为 5.2.1：

```
distributionUrl=https\://services.gradle.org/distributions/gradle-5.2.1-bin.zip
```

内容很简单，有个简单的类：

```
package top.wisely;

public class GreetingService {
```

```
    public String greeting(String user, String greetings){
        return "Hi " + user + "," + greetings;
    }
}
```

build.gradle 的定义如下：

```
plugins {
    id 'java'
    id 'maven-publish'
}

group 'top.wisely'
version '1.0-SNAPSHOT'

sourceCompatibility = 1.8

repositories {
    mavenCentral()
}

dependencies {
}

publishing {
    repositories {
        maven { //此处配置上传库的地址
            credentials {
                username "myMavenRepo"
                password "test"
            }
            url "https://mymavenrepo.com/repo/XL62J2nMpLk4ILr5VES7/"
        }
    }

    publications {
        maven(MavenPublication) {
            from components.java
        }
    }
}
```

https://mymavenrepo.com 是一个免费 Maven 私服（读者请自行注册），通过插件 maven-publish 可将技术库 hello 发布到 https://mymavenrepo.com 上。

发布命令如下：

```
$ ./gradlew publish
```

发布成功后如图 4-31 所示。

```
https://mymavenrepo.com/repo/XL62J2nMpLk4ILr5VES7/top/wisely/hello/1.0-SNAPSHOT/

Index of /repo/UwLSTfx0ltwDBZiHWbIL/top/wisely/hell

../
hello-1.0-20190415.161651-1.jar            15-Apr-2019 16:32     1146
hello-1.0-20190415.161651-1.jar.md5        15-Apr-2019 16:32       32
hello-1.0-20190415.161651-1.jar.sha1       15-Apr-2019 16:32       40
hello-1.0-20190415.161651-1.pom            15-Apr-2019 16:32      395
hello-1.0-20190415.161651-1.pom.md5        15-Apr-2019 16:32       32
hello-1.0-20190415.161651-1.pom.sha1       15-Apr-2019 16:32       40
maven-metadata.xml                         15-Apr-2019 16:32      759
maven-metadata.xml.md5                     15-Apr-2019 16:32       32
maven-metadata.xml.sha1                    15-Apr-2019 16:32       40
```

图 4-31

在 Spring Boot 的自动配置中做两件事情：

◎　从外部配置获取 user 和 greeting 的值。

◎　自动注册 GreetingService 的 Bean。

4.7.2　Starter 的结构

新建 Gradle 项目，不选择额外的库和框架，如图 4-32 和图 4-33 所示。

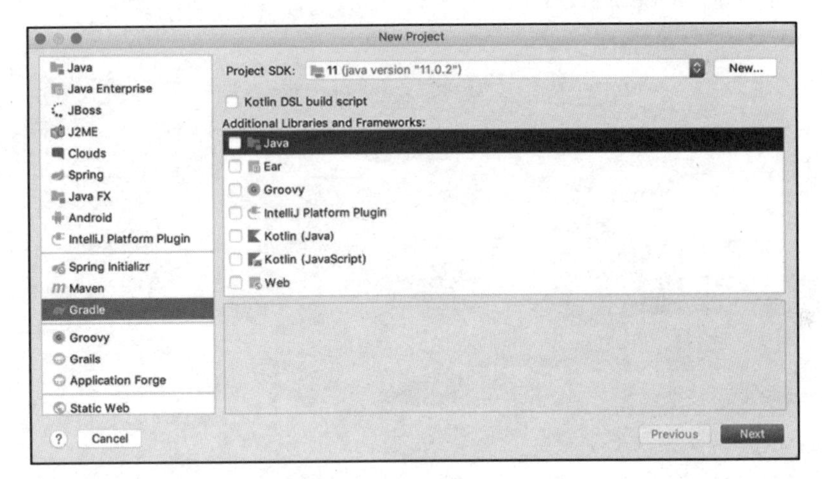

图 4-32

New Project

GroupId　top.wisely

ArtifactId　hello-starter

Version　1.0-SNAPSHOT

图 4-33

修改 Gradle Wrapper 的版本为 5.2.1，build.gradle 的定义如下：

```
buildscript {
    repositories {
        mavenCentral()
        maven { url 'https://repo.spring.io/snapshot' }
        maven { url 'https://repo.spring.io/milestone' }

    }
    dependencies {
        classpath "org.springframework.boot:spring-boot-gradle-plugin:2.2.X.RELEASE"
    }
}

allprojects {
    repositories {
        mavenCentral()
        maven { url 'https://repo.spring.io/snapshot' }
        maven { url 'https://repo.spring.io/milestone' }
        maven { // 此处配置下载库的地址
            credentials {
                username "myMavenRepo"
                password "test"
            }
            url "https://mymavenrepo.com/repo/XL62J2nMpLk4ILr5VES7/"
        }
    }
    group = "top.wisely"
    version = "1.0-SNAPSHOT"

    apply plugin: 'java'
    apply plugin: "io.spring.dependency-management"
    apply plugin: 'maven-publish'

    sourceCompatibility = 1.8
    targetCompatibility = 1.8

    dependencyManagement {
        imports {
            mavenBom
org.springframework.boot.gradle.plugin.SpringBootPlugin.BOM_COORDINATES
        }

    }
}
```

4.7.3　autoconfigure 模块

在“hello-starter”上单击鼠标右键，在弹出的快捷菜单上单击“New”→“Module”选项，显示如图 4-34 和图 4-35 所示。

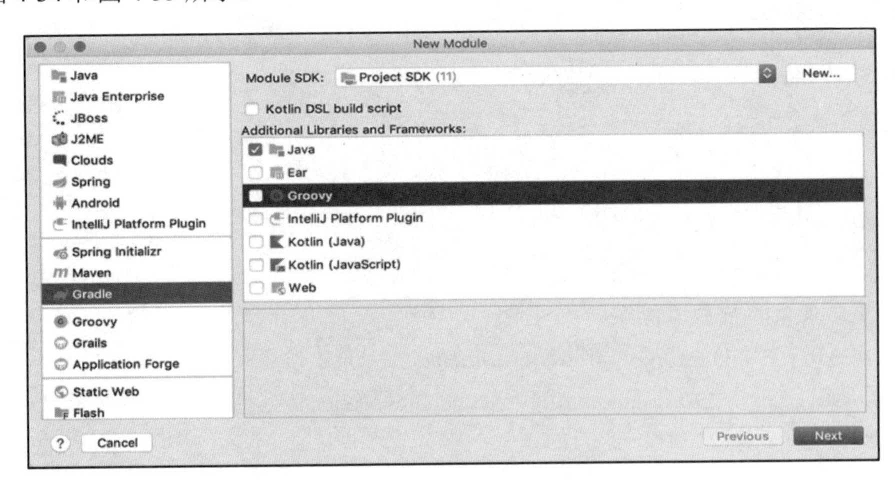

图 4-34

图 4-35

build.gradle 的定义如下。

```
repositories {
   dependencies {
      compileOnly 'top.wisely:hello:1.0-SNAPSHOT' //a
      annotationProcessor
'org.springframework.boot:spring-boot-configuration-processor' // a
      annotationProcessor
'org.springframework.boot:spring-boot-autoconfigure-processor' // b
      compileOnly 'org.springframework.boot:spring-boot-autoconfigure'
   }
}

publishing {
   repositories {
```

```
        maven { // 此处配置上传库的地址
            credentials {
                username "myMavenRepo"
                password "test"
            }
            url "https://mymavenrepo.com/repo/XL62J2nMpLk4ILr5VES7/"
        }
    }

    publications {
        maven(MavenPublication) {
            from components.java
        }
    }
}
```

a. 使用自定义的技术组件库。

b. 生成 META-INF/spring-configuration-metadata.json，该文件能帮助 IDE 给我们提供自动提示。生成 META-INF/spring-autoconfigure-metadata.properties，帮助应用在启动过程中过滤自动配置，加快启动速度。

属性绑定文件：

```
@ConfigurationProperties(prefix = "greeting")
public class GreetingProperties {
    private String user;
    private String greetings;

    public String getUser() {
        return user;
    }

    public void setUser(String user) {
        this.user = user;
    }

    public String getGreetings() {
        return greetings;
    }

    public void setGreetings(String greetings) {
        this.greetings = greetings;
    }
}
```

自动配置文件：

```
@Configuration
@ConditionalOnClass({GreetingService.class})
@EnableConfigurationProperties({GreetingProperties.class})//此处必须使用，当前类的包
```

```
//路径一般不会和使用者的包扫描路径相同
public class GreetingAutoConfiguration {

    private final GreetingProperties greetingProperties;

    public GreetingAutoConfiguration(GreetingProperties greetingProperties) {
        this.greetingProperties = greetingProperties;
    }

    @Bean
    @ConditionalOnMissingBean
    public GreetingService greetingService(){
        return new
GreetingService(greetingProperties.getUser(),greetingProperties.getGreetings());
    }
}
```

在 META-INF/spring.factories 中定义配置：

```
org.springframework.boot.autoconfigure.EnableAutoConfiguration=\
top.wisely.autoconfigure.GreetingAutoConfiguration
```

发布到 maven 私服：

```
$ ./gradlew hello-spring-boot-autoconfigure:publish
```

4.7.4 Starter 模块

以同样的方式新建 hello-spring-boot-starter 模块，Starter 模块其实是一个空模块，它只为 autoconfigure 提供需要的依赖。其中，build.gradle 中的内容如下。

```
dependencies {
    compile 'top.wisely:hello:1.0-SNAPSHOT'
}

publishing {
    repositories {
        maven {
            credentials {
                username "myMavenRepo"
                password "test"
            }
            url "https://mymavenrepo.com/repo/XL62J2nMpLk4ILr5VES7/"
        }
    }

    publications {
        maven(MavenPublication) {
            from components.java
        }
```

```
    }
}
```

在 hello-spring-boot-starter 模块中没有任何内容，只声明了对 hello-spring-boot-autoconfigure 模块和 hello 库的依赖。

同样发布到 maven 私服：

```
$ ./gradlew hello-spring-boot-starter:publish
```

至此，私服上有 hello 库、hello-spring-boot-autoconfigure 模块和 hello-spring-boot-starter 模块三个，如图 4-36 所示。

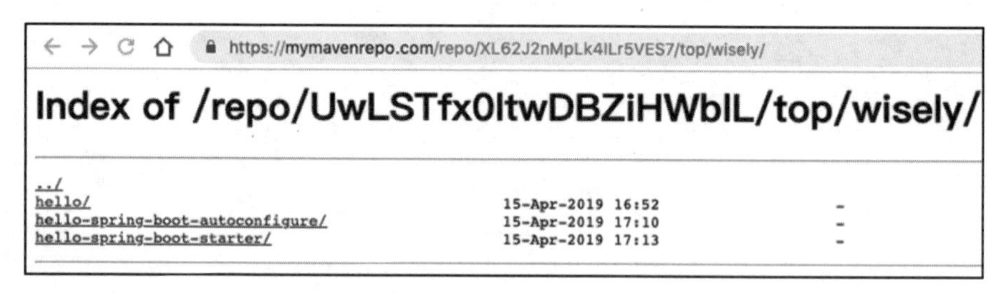

图 4-36

4.7.5 使用 Starter

通过 Spring Initializr 生成一个 Spring Boot 项目。

Group：io.github.wiselyman。

Artifact：hello-client。

在 build.gradle 中添加我们自定义的 Starter 依赖。

```
repositories {
    mavenCentral()
    maven { url 'https://repo.spring.io/snapshot' }
    maven { url 'https://repo.spring.io/milestone' }
    maven { //设置下载库地址
        credentials {
            username "myMavenRepo"
            password "test"
        }
        url "https://mymavenrepo.com/repo/XL62J2nMpLk4ILr5VES7/"
    }
}

dependencies {
    implementation 'org.springframework.boot:spring-boot-starter'
    implementation 'top.wisely:hello-spring-boot-starter:1.0-SNAPSHOT' //Starter
//依赖
  implementation 'top.wisely:hello-spring-boot-autoconfigure:1.0-SNAPSHOT'//自动配置
```

```
    testImplementation 'org.springframework.boot:spring-boot-starter-test'
}
```

hello-spring-boot-starter 可自动导入相关依赖，如图 4-37 所示。

hello-client
Source Sets
main
Dependencies
org.springframework.boot:spring-boot-starter:2.2.0.M2 (Compile)
top.wisely:hello-spring-boot-starter:1.0-SNAPSHOT (Compile)
top.wisely:hello:1.0-SNAPSHOT (Compile)
top.wisely:hello-spring-boot-autoconfigure:1.0-SNAPSHOT (Compile)

图 4-37

在 application.yml 中配置 greeting 信息。

```
greeting:
  user: wyf
  greetings: 祝你幸福
```

通过 CommandLineRunner 进行验证。

```
    @Bean
    CommandLineRunner commandLineRunner(GreetingService greetingService){
        return args -> {
            System.out.println(greetingService.greeting());
        };
    }
}
```

hello-spring-boot-autoconfigure 已经自动配置了 GreetingService 的 Bean，正确的输出信息如图 4-38 所示。

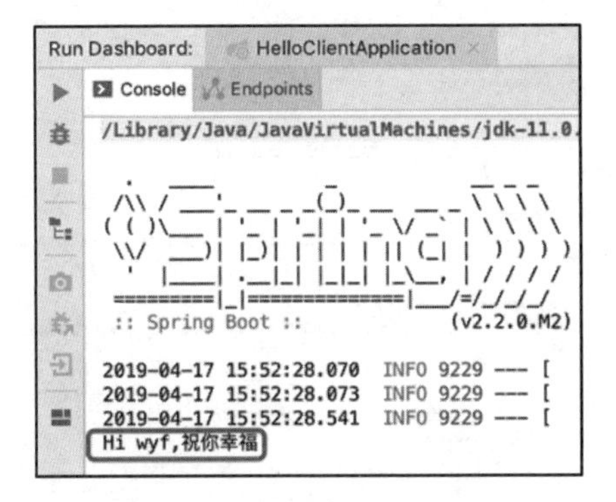

图 4-38

4.8 Spring Boot Actuator

Spring Boot Actuator 提供了在生产环境下所需要的一些特性，可以使用 Spring Boot Actuator 来监控应用的行为。

下面新建一个应用作为本节的演示示例。

Group：top.wisely。

Artifact：learning-spring-boot-actuator。

Dependencies：Spring Boot Actuator、Spring Web Starter 和 Lombok。

具体依赖如下。

```
dependencies {
  implementation 'org.springframework.boot:spring-boot-starter-actuator'
  implementation 'org.springframework.boot:spring-boot-starter-web'
  compileOnly 'org.projectlombok:lombok'
  annotationProcessor 'org.projectlombok:lombok'
    //……
}
```

4.8.1 常用端点

1. actuator

/actuator 是所有端点的前缀，访问 http://localhost:8080/actuator，可显示所有地址已暴露且功能已开启的端点访问信息，如图 4-39 所示。

```
localhost:8080/actuator
{
 - _links: {
   - self: {
         href: "http://localhost:8080/actuator",
         templated: false
     },
   - health: {
         href: "http://localhost:8080/actuator/health",
         templated: false
     },
   - health-component: {
         href: "http://localhost:8080/actuator/health/{component}",
         templated: true
     },
   - health-component-instance: {
         href: "http://localhost:8080/actuator/health/{component}/{instance}",
         templated: true
     },
   - info: {
         href: "http://localhost:8080/actuator/info",
         templated: false
     }
   }
}
```


图 4-39

2. health

/actuator/health 可显示应用的整体健康情况，访问 http://localhost:8080/actuator/health，如图

4-40 所示。

```
localhost:8080/actuator/health
{
    status: "UP"
}
```

图 4-40

想要看到健康信息的明细，可在 application.properties 中设置：

```
management.endpoint.health.show-details=always
```

重启应用后，访问 http://localhost:8080/actuator/health，如图 4-41 所示。

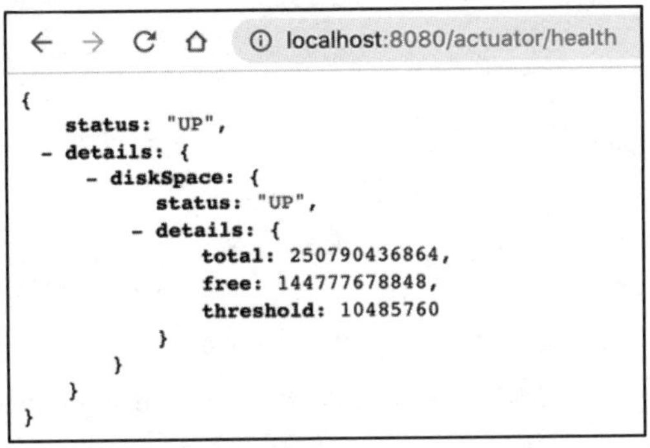

```
localhost:8080/actuator/health
{
    status: "UP",
  - details: {
      - diskSpace: {
            status: "UP",
          - details: {
                total: 250790436864,
                free: 144777678848,
                threshold: 10485760
            }
        }
    }
}
```

图 4-41

在本例中没有其他组件，因而只显示磁盘空间的健康情况。我们可以通过 details 下的健康名称（如 diskSpace）来专门查看相关信息。访问 http://localhost:8080/actuator/health/diskSpace，如图 4-42 所示。

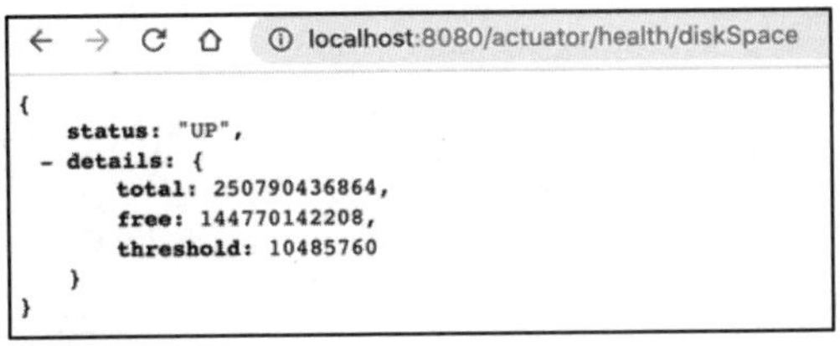

```
localhost:8080/actuator/health/diskSpace
{
    status: "UP",
  - details: {
        total: 250790436864,
        free: 144770142208,
        threshold: 10485760
    }
}
```

图 4-42

3. info

/actuator/info 可显示应用信息，在 application.properties 中可添加应用信息，使用 info.* 可设置任意信息。

```
info.app.name=learning spring boot actuator
info.app.author=wang yun fei
info.some=some info
```

重启应用，访问 http://localhost:8080/actuator/info，如图 4-43 所示。

```
localhost:8080/actuator/info
{
  - app: {
        name: "learning spring boot actuator",
        author: "wang yun fei"
    },
    some: "some info"
}
```

图 4-43

4．shutdown

/actuator/shutdown 可优雅地关闭 Spring Boot 应用。该节点默认路径没有暴露，且功能没有开启。

```
management.endpoints.web.exposure.include=shutdown,health,info # a
management.endpoint.shutdown.enabled=true # b
```

a. 配置需要暴露的端点，多个端点暴露用“,”隔开。

b. 开启 shutdown 端点，端点的开启和关闭可以通过 management.endpoint.端点名.enabled 来设置。

在 Postman 中，用 Post 方法调用 http://localhost:8080/shutdown，可优雅地关闭当前应用，如图 4-44 所示。

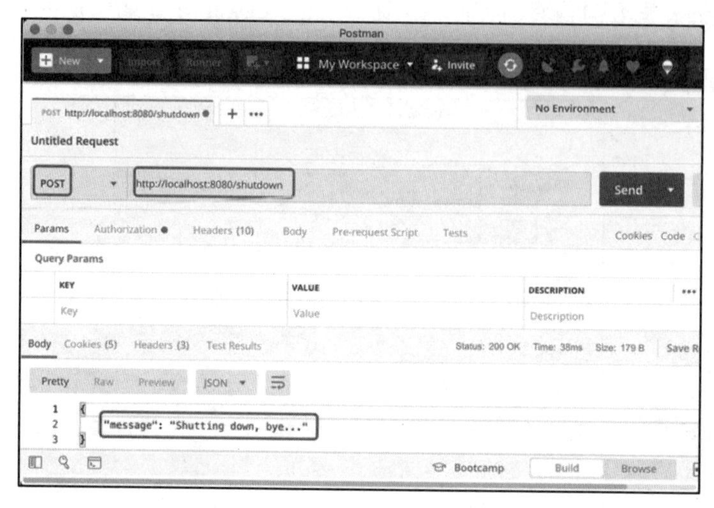

图 4-44

5．env

/actuator/env 可获得应用的所有 Environment 信息，包含 Profile、系统环境变量和应用的 properties 信息。

暴露该端点：

```
management.endpoints.web.exposure.include=shutdown,health,info,env
```

重启应用，访问 http://localhost:8080/actuator/env，如图 4-45 所示。

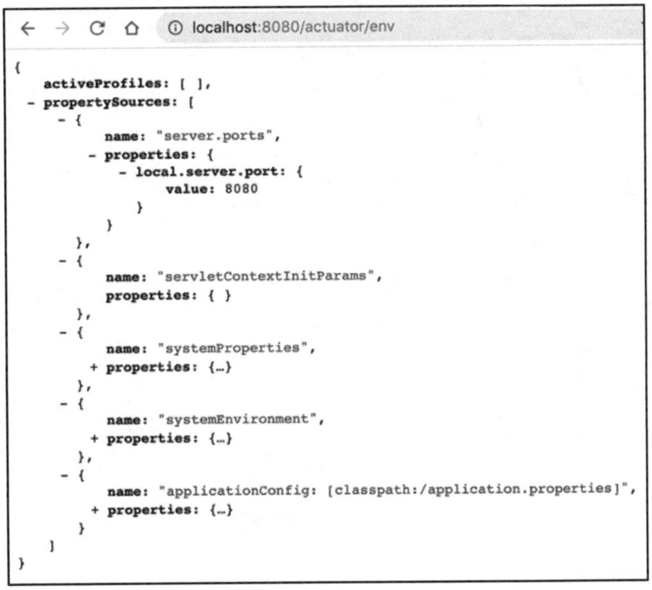

```
localhost:8080/actuator/env
{
   activeProfiles: [ ],
 - propertySources: [
   - {
         name: "server.ports",
       - properties: {
         - local.server.port: {
               value: 8080
           }
         }
     },
   - {
         name: "servletContextInitParams",
         properties: { }
     },
   - {
         name: "systemProperties",
       + properties: {…}
     },
   - {
         name: "systemEnvironment",
       + properties: {…}
     },
   - {
         name: "applicationConfig: [classpath:/application.properties]",
       + properties: {…}
     }
   ]
}
```

图 4-45

6．beans

/actuator/beans 可显示当前应用的所有 Bean，暴露该端点：

```
management.endpoints.web.exposure.include=shutdown,health,info,env,beans
```

重启应用，访问 http://localhost:8080/actuator/beans，如图 4-46 所示。

7．conditions

/actuator/conditions 可显示应用的自动配置报告，包含匹配的自动配置类（positiveMatches）、不匹配的自动配置类（negativeMatches）和非条件配置类（unconditionalClasses）。暴露该端点：

```
management.endpoints.web.exposure.include=shutdown,health,info,env,beans,conditions
```

重启应用，访问 http://localhost:8080/actuator/conditions，如图 4-47 所示。

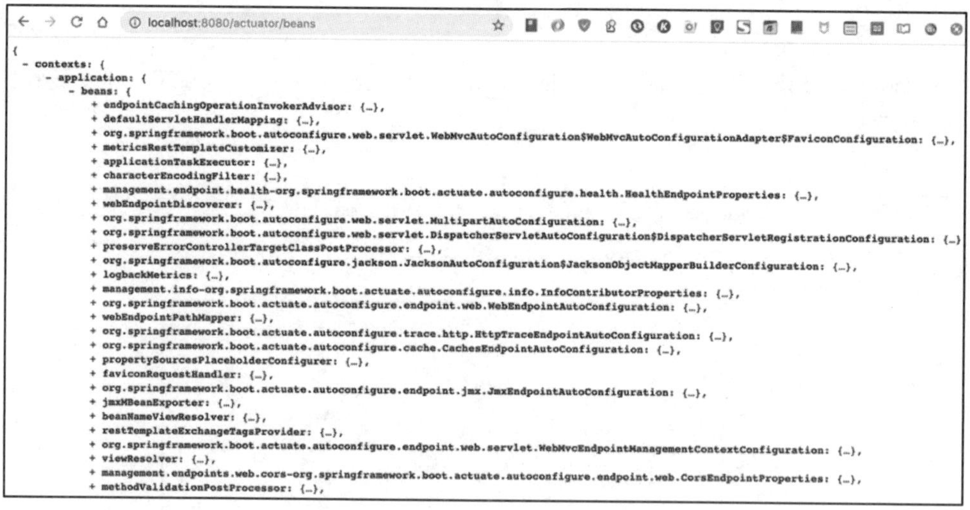

localhost:8080/actuator/beans

{
  - contexts: {
    - application: {
      - beans: {
        + endpointCachingOperationInvokerAdvisor: {…},
        + defaultServletHandlerMapping: {…},
        + org.springframework.boot.autoconfigure.web.servlet.WebMvcAutoConfiguration$WebMvcAutoConfigurationAdapter$FaviconConfiguration: {…},
        + metricsRestTemplateCustomizer: {…},
        + applicationTaskExecutor: {…},
        + characterEncodingFilter: {…},
        + management.endpoint.health-org.springframework.boot.actuate.autoconfigure.health.HealthEndpointProperties: {…},
        + webEndpointDiscoverer: {…},
        + org.springframework.boot.autoconfigure.web.servlet.MultipartAutoConfiguration: {…},
        + org.springframework.boot.autoconfigure.web.servlet.DispatcherServletAutoConfiguration$DispatcherServletRegistrationConfiguration: {…}
        + preserveErrorControllerTargetClassPostProcessor: {…},
        + org.springframework.boot.autoconfigure.jackson.JacksonAutoConfiguration$JacksonObjectMapperBuilderConfiguration: {…},
        + logbackMetrics: {…},
        + management.info-org.springframework.boot.actuate.autoconfigure.info.InfoContributorProperties: {…},
        + org.springframework.boot.actuate.autoconfigure.endpoint.web.WebEndpointAutoConfiguration: {…},
        + webEndpointPathMapper: {…},
        + org.springframework.boot.actuate.autoconfigure.trace.http.HttpTraceEndpointAutoConfiguration: {…},
        + org.springframework.boot.actuate.autoconfigure.cache.CachesEndpointAutoConfiguration: {…},
        + propertySourcesPlaceholderConfigurer: {…},
        + faviconRequestHandler: {…},
        + org.springframework.boot.actuate.autoconfigure.endpoint.jmx.JmxEndpointAutoConfiguration: {…},
        + jmxMBeanExporter: {…},
        + beanNameViewResolver: {…},
        + restTemplateExchangeTagsProvider: {…},
        + org.springframework.boot.actuate.autoconfigure.endpoint.web.servlet.WebMvcEndpointManagementContextConfiguration: {…},
        + viewResolver: {…},
        + management.endpoints.web.cors-org.springframework.boot.actuate.autoconfigure.endpoint.web.CorsEndpointProperties: {…},
        + methodValidationPostProcessor: {…},

图 4-46

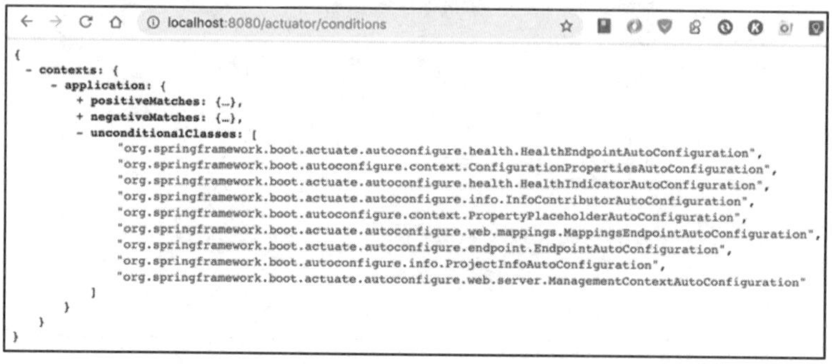

localhost:8080/actuator/conditions

{
  - contexts: {
    - application: {
      + positiveMatches: {…},
      + negativeMatches: {…},
      - unconditionalClasses: [
          "org.springframework.boot.actuate.autoconfigure.health.HealthEndpointAutoConfiguration",
          "org.springframework.boot.autoconfigure.context.ConfigurationPropertiesAutoConfiguration",
          "org.springframework.boot.actuate.autoconfigure.health.HealthIndicatorAutoConfiguration",
          "org.springframework.boot.actuate.autoconfigure.info.InfoContributorAutoConfiguration",
          "org.springframework.boot.autoconfigure.context.PropertyPlaceholderAutoConfiguration",
          "org.springframework.boot.actuate.autoconfigure.web.mappings.MappingsEndpointAutoConfiguration",
          "org.springframework.boot.actuate.autoconfigure.endpoint.EndpointAutoConfiguration",
          "org.springframework.boot.autoconfigure.info.ProjectInfoAutoConfiguration",
          "org.springframework.boot.actuate.autoconfigure.web.server.ManagementContextAutoConfiguration"
        ]
      }
    }
}

图 4-47

8. httptrace

/actuator/httptrace 可显示 HTTP 请求的追踪信息，当设置暴露大量的端点较为烦琐时，可以用“*”表示所有。

```
management.endpoints.web.exposure.include=*
```

从 Spring 2.2.x 开始，我们必须定制实现 HttpTraceRepository 的 Bean，让它来实现存储追踪和查询信息的功能，这样才能启用此端点。

```
@Component
public class MyHttpTraceRepository implements HttpTraceRepository {
   private static List< HttpTrace> traceList = new ArrayList<>();
   @Override
   public List<HttpTrace> findAll() {
      return traceList;
```

```
    }

    @Override
    public void add(HttpTrace trace) {
        traceList.add(trace);
    }
}
```

显然，把追踪信息存储在内存中是不合适的，读者在学习完数据访问相关技术后，可以将追踪信息存储在合适的数据库中。

启动应用，访问 http://localhost:8080/actuator/httptrace，如图 4-48 所示。

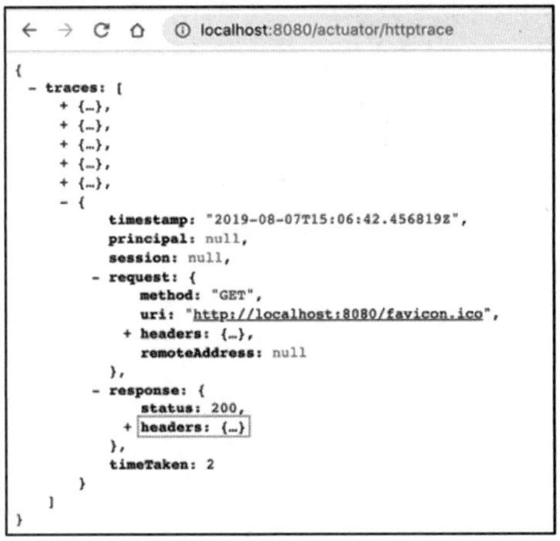

```
localhost:8080/actuator/httptrace

{
  - traces: [
      + {…},
      + {…},
      + {…},
      + {…},
      + {…},
      - {
            timestamp: "2019-08-07T15:06:42.456819Z",
            principal: null,
            session: null,
          - request: {
                method: "GET",
                uri: "http://localhost:8080/favicon.ico",
              + headers: {…},
                remoteAddress: null
            },
          - response: {
                status: 200,
              + headers: {…}
            },
            timeTaken: 2
        }
    ]
}
```

图 4-48

9．configprops

/actuator/configprops 列出了所有注解 @ConfigurationProperties 的 Bean，访问 http://localhost:8080/actuator/configprops，如图 4-49 所示。

```
localhost:8080/actuator/configprops

{
  - contexts: {
      - application: {
          - beans: {
              + management.trace.http-org.springframework.boot.actuate.autoconfigure.trace.http.HttpTraceProperties: {…},
              + management.endpoints.web-org.springframework.boot.actuate.autoconfigure.endpoint.web.WebEndpointProperties: {…},
              + management.endpoint.configprops-org.springframework.boot.actuate.autoconfigure.context.properties.ConfigurationPropertiesRepo
              + spring.security.oauth2.resourceserver-org.springframework.boot.autoconfigure.security.oauth2.resource.OAuth2ResourceServerPro
              + spring.jackson-org.springframework.boot.autoconfigure.jackson.JacksonProperties: {…},
              + spring.http-org.springframework.boot.autoconfigure.http.HttpProperties: {…},
              + spring.info-org.springframework.boot.autoconfigure.info.ProjectInfoProperties: {…},
              + management.endpoints.jmx-org.springframework.boot.actuate.autoconfigure.endpoint.jmx.JmxEndpointProperties: {…},
              + spring.task.scheduling-org.springframework.boot.autoconfigure.task.TaskSchedulingProperties: {…},
              + management.endpoint.health-org.springframework.boot.actuate.autoconfigure.health.HealthEndpointProperties: {…},
              + management.endpoints.web.cors-org.springframework.boot.actuate.autoconfigure.endpoint.web.CorsEndpointProperties: {…},
              + spring.resources-org.springframework.boot.autoconfigure.web.ResourceProperties: {…},
              + management.metrics.export.simple-org.springframework.boot.actuate.autoconfigure.metrics.export.simple.SimpleProperties: {…},
              + management.metrics-org.springframework.boot.actuate.autoconfigure.metrics.MetricsProperties: {…},
              + management.endpoint.env-org.springframework.boot.actuate.autoconfigure.env.EnvironmentEndpointProperties: {…},
              + management.endpoint.logfile-org.springframework.boot.actuate.autoconfigure.logging.LogFileWebEndpointProperties: {…},
              - spring.mvc-org.springframework.boot.autoconfigure.web.servlet.WebMvcProperties: {
```

图 4-49

10．threaddump

/actuator/threaddump 可显示运行应用的 Java 虚拟机线程信息，访问 http://localhost:8080/actuator/threaddump，如图 4-50 所示。

```
localhost:8080/actuator/threaddump
{
 - threads: [
    - {
          threadName: "Reference Handler",
          threadId: 2,
          blockedTime: -1,
          blockedCount: 6,
          waitedTime: -1,
          waitedCount: 0,
          lockName: null,
          lockOwnerId: -1,
          lockOwnerName: null,
          daemon: true,
          inNative: false,
          suspended: false,
          threadState: "RUNNABLE",
          priority: 10,
        - stackTrace: [
            - {
                  classLoaderName: null,
                  moduleName: "java.base",
                  moduleVersion: "11.0.2",
                  methodName: "waitForReferencePendingList",
                  fileName: "Reference.java",
                  lineNumber: -2,
                  className: "java.lang.ref.Reference",
                  nativeMethod: true
              },
```


图 4-50

11．loggers

/actuator/loggers 可显示应用中所有的 logger，访问 http://localhost:8080/actuator/loggers，如图 4-51 所示。

```
localhost:8080/actuator/loggers
{
 - levels: [
       "OFF",
       "ERROR",
       "WARN",
       "INFO",
       "DEBUG",
       "TRACE"
   ],
 - loggers: {
     - ROOT: {
           configuredLevel: "INFO",
           effectiveLevel: "INFO"
       },
     - io: {
           configuredLevel: null,
           effectiveLevel: "INFO"
       },
     - io.micrometer: {
           configuredLevel: null,
           effectiveLevel: "INFO"
       },
     - io.micrometer.core: {
           configuredLevel: null,
           effectiveLevel: "INFO"
       },
     - io.micrometer.core.instrument: {
           configuredLevel: null,
           effectiveLevel: "INFO"
       },
```


图 4-51

访问某个 logger，可通过包名称来访问。例如，设置一下 logger。

```
logging.level.top.wisely=debug
```

访问 http://localhost:8080/actuator/loggers/top.wisely，如图 4-52 所示。

```
{
    configuredLevel: "DEBUG",
    effectiveLevel: "DEBUG"
}
```

图 4-52

12．mappings

/actuator/mappings 可显示应用中所有的@RequestMapping，下面定义一个请求映射：

```
@SpringBootApplication
@RestController
public class LearningSpringBootActuatorApplication {

    @GetMapping("/")
    public String hello(){
        return "Hello Spring Boot Actuator";
    }

    public static void main(String[] args) {
        SpringApplication.run(LearningSpringBootActuatorApplication.class, args);
    }

}
```

启动应用，访问 http://localhost:8080/actuator/mappings，如图 4-53 所示。

```
- {
    handler: "top.wisely.learningspringbootactuator.LearningSpringBootActuatorApplication#hello()",
    predicate: "{GET /}",
  - details: {
      - handlerMethod: {
            className: "top.wisely.learningspringbootactuator.LearningSpringBootActuatorApplication",
            name: "hello",
            descriptor: "()Ljava/lang/String;"
        },
      - requestMappingConditions: {
            consumes: [ ],
            headers: [ ],
          - methods: [
                "GET"
            ],
            params: [ ],
          - patterns: [
                "/"
            ],
            produces: [ ]
        }
    }
},
```

图 4-53

13. metrics

/actuator/metrics 可显示当前应用的指标信息，包含内存使用情况等。访问 http://localhost:8080/actuator/metrics，如图 4-54 所示。

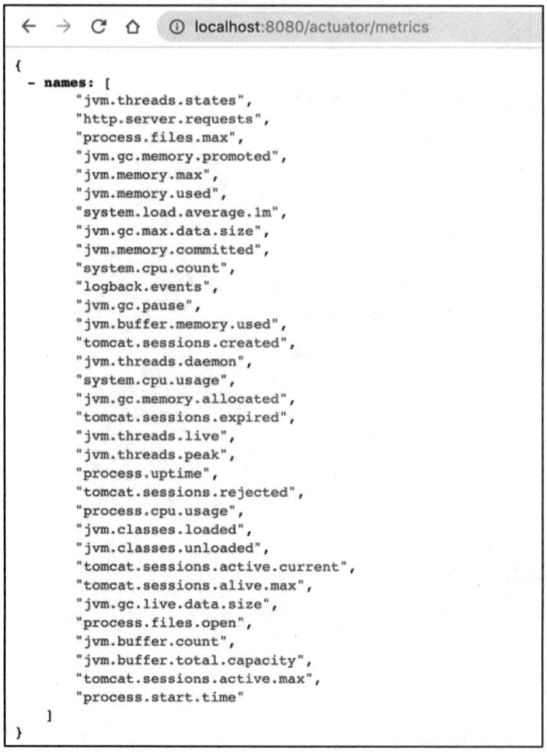

```
localhost:8080/actuator/metrics
{
  - names: [
        "jvm.threads.states",
        "http.server.requests",
        "process.files.max",
        "jvm.gc.memory.promoted",
        "jvm.memory.max",
        "jvm.memory.used",
        "system.load.average.1m",
        "jvm.gc.max.data.size",
        "jvm.memory.committed",
        "system.cpu.count",
        "logback.events",
        "jvm.gc.pause",
        "jvm.buffer.memory.used",
        "tomcat.sessions.created",
        "jvm.threads.daemon",
        "system.cpu.usage",
        "jvm.gc.memory.allocated",
        "tomcat.sessions.expired",
        "jvm.threads.live",
        "jvm.threads.peak",
        "process.uptime",
        "tomcat.sessions.rejected",
        "process.cpu.usage",
        "jvm.classes.loaded",
        "jvm.classes.unloaded",
        "tomcat.sessions.active.current",
        "tomcat.sessions.alive.max",
        "jvm.gc.live.data.size",
        "process.files.open",
        "jvm.buffer.count",
        "jvm.buffer.total.capacity",
        "tomcat.sessions.active.max",
        "process.start.time"
    ]
}
```

图 4-54

14. 修改端点地址

```
management.endpoints.web.base-path=/ #a
management.endpoints.web.path-mapping.health=check-health #b
```

a. 将 Actuator 前缀由/actuator 修改为/。

b. 将健康端点的路径由/health 修改为/check-health，若需要修改成其他端点的路径，则可使用：

```
management.endpoints.web.path-mapping.端点名=路径
```

在此例中，我们访问的健康端点地址由 http://localhost:8080/actuator/health 修改为了 http://localhost:8080/check-health。

4.8.2 自定义

1. 自定义端点

在 Bean 上注解@Endpoint、 @WebEndpoint 或 @EndpointWebExtension，可以将 Bean 通过 HTTP 暴露为端点。

自定义端点支持以下三种操作。

◎ @ReadOperation：GET（查询请求）。

◎ @WriteOperation：POST（保存请求）。

◎ @DeleteOperation：DELETE（删除请求）。

下面进行简单的演示，端点类似于一个常规的 RESTful 服务。

```
//……
import org.springframework.boot.actuate.endpoint.annotation.*;
//……
@Component
@Endpoint(id = "my") //a
public class MyEndpoint {
   public static Map<String, Boolean> status = new HashMap<>(); //b

   @ReadOperation //c
   public Map<String, Boolean> findAll() {
      return status;
   }
   @ReadOperation
   public Boolean findOne(@Selector String id) { //d
      return status.get(id);
   }
   @WriteOperation //e
   public String save(@Selector String id) {
      status.put(id, true);
      return "保存成功";
   }
   @DeleteOperation //f
   public String delete(@Selector String id) {
      status.remove(id);
      return "删除成功";
   }
}
```

a. 使用@Endpoint 注解 Bean，当前端点的名称（id）为 my。

b. 使用一个 Map 来存储状态信息。

c. 使用@ReadOperation 执行查询操作，这里查询整个状态信息。

d. 使用@Selector 接收参数。

e. 使用@WriteOperation 保存状态。

f. 使用@DeleteOperation 删除状态。

使用 Postman，以 POST 方式添加两个状态，分别为 http://localhost:8080/actuator/my/1 和 http://localhost:8080/actuator/my/2，这里的 1 和 2 为 ID 参数，如图 4-55 所示。

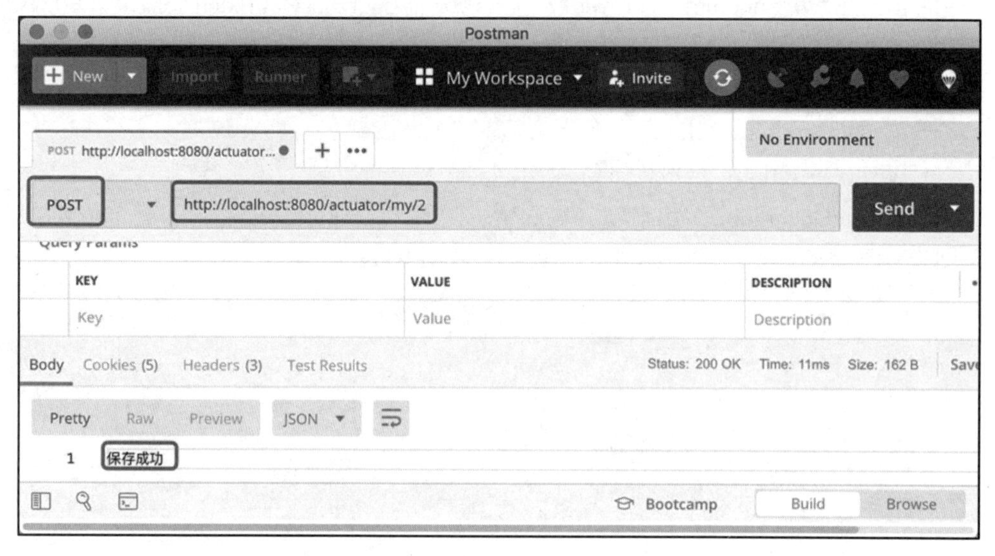

图 4-55

通过 Chrome 浏览器查看所有的状态（http://localhost:8080/actuator/my），如图 4-56 所示。

图 4-56

访问某个状态（http://localhost:8080/actuator/my/2），如图 4-57 所示。

图 4-57

在 Postman 中，用 DELETE 方法删除一个状态，如图 4-58 所示。

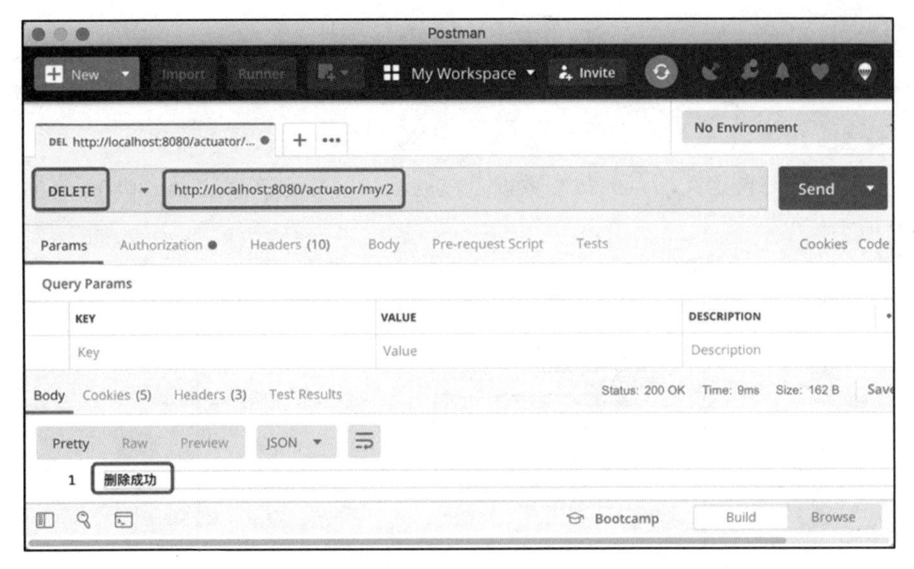

图 4-58

此时获取全部的状态（http://localhost:8080/actuator/my），如图 4-59 所示。

图 4-59

2. 自定义健康指示器

除 Spring Boot Actuator 提供的健康指示器外，还可以自定义健康指示器。只需使用 HealthIndicator 来定义健康指示器即可。

把前面端点的状态数据作为健康依据。如果状态不为空，则为健康（UP）；如果状态为空，则不健康（DOWN）。

```
//……
import org.springframework.boot.actuate.health.Health;
import org.springframework.boot.actuate.health.HealthIndicator;
//……
@Component
public class MyHealthIndicator implements HealthIndicator {
   @Override
   public Health health() {
      if(MyEndpoint.status.isEmpty()){
         return Health.down().build();
```

```
        }else {
            return Health.up().build();
        }
    }
}
```

启动应用，这时自定义端点的状态数据是空的，访问 http://localhost:8080/actuator/health，如图 4-60 所示。

```
localhost:8080/actuator/health
{
    status: "DOWN",
  - details: {
      - my: {
            status: "DOWN"
        },
      - diskSpace: {
            status: "UP",
          - details: {
                total: 250790436864,
                free: 160557846528,
                threshold: 10485760
            }
        }
    }
}
```

图 4-60

在 Postman 中添加状态信息，再次访问 http://localhost:8080/actuator/health，如图 4-61 所示。

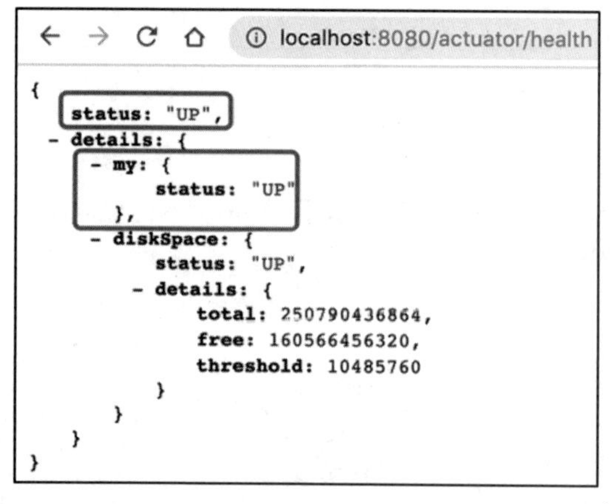

图 4-61

3. 自定义指标

我们可以自定义指标来扩展对应用的监控。Spring Boot Actuator 提供了基于“Micrometer”的基本指标和自动配置。

通过实现 MeterBinder 接口和自动配置的 MeterRegistry 来注册指标数据，示例如下。

```
//……
import io.micrometer.core.instrument.Gauge;
import io.micrometer.core.instrument.MeterRegistry;
import io.micrometer.core.instrument.binder.MeterBinder;
//……
@Component
public class MyMetrics implements MeterBinder{

   @Override
   public void bindTo(MeterRegistry registry) {
      Gauge.builder("top.wisely.size", () -> MyEndpoint.status.size())
            .baseUnit("个")
         .description("获取自定义端点中的状态数量")
         .register(registry);
   }
}
```

更多关于“Micrometer”的用法请参考官网。

现在启动应用，访问 http://localhost:8080/actuator/metrics，如图 4-62 所示。

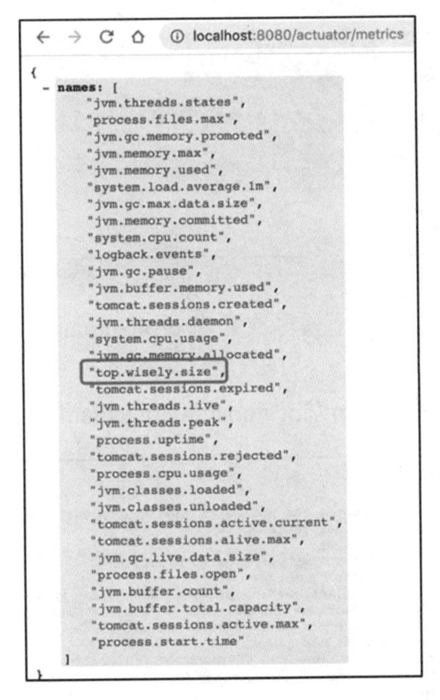

图 4-62

在 Postman 中，为自定义端点添加两个状态。通过 http://localhost:8080/actuator/metrics/

top.wisely.counter 访问某个指标，如图 4-63 所示。

```
localhost:8080/actuator/metrics/top.wisely.size
{
    name: "top.wisely.size",
    description: "获取自定义端点中状态数量",
    baseUnit: "个",
  - measurements: [
      - {
            statistic: "VALUE",
            value: 2
        }
    ],
    availableTags: [ ]
}
```

图 4-63

4.8.3 使用 Prometheus 和 Grafana 监控指标

监控指标是应用的重要功能，我们可以使用 Prometheus 存储指标数据，使用 Grafana 对数据进行可视化展示。Micrometer 可以将指标数据输出到不同的监控平台，包括 Prometheus。下面演示使用 Prometheus 和 Grafana 来监控指标数据。

（1）添加依赖：

```
dependencies {
    //……
  implementation 'io.micrometer:micrometer-registry-prometheus'
    //……
}
```

（2）开启端点：

```
management.endpoints.web.exposure.include=prometheus
```

因为前面已配置了“*”，所以无须再次开启 prometheus 端点。

（3）访问端点。

（4）启动应用，访问端点地址 http://localhost:8080/actuator/promctheus，显示如图 4-64 所示。

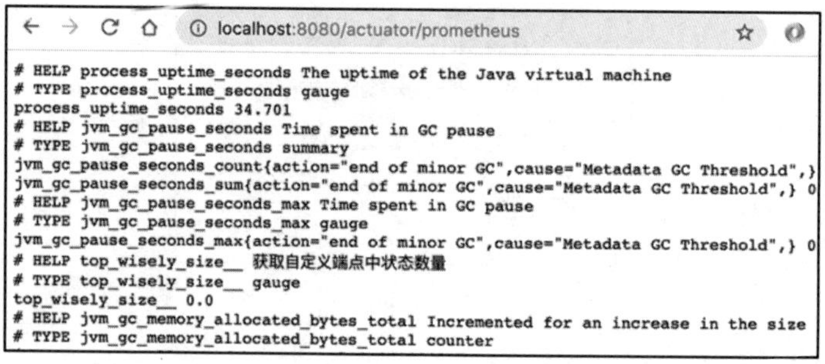

```
localhost:8080/actuator/prometheus
# HELP process_uptime_seconds The uptime of the Java virtual machine
# TYPE process_uptime_seconds gauge
process_uptime_seconds 34.701
# HELP jvm_gc_pause_seconds Time spent in GC pause
# TYPE jvm_gc_pause_seconds summary
jvm_gc_pause_seconds_count{action="end of minor GC",cause="Metadata GC Threshold",}
jvm_gc_pause_seconds_sum{action="end of minor GC",cause="Metadata GC Threshold",} 0
# HELP jvm_gc_pause_seconds_max Time spent in GC pause
# TYPE jvm_gc_pause_seconds_max gauge
jvm_gc_pause_seconds_max{action="end of minor GC",cause="Metadata GC Threshold",} 0
# HELP top_wisely_size__ 获取自定义端点中状态数量
# TYPE top_wisely_size__ gauge
top_wisely_size__ 0.0
# HELP jvm_gc_memory_allocated_bytes_total Incremented for an increase in the size
# TYPE jvm_gc_memory_allocated_bytes_total counter
```

图 4-64

（5）编写 docker compose 文件，把它放置到应用根目录下的 stack.yml 中。

```
version: "3"
services:
 prom:
  image: prom/prometheus:v2.11.1
  volumes:
   - ./config/prometheus.yml:/etc/prometheus/prometheus.yml # a
  ports:
   - 9090:9090 # b
  networks:
   metrics:
    aliases:
     - grafana # c
 grafana:
  image: grafana/grafana:6.3.2
  ports:
   - 3000:3000 # b
  environment:
   - GF_SECURITY_ADMIN_USER=wisely # d
   - GF_SECURITY_ADMIN_PASSWORD=zzzzzz
  networks:
   metrics:
    aliases:
     - prometheus # c

networks:
 metrics: # c
```

a. 把应用根目录下的 config/prometheus.yml 作为 prometheus 的配置文件。

b. 将容器端口号暴露给当前开发主机。

c. 定义网络 metrics，让 prometheus 和 grafana 在同一个网络下。

d. 设置 grafana 的账号和密码。

（6）配置 prometheus，位置在源码根目录的 config 目录下。

```
scrape_configs:
 - job_name: 'prometheus'
   scrape_interval: 1m
   static_configs:
    - targets: ['localhost:9090']
 - job_name: 'learning-spring-boot-actuator' #a
   scrape_interval: 1m #b
   metrics_path: '/actuator/prometheus' #c
   static_configs:
    - targets: ['192.168.31.199:8080'] #d
 - job_name: 'grafana'
   scrape_interval: 1m
   metrics_path: '/metrics'
```

```
    static_configs:
      - targets: ['grafana:3000']
```

a. 此处配置的是监控 learning-spring-boot-actuator 应用的任务。

b. 每隔一分钟获取一次/actuator/prometheus 数据。

c. 指定获取数据的端点地址/actuator/prometheus。

d. 此处设置的是应用的地址，这里的 IP 地址是当前开发主机的地址。

（7）启动容器。

执行命令：

```
$ docker-compose -f stack.yml up -d
```

（8）检查 prometheus 任务状态。

访问 http://localhost:9090/targets，查看 prometheus 的任务状态。正常情况下，所有的状态都显示“UP”，如图 4-65 所示。

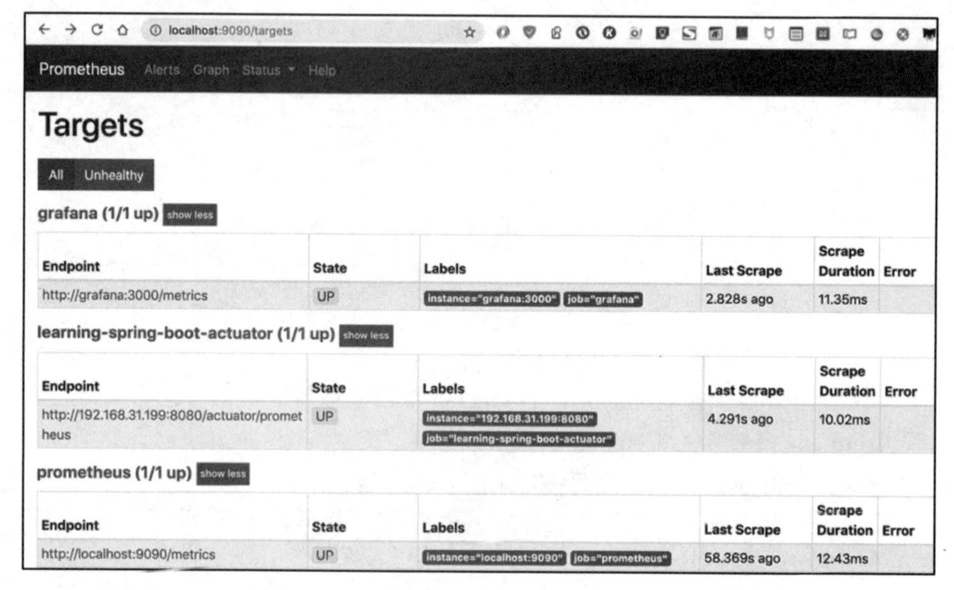

图 4-65

（9）访问 grafana。

访问 http://localhost:3000，使用账号和密码登录，如图 4-66 所示。

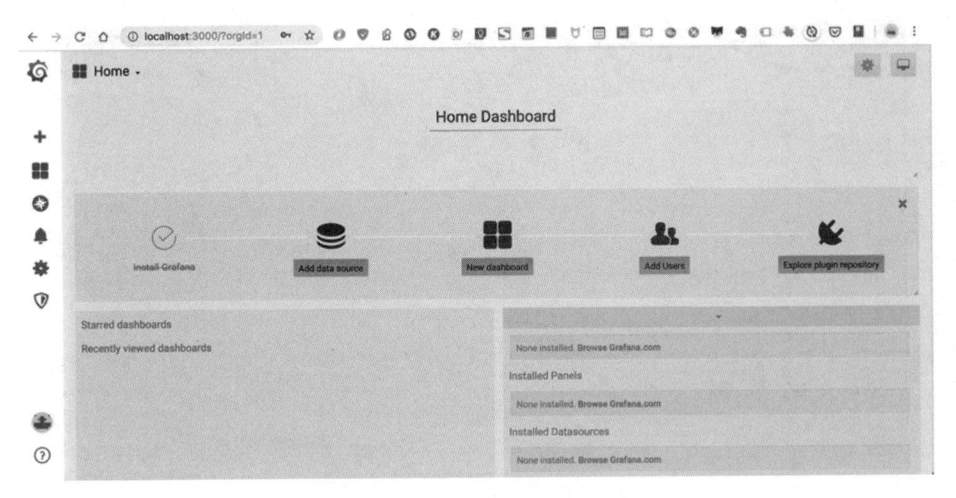

图 4-66

（10）添加 prometheus 数据源。

单击首页上的“Add data source”按钮，在筛选过滤界面选择“Prometheus”选项，设置 Prometheus 的地址为 http://prometheus:9090，如图 4-67 所示。填写完成后单击“Save & Test”按钮，显示“Data source is working”，代表可正常工作

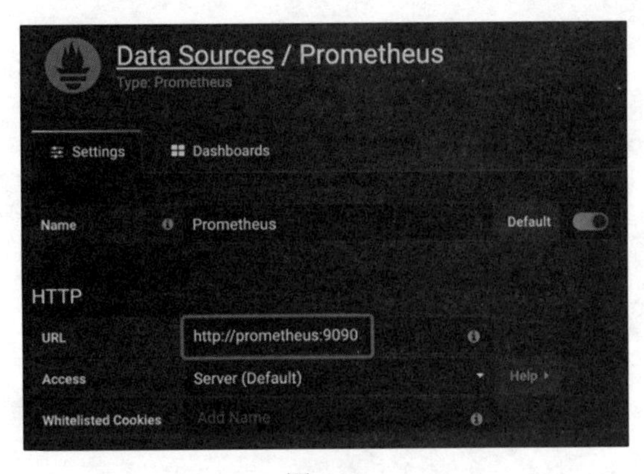

图 4-67

（11）创建监控 Dashboard。

单击首页上的“New dashboard”按钮，在弹出的 New dashboard 界面单击“Add Query”选项：

在“Query”选择前面配置的数据源 Prometheus。

在“Metrics”搜索所需要监控的指标，如输入“top_wisely_size”。

单击上方保存图标，保存 Dashboard。如图 4-68 所示。

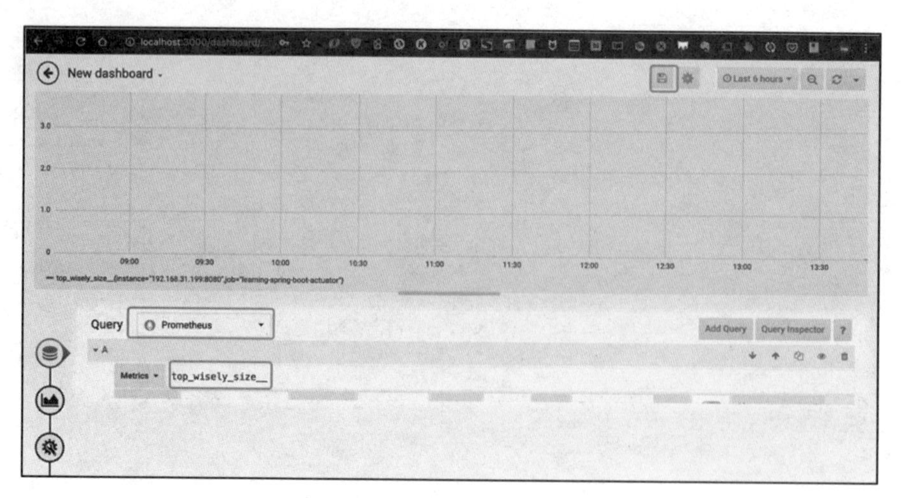

图 4-68

（12）监控指标。

向自定义端点 http://localhost:8080/actuator/my 添加数据，观察 Dashboard 的变化，可得到如图 4-69 所示监控。

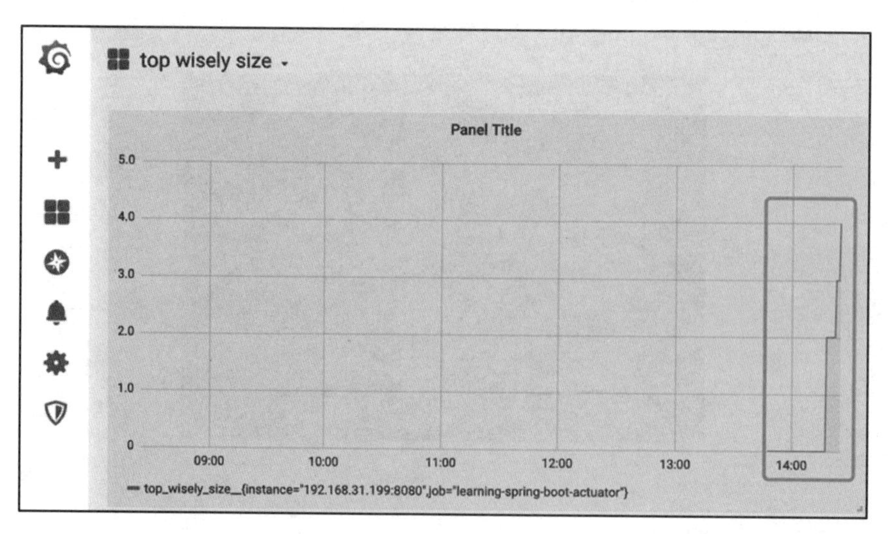

图 4-69

4.9 小结

本章深入介绍了 Spring Boot 的相关知识，包括 Spring Boot 的原理和 Spring Boot 的配置。如果在阅读本章时，觉得某些地方难以理解，千万不要害怕，请继续学习本书后面的知识。

第 5 章 Spring Web MVC

5.1 Spring Web MVC 简介

Spring Web MVC 通常简称为 Spring MVC，它是一个基于 MVC 的 Web 开发框架。从 Spring 5.0 开始，Spring 引入了响应式 Web 框架 Spring WebFlux。Spring MVC 和 Spring WebFlux 使用相同的编程模型。

Spring MVC 是基于 Servlet API 构建的，Spring MVC 的 Servlet 实现为 DispatcherServlet，它使用了一个 Spring 容器（WebApplicationContext），从而让 Servlet 和 Spring 容器结合在一起。

基于 Spring Boot 的 Spring MVC 应用内嵌了 Servlet 容器，我们可以直接运行 Spring Boot 来开发 Web 应用。在 Web 应用下，Spring Boot 自动新建了一个类型为 AnnotationConfigServletWebServerApplicationContext 的 Spring IoC 容器。

5.2 用 Spring Boot 学习 Web MVC

下面新建一个应用，作为本章的演示示例。

Group：top.wisely。

Arifact：learning-spring-mvc。

Dependencies：Spring Web Starter 和 Lombok。

5.2.1 核心注解

Spring MVC 的核心注解如下。

（1）@Controller：

◎ 它组合了@Component 注解，被注解的类是一个 Bean。

◎ 它是一个控制器。

（2）@RequestMapping：将 Web 请求路径映射到被注解的类或者方法上。

```
@Controller
@RequestMapping("/demo")
public class DemoController {

   @RequestMapping("/hello")
   @ResponseBody
   public String hello(){
     return "Hello World !!!";
   }

}
```

启动应用，在 Postman 中访问 http://localhost:8080/demo/hello，如图 5-1 所示。

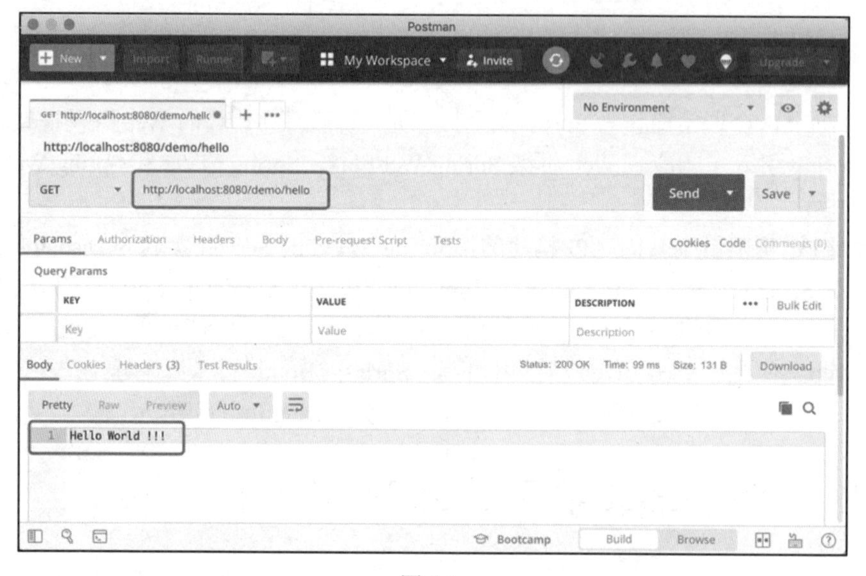

图 5-1

5.2.2 RESTful 服务

REST（Representational State Transfer，表述性状态传递）是目前基于 HTTP 的 Web 服务中最常用的一种架构风格。遵循 REST 架构规范的 Web 服务称之为 RESTful Web 服务，简称 RESTful 服务。它区别于以方法调用为导向的 Web 服务（SOAP），强调基于网络资源的 Web 服务。本书只关注 RESTful 服务，因为目前应用的客户端大多基于 Android、iOS 或者浏览器，而基于浏览器的应用会和服务端剥离，所以本书不讨论 Spring MVC 的模板引擎（如 Thymeleaf、FreeMarker、Jsp 等）部分。

当访问 RESTful 的网络资源时，将得到如 JSON、XML 或 HTML 格式的数据。一次完整的 Web 交互由两部分组成。

（1）请求：代表客户端向服务端发送一次 HTTP Web 请求。由 RequestEntity<T>（HttpEntity 的子类）来代表访问请求的所有内容。

◎ header 为元数据信息，可由@RequestHeader 注解来获取。
  ➢ 内容类型：在请求头信息里有一项重要的属性叫作内容类型（Content-Type），它用来指定 body 请求体的媒体类型（如 application/json）。我们可以使用@RestController 的 consumes 属性来限制控制器方法支持的内容类型。
  ➢ Accept：客户端可接收的返回媒体类型，如 Accept:application/json 或 text/html。若不设置，则默认为 Accept:*/*。我们可以使用@RestController 的 produces 属性来限制控制器方法支持的返回媒体类型。默认情况下，客户端可接收所有的返回媒体类型。

◎ body 为请求体数据，可由@RequestBody 注解来获取。
◎ 参数数据：包括 URL 路径上的数据、请求参数@RequestParam 和路径变量@PathVariable 等。
◎ 请求方法：包括 GET、POST、HEAD、OPTIONS、PUT、PATCH、DELETE 和 TRACE。

（2）返回：response，由 status、header 和 body 组成，可由 ResponseEntity<T>（HttpEntity 的子类，封装了公共的 header 和 body 部分）来设置。

◎ header 为元数据信息。
  ➢ 内容类型：返回头中也有内容类型，它可指定 body 返回体的媒体类型。
◎ body 为返回数据，可由@ResponseBody 来设置。
◎ status 状态信息，可由@ResponseStatus 来设置。状态主要分为以下几类。
  ➢ 1xx：信息反馈。
  ➢ 2xx：成功反馈。
  ➢ 3xx：重定向。
  ➢ 4xx：客户端导致的错误。
  ➢ 5xx：服务端导致的错误。

Spring MVC 提供了@RestController 来开发 RESTful 服务，它组合了@Controller 和@ResponseBody，意味着需要通过返回体来返回数据。

Spring MVC 还针对不同的 HTTP 动作制定了语义化的@RequestMapping，下面先来了解 RESTful 服务支持的 HTTP 请求方法。

◎ GET：从返回体重新获取资源。使用@GetMapping 等同于使用@RequestMapping(method = RequestMethod.GET)。

◎ POST：通过请求体创建新的资源。使用@PostMapping 等同于使用@RequestMapping(method = RequestMethod.POST)。

◎ PUT：通过请求体替换资源，如果资源不存在，则创建资源。使用@PutMapping 等同于使用@RequestMapping(method = RequestMethod.PUT)。

◎ PATCH：通过请求体更新资源，若资源不存在，则创建资源。使用@PatchMapping 等同于使用@RequestMapping(method = RequestMethod.PATCH)。

◎ DELETE：删除资源。使用@DeleteMapping 等同于使用@RequestMapping(method = RequestMethod.DELETE)。

下面使用 Spring MVC 开发一个标准的 RESTful 服务作为示例。我们的资源是一个叫作 Person 的领域模型。

```
@Getter
@Setter
@AllArgsConstructor
@NoArgsConstructor
@EqualsAndHashCode(exclude = {"name", "age"}) //用 equals 比较两个对象，若 id 相同即 equal
@ToString
public class Person {
   private Long id;
   private String name;
   private Integer age;
}
```

控制器声明如下：

```
@RestController //组合了@Controller 和@ResponseBody，类中所有方法的返回值都是通过返回体返回的
@RequestMapping("/people") //资源访问的路径为"/people"
public class PeopleController {
   private PersonRepository personRepository; // 参照本节最后“数据操作代码”部分

   public PeopleController(PersonRepository personRepository) { //注入数据操作的 Bean
      this.personRepository = personRepository;
   }
}
```

1. 新增资源

```
@PostMapping
@ResponseStatus(HttpStatus.CREATED) // 指定状态为资源已创建 201，若不指定，则皆为
//HttpStatus.OK:200
public Person save(@RequestBody Person person){ //从请求体获取数据
   return personRepository.save(person); //把返回的数据写入返回体
}
```

这里使用了@RequestBody 和@ResponseBody 组合，下面使用 RequestEntity 和 ResponseEntity 组合。

```
@PostMapping
public ResponseEntity<Person> save(RequestEntity<Person> personRequestEntity){
   Person p = personRepository.save(personRequestEntity.getBody());//使用
//RequestEntity 获得请求体信息
   return new ResponseEntity<Person>(p, HttpStatus.CREATED); //使用 ResponseEntity
//设置返回信息
}
```

执行结果如图 5-2 所示。

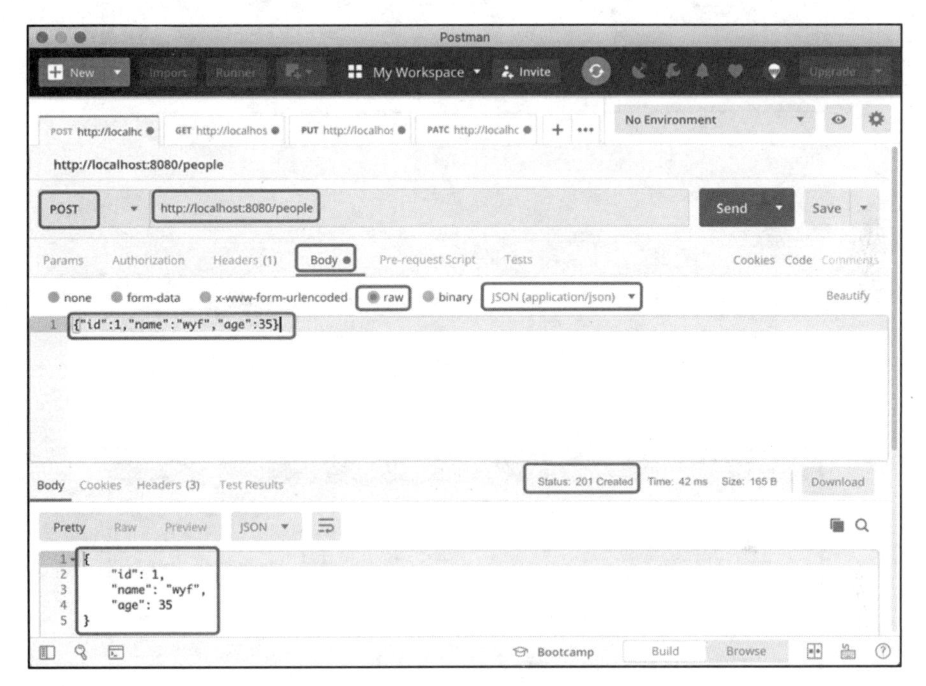

图 5-2

2. 获取指定的资源

```
@GetMapping("/{id}")
public Person getOne(@PathVariable Long id){ //使用@PathVariable 获取路径中的{id}的值
   return personRepository.findOne(id);
}
```

执行结果如图 5-3 所示。

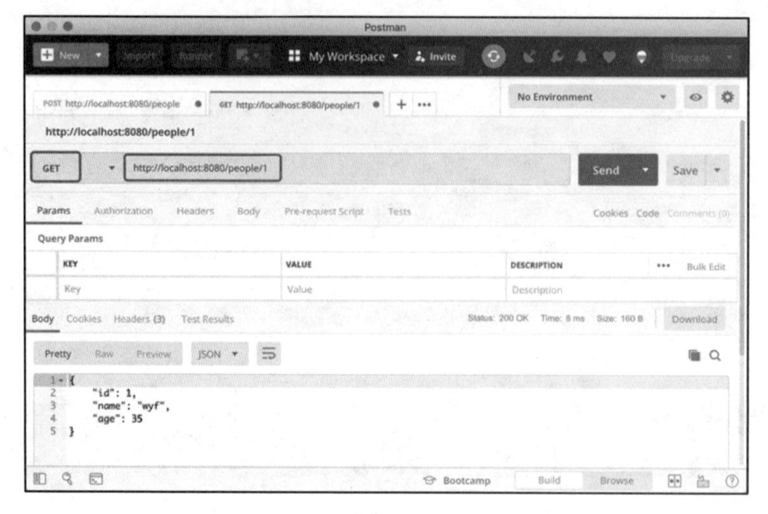

图 5-3

3. 替换更新资源

```
@PutMapping("/{id}")
public Person replace(@PathVariable Long id, @RequestBody Person person){
  return personRepository.replace(id, person); //整个对象替换更新
}
```

执行结果如图 5-4 所示。

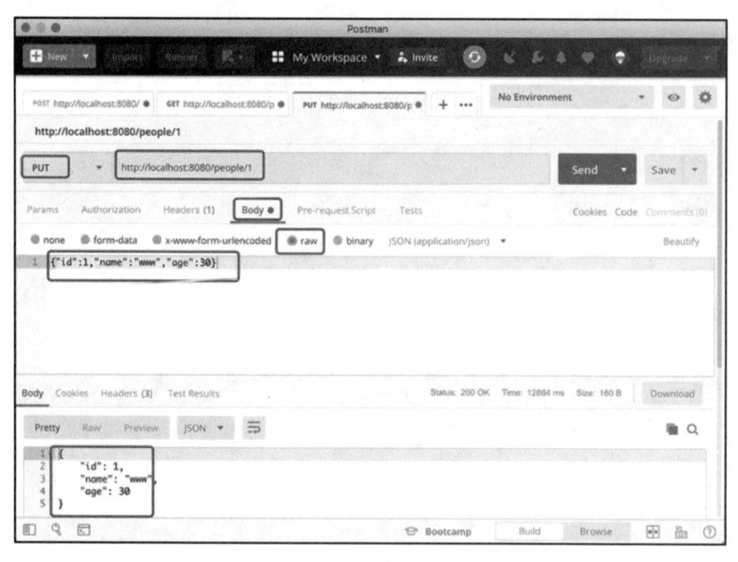

图 5-4

4. 更新资源

```
@PatchMapping("/{id}")
public Person patch(@PathVariable Long id, @RequestBody Person person) {
   return personRepository.patched(id, person); //只更新修改的部分
}
```

执行结果如图 5-5 所示。

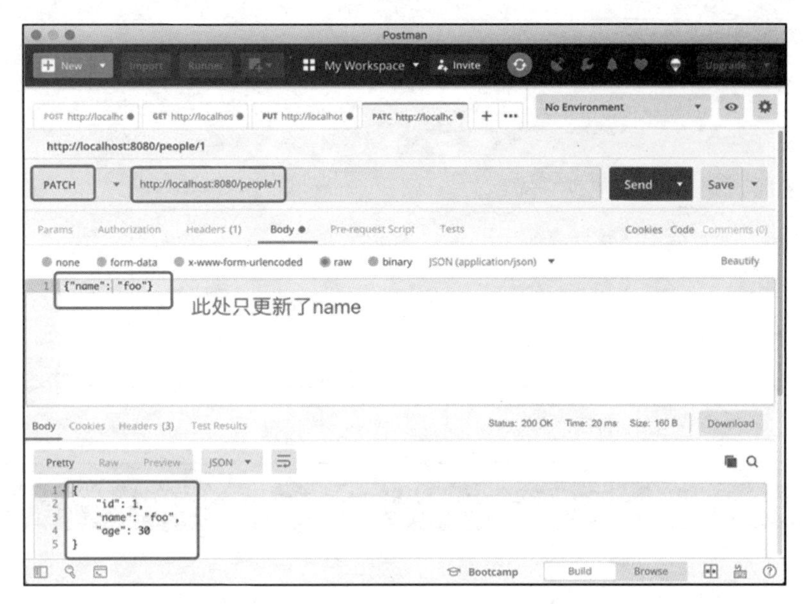

图 5-5

5. 查找资源

```
@GetMapping("/findByName")
public Person findByName(@RequestParam String name){ //通过@RequestParam 获取?后的
//参数内容
   return personRepository.findByName(name);
}
```

执行结果如图 5-6 所示。

6. 删除资源

```
@DeleteMapping("/{id}")
@ResponseStatus(HttpStatus.NO_CONTENT) //状态修改成无内容：204
public void delete(@PathVariable Long id){
   personRepository.delete(id);
}
```

执行结果如图 5-7 所示。

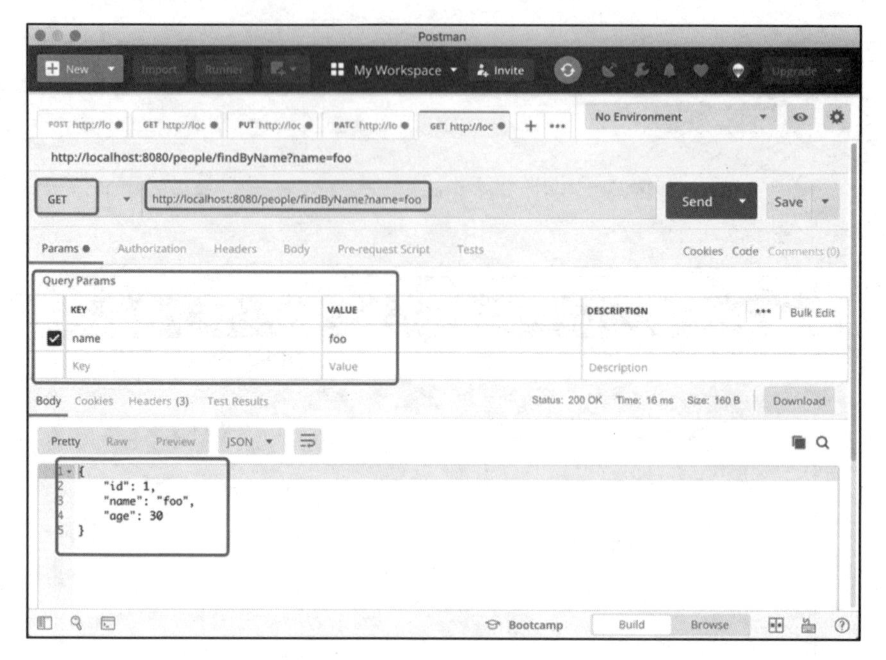
Postman
New
My Workspace
Invite
No Environment
http://localhost:8080/people/findByName?name=foo
GET
http://localhost:8080/people/findByName?name=foo
Send
Save
Params
Authorization
Headers
Body
Pre-request Script
Tests
Cookies
Code
Query Params
KEY
VALUE
DESCRIPTION
Bulk Edit
name
foo
Key
Value
Description
Body
Cookies
Headers (3)
Test Results
Status: 200 OK
Time: 16 ms
Size: 160 B
Download
Pretty
Raw
Preview
JSON
{
"id": 1,
"name": "foo",
"age": 30
}
Bootcamp
Build
Browse

图 5-6

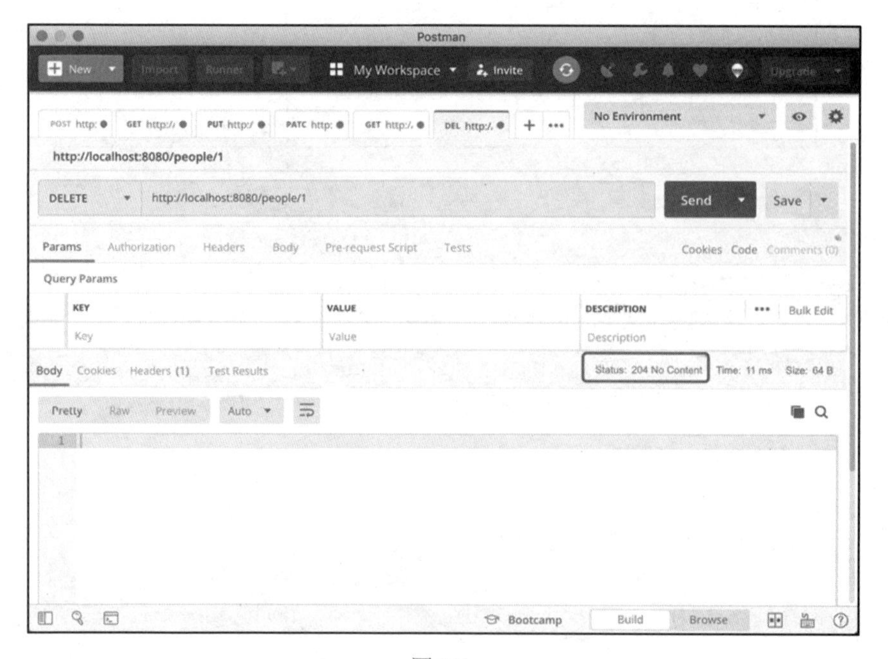
Postman
New
My Workspace
Invite
No Environment
http://localhost:8080/people/1
DELETE
http://localhost:8080/people/1
Send
Save
Params
Authorization
Headers
Body
Pre-request Script
Tests
Cookies
Code
Query Params
KEY
VALUE
DESCRIPTION
Bulk Edit
Key
Value
Description
Body
Cookies
Headers (1)
Test Results
Status: 204 No Content
Time: 11 ms
Size: 64 B
Pretty
Raw
Preview
Auto
Bootcamp
Build
Browse

图 5-7

7. 获取请求头

```
@GetMapping("/getRequestHeaders")
public String[] getHeaders(@RequestHeader HttpHeaders httpHeaders, //a
                           @RequestHeader Map<String, String> headerMap, //a
                           @RequestHeader MultiValueMap<String, String> multiValueMap, //a
                           @RequestHeader("accept") String accept, //b
                           RequestEntity requestEntity) { //c
    String userAgent = httpHeaders.getFirst("user-agent");
    String host = headerMap.get("host");
    String cacheControl = multiValueMap.getFirst("cache-control");
    HttpHeaders sameHttpHeaders = requestEntity.getHeaders();
    return new String[]{userAgent, host, cacheControl, accept};
}
```

a. 使用@RequestHeader 注解在 HttpHeaders、Map<String, String>和 MultiValueMap<String, String>上来获取请求头信息。

b. 通过@RequestHeader 设置请求头的 key，获得指定的请求头的值。

c. 在方法参数里注入 RequestEntity 对象，从而获得请求头的信息。

执行结果如图 5-8 所示。

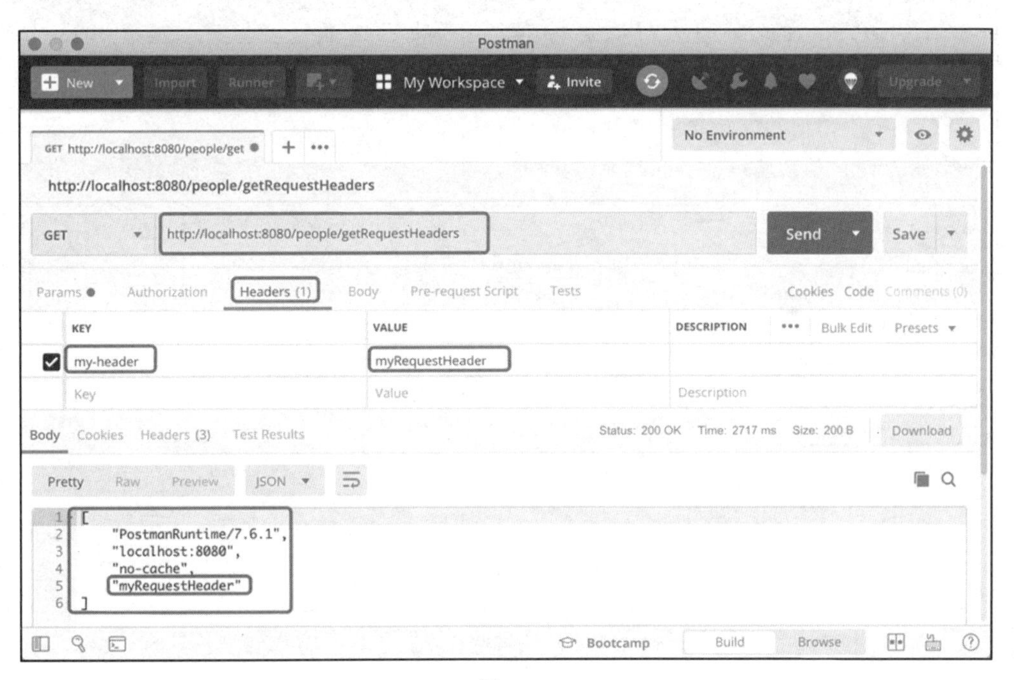

图 5-8

8. ResponseEntity

ResponseEntity 可以定制整个返回，包含 status、header 和 body。

```
@GetMapping("/responseEntityShowcase")
public ResponseEntity<Person> responseEntityShowcase(){
    HttpHeaders responseHeaders = new HttpHeaders();
    responseHeaders.set("my-header", "hello header");
    return new ResponseEntity<Person>(new Person(12341,"bar", 22),
            responseHeaders,
            HttpStatus.OK);
}
```

ResponseEntity 还提供了建造者模式的 Fluent API，下面的代码与上面的代码等同。

```
return ResponseEntity.ok() //中间操作，设置 status
        .header("my-header", "hello header") //中间操作，设置 header
        .body(new Person(12341,"bar", 22)); //终结操作，设置 body
```

执行结果如图 5-9 所示。

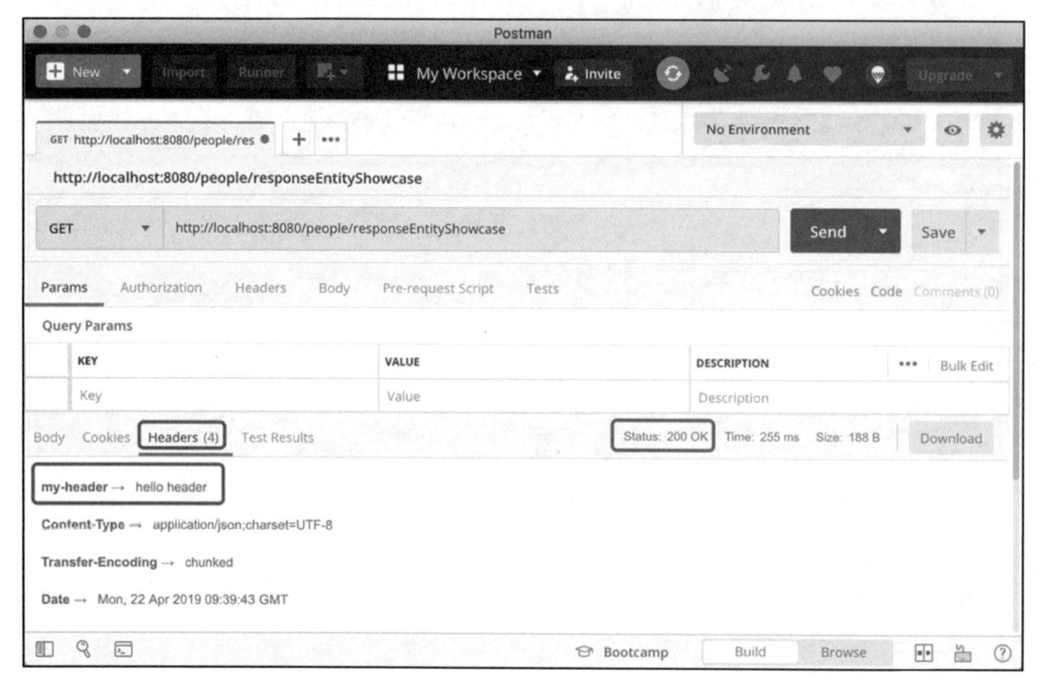

图 5-9

更多关于 ResponseEntity 提供的方法，可参考 ResponseEntity 的 API。

9.获取 cookie 信息

可以使用@CookieValue 来获得客户端 cookie 的值。

```
@GetMapping("/getCookie")
public String[] getCookie(@CookieValue("myCookie") String myCookie,
                 @CookieValue("anotherCookie") String anotherCookie) {
   return new String[]{myCookie, anotherCookie};
}
```

执行结果如图 5-10 所示。

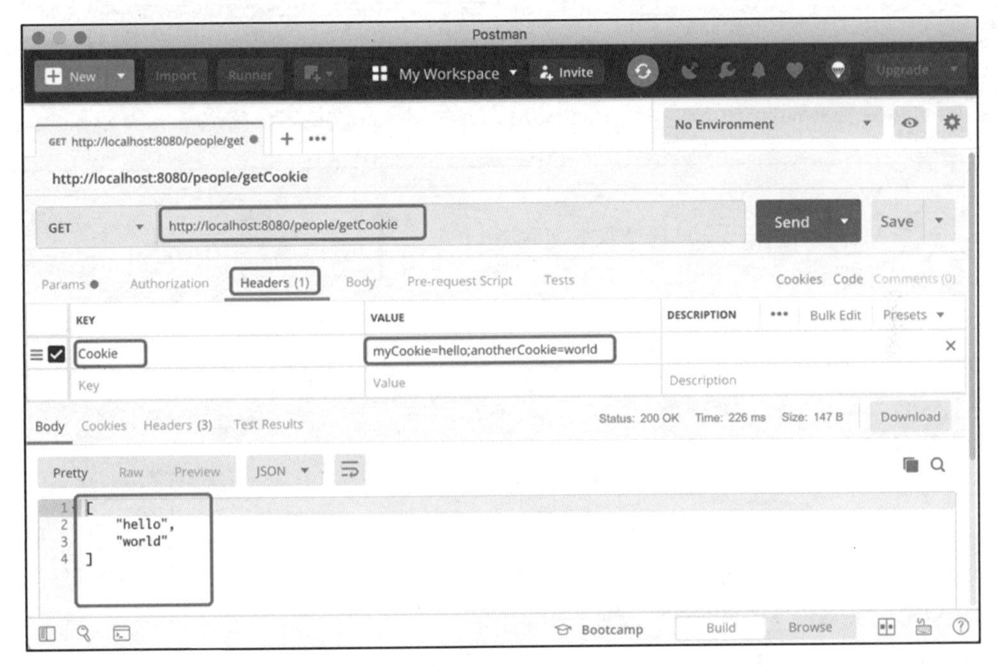

图 5-10

10. 方法参数

Spring MVC 支持注入一些参数到方法中：

```
@GetMapping("/methodArguments")
public void methodArguments(WebRequest request,//a
                    NativeWebRequest nativeWebRequest,//b
                    ServletRequest servletRequest, //c
                    ServletResponse servletResponse,//d
                    HttpSession httpSession, //e
                    Locale locale, //f
                    TimeZone timeZone,//g
                    ZoneId zoneId, //h
                    DemoService demoService //i) throws Exception{
```

```
    System.out.println(request);
    System.out.println(nativeWebRequest);
    System.out.println(servletRequest);
    System.out.println(servletResponse);
    System.out.println(httpSession);
    System.out.println(locale);
    System.out.println(timeZone);
    System.out.println(zoneId);
    System.out.println(demoService.sayHello());
}
```

```
ServletWebRequest: uri=/people/methodArguments;client=0:0:0:0:0:0:0:1;session=90AEE4BD229225C7BC950D7511AFD41C
ServletWebRequest: uri=/people/methodArguments;client=0:0:0:0:0:0:0:1;session=90AEE4BD229225C7BC950D7511AFD41C
org.apache.catalina.connector.RequestFacade@140bc84e
org.apache.catalina.connector.ResponseFacade@2400da03
org.apache.catalina.session.StandardSessionFacade@45fd6ddb
zh_CN_#Hans
sun.util.calendar.ZoneInfo[id="Asia/Shanghai",offset=28800000,dstSavings=0,useDaylight=false,transitions=29,lastRule=null]
Asia/Shanghai
Hello World
```

a. 通用 Web 请求接口。

b. 继承 WebRequest，通用的请求和返回接口。

c. Servlet 请求。

d. Servlet 返回。

e. HTTP 会话。

f. 本地信息。

g. 时区信息。

h. 时区 ID。

i. 新建一个对象（非注入 Bean）。

这些值都不为空，都成功注入了。还有其他一些参数如 PushBuilder、Principle 等也是可以直接注入的，将在本书对应的章节进行讲解。

DemoService 的代码如下：

```
@Service
public class DemoService {
    public String sayHello(){
        return "Hello World";
    }
}
```

11. 文件上传和返回文件

在 Spring MVC 中，可以使用 MultipartFile 来接收上传的文件，同样，也可以使用 Servlet 3.0 的 javax.servlet.http.Part 来接收上传的文件。

```
@PostMapping("/upload")
public ResponseEntity<Resource> upload(MultipartFile file, //a
                                       @RequestPart("file") MultipartFile file2, //a
```

```
                              @RequestParam("file") MultipartFile file3, //a
                              @RequestParam("file")Part part //a
                              ) throws Exception{

   System.out.println(file.equals(file2)); //b
   System.out.println(file.equals(file3)); //b
   System.out.println(file.getSize() == part.getSize());//b

   Resource imageResource = new InputStreamResource(file.getInputStream()); //c

   return ResponseEntity.ok()
         .contentType(MediaType.IMAGE_PNG) //d
         .body(imageResource); //e

}
```

a. 可以用四种形式来接收 multipart/form-data 上传的文件。

b. 通过这几句可以看出，得到的是相同的文件。

c. Resource 是 Spring 对资源的抽象，它的实现可以是文件资源、网址资源、类路径资源或输入流资源等。

d. 设置返回值类型为 MediaType.IMAGE_PNG，即指定返回格式是 PNG 图片。

e. 将图片资源作为返回体。

执行结果如图 5-11 所示。

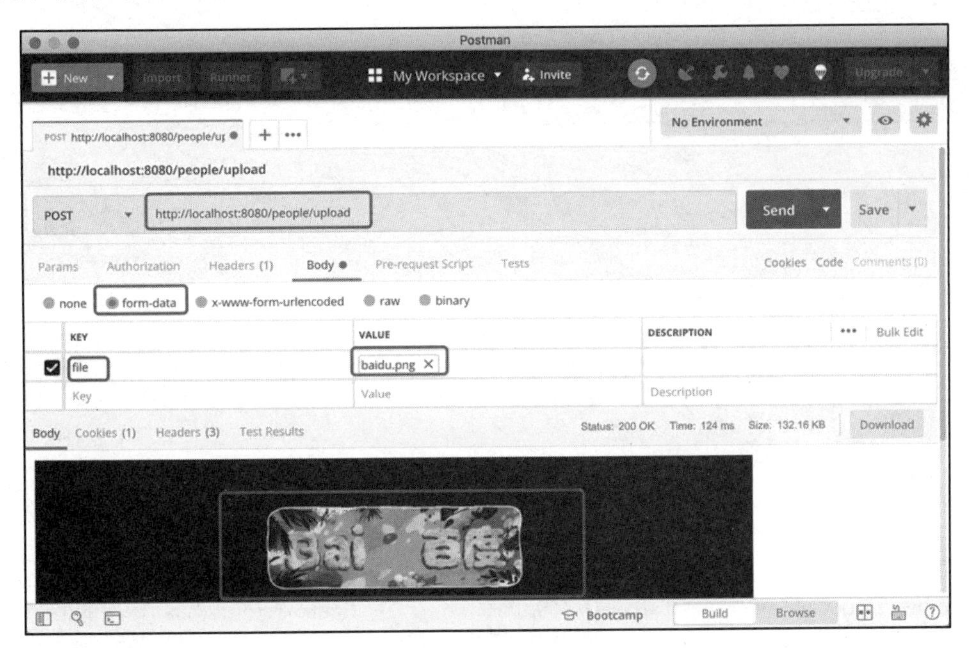

图 5-11

若需要上传多个文件，则只需使用@RequestParam 注解 Map<String, MultipartFile> 或 MultiValueMap<String, MultipartFile>即可。

```
@PostMapping("/multipleUpload")
public String uploads(@RequestParam Map<String, MultipartFile> fileMap){
   StringBuilder info = new StringBuilder();
   fileMap.forEach((key, file) -> {
    info.append(file.getName()).append("'s length is
").append(file.getSize()).append("\n");
   });
   return info.toString();
}
```

执行结果如图 5-12 所示。

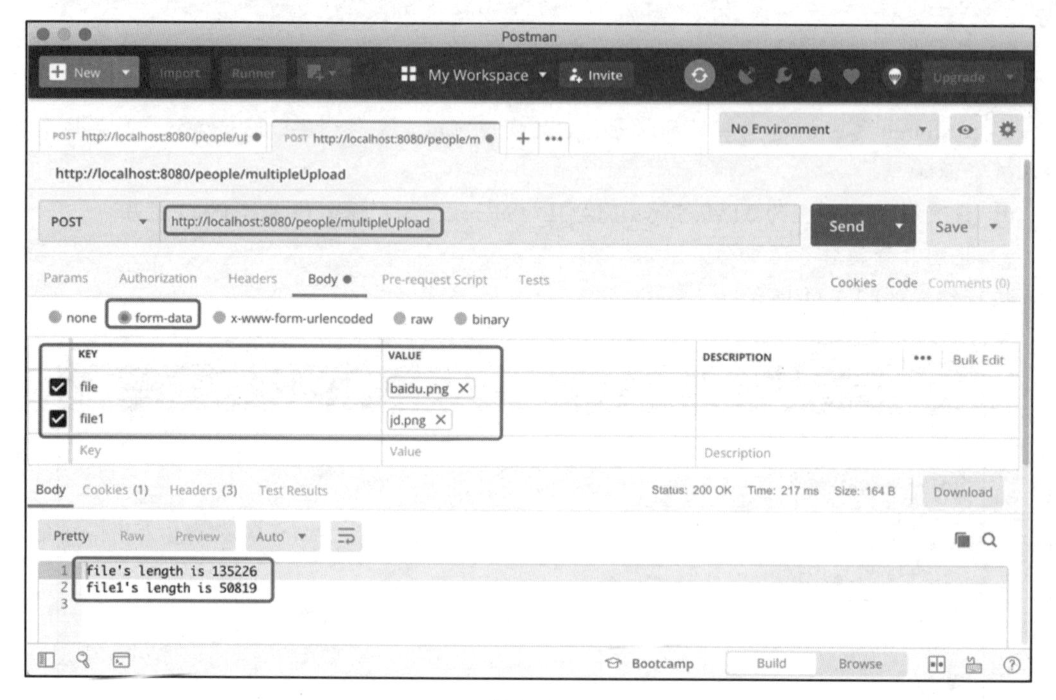

图 5-12

在外部配置中可以使用 spring.servlet.multipart.*来配置文件上传。

```
spring.servlet.multipart.max-file-size: 10MB #文件最大为10MB
```

12. 参数校验

Spring Boot 支持基于 JSR-303/349/380 规范的 Bean 校验 API，功能实现者为 hibernate-validator。

JSR（Java Specification Requests，Java 规范请求）是对 Java 新功能的请求，是 JCP 组织的一部分。Java 社区的参与者们通过 JCP 组织，利用自己的创意来影响 Java 语言的发展。在 JSR 中，jar 的包名一般以 javax 开头，如 javax.validation。

JSR 只提供功能规范定义，不提供实现。Spring Boot 的 Web 依赖添加了 spring-boot-starter-validation，它添加了规范包 jakarta.validation-api-2.0.1.jar 和实现包 hibernate-validator-6.0.16.Final.jar，

JSR-303：Bean Validation API 1.0。

JSR-349：Bean Validation API 1.1。

JSR-380：Bean Validation API 2.0。

下面给 Person 类加上校验注解。

```
@NotNull(message = "id 不能为空")
private Long id;
@Size(min = 3, max = 5, message = "name 在 3 到 5 个字符之间")
private String name;
@Min(value = 18, message = "age 不能低于 18 岁")
private Integer age;
```

JSR 校验注解在 javax.validation.constraints 包中，Hibernate 提供的更多校验注解在 org.hibernate.validator.constraints 中。

在参数前使用@Valid 注解即可校验该参数。

```
@PostMapping("/validate")
public ResponseEntity validate(@Valid @RequestBody Person person, BindingResult
result){ //a
   if (result.hasErrors()){ //b
      return ResponseEntity.badRequest()
            .body(result.getAllErrors()); //c
   }
   return ResponseEntity.ok()
         .body("everything is fine");
}
```

a. @Valid 注解是 JSR 校验注解的一种。使用@RequestBody 意味着数据来源于请求体，BindingResult 可直接作为参数注入，从而获得校验的错误。

b. 可以获得 BindingResult 是否有错误。

c. 当 BindingResult 中有错误时，向返回体写入所有错误。

执行结果如图 5-13 所示。

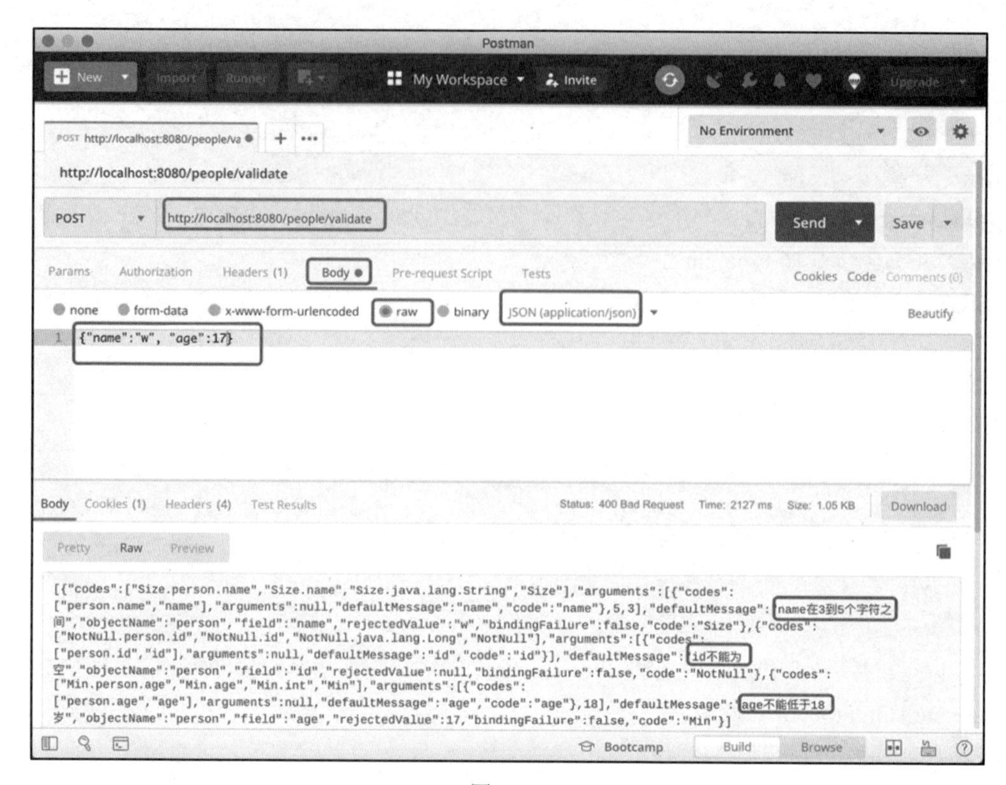

图 5-13

13. 数据操作代码

前面例子的数据操作代码在 PersonRepository 类中：

```
@Repository
public class PersonRepository implements CommonRepository{

   private static Set<Person> people = new HashSet<>();

   @Override
   public Person save(Person person) {
      people.add(person);
      return person;
   }

   @Override
   public Person findOne(Long id) {
      Person person = people.stream()
             .filter(p -> p.getId().equals(id))
             .findFirst()
             .get();
      return person;
```

```
    }

    @Override
    public Person replace(Long id, Person person) {
        Person oldPerson = people.stream()
                .filter(p -> p.getId().equals(id))
                .findFirst().get();
        people.remove(oldPerson);
        people.add(person);
        return person;
    }

    @Override
    public Person patched(Long id, Person person) {
        Person patchedPerson = people.stream()
                .filter(p -> p.getId().equals(id))
                .map(p -> {
                    String[] ignoredNullPropertyNames =
MyBeanUtils.ignoredNullPropertyNames(person);
                    BeanUtils.copyProperties(person, p, ignoredNullPropertyNames);
                    return p ;
                })
                .findFirst().get();
        return patchedPerson;
    }

    @Override
    public void delete(Long id) {
        people.removeIf(p -> p.getId().equals(id));
    }

    @Override
    public Person findByName(String name) {
        Person person = people.stream()
                .filter(p -> p.getName().equals(name))
                .findFirst().get();
        return person;
    }
}
```

操作规范接口：

```
public interface CommonRepository {
    Person save(Person person);
    Person findOne(Long id);
    Person replace(Long id, Person person);
    Person patched(Long id, Person person);
    void delete(Long id);
    Person findByName(String name);
}
```

找出对象中属性为空的工具方法：

```
public class MyBeanUtils {

   public static String[] ignoredNullPropertyNames(Object source){
      final BeanWrapper wrappedSource = new BeanWrapperImpl(source);
      return Stream.of(wrappedSource.getPropertyDescriptors())
              .map(FeatureDescriptor::getName)
              .filter(propertyName -> wrappedSource.getPropertyValue(propertyName) ==
null)
              .toArray(String[]::new);
   }

}
```

5.2.3 @ControllerAdvice

@ControllerAdvice 注解是一个特殊的@Component，顾名思义，它负责所有控制器共享的功能，如异常处理（配合@ExceptionHandler）、数据绑定（配合@InitBinder）等。@ControllerAdvice 通过属性指定生控制器的生效范围。

通过注解限制：

```
@ControllerAdvice(annotations = RestController.class)
public class someControllerAdvice{}
```

通过包名指定：

```
@ControllerAdvice("top.wisely.learningspringmvc.controller")
public class someControllerAdvice{}
```

指定控制器：

```
@ControllerAdvice(assignableTypes = {PeopleController.class, DemoController.class})
public class someControllerAdvice{}
```

1. 异常处理

使用@ExceptionHandler 组合@ControllerAdvice 可处理所有控制器的异常。当然，@ExceptionHandler 也可以注解在@Controller（包含@RestController）内，异常处理只对当前控制器有效。

自定义一个异常：

```
@Getter
@Setter
@AllArgsConstructor
@NoArgsConstructor
public class PersonNameNotFoundException extends RuntimeException{
   private String name;
}
```

定义异常处理类：

```
@ControllerAdvice
public class ExceptionHandlerAdvice {

   @ExceptionHandler(PersonNameNotFoundException.class)
   public ResponseEntity<String> customExceptionHandler(PersonNameNotFoundException
exception){
       return ResponseEntity.status(HttpStatus.NOT_FOUND)
              .body(exception.getName() + "没有找到! ");
   }

}
```

在控制器中，手动抛出异常：

```
@GetMapping("/exceptions")
public void exceptions(String name)  {
   throw new PersonNameNotFoundException(name);
}
```

异常被 ExceptionHandlerAdvice 的 exceptionHandler 方法处理，如图 5-14 所示。

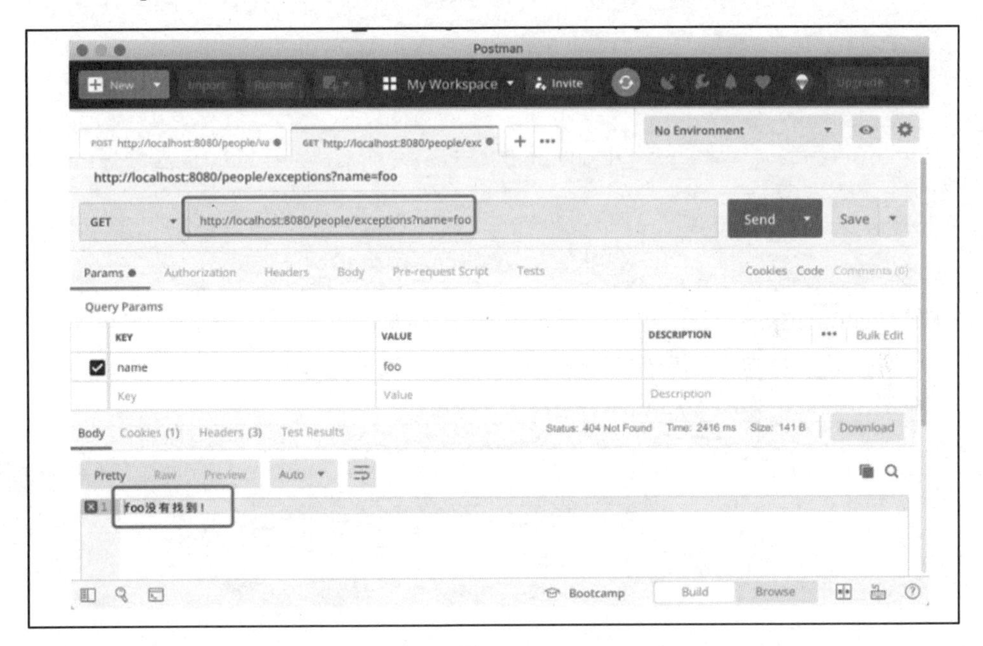

图 5-14

使用@InitBinder 注解组合@ControllerAdvice，可实现初始化数据绑定。所谓初始化数据绑定，即在 Web 请求进入控制器方法处理之前，对 Web 请求参数（不包含请求体@RequestBody，它由后面的 RequestBodyAdvice 进行处理）进行预先的初始化处理，这个处理是通过

WebDataBinder 对象来做的。这些请求参数来自@RequestParam、@PathVariable、ServletRequest 和 ServletReponse 等。也就是说，这些参数可以在@InitBinder 注解的方法中先进行处理，该方法没有返回值。例如，将参数 1-wyf-35 转换成一个 Person 对象。@InitBinder 也可以用在@Controller 内，只对使用的控制器有效。

定义一个属性编辑器，将接收到的格式为 id-name-age 的参数转换成对象：

```
public class PersonEditor extends PropertyEditorSupport {
   @Override
   public void setAsText(String text) throws IllegalArgumentException {
      String[] personStr = text.split("-"); // 将1-wyf-35分割成字符串数组
      Long id = Long.valueOf(personStr[0]);
      String name = personStr[1];
      Integer age = Integer.valueOf(personStr[2]);
      setValue(new Person(id, name, age)); //利用字符串数组建立Person对象
   }

}
```

使用@InitBinder 来注册这个属性编辑器：

```
@ControllerAdvice
@Slf4j
public class InitBinderAdvice {

   @InitBinder
   public void registerPersonEditor(WebDataBinder binder, @RequestBody String person){
      log.info("在InitBinder中为字符串：" + person);
      binder.registerCustomEditor(Person.class, new PersonEditor());
   }
}
```

在未经 WebDataBinder 注册的 PersonEditor 转换前，从请求参数里拿到的只是字符串 person。此处的字符串参数只是为了帮助读者加深理解，在开发时不需要这个参数。

用控制器来验证，如图 5-15 所示。

```
@GetMapping("/propertyEditor")
public Person propertyEditor(@RequestParam Person person){//支持的参数类型@RequetParam
   log.info("经过InitBinder注册的PropertyEditor转换后为对象：" + person);
   return person;
   return person;
}
```

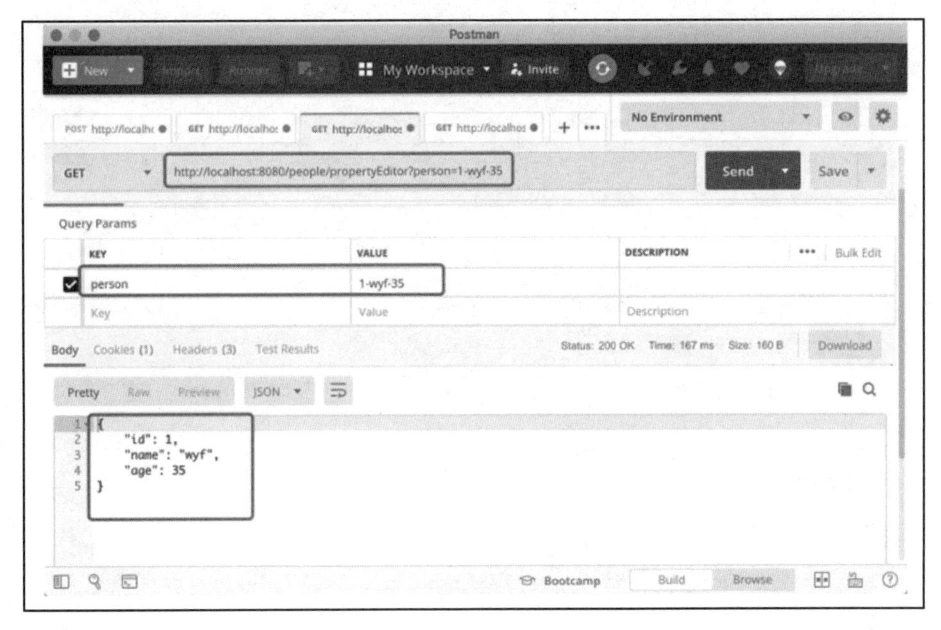

图 5-15

控制台显示如图 5-16 所示。

```
在InitBinder中为字符串：1-wyf-35
经过InitBinder注册的PropertyEditor转换后为对象：Person(id=1, name=wyf, age=35)
```

图 5-16

我们成功地将字符串 1-wyf-35 转换成了 Person 对象。

3. 自定义 Validator

在 5.2.12 节演示了默认的校验方式，若有特殊的校验需求，则可以通过实现 org.springframework.validation.Validator 接口来完全控制校验行为。

```
public class PersonValidator implements Validator {
   @Override
   public boolean supports(Class<?> clazz) {
      return Person.class.isAssignableFrom(clazz); //只支持 Person 类
   }

   @Override
   public void validate(Object target, Errors errors) {
      Person person = (Person) target;
      validateId(person, errors); //校验 Id
      validateName(person, errors); //校验 Name
      validateAge(person, errors); //校验 Age
```

```
    }
    private void validateId(Person person, Errors errors){
        ValidationUtils.rejectIfEmpty(errors,"id", "person.code", "id 不能为空-自定义");
    }
    private void validateName(Person person, Errors errors){
        int nameLength = person.getName().length();
        if (nameLength < 3 || nameLength > 5)
            errors.rejectValue("name", "person.name", "name 在 3 到 5 个字符之间-自定义");
    }
    private void validateAge(Person person, Errors errors){
        if (person.getAge() < 18)
            errors.rejectValue("age", "person.age", "age 不能低于 18 岁-自定义");
    }
}
```

在 InitBinderAdvice 类的@InitBinder 方法中注册：

```
@ControllerAdvice
public class InitBinderAdvice {
    @InitBinder
    public void setPersonValidator(WebDataBinder binder){
        binder.setValidator(new PersonValidator());
    }
}
```

还是使用前面的控制器来验证，如图 5-17 所示。

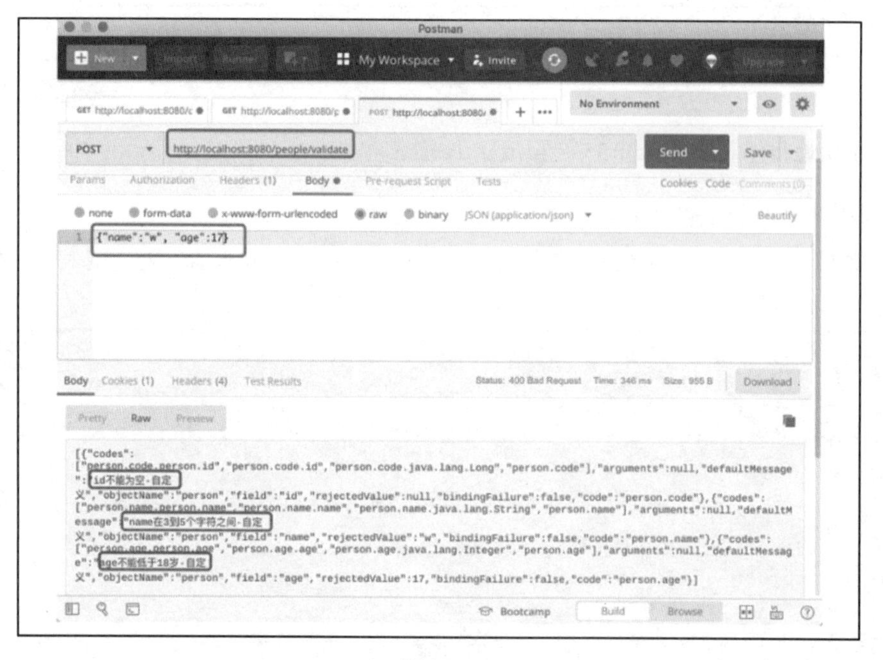

图 5-17

5.2.4 @RestControllerAdvice

@RestControllerAdvice 是组合注解，它组合了@ControllerAdvice 和@ResponseBody，它的功能和@ControllerAdvice 一样，主要用于对 RESTful 的请求体和返回体进行定制处理。

1. 先处理请求体与后处理返回体

对请求体进行定制处理用 RequestBodyAdvice 接口，它会在请求体进入控制器方法之前对请求体进行处理；对返回体进行定制处理用 ResponseBodyAdvice 接口，它会在控制器方法返回值确定之后再对返回值进行处理。它们需要和@RestControllerAdvice 一起使用。

定义一个注解，这个注解将作为使用我们定制功能的标记：

```
@Target({ElementType.PARAMETER, ElementType.METHOD}) //支持注解在方法参数和方法上
@Retention(RetentionPolicy.RUNTIME)
@Documented
public @interface ProcessTag {
}
```

定制请求体的建言：

```
@RestControllerAdvice
public class CustomRequestBodyAdvice implements RequestBodyAdvice {
   @Override
   public boolean supports(MethodParameter methodParameter, Type targetType, Class<?
extends HttpMessageConverter<?>> converterType) {
      return methodParameter.getParameterAnnotation(ProcessTag.class) != null; //a

   }

   @Override
   public HttpInputMessage beforeBodyRead(HttpInputMessage inputMessage,
MethodParameter parameter, Type targetType, Class<? extends HttpMessageConverter<?>>
converterType) throws IOException {
       return inputMessage; //b
   }

   @Override
   public Object afterBodyRead(Object body, HttpInputMessage inputMessage,
MethodParameter parameter, Type targetType, Class<? extends HttpMessageConverter<?>>
converterType) {
       if (body instanceof Person) {
          Person person = (Person) body;
          String upperCaseName = person.getName().toUpperCase();
          return new Person(person.getId(), upperCaseName, person.getAge());
       }
       return body; //c
   }
```

```
    @Override
    public Object handleEmptyBody(Object body, HttpInputMessage inputMessage,
MethodParameter parameter, Type targetType, Class<? extends HttpMessageConverter<?>>
converterType) {
        if(Person.class.isAssignableFrom((Class<?>) targetType)){
            return new Person(new Random().nextLong(),"Nobody",-1);
        }
        return body; //d
    }
}
```

a. 该请求体建言生效的条件。本例以请求体参数标记了@ProcessTag 注解为条件。

b. 在读取请求体之前，未做任何处理。

c. 在读取请求体之后，如果请求体类型是 Person，则将 name 转为大写；如果请求体类型不是 Person，则保持不变。

d. 当处理的请求体为空时，如果请求体类型是 Person，则新建一个对象，否则保持不变。

定制的返回体处理建言。

```
@RestControllerAdvice
public class CustomResponseBodyAdvice implements ResponseBodyAdvice {
    @Override
    public boolean supports(MethodParameter returnType, Class converterType) {
        return returnType.hasMethodAnnotation(ProcessTag.class); //a
    }

    @Override
    public Object beforeBodyWrite(Object body, MethodParameter returnType, MediaType
selectedContentType, Class selectedConverterType, ServerHttpRequest request,
ServerHttpResponse response) {
        if (body instanceof Person){
            Map<String, Object> map = new HashMap<>();
            map.put("person", body);
            map.put("extra-response-body", "demo-body");
            return map;
        }
        return body; //b
    }
}
```

a. 该返回体建言生效的条件，本例以在控制器方法上标记了@ProcessTag 注解为案件。

b. 在写返回体之前，将额外的信息和 body 封装到 map 里返回；若不是 Person 类，则保持不变。

在控制里验证这两个建言：

```
@GetMapping("/modifyBodies")
@ProcessTag //标记定制返回体
public Person modifyRequestBody(@ProcessTag @RequestBody Person person){ //标记定制
//请求体
    return person;
}
```

当请求体不为空时，请求结果如图 5-18 所示。

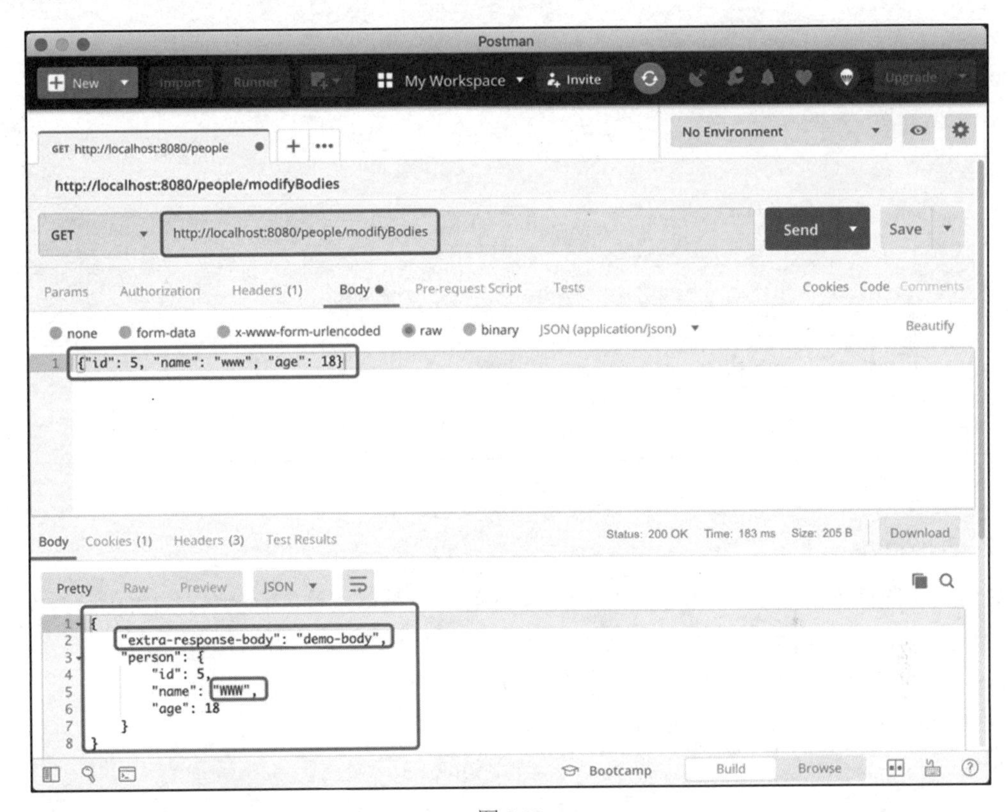

图 5-18

当请求体为空时，请求结果如图 5-19 所示。

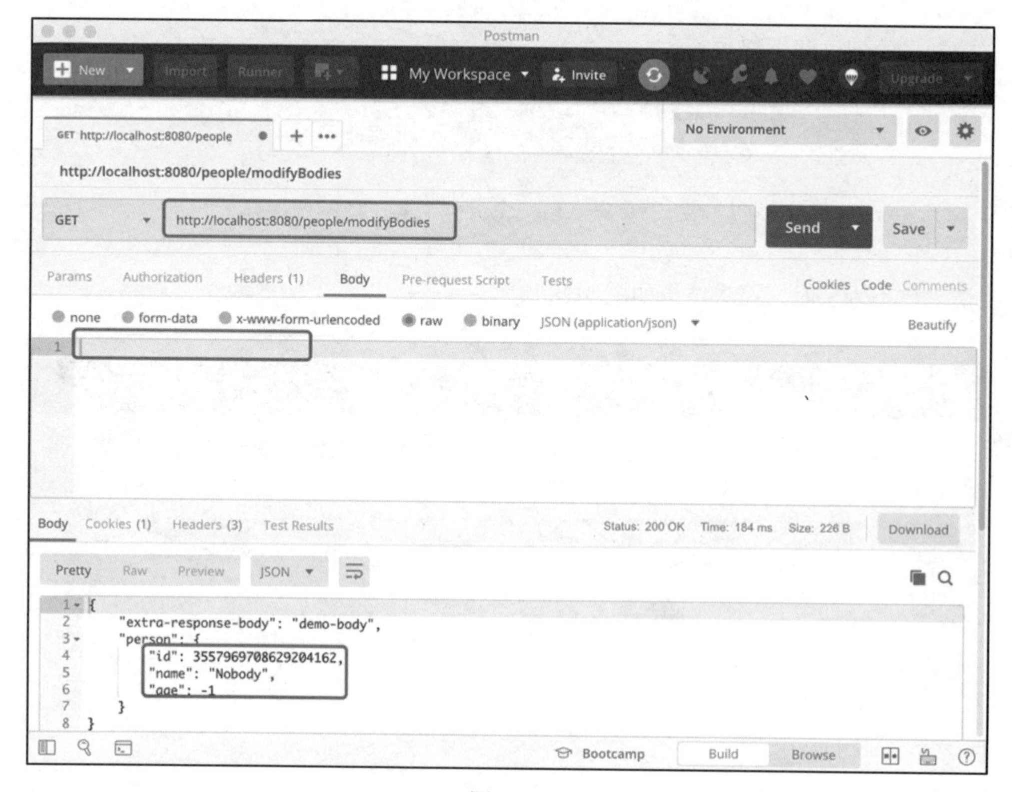

图 5-19

2. @JsonView

Spring MVC 支持用 Jackson 的@JsonView 注解来定制请求体和返回体的 JSON。这个功能的实现和前面类似，也是通过 JsonViewRequestBodyAdvice 和 JsonViewResponseBodyAdvice 来实现的。它们是自动添加到 IoC 容器，而不是通过@RestControllerAdvice 注册的。

通过@JsonView 注解在 Person 类中定义视图：

```
@Setter
@AllArgsConstructor
@NoArgsConstructor
@EqualsAndHashCode(exclude = {"name", "age"})
public class Person {
   public interface WithoutIdView {}; //a
   public interface WithIdView extends WithoutIdView {}; //a

   @NotNull(message = "id不能为空")
   private Long id;
   @Size(min = 3, max = 5, message = "name在3到5个字符之间")
   private String name;
```

```
    @Min(value = 18, message = "age 不能低于 18 岁")
    private Integer age;

    @JsonView(WithIdView.class) //b
    public Long getId() {
       return id;
    }

    @JsonView(WithoutIdView.class)//b
    public String getName() {
       return name;
    }
    @JsonView(WithoutIdView.class)//b
    public Integer getAge() {
       return age;
    }
}
```

a. 定义两个接口作为视图，WithIdView 继承 WithoutId 视图。

b. 定制两个视图对应需要展示的属性。

在控制器中进行检验，结果如图 5-20 所示。

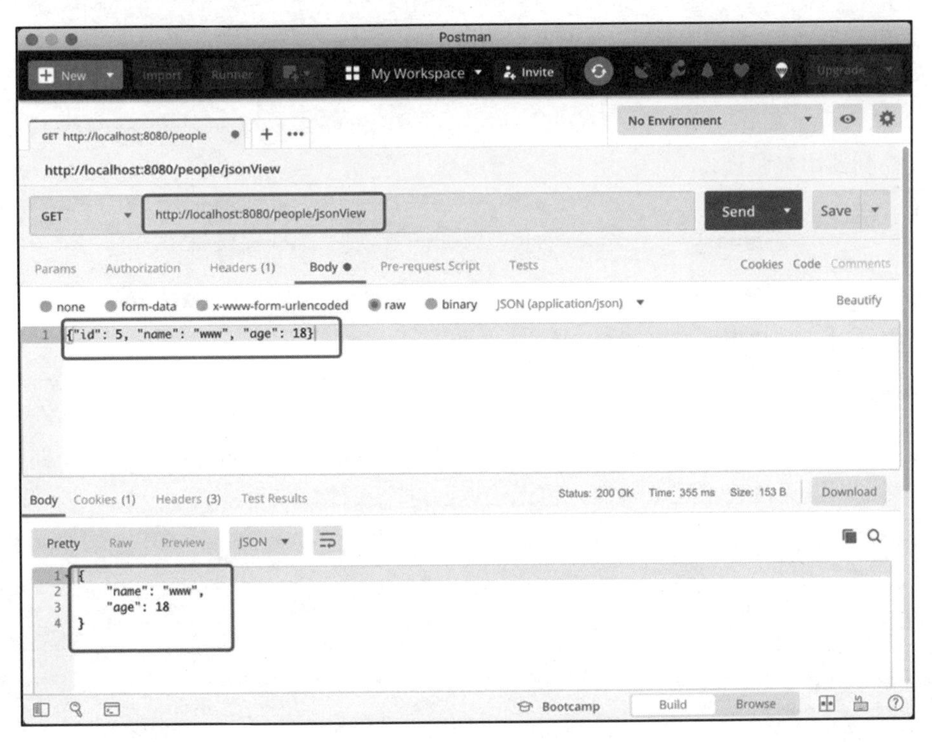

图 5-20

```
@GetMapping("/jsonView")
@JsonView(Person.WithoutIdView.class) //使用无 Id 视图
public Person jsonView(@RequestBody Person person){
   return person;
}
```

5.2.5 JSON 定制

前面使用了 Jackson 的注解@JsonView 来定制返回视图，本节使用 Jackson 提供的注解对序列化（将 Java 对象转换成 JSON，用于返回体的定制）和反序列化（将 JSON 转成 Java 对象，用于请求体的定制）进行更细节的设置。

定义一个 Java 类来作为演示：

```
@Getter
@Setter
@AllArgsConstructor
@NoArgsConstructor
@ToString
public class SecondPerson {
   private Long id;
   private String name;
   private Integer age;
    private Float height;
   private Date birthday;
}
```

定义控制器用来验证：

```
@GetMapping("/json")
public SecondPerson jsonOut(@RequestBody SecondPerson person){
   return person;
}
```

本节将针对一些常用的注解进行讲解。

1. 忽略属性

我们可以用@JsonIgnore 忽略某个属性，或者用@JsonIgnoreProperties 忽略多个属性。属性一旦被忽略，则既无法请求这个属性，也无法返回这个属性。

```
@Getter
@Setter
@AllArgsConstructor
@NoArgsConstructor
@ToString
@JsonIgnoreProperties({"name","age"})
public class SecondPerson {
   private Long id;
   private String name;
```

```
    private Integer age;
    @JsonIgnore
    private Float height;
    @JsonIgnore
    private Date birthday;
}
```

执行结果如图 5-21 所示。

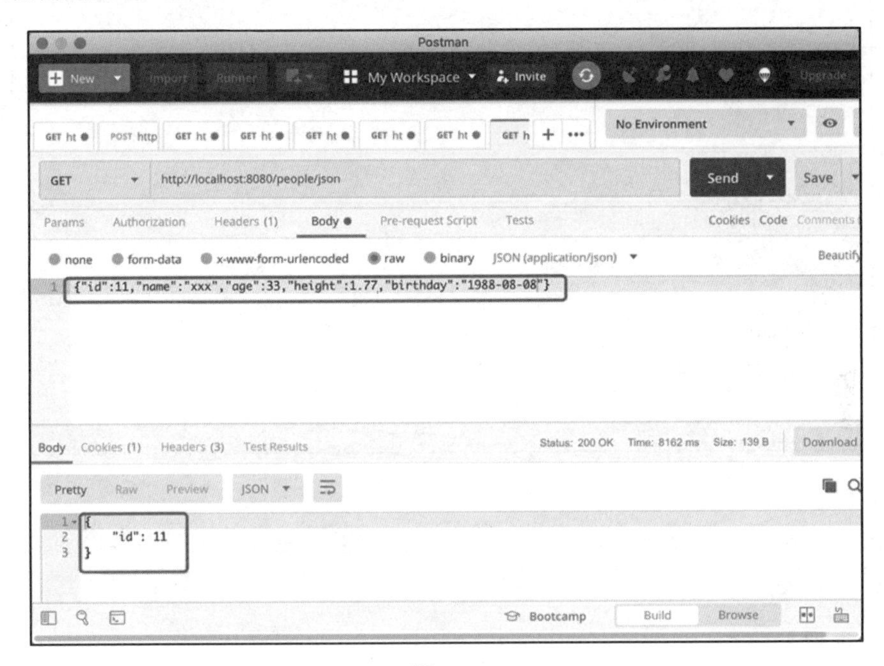

图 5-21

2. 格式定制

可以使用@JsonFormat 定制时间格式等，如图 5-22 所示。

```
@Getter
@Setter
@AllArgsConstructor
@NoArgsConstructor
@ToString
public class SecondPerson {
    private Long id;
    private String name;
    private Integer age;
    private Float height;
    @JsonFormat(pattern = "yyyy-MM-dd")
    private Date birthday;
}
```

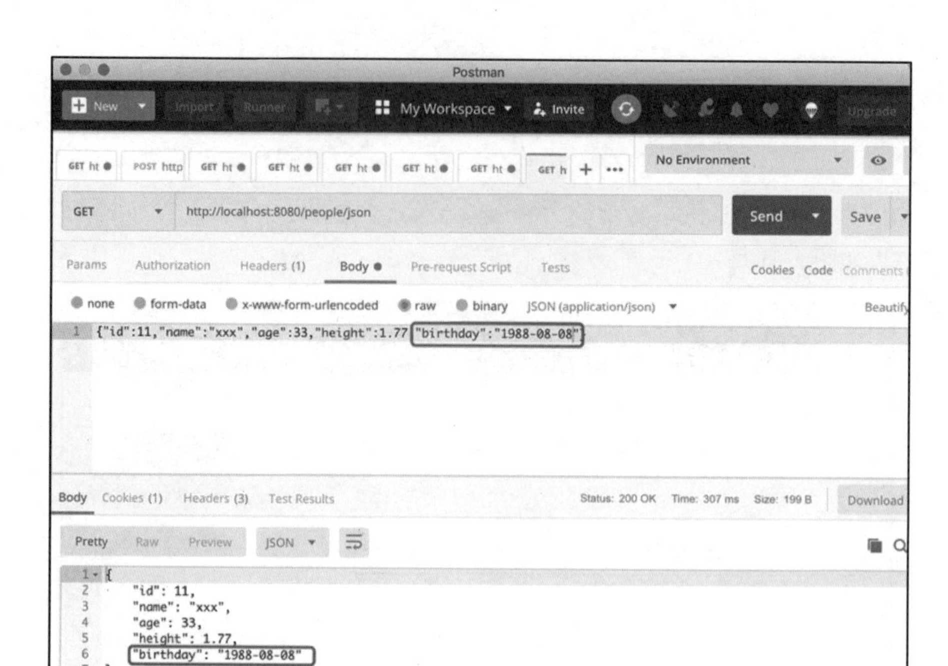

图 5-22

在 Spring Boot 中，可以通过外部配置来全局设置 JSON 数据中的日期格式。

```
spring.jackson.date-format: yyyy-MM-dd
```

3. key 定制

可以使用@JsonProperty 来定制 JSON 中的 key，如图 5-23 所示。

```
@Getter
@Setter
@AllArgsConstructor
@NoArgsConstructor
@ToString
public class SecondPerson {
   @JsonProperty("person-id")
   private Long id;
   @JsonProperty("person-name")
   private String name;
   @JsonProperty("person-age")
   private Integer age;
   private Float height;
   @JsonFormat(pattern = "yyyy-MM-dd")
   private Date birthday;
}
```

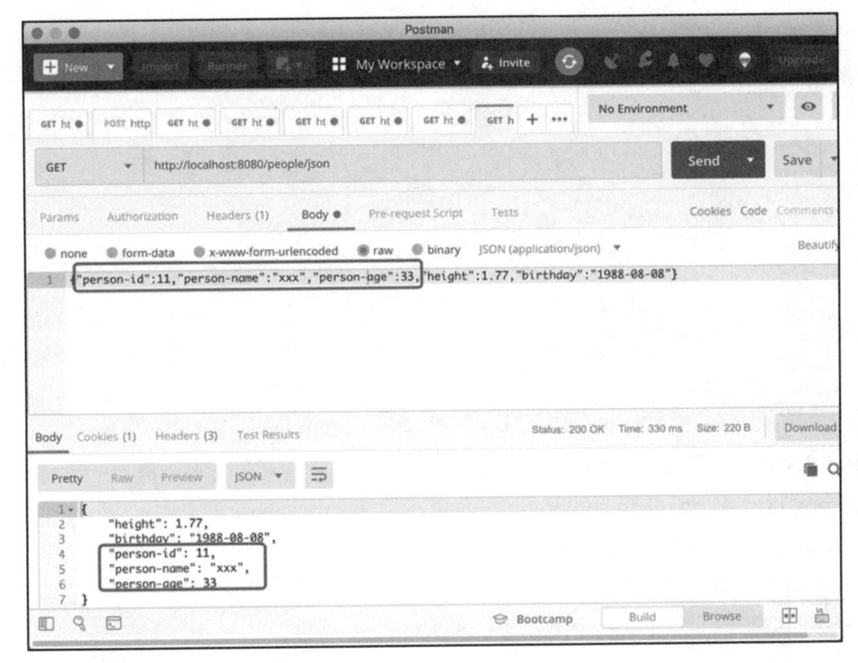

图 5-23

5.2.6 RestTemplate

可以使用 RestTemplate 作为客户端来访问其他应用提供的 RESTful 服务。在前面的演示中，使用的是 Postman 或者浏览器，这里使用 RestTemplate 作为客户端完成同样的功能。

新建客户端类 PersonClient，内容如下。

```
@Component
public class PersonClient {
   private RestTemplate restTemplate;
   public PersonClient(RestTemplateBuilder builder) {
      this.restTemplate = builder.build(); //a
   }

   public void save(){
      HttpEntity<Person> request = new HttpEntity<>(new Person(11, "wyf", 35));
      ResponseEntity<Person> savedPersonEntity = restTemplate
            .postForEntity("http://localhost:8080/people", request, Person.class); //b
      System.out.println(savedPersonEntity.getBody());
   }

   public void getOne(){
      Map<String, String> params = new HashMap<>();
      params.put("id", "1");
      Person person = restTemplate
```

```
            .getForObject("http://localhost:8080/people/{id}",Person.class,
params);//c
        System.out.println(person);
    }

    public void findByName(){
        UriComponentsBuilder builder = UriComponentsBuilder
                .fromHttpUrl("http://localhost:8080/people/findByName")
                .queryParam("name", "wyf");
        Person person = restTemplate
                .getForObject(builder.toUriString(),Person.class);//d
        System.out.println(person);
    }

}
```

a. Spring Boot 自动配置了 RestTemplateBuilder，我们可以通过它来获得 RestTemplate。

b. 使用 postForEntity()来保存 Person，方法参数 Person.class 用来指定返回值的类型。这里服务端的代码针对的是下面的内容。

```
@PostMapping
public ResponseEntity<Person> save(RequestEntity<Person> personRequestEntity){
  //……
}
```

c. 使用 getForObject()获取某个 Person，路径变量通过 params 来传递。

d. 同样，通过 getForObject()按请求参数查询 Person，查询参数通过 UriComponentsBuilder 的 queryParam()方法来构造。

在 LearningSpringMvcApplication 入口类添加 CommandLineRunner，自动执行上面的客户端。

```
@Bean
  CommandLineRunner personClientClr(PersonClient personClient){
     return args -> {
        personClient.save();
        personClient.getOne();
        personClient.findByName();
     };
  }
}
```

运行应用，控制台输出如图 5-24 所示。

```
Person(id=1, name=wyf, age=35)
Person(id=1, name=wyf, age=35)
Person(id=1, name=wyf, age=35)
```

图 5-24

更多关于 RestTemplate 的用法请参考：*Spring Pramework JavaDocAPI*。

5.3 Web MVC 配置

5.3.1 Spring MVC 的工作原理

Spring MVC 对 REST 的支持是通过 DispatchServlet 开始的触发。在 Spring Boot 中，通过 DispatcherServletAutoConfiguration 来定义 DispatcherServlet 的 Bean，并通过 DispatcherServletRegistrationBean 的 Bean 来注册 DispatcherServlet 到 Servlet 容器中。

DispatcherServlet 在处理（doDispatch 方法）时，将主要的功能都代理给了下面的两个 Bean。

（1）HandlerMapping：RequestMappingHandlerMapping。

◎ 它主要负责获取 Web 请求与 Handler（控制器方法）之间的映射。当前使用的实现为 RequestMappingHandlerMapping，它获取 RequestMappingInfo 作为请求和 Handler 方法之间的映射，RequestMappingInfo 信息是从@RequestMapping 注解中获取的。

◎ 负责 PathMatchConfigurer（路径匹配）的设置。

◎ 负责 Interceptor（拦截器）的设置。

◎ 负责 ContentNegotiationManager（内容协商管理器）的设置。

◎ 负责 CORS（跨域资源共享）的设置。

（2）HandlerAdapter：RequestMappingHandlerAdapter。

◎ 它主要负责从映射中获取 Handler 并调用。当前使用的实现为 RequestMappingHandlerAdapter，它通过 HandlerMapping 的 getHandler 方法从 RequestMappingInfo 中获取并执行 HandlerMethod，HandlerMethod 即我们实际要执行的控制器方法。

◎ 负责 HttpMessageConverter 的设置。

◎ 负责 WebBindingInitializer 的设置。

◎ 负责 HandlerMethodArgumentResolver（控制器方法的参数）的设置。

◎ 负责 HandlerMethodReturnValueHandler（控制器方法返回值）的设置。

DispatcherServlet 支持多个 HandlerMapping，从 HandlerMapping 的 getHandler 方法可以获得不同类型的 Handler。HandlerAdapter 的 supports 方法声明只处理符合它支持的 Handler 类型；RequestMappingHandlerAdapter 只支持类型为 HandlerMethod 的 Handler。

5.3.2 配置 MVC

前面 HandlerMapping 和 HandlerAdapter 负责的内容都是需要做配置的内容，幸运的是，Spring MVC 提供了@EnableWebMvc 注解，它通过导入 DelegatingWebMvcConfiguration 中的配

置，做好了默认配置。Spring Boot 虽然没有直接使用@EnableWebMvc，但是在 Spring Boot 的 WebMvcAutoConfiguration.EnableWebMvcConfiguration（继承 DelegatingWebMvcConfigurations）中做了等同于@EnableWebMvc 的配置。除 WebMvcAutoConfiguration.EnableWebMvcConfiguration 中的配置外，WebMvcAutoConfiguration 还做了非常多的配置，在本书后面会有针对讲解。

可以通过实现 WebMvcConfigurer 接口来定制 HandlerMapping 和 HandlerAdapter。

```
@Configuration
public class WebConfiguration implements WebMvcConfigurer {
  //……
}
```

覆写 WebMvcConfigurer 接口的方法，实现对 Spring MVC 配置的定制。如果完全不想使用 Spring Boot 提供的自动配置，则只需在配置类上加上@EnableWebMvc 即可，但一般不需要这样做。

自 Spring 5.0 开始，Spring 全面支持 Java 8，因而 WebMvcConfigurerAdapter 的实现内容已经被 WebMvcConfigurer 的默认方法替代，WebMvcConfigurerAdapter 将被废弃。

当然，除上面的配置形式外，有些配置是可以通过配置 Bean 来实现的，在本书后面会有针对讲解。

5.3.3 Interceptor

Servlet 提供了 filter（过滤器）来预处理和后处理每一个 Web 请求，Spring MVC 提供了 Interceptor（拦截器）来预处理和后处理每一个 Web 请求，它的优势是能使用 IoC 容器的一些功能。Interceptor 接口有 3 个方法。

◎ preHandle：在 Handler 执行前。

◎ postHandle：在 Handler 执行后 。

◎ afterCompletion：在整个请求完成后。

只要实现接口 HandlerInterceptor 或者继承 HandlerInterceptorAdapter，就能定义拦截器。

```
@Slf4j
public class CustomInterceptor implements HandlerInterceptor {
   private final static String START_TIME = "startTime";
   private final static String PROCESS_TIME = "processTime";

   @Override
   public boolean preHandle(HttpServletRequest request, HttpServletResponse response,
Object handler) throws Exception {
       request.setAttribute(START_TIME, System.currentTimeMillis());
       log.info("preHandle 处理中...");
       return true;
   }
```

```
    @Override
    public void postHandle(HttpServletRequest request, HttpServletResponse response,
Object handler, ModelAndView modelAndView) throws Exception {
        long startTime = (long) request.getAttribute(START_TIME);
        long endTime = System.currentTimeMillis();
        request.removeAttribute(START_TIME);
        request.setAttribute(PROCESS_TIME, endTime - startTime);
        log.info("postHandle 处理中...");
    }

    @Override
    public void afterCompletion(HttpServletRequest request, HttpServletResponse response,
Object handler, Exception ex) throws Exception {
        log.info("请求处理时间为：" + request.getAttribute(PROCESS_TIME) + "毫秒");
        request.removeAttribute(PROCESS_TIME);
    }
}
```

在 WebConfiguration 中，通过覆写 WebMvcConfigurer 接口的方法 InterceptorRegistry 来注册 Interceptor：

```
@Configuration
public class WebConfiguration implements WebMvcConfigurer {
    @Override
    public void addInterceptors(InterceptorRegistry registry) {
        registry.addInterceptor(customInterceptor());
    }

    @Bean
    public CustomInterceptor customInterceptor(){
        return new CustomInterceptor();
    }
}
```

在 Postman 中执行任意请求，控制台输出如图 5-25 所示。

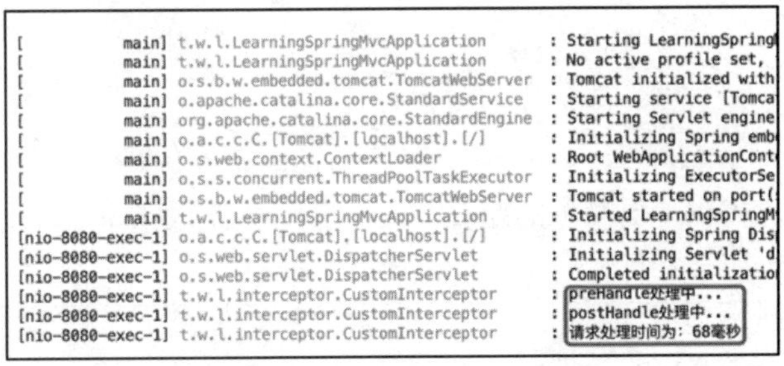

```
[           main] t.w.l.LearningSpringMvcApplication       : Starting LearningSpring
[           main] t.w.l.LearningSpringMvcApplication       : No active profile set,
[           main] o.s.b.w.embedded.tomcat.TomcatWebServer  : Tomcat initialized with
[           main] o.apache.catalina.core.StandardService   : Starting service [Tomca
[           main] org.apache.catalina.core.StandardEngine  : Starting Servlet engine
[           main] o.a.c.c.C.[Tomcat].[localhost].[/]       : Initializing Spring emb
[           main] o.s.web.context.ContextLoader            : Root WebApplicationCont
[           main] o.s.s.concurrent.ThreadPoolTaskExecutor  : Initializing ExecutorSe
[           main] o.s.b.w.embedded.tomcat.TomcatWebServer  : Tomcat started on port(
[           main] t.w.l.LearningSpringMvcApplication       : Started LearningSpringM
[nio-8080-exec-1] o.a.c.c.C.[Tomcat].[localhost].[/]       : Initializing Spring Dis
[nio-8080-exec-1] o.s.web.servlet.DispatcherServlet        : Initializing Servlet 'd
[nio-8080-exec-1] o.s.web.servlet.DispatcherServlet        : Completed initializatio
[nio-8080-exec-1] t.w.l.interceptor.CustomInterceptor      : preHandle处理中...
[nio-8080-exec-1] t.w.l.interceptor.CustomInterceptor      : postHandle处理中...
[nio-8080-exec-1] t.w.l.interceptor.CustomInterceptor      : 请求处理时间为：68毫秒
```

图 5-25

5.3.4 Formatter

org.springframework.format.Formatter 是 Spring 提供用来替代 PropertyEditor 的，PropertyEditor 不是线程安全的，每个 Web 请求都会通过 WebDataBinder 注册创建新的 PropertyEditor 实例；而 Formatter 是线程安全的。

定义一个类似于 PersonEditor 的 PersonFormmater，它可以实现 Formatter 接口：

```
public class PersonFormatter implements Formatter<Person> {
    //处理逻辑和 PersonEditor 一致
    @Override
    public Person parse(String text, Locale locale) throws ParseException {
        String[] personStr = text.split("\\|");
        Long id = Long.valueOf(personStr[0]);
        String name = personStr[1];
        Integer age = Integer.valueOf(personStr[2]);
        return new Person(id, name, age);
    }

    @Override
    public String print(Person object, Locale locale) {
        return object.toString();
    }
}
```

在 Spring MVC 下，通过 WebConfiguration 覆写 WebMvcConfigurer 接口的 addFormatters 方法来注册 Formatter。

```
@Configuration
public class WebConfiguration implements WebMvcConfigurer {

    @Override
    public void addFormatters(FormatterRegistry registry) {
        registry.addFormatter(personFormatter());
    }

    @Bean
    public Formatter personFormatter(){
        return new PersonFormatter();
    }
}
```

Spring Boot 为我们提供了更简单的配置方式，只要注册 Formatter 的 Bean 即可，无须使用 addFormatters 方法。

```
@Bean
public Formatter personFormatter(){
    return new PersonFormatter();
}
```

此时还要注释掉 InitBinderAdvice 中关于代码注册（registerPersonEditor）的代码，否则 PersonEditor 和 PersonFormatter 会产生冲突。

除用 addFormatters 方法注册外，还可以通过 InitBinderAdvice 中@InitBinder 注解的方法来注册 Formatter。

```
@ControllerAdvice
public class InitBinderAdvice {
   @InitBinder
   public void addPersonFormatter(WebDataBinder binder){
     binder.addCustomFormatter(new PersonFormatter());
   }
}
```

用控制器验证，结果如图 5-26 所示。

```
@GetMapping("/formatter")
public Person formatter(@RequestParam Person person){
   return person;
}
```

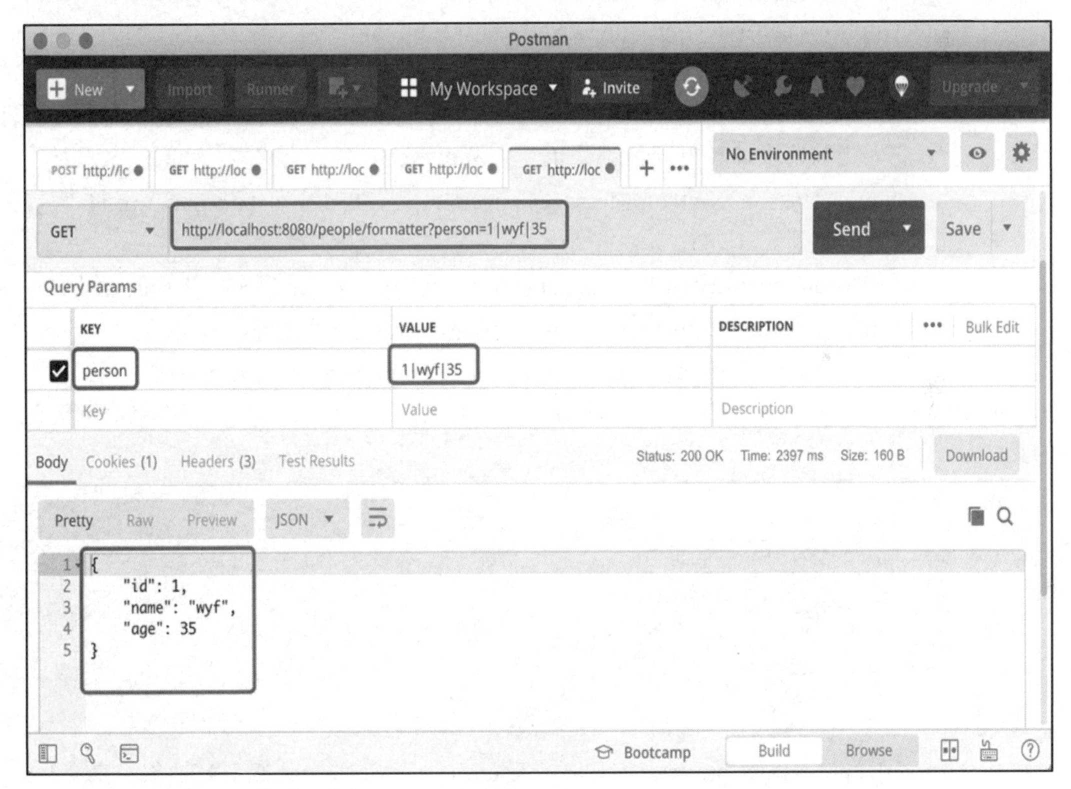

图 5-26

5.3.5 HttpMessageConverter

在 Spring MVC 中，请求（@RequestBody、RequestEntity 等）和返回（@Responsebody、ResponseEntity 等）都是通过 HttpMessageConverter 来实现数据转换的。

外部的请求数据通过 HttpMessageConverter 可以转换成 Java 对象，Java 对象通过 HttpMessageConverter 又可以转换成返回数据。在前面的例子中，Web 请求体中的 JSON 数据通过 MappingJackson2HttpMessageConverter 转换成 Java 对象，而返回的 Java 对象又通过 MappingJackson2HttpMessageConverter 转换成 JSON 数据。

HttpMessageConverter 会根据请求或返回的内容类型（Content-Type 如 application/json）来选择对应的 HttpMessageConverter 对数据进行转换。

Spring MVC 自动注册了下列 HttpMessageConverter。

- ◎ ByteArrayHttpMessageConverter：二进制数组转换。
- ◎ StringHttpMessageConverter：字符串转换，默认支持所有的媒体类型。
- ◎ ResourceHttpMessageConverter：org.springframework.core.io.Resource 类型转换。
- ◎ SourceHttpMessageConverter：javax.xml.transform.Source 类型转换。
- ◎ 各种 JSON 库的 HttpMessageConverter。
  - ➢ MappingJackson2HttpMessageConverter：当 jackson-databind jar 包在类路径时注册，当前请求体和返回体都是用它来做数据转换的。
  - ➢ MappingJackson2XmlHttpMessageConverter：当 jackson-dataformat-xml jar 包在类路径时注册。
  - ➢ Jaxb2RootElementHttpMessageConverter：当 jaxb-api jar 包在类路径时注册。
  - ➢ GsonHttpMessageConverter：当 gson jar 包在类路径时注册。
  - ➢ JsonbHttpMessageConverter：当 javax.json.bind-api jar 包在类路径时注册。

下面定义一个 HttpMessageConverter 并演示它的功能。首先定义一个专门用来演示数据转换的类：

```
@Getter
@Setter
@AllArgsConstructor
@NoArgsConstructor
public class AnotherPerson {
   private Long id;
   private String name;
   private Integer age;
}
```

继承 AbstractHttpMessageConverter（实现了 HttpMessageConverter 接口）抽象类来定制：

```
public class AnotherPersonHttpMessageConverter extends
AbstractHttpMessageConverter<AnotherPerson> {

   public AnotherPersonHttpMessageConverter() {
      super(new MediaType("application","another-person",
Charset.defaultCharset()));//a
   }

   @Override
   protected boolean supports(Class<?> clazz) {
      return AnotherPerson.class.isAssignableFrom(clazz); //b
   }

   @Override
   protected AnotherPerson readInternal(Class<? extends AnotherPerson> clazz,
HttpInputMessage inputMessage) throws IOException, HttpMessageNotReadableException {
      String body = StreamUtils.copyToString(inputMessage.getBody(),
Charset.defaultCharset());
      String[] personStr = body.split("-");
      return new AnotherPerson(Long.valueOf(personStr[0]), personStr[1],
Integer.valueOf(personStr[2])); //c
   }

   @Override
   protected void writeInternal(AnotherPerson anotherPerson, HttpOutputMessage
outputMessage) throws IOException, HttpMessageNotWritableException {
      String out = "Hello:" + anotherPerson.getId() + "-" +
            anotherPerson.getName() + "-" +
            anotherPerson.getAge();
      StreamUtils.copy(out, Charset.defaultCharset(), outputMessage.getBody()); //d
   }
}
```

a. 在构造器中自定义媒体类型 application/another-person。

b. 当前 HttpMessageConverter 支持转换的类型是 AnotherPerson。

c. 将请求体内的字符串 6-foo-28 转换成 AnotherPerson 对象。

d. 将 AnotherPerson 对象转换成对应的字符串形式。

注册 HttpMessageConverter 的方式有很多种，举例如下。

（1）直接定义 Bean。

```
@Bean
HttpMessageConverter anotherPersonHttpMessageConverter(){
   return new AnotherPersonHttpMessageConverter();
}
```

（2）定义 HttpMessageConverters 的 Bean，这是 Spring Boot 专门提供的，推荐使用。

```
@Bean
HttpMessageConverters httpMessageConverters(){
```

```
    return new HttpMessageConverters(new AnotherPersonHttpMessageConverter());
}
```

（3）覆写 WebMvcConfigurer 的方法 configureMessageConverters。

```
@Override
public void configureMessageConverters(List<HttpMessageConverter<?>> converters) {
    converters.add(new AnotherPersonHttpMessageConverter());
}
```

当使用前两种注册方式注册时，控制器中的内容如下：

```
@GetMapping("/converter")
public AnotherPerson converter(@RequestBody AnotherPerson person){
    return person;
}
```

当使用第三种注册方式注册时，因为 Spring MVC 默认使用 MappingJackson2HttpMessageConverter 对请求体和返回体进行处理，所以需要指定返回体的媒体类型。

```
@GetMapping(value = "/converter", produces = {"application/another-person"})
public AnotherPerson converter(@RequestBody AnotherPerson person){
    return person;
}
```

当在 Postman 中进行请求时，需要将请求头的 Content-Type 设置为 application/another-person，如图 5-27 所示。

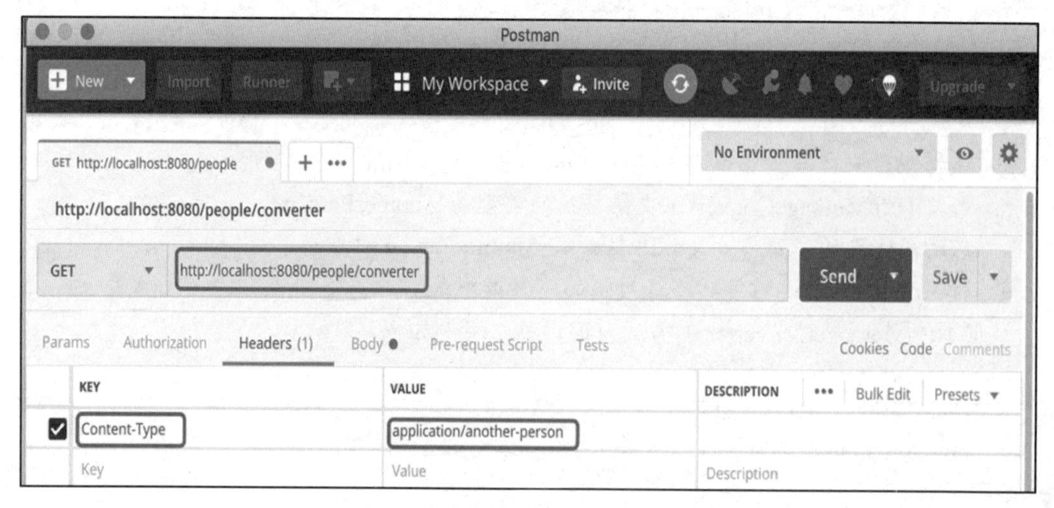

图 5-27

请求的格式也是用“-”隔开的，如图 5-28 所示。

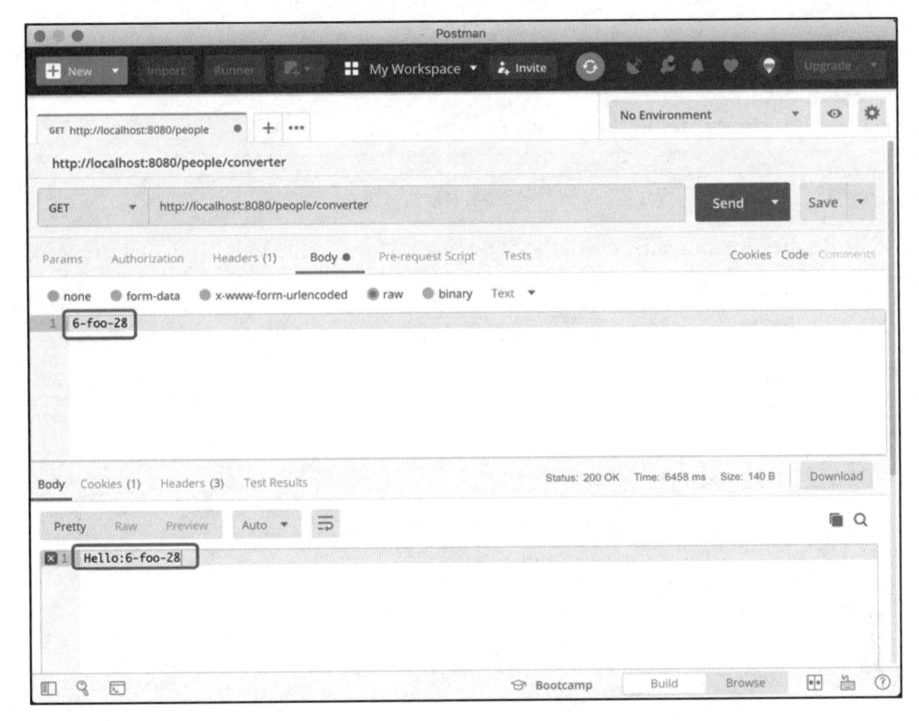

图 5-28

5.3.6 方法参数和返回值处理设置

在控制器方法里，可以使用@RequestBody、RequestEntity 获得请求体里的数据，使用@ResponseBody、ResponseEntity 对返回体进行处理。

在 Spring MVC 中，可以通过实现 HandlerMethodArgumentResolver 接口对控制器方法参数进行处理，通过实现 HandlerMethodReturnValueHandler 对方法的返回值进行处理。

若同时需要对请求参数和返回返回体进行处理，则继承 AbstractMessageConverterMethodProcessor 抽象类，它分别实现了 HandlerMethodArgumentResolver 和 HandlerMethodReturnValueHandler。

Spring MVC 的@RequestBody 和@ResponseBody 就是通过 RequestResponseBodyMethod Processor 对请求体和返回体进行处理的；而 RequestEntity 和 ResponseEntity 是通过 HttpEntityMethodProcessor 生效的。

◎ @RequestParam：RequestParamMethodArgumentResolver。

◎ @PathVariable：PathVariableMethodArgumentResolver。

◎ @RequestPart：RequestPartMethodArgumentResolver。

◎ @RequestHeader：RequestHeaderMethodArgumentResolver。

◎ ServletRequest：ServletRequestMethodArgumentResolver。

◎ ServletResponse：ServletResponseMethodArgumentResolver。

RequestResponseBodyMethodProcessor 的构造如下：

```
public RequestResponseBodyMethodProcessor(List<HttpMessageConverter<?>> converters,
    @Nullable List<Object> requestResponseBodyAdvice) {}
```

这意味着 RequestResponseBodyMethodProcessor 负责用 HttpMessageConverter 将请求体中的外部数据转换为 Java 对象，并将 Java 对象转换成返回体中的外部数据（让@RequestBody 和@ResponseBody 生效）。同时，请求体数据的先处理 RequestBodyAdvice 和返回体的后处理 ResponseBodyAdvice 也是由它处理的。

1. 示例 1

在 5.2.10 节的方法参数中，有如下代码：

```
@GetMapping("/methodArguments")
public void methodArguments(WebRequest request,
                            NativeWebRequest nativeWebRequest,
                            ServletRequest servletRequest,
                            ServletResponse servletResponse,
                            HttpSession httpSession,
                            Locale locale,
                            TimeZone timeZone,
                            ZoneId zoneId,
                            DemoService demoService) throws Exception{
   System.out.println(request);
   System.out.println(nativeWebRequest);
   System.out.println(servletRequest);
   System.out.println(servletResponse);
   System.out.println(httpSession);
   System.out.println(locale);
   System.out.println(timeZone);
   System.out.println(zoneId);
   System.out.println(demoService.sayHello());
}
```

此处的 DemoService demoService 不是注入的，而是新建的，如果想将其变成注入的而非新建的，那么我们要做的就是定制方法参数，即用 HandlerMethodArgumentResolver 接口去定制。

下面定义一个控制器方法来检验：

```
@Controller
@RequestMapping("/demo")
public class DemoController {

   private DemoService demoService;

   public DemoController(DemoService demoService) {
```

```
        this.demoService = demoService;
    }

    @GetMapping("/bean-argument")
    public void argument(DemoService demoService){
        System.out.println("this.demoService 和方法参数 demoService 是否相等：" +
this.demoService.equals(demoService));
    }


}
```

此时，一个注入的 demoService 和一个新建的 DemoService 肯定是不相等的，访问 http://localhost:8080/demo/bean-argument，控制台显示如图 5-29 所示。

```
2019-05-05 19:46:32.757  INFO 62005 --- [nio-8080-exec-2]
2019-05-05 19:46:32.768  INFO 62005 --- [nio-8080-exec-2]
this.demoService和方法参数demoService是否相等：false
2019-05-05 19:46:34.900  INFO 62005 --- [nio-8080-exec-2]
2019-05-05 19:46:34.932  INFO 62005 --- [nio-8080-exec-2]
2019-05-05 19:46:34.935  INFO 62005 --- [nio-8080-exec-2]
```

图 5-29

下面通过 HandlerMethodArgumentResolver 接口来定制：

```
public class BeanArgumentResolver implements HandlerMethodArgumentResolver,
BeanFactoryAware {
    private BeanFactory beanFactory;
    @Override
    public void setBeanFactory(BeanFactory beanFactory) throws BeansException {
        this.beanFactory = beanFactory; //a
    }

    @Override
    public boolean supportsParameter(MethodParameter parameter) {
        return parameter.getParameterType().equals(DemoService.class); //b
    }

    @Override
    public Object resolveArgument(MethodParameter parameter, ModelAndViewContainer
mavContainer, NativeWebRequest webRequest, WebDataBinderFactory binderFactory) throws
Exception {
        return beanFactory.getBean(DemoService.class); //c
    }
}
```

a 通过实现 BeanFactoryAware 可获得 BeanFactory。

b 当前参数只支持 DemoService 类。

c 获得支持类型的 Bean。

这个示例并不通用，当然，可以很容易地将它改造成通用的示例，读者可以自己尝试。

下面通过 WebMvcConfigurer 接口的 addArgumentResolvers 方法来注册：

```
@Bean
HandlerMethodArgumentResolver beanArgumentResolver(){
   return new BeanArgumentResolver();
}

@Override
public void addArgumentResolvers(List<HandlerMethodArgumentResolver> resolvers) {
   resolvers.add(beanArgumentResolver());
   resolvers.add(requestResponsePersonMethodProcessor());
}
```

此时再次访问 http://localhost:8080/demo/bean-argument，控制台显示如图 5-30 所示。

```
2019-05-05 19:55:57.102  INFO 62084 --- [nio-8080-exec-1]
2019-05-05 19:55:57.113  INFO 62084 --- [nio-8080-exec-1]
2019-05-05 19:55:57.127  INFO 62084 --- [nio-8080-exec-1]
2019-05-05 19:55:57.138  INFO 62084 --- [nio-8080-exec-1]
this.demoService和方法参数demoService是否相等: true
2019-05-05 19:56:02.961  INFO 62084 --- [nio-8080-exec-1]
2019-05-05 19:56:02.991  INFO 62084 --- [nio-8080-exec-1]
```

图 5-30

2. 示例 2

下面模仿@RequestBody 和@ResponseBody 的实现做一个简单的示例。

请求体注解：

```
@Target({ElementType.PARAMETER})
@Retention(RetentionPolicy.RUNTIME)
@Documented
public @interface RequestPerson {
}
```

返回体注解：

```
@Target({ElementType.METHOD})
@Retention(RetentionPolicy.RUNTIME)
@Documented
public @interface ResponsePerson {
}
```

针对 Person 的请求处理器和返回处理器：

```
public class RequestResponsePersonMethodProcessor extends
AbstractMessageConverterMethodProcessor {

   public RequestResponsePersonMethodProcessor(List<HttpMessageConverter<?>>
converters) {
      super(converters);
   }
```

```
    @Override //HandlerMethodArgumentResolver 的覆写方法
    public boolean supportsParameter(MethodParameter parameter) {
        return  parameter.hasMethodAnnotation(ResponsePerson.class) &&
              parameter.getParameterType().equals(AnotherPerson.class); //a
    }

    @Override //HandlerMethodArgumentResolver 的覆写方法
    public Object resolveArgument(MethodParameter parameter, ModelAndViewContainer
mavContainer, NativeWebRequest webRequest, WebDataBinderFactory binderFactory) throws
Exception {
        Object object = readWithMessageConverters(webRequest, parameter,
parameter.getNestedGenericParameterType());
        return object; //b
    }

    @Override //HandlerMethodReturnValueHandler 的覆写方法
    public boolean supportsReturnType(MethodParameter returnType) {
        return (AnnotatedElementUtils.hasAnnotation(returnType.getContainingClass(),
ResponsePerson.class) ||
                returnType.hasMethodAnnotation(ResponsePerson.class)) &&
                returnType.getParameterType().equals(AnotherPerson.class); //c
    }

    @Override //HandlerMethodReturnValueHandler 的覆写方法
    public void handleReturnValue(Object returnValue, MethodParameter returnType,
ModelAndViewContainer mavContainer, NativeWebRequest webRequest) throws Exception {
        AnotherPerson person = (Person) returnValue;
        person.setName(person.getName().toUpperCase());
        writeWithMessageConverters(person, returnType, webRequest); //d
    }
}
```

a. 参数只有注解了@RequestPerson 且类型为 AnotherPerson 才有效。

b. 使用 HttpMessageConverter 将请求体内的数据（6-foo-28）转换成 AnotherPerson 类，readWithMessageConverters 方法使用匹配的 HttpMessageConverter，将请求体中的内容转换为 Java 对象，它是 AbstractMessageConverterMethodProcessor 的父类 AbstractMessageConverterMethodArgumentResolver 的方法。

c. 只有从方法返回上注解了@ResponsePerson 且返回值类型为 AnotherPerson 才有效。

d. 使用 HttpMessageConverter 将对象转换成数据展示给客户端，writeWithMessageConverters 方法使用匹配的 HttpMessageConverter 将对象转换成对应的数据类型，它同样是 AbstractMessageConverterMethodArgumentResolver 的方法。

前面通过自定义方法参数返回值处理器了解了 HttpMessageConverter 的工作原理。下面注册这个处理器，这个处理器分成两个部分：HandlerMethodArgumentResolver 和 HandlerMethod

ReturnValueHandler。我们要分别覆写 WebMvcConfigurer 接口的 addArgumentResolvers 和 addReturnValueHandlers 方法。

```
@Configuration
public class WebConfiguration implements WebMvcConfigurer {
   @Bean
   HttpMessageConverters httpMessageConverters(){
      return new HttpMessageConverters(new AnotherPersonHttpMessageConverter());
   }

   @Bean
   RequestResponsePersonMethodProcessor requestResponsePersonMethodProcessor(){
      return new
RequestResponsePersonMethodProcessor(httpMessageConverters().getConverters());
   }

   @Override
   public void addArgumentResolvers(List<HandlerMethodArgumentResolver> resolvers) {
      resolvers.add(requestResponsePersonMethodProcessor());
   }

   @Override
   public void addReturnValueHandlers(List<HandlerMethodReturnValueHandler> handlers)
{
      handlers.add(requestResponsePersonMethodProcessor());
   }

}
```

使用前面例子中定义的 AnotherPersonHttpMessageConverter 作为本例演示所用的 HttpMessageConverter，RequestResponsePersonMethodProcessor 所需的 converters 可从 HttpMessageConverters 的 Bean 中获得。

本例使用@ResponsePerson 注解方法返回值，而@RestController 的控制器默认每个方法都注解了@ResponseBody，这会让@ResponsePerson 失效，所以需要在 DemoController 控制器进行检验。

```
@Controller
@RequestMapping("/demo")
public class DemoController {
   @GetMapping("/argument")
   @ResponsePerson
   public AnotherPerson argument(@RequestPerson AnotherPerson person){
      return person;
   }
}
```

为了使 AnotherPersonHttpMessageConverter 生效，需设置请求头信息，如图 5-31 所示。

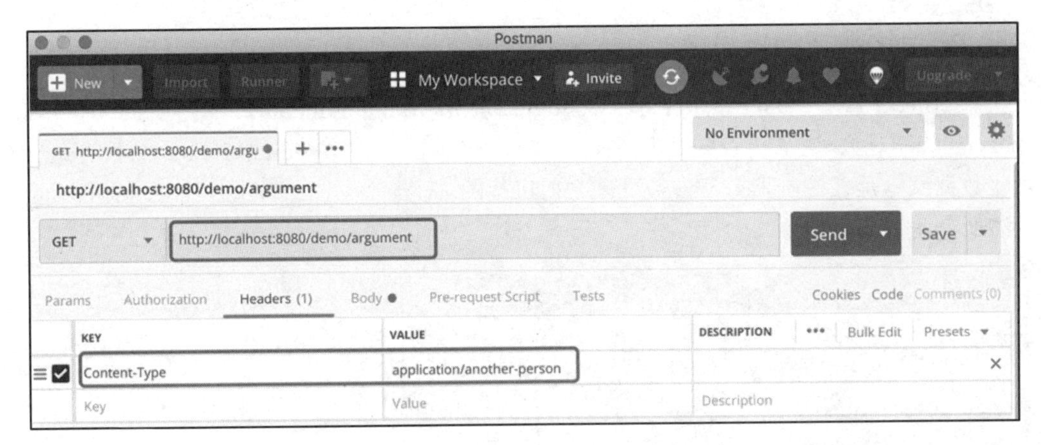

图 5-31

使用 AnotherPersonHttpMessageConverter 正确解析请求体的内容，正确地生成返回体，方法返回值处理器还额外将 name 改成了大写，如图 5-32 所示。

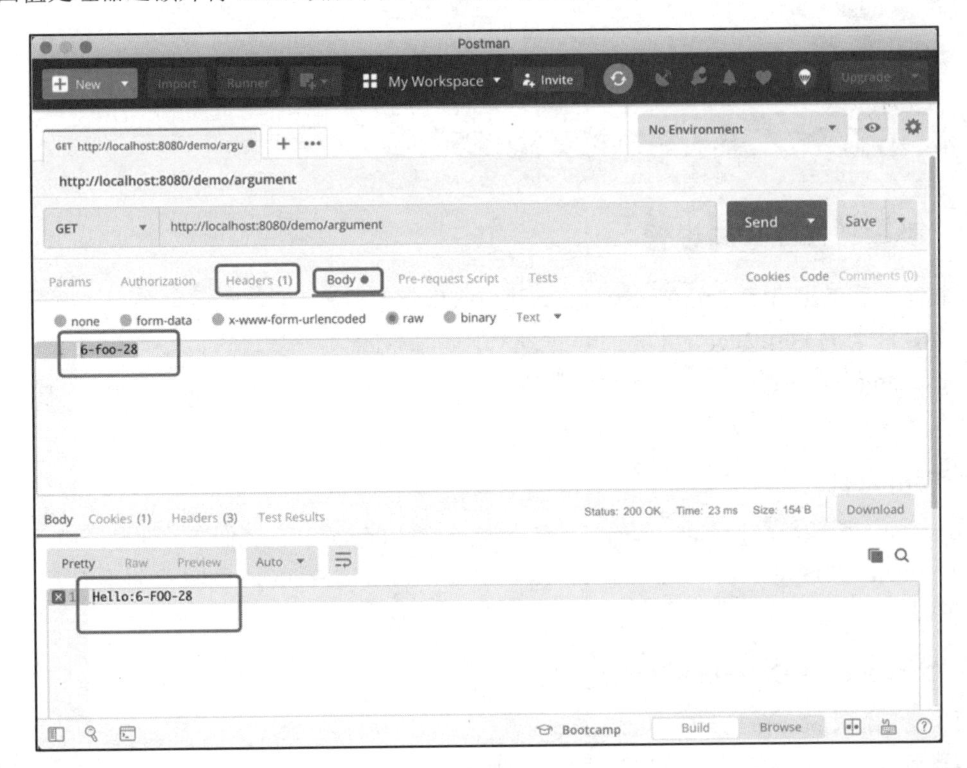

图 5-32

5.3.7 初始化数据绑定设置

在 Spring MVC 中，可以配置 ConfigurableWebBindingInitializer 的 Bean 来初始化 WebDataBinder 对象，这意味着可以用现在这种方式来设置 PropertyEditor、Formatter 和 Validator。初始化数据绑定是在控制器方法参数处理之前进行的。

注册 PropertyEditor：

```
@Bean
ConfigurableWebBindingInitializer
ConfigurableWebBindingInitializer(FormattingConversionService
mvcConversionService){ //Spring Boot 已经自动注册了 Bean，可直接注入
   ConfigurableWebBindingInitializer initializer = new
ConfigurableWebBindingInitializer();
   initializer.setPropertyEditorRegistrar(registry ->
         registry.registerCustomEditor(Person.class, new PersonEditor()));
   return initializer;
}
```

设置 Validator：

```
initializer.setValidator(new PersonValidator());
```

设置 Formatter：

```
mvcConversionService.addFormatter(new PersonFormatter());
initializer.setConversionService(mvcConversionService);
```

5.3.8 类型转换原理与设置

Spring 为我们提供了 ConversionService 接口来做类型转换，它是 Spring 类型转换系统的入口。例如，注册的 Formatter 的 FormattingConversionService 类就是它的实现类。

FormattingConversionService 支持注册 Formatter、Converter 和 AnnotationFormatterFactory，它属于配置初始化数据绑定的一部分。

1. Formatter

Formatter 用于格式化（id|name|age）对象类型（如 Person）。我们在前面已经演示了自定义的 Formatter，这里不再赘述了。在 AnnotationFormatterFactory 中我们将会使用大量的 Formatter。

2. Converter 或 ConverterFactory

（1）常用的 Converter 或 ConverterFactory。

将源对象类型 S 转换成目标对象类型 T，是 Sping 和 Spring MVC 对象之间类型转换的重要功能，若外部的对象类型是字符串“1”，则 Spring MVC 也能将它正确地转换成对应的数字类型。

下面对控制器做如下检验，检验结果如图 5-33 所示。

```
@GetMapping("/{id}/convert")
public Map<String, Object> convert(@PathVariable Long id,
                                   @RequestParam Boolean testBool,
                                   @RequestBody Person person){
    Map<String, Object> map = new HashMap<>();
    map.put("id", id);
    map.put("date", testBool);
    return map;
}
```

请求地址为 http://localhost:8080/people/8/convert?testBool=true，字符串“8”可以通过 StringToNumberConverterFactory.StringToNumber 转换成数字；同样，字符串“true”也可以通过 StringToBooleanConverter 转换成布尔值。这里放一个@RequestBody 是为了告诉我们，请求体的转换是通过 HttpMessageConverter 来做的，且初始化数据绑定是先处理的，对@ResponseBody 的控制器方法参数的处理是后处理的。

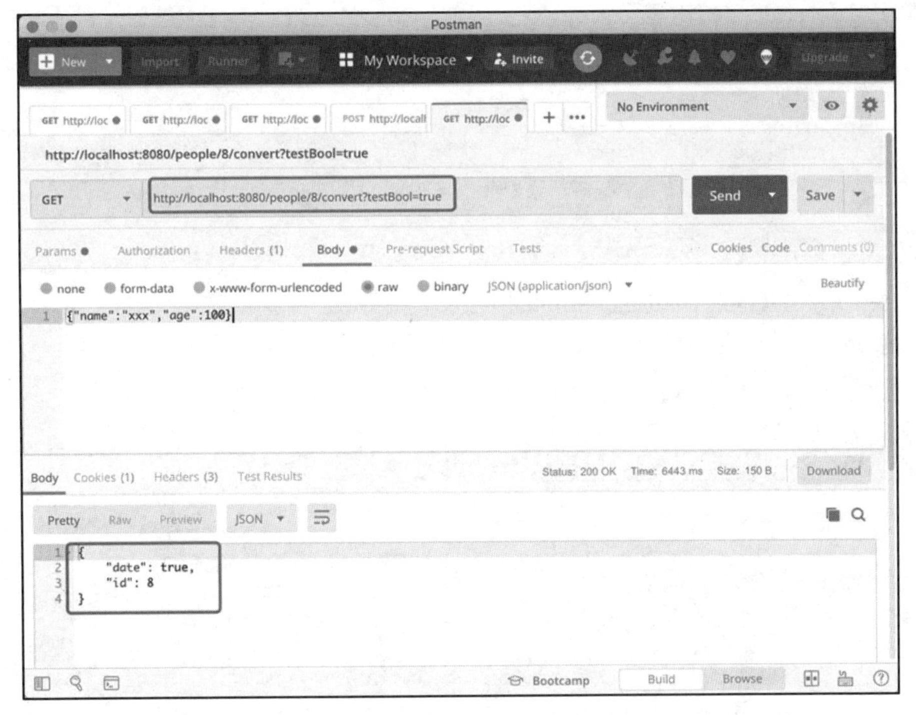

图 5-33

Spring MVC 内置了很多 Converter 或 ConverterFactory，它们主要存放在 org.springframework.core.convert 包中。

（2）自定义 Converter。

下面简单定义一个 Converter，用来演示将字符串转换成字符串长度：

```
//……
import org.springframework.core.convert.converter.Converter;
//..
public class StringToLengthConverter implements Converter<String, Integer> ,
ConditionalConverter {
    @Override
    public Integer convert(String source) {
        return Integer.valueOf(source.length());
    }

    @Override
    public boolean matches(TypeDescriptor sourceType, TypeDescriptor targetType) {
        return targetType.hasAnnotation(StrLength.class);
    }
}
```

除实现 Converter 接口进行转换外，还要实现 ConditionalConverter 接口。声明在何种情况下该 Converter 才会被调用，本例是在方法参数前注解了@StrLength。

标记注解的声明：

```
@Target({ElementType.PARAMETER})
@Retention(RetentionPolicy.RUNTIME)
@Documented
public @interface StrLength {
}
```

定义 ConfigurableWebBindingInitializer 的 Bean 配置，通过注入的 FormattingConversionService 实例的 addConverter 方法注册：

```
    @Bean
    ConfigurableWebBindingInitializer
ConfigurableWebBindingInitializer(FormattingConversionService mvcConversionService){
        ConfigurableWebBindingInitializer initializer = new
ConfigurableWebBindingInitializer();

        mvcConversionService.addConverter(new StringToLengthConverter());
        initializer.setConversionService(mvcConversionService);
        return initializer;

    }
```

当然，在 Spring MVC 下还可以通过 WebMvcConfigurer 接口的 addFormatters 来注册：

```
@Configuration
public class WebConfiguration implements WebMvcConfigurer {
    @Override
    public void addFormatters(FormatterRegistry registry) {
```

```
        registry.addConverter(new StringToLengthConverter());
    }
}
```

在 Spring Boot 下更简单，我们只需注册 Converter 的 Bean 即可：

```
@Bean
Converter stringToLengthConverter(){
    return new StringToLengthConverter();
}
```

通过控制器进行检验，结果如图 5-34 所示。

```
@GetMapping("/convert")
public Integer convert(@StrLength Integer person){
    return person;
}
```

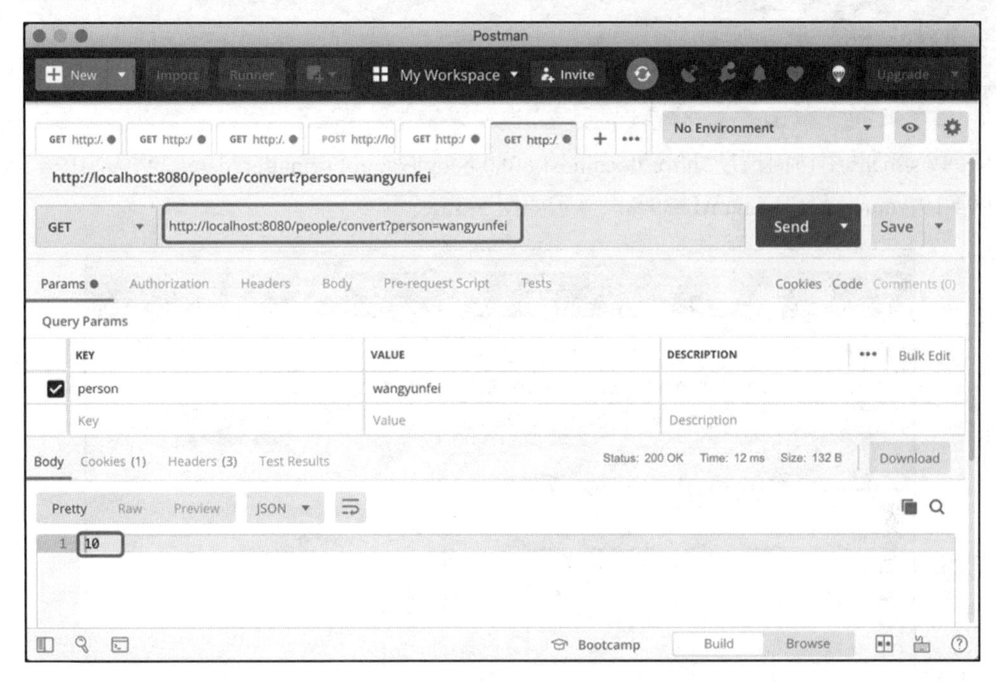

图 5-34

3. AnnotationFormatterFactory

AnnotationFormatterFactory 创建 Formatter，用来格式化标记特定注解的属性或参数值。

◎ NumberFormatAnnotationFormatterFactory 使用@NumberFormat 注解创建 NumberStyleFormatter、CurrencyStyleFormatter 和 PercentStyleFormatter 这些 Formatter 来格式化注解的属性。

◎ DateTimeFormatAnnotationFormatterFactory 使用@DateTimeFormat 注解创建 DateFormatter 的来格式化日期。

（1）@DateTimeFormat 和@NumberFormat。

下面演示一下@DateTimeFormat 和@NumberFormat 的使用，定义控制器方法：

```
@GetMapping("/annoFormatter")
public Map<String, Object> annoFormatter(@DateTimeFormat(pattern = "dd/MM/yyyy")Date
date,
                                 @DateTimeFormat(pattern = "yyyy-MM-dd") Date date1,
                                 @NumberFormat(style = NumberFormat.Style.CURRENCY) BigDecimal num,
                                 @NumberFormat(style = NumberFormat.Style.PERCENT) BigDecimal num1
                    ){
    Map<String, Object> map = new HashMap<>();
    map.put("date", date);
    map.put("date1", date1);
    map.put("number", num);
    map.put("number1", num1);
    return map;
}
```

在 Postman 中访问地址 http://localhost:8080/people/annoFormatter?date=22/04/2019&date1=2019-04-19&num=￥123&num1=39%，显示如图 5-35 所示。

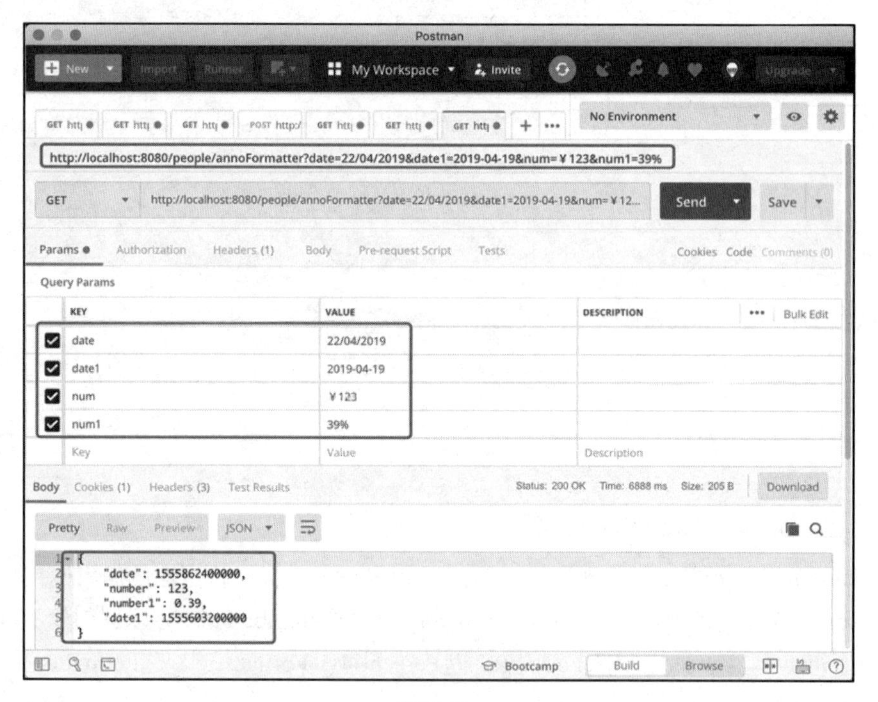

图 5-35

在 Spring Boot 中，可以通过外部配置来全局设置参数中的日期格式。

```
spring.mvc.date-format: dd/MM/yyyy
```

（2）自定义 AnnotationFormatterFactory。

同样自定义一个 AnnotationFormatterFactory 来作为演示。因为前面已经有了一个 PersonFormatter，所以再定义一个 Formatter 作为条件选择使用。

```
public class AnotherPersonFormatter implements Formatter<Person> {

   @Override
   public Person parse(String text, Locale locale) throws ParseException {
      String[] personStr = text.split("-");
      Long id = Long.valueOf(personStr[0]);
      String name = personStr[1];
      Integer age = Integer.valueOf(personStr[2]);
      return new Person(id, name, age);
   }

   @Override
   public String print(Person object, Locale locale) {
      return object.toString();
   }
}
```

定义注解作为选择时的条件：

```
@Target({ElementType.PARAMETER})
@Retention(RetentionPolicy.RUNTIME)
@Documented
public @interface PersonFormat {
   boolean style() default true; //根据 style 选择格式
}
```

定义 PersonFormatAnnotationFormatterFactory 实现 AnnotationFormatterFactory：

```
      implements AnnotationFormatterFactory<PersonFormat> {
   @Override
   public Set<Class<?>> getFieldTypes() {
      Set<Class<?>> fieldTypes = new HashSet<>(1);
      fieldTypes.add(Person.class);
      return  Collections.unmodifiableSet(fieldTypes); //支持的类型是 Person
   }

   @Override
   public Parser<?> getParser(PersonFormat annotation, Class<?> fieldType) {
      if(annotation.style())
         return new PersonFormatter(); //当 style 为 true 时使用 PersonFormatter 格式化
      else
         return new AnotherPersonFormatter(); //为 false 时使用
```

```
                                        //AnotherPersonFormatter 格式化
    }

    @Override
    public Printer<?> getPrinter(PersonFormat annotation, Class<?> fieldType) {
        if(annotation.style())
            return new PersonFormatter();
        else
            return new AnotherPersonFormatter();
    }
}
```

在 ConfigurableWebBindingInitializer Bean 的定义中，通过注入的 FormattingConversionService 实例的 addFormatterForFieldAnnotation 方法注册：

```
    @Bean
    ConfigurableWebBindingInitializer
ConfigurableWebBindingInitializer(FormattingConversionService mvcConversionService){
        ConfigurableWebBindingInitializer initializer = new
ConfigurableWebBindingInitializer();

        mvcConversionService.addFormatterForFieldAnnotation(new
PersonFormatAnnotationFormatterFactory());
        initializer.setConversionService(mvcConversionService);

        return initializer;

    }
```

在 Spring MVC 下还可以通过 WebMvcConfigurer 接口的 addFormatter 方法注册：

```
@Configuration
public class WebConfiguration implements WebMvcConfigurer {

    @Override
    public void addFormatters(FormatterRegistry registry) {
        registry.addFormatterForFieldAnnotation(new
PersonFormatAnnotationFormatterFactory());
    }
}
```

使用控制器验证，结果如图 5-36 所示。

```
@GetMapping("/customAnnoFormatter")
public Map<String, Person> customAnnoFormatter(@PersonFormat @RequestParam Person person,
                        @PersonFormat(style = false) @RequestParam Person person1){
    Map<String, Person> map = new HashMap<>();
    map.put("person", person);
    map.put("person1", person1);
```

```
    return map;
}
```

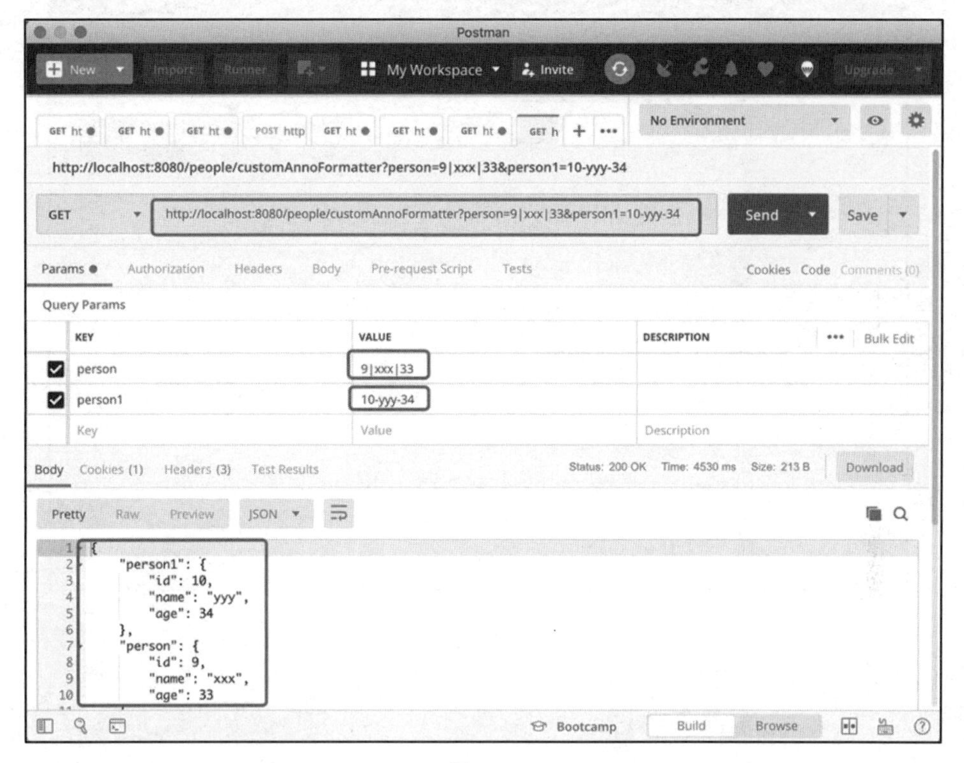

图 5-36

5.3.9　路径匹配和内容协商

在 Spring MVC 中，可以通过覆写 WebMvcConfigurer 接口的 configurePathMatch 方法来设置路径匹配。Spring MVC 提供了 PathMatchConfigurer 来进行路径匹配配置。

```
public void configurePathMatch(PathMatchConfigurer configurer) {
}
```

1. 后缀匹配

使用 PathMatchConfigurer.setUseSuffixPatternMatch(Boolean suffixPatternMatch)可以设置是否使用后缀匹配。若设置为 true，则路径/xx 和/xx.*是等效的；在 Spring Boot 下，默认是 false，如图 5-37 所示。

```
configurer.setUseSuffixPatternMatch(true);
```

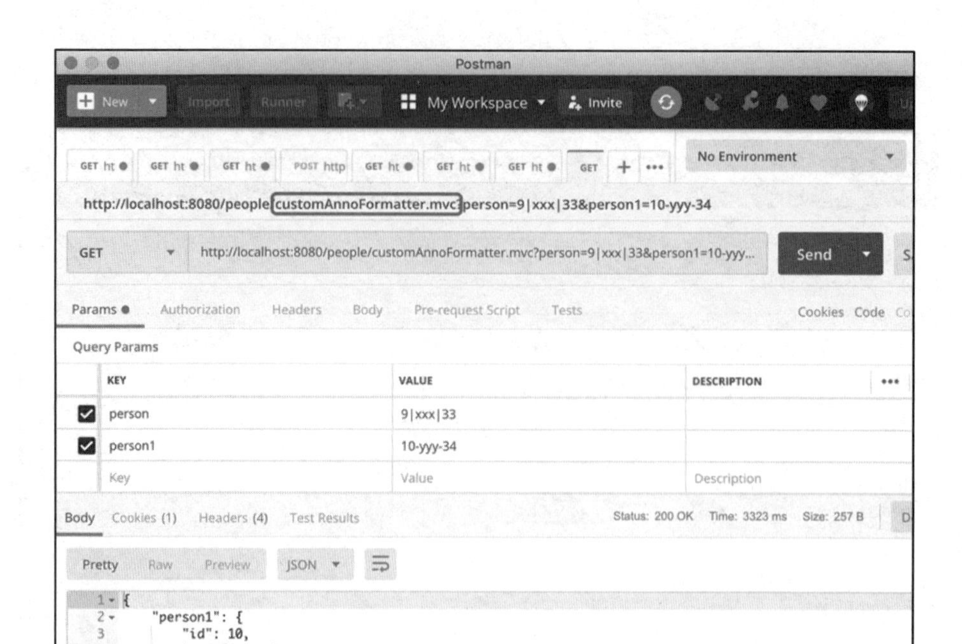

图 5-37

我们还可以在外部快捷配置 application.yml，效果与前面相同：

```
spring.mvc.pathmatch.use-suffix-pattern: true //在 Spring Boot 中，默认是 false
```

2. 斜线匹配

使用 PathMatchConfigurer.setUseTrailingSlashMatch(Boolean trailingSlashMatch)可以设置是否使用尾随斜线匹配。若设置为 true，则路径/xx 和/xx/是等效的，在 Spring MVC 中，默认是开启的，如需关闭，则设置为 false。

```
configurer.setUseTrailingSlashMatch(false);
```

在默认情况下效果如图 5-38 所示。

3. 路径前缀

通过 PathMatchConfigurer.addPathPrefix(String prefix, Predicate<Class<?>> predicate)可以设置路径前缀。

Prefix 用于设置路径的前缀，predicate 用于设置匹配生效的控制器类型。本例为对@RestController 生效，如图 5-39 所示。

```
configurer.addPathPrefix("/api",
HandlerTypePredicate.forAnnotation(RestController.class));
```

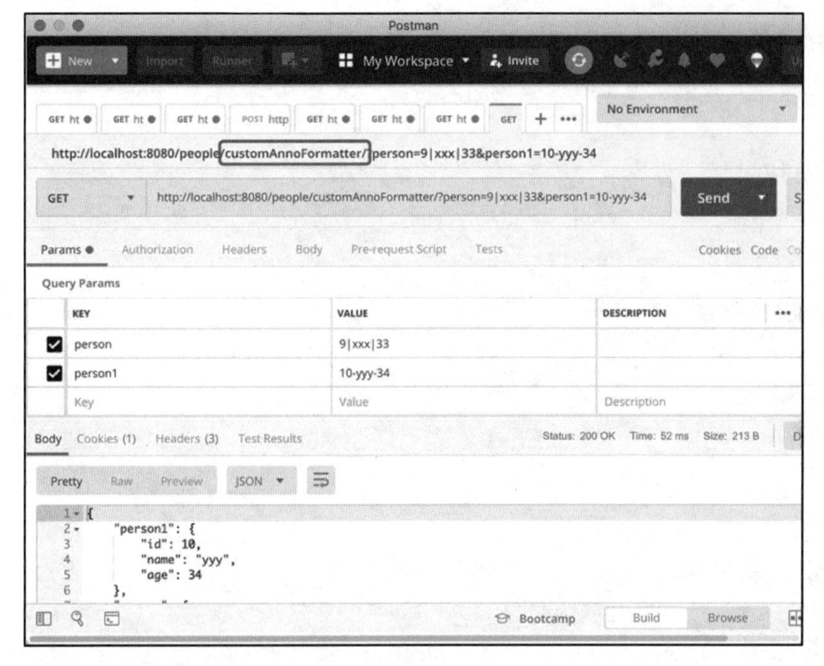

Postman
New
My Workspace
Invite
No Environment
http://localhost:8080/people/customAnnoFormatter/?person=9|xxx|33&person1=10-yyy-34
GET
http://localhost:8080/people/customAnnoFormatter/?person=9|xxx|33&person1=10-yyy-34
Send
Params
Authorization
Headers
Body
Pre-request Script
Tests
Cookies
Code
Query Params
KEY
VALUE
DESCRIPTION
person
9|xxx|33
person1
10-yyy-34
Key
Value
Description
Body
Cookies (1)
Headers (3)
Test Results
Status: 200 OK
Time: 52 ms
Size: 213 B
Pretty
Raw
Preview
JSON
{
"person1": {
"id": 10,
"name": "yyy",
"age": 34
},
Bootcamp
Build
Browse

图 5-38

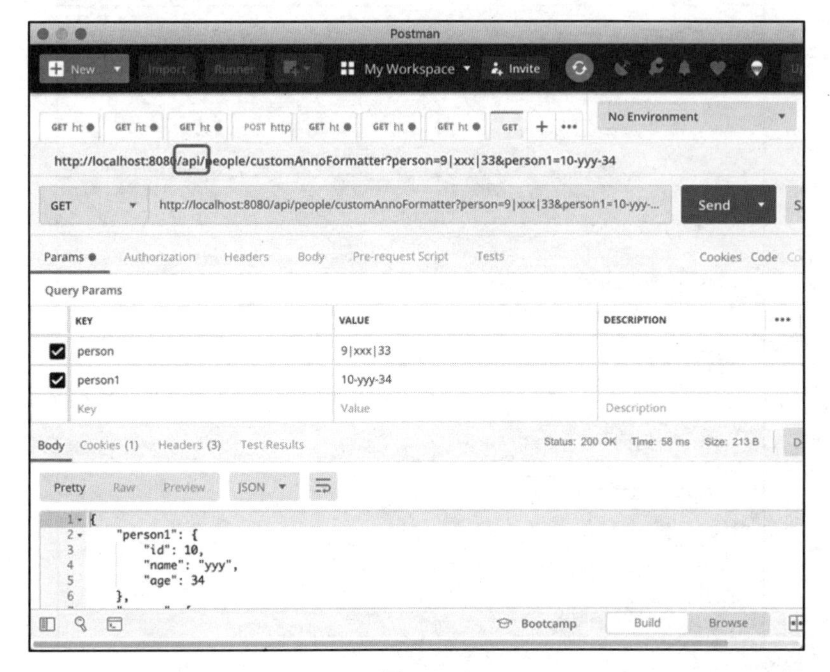

Postman
New
My Workspace
Invite
No Environment
http://localhost:8080/api/people/customAnnoFormatter?person=9|xxx|33&person1=10-yyy-34
GET
http://localhost:8080/api/people/customAnnoFormatter?person=9|xxx|33&person1=10-yyy-...
Send
Params
Authorization
Headers
Body
Pre-request Script
Tests
Cookies
Code
Query Params
KEY
VALUE
DESCRIPTION
person
9|xxx|33
person1
10-yyy-34
Key
Value
Description
Body
Cookies (1)
Headers (3)
Test Results
Status: 200 OK
Time: 58 ms
Size: 213 B
Pretty
Raw
Preview
JSON
{
"person1": {
"id": 10,
"name": "yyy",
"age": 34
},
Bootcamp
Build
Browse

图 5-39

4. 内容协商

“内容协商”指的是在请求时使用不同的条件可获得不同的返回体的类型，如 JSON、XML 等。

在 Spring Boot 中，可以在外部配置“路径+.+扩展名”获得不同类型的返回体。

```
spring.mvc.contentnegotiation.favor-path-extension: true //偏好使用路径扩展名，如
//content.json
spring.mvc.pathmatch.use-registered-suffix-pattern: true //后缀只能是已注册扩展
//名，content.xx 无效
```

Spring MVC 已经注册了媒体类型 JSON。下面添加 XML 的支持 jar 到 build.gradle。

```
dependencies {
  implementation 'org.springframework.boot:spring-boot-starter-web'
  implementation 'com.fasterxml.jackson.dataformat:jackson-dataformat-xml'
}
```

用控制器验证，结果如图 5-40 所示。

```
@GetMapping("/content")
public AnotherPerson content(@RequestBody AnotherPerson person){
   return person;
}
```

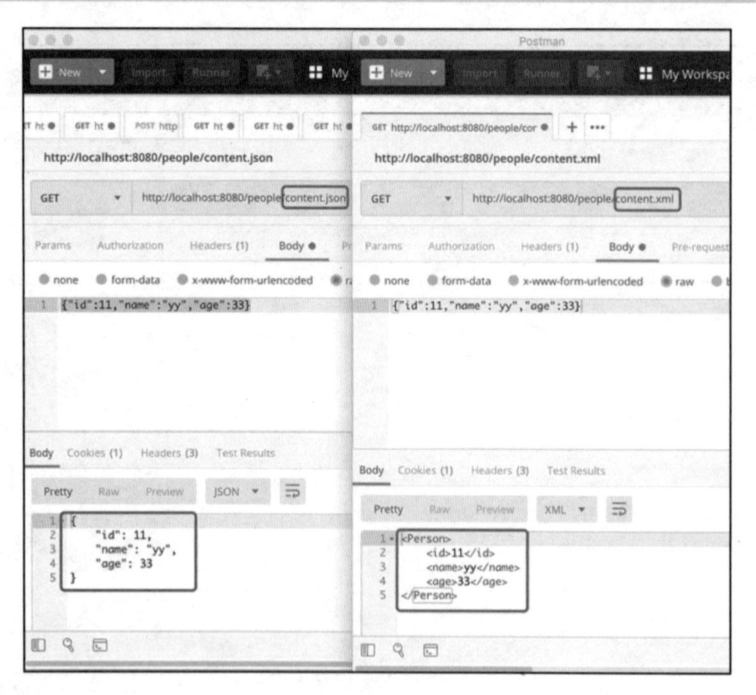

图 5-40

5. 添加新的媒体类型

媒体类型的功能是由 HttpMessageConverter 提供的，下面将未在前面注册的 AnotherPersonHttpMessageConverter 支持的媒体类型注册到内容协商。

可以通过覆写 WebMvcConfigurer 接口的 configureContentNegotiation 方法，同时使用 Spring MVC 提供的 ContentNegotiationConfigurer 来配置内容协商。

```
@Override
public void configureContentNegotiation(ContentNegotiationConfigurer configurer) {
   configurer.mediaType("ap",
         new MediaType("application","another-person", Charset.defaultCharset()));
}
```

或者通过外部配置来配置内容协商：

```
spring.mvc.contentnegotiation.media-types.ap: application/another-person
```

在 spring.mvc.contentnegotiation.media-types 后支持一个 Map 类型，ap 是 key，application/another-person 是 MediaType 类型的 value。

访问 http://localhost:8080/people/content.ap，使用 AnotherPersonHttpMessageConverter 处理并生成返回体，如图 5-41 所示。

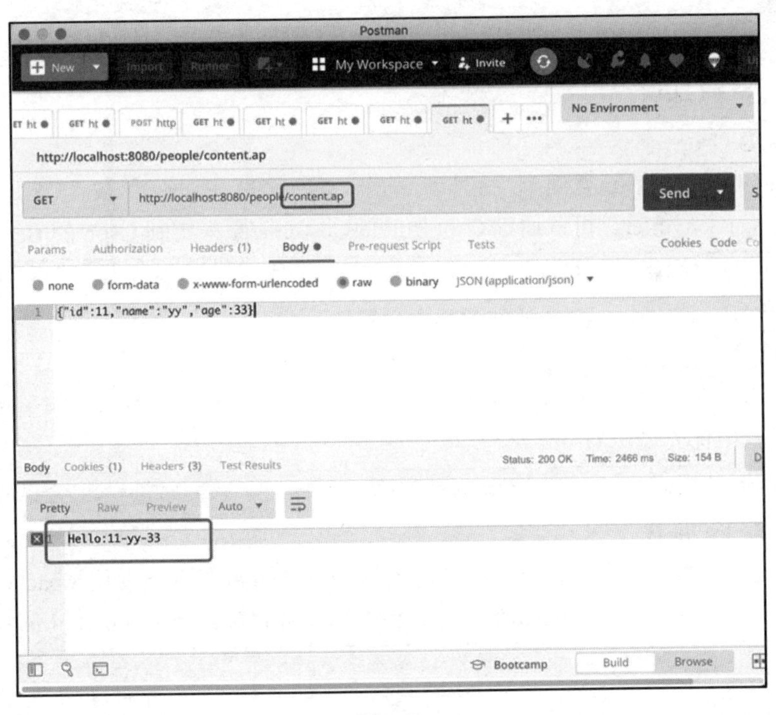

图 5-41

5.3.10 JSON

Spring MVC 提供了多种 JSON 的实现，包含 Jackson、Gson 和 JSON-B。

◎ Jackson 使用的是 MappingJackson2HttpMessageConverter，Spring Boot 提供的自动配置为 JacksonAutoConfiguration。

◎ Gson 使用的是 GsonHttpMessageConverter，Spring Boot 提供的自动配置为 GsonAutoConfiguration。

◎ JSON-B 使用的是 JsonbHttpMessageConverter，Spring Boot 提供的自动配置为 JsonbAutoConfiguration。

Spring MVC 默认使用的是 Jackson，即使用 MappingJackson2HttpMessageConverter。

1. Jackson 的 ObjectMapper 配置

在第 4 章中，我们是通过配置 ObjectMapper 的 Bean 或者实现 Jackson2ObjectMapperBuilderCustomizer 接口来配置 ObjectMapper 的示例的。

ObjectMapper 还可以通过前缀 spring.jackson.*在外部配置中配置。例如，配置缩进可使用下面的代码：

```
spring.jackson.serialization.indent_output: true
spring.jackson.date-format: dd/MM/yyyy
```

2. 切换 JSON 实现

（1）切换为 Gson。

当 Gson 的 jar 包在类路径中时，一个 Gson 的 Bean 会被配置，并自动注册 GsonHttpMessageConverter。可以通过 GsonBuilderCustomizer 来定制 Gson 的 Bean：

```
implementation 'com.google.code.gson:gson'
```

可以通过前缀 spring.gson.*在外部配置中配置。

```
spring.gson.pretty-printing: true
spring.gson.date-format: dd/MM/yyyy
```

在外部配置中优先使用 Gson。

```
spring.http.converters.preferred-json-mapper: gson
```

（2）切换为 JSON-B。

JSON-B 是 JSR-367 提供的 JSON-B API 规范，Spring Boot 偏向于使用 Apache Johnzon 作为实现。当 jar 包在类路径时，一个 Jsonb 的 Bean 会被配置，并自动注册 JsonbHttpMessageConverter。

```
implementation 'javax.json.bind:javax.json.bind-api'
implementation 'org.apache.geronimo.specs:geronimo-json_1.1_spec:1.2'
implementation 'org.apache.johnzon:johnzon-jsonb'
```

在 JSON-B 下还可以：

◎　使用@JsonbProperty 来配置 JSON 的 key。

◎　使用@JsonbDateFormat 来定制时间格式。

◎　使用@JsonbNumberFormat 来定制数字格式。

◎　使用@JsonbTransient 来忽略字段。

在外部配置中配置时优先使用 JSON-B：

```
spring.http.converters.preferred-json-mapper: jsonb
```

5.3.11 其他外部属性配置

1. 静态文件目录

在 Web 开发中有很多静态文件，如 HTML、JS、CSS、图片等。在 Spring Boot 中，下列目录都可放置静态文件：

◎　classpath:/META-INF/resources/。

◎　classpath:/resources/。

◎　classpath:/static/。

◎　classpath:/public/。

当前的 classpath 类路径是 Spring Boot 生成的 src/main/resources 目录。我们还可以通过 spring.resources.static-locations 来覆盖默认的静态目录。

```
spring.resources.static-locations:
classpath:/my-static/,classpath:/META-INF/resources/,
classpath:/resources/,classpath:/static/, classpath:/public/
```

当前所有的静态文件目录如图 5-42 所示。

可以直接访问这些静态文件，如图 5-43 所示。

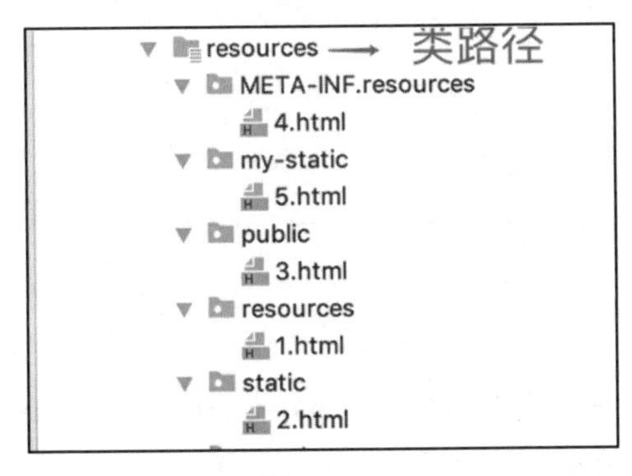

图 5-42

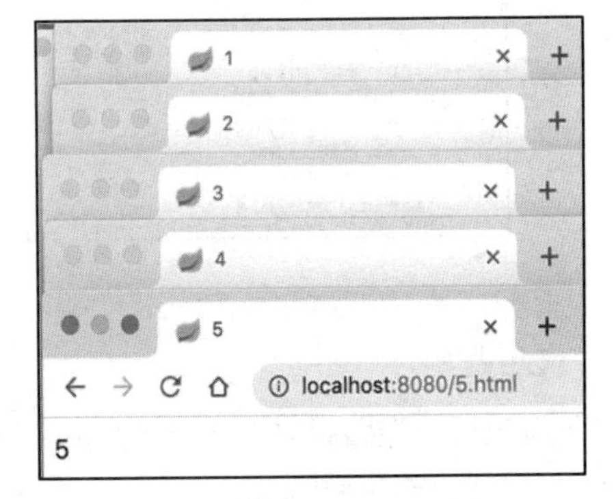

图 5-43

Spring Boot 还对 webjar（将前端开发相关库放入 jar 包，读者可访问 WebJars 官网了解更多）中的静态文件（classpath: /META-INF/resources/webjars/映射成/webjars/**）提供了支持，如添加 bootstrap 的 webjar 到 build.gradle，如图 5-44 所示。

```
implementation 'org.webjars:bootstrap:4.3.1'
```

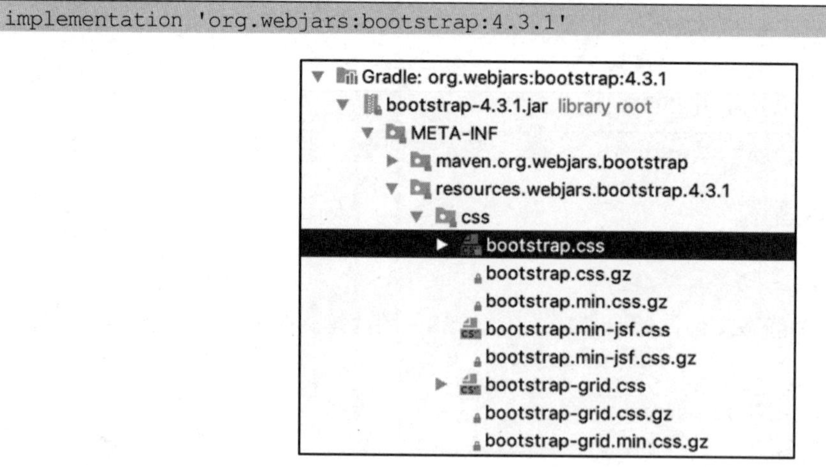

图 5-44

可以通过 http://localhost:8080/webjars/bootstrap/4.3.1/css/bootstrap.css 来访问，如图 5-45 所示。

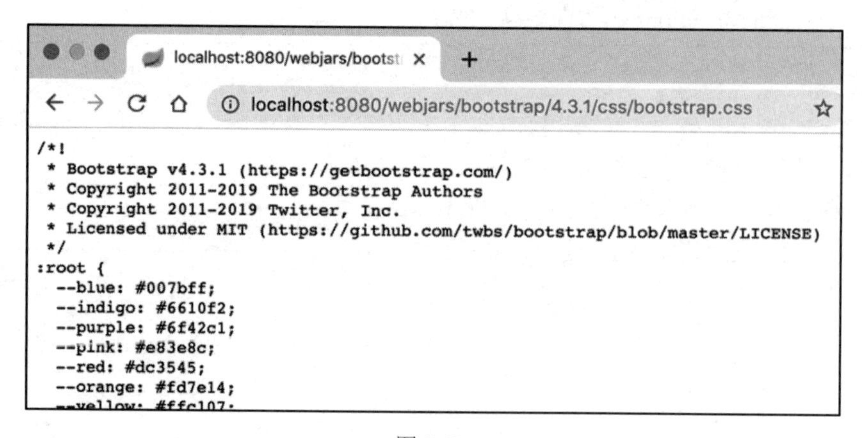

图 5-45

2. 欢迎页

当访问应用的根路径时，Spring Boot 会在静态文件目录中寻找 index.html 文件作为欢迎页。index.html 文件应放置到 static 目录下，如图 5-46 所示。

访问 http://localhost:8080/，即可打开 index.html，如图 5-47 所示。

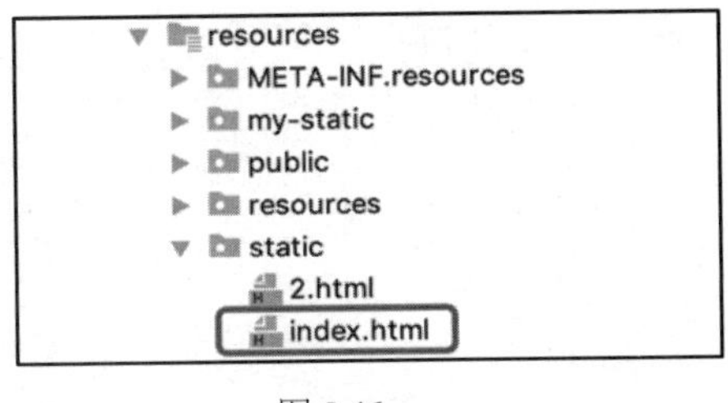

图 5-46

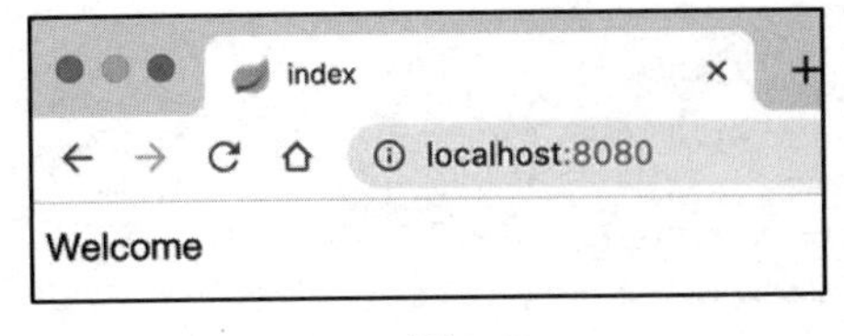

图 5-47

3. favicon ico 文件

Spring Boot 会寻找静态文件目录文件下的 favicon.ico 文件。我们在线生成（读者可自行检索此类网站）一个 favicon.ico 文件放置在 static 目录下，如图 5-48 所示。

这时浏览器 Tab 页的小图标就变成了我们定制的图标，如图 5-49 所示。

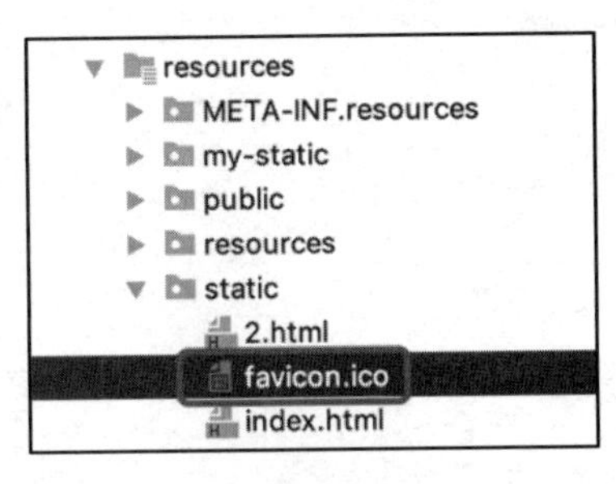

图 5-48

图 5-49

5.4　Servlet 容器

Spring Boot 使用的是内嵌式的 Servlet 容器，支持 Tomcat、Jetty 和 Undertow。

5.4.1　注册 Servlet、Filter 和 Listener

在 Spring Boot 下，可以用多种形式注册 Servlet、Filter 和 Listener，下面先定义 3 个简单的示例。

Servlet：

```
@Slf4j
@Getter
@Setter
public class CustomServlet extends HttpServlet {
   private String msg;

   public CustomServlet(String msg) {
      this.msg = msg;
   }
```

```
    @Override
    public void service(ServletRequest req, ServletResponse res) throws ServletException, IOException {
        PrintWriter writer = res.getWriter();
        writer.println("Welcome " + msg);
        writer.close();
        log.info("--CustomServlet/service---");
    }

}
```

Filter：

```
@Slf4j
public class CustomFilter extends HttpFilter {
    @Override
    public void doFilter(ServletRequest request, ServletResponse response, FilterChain chain) throws IOException, ServletException {
        log.info("--CustomFilter/doFilter---");
        chain.doFilter(request, response);
    }
}
```

Listener：

```
@Slf4j
public class CustomListener implements ServletContextListener {

    @Override
    public void contextInitialized(ServletContextEvent sce) {
        log.info("--CustomListener/contextInitialized---");
    }

    @Override
    public void contextDestroyed(ServletContextEvent sce) {
        log.info("--CustomListener/contextDestroyed---");
    }
}
```

1. 以默认形式注册 Bean

在 Spring Boot 下，可以直接注册 Servlet、Filter 和 Listener 的 Bean，默认如下。

◎ Servlet：多个 Servlet Bean 映射的路径为 Bean 的名称。

◎ Filter：默认映射到/*。

```
@Bean
public CustomServlet customServlet(){
    return new CustomServlet("Custom Servlet");
}
```

```
@Bean
public CustomServlet customServlet2(){
    return new CustomServlet("Custom Servlet2");
}

@Bean
public CustomFilter customFilter(){
    return new CustomFilter();
}

@Bean
public CustomListener customListener(){
    return new CustomListener();
}
```

执行结果如图 5-50 所示。

图 5-50

如果默认注册的 Bean 无法满足要求，则需要使用*RegistrationBean 来注册 Bean。

2. 用*RegistrationBean 注册 Bean

用*RegistrationBean 来注册 Servlet、Filter 和 Listener 的 Bean，通过 ServletRegistrationBean、FilterRegistrationBean 和 ServletListenerRegistrationBean 的方法对 Servlet、Filter 和 Listener 进行完全控制。

```
@Bean
public ServletRegistrationBean servletRegistrationBean(){
    ServletRegistrationBean registrationBean = new ServletRegistrationBean();
    registrationBean.setServlet(new CustomServlet("Custom Servlet by
ServletRegistrationBean"));
    registrationBean.setUrlMappings(Arrays.asList("/custom-servlet"));
    return registrationBean;
}

@Bean
public FilterRegistrationBean filterRegistrationBean(){
```

```
    FilterRegistrationBean registrationBean = new FilterRegistrationBean();
    registrationBean.setFilter(new CustomFilter());
    registrationBean.setUrlPatterns(Arrays.asList("/*"));
    return registrationBean;
}

@Bean
public ServletListenerRegistrationBean servletListenerRegistrationBean(){
    ServletListenerRegistrationBean registrationBean = new
ServletListenerRegistrationBean();
    registrationBean.setListener(new CustomListener());
    return registrationBean;
}
```

执行结果如图 5-51 所示。

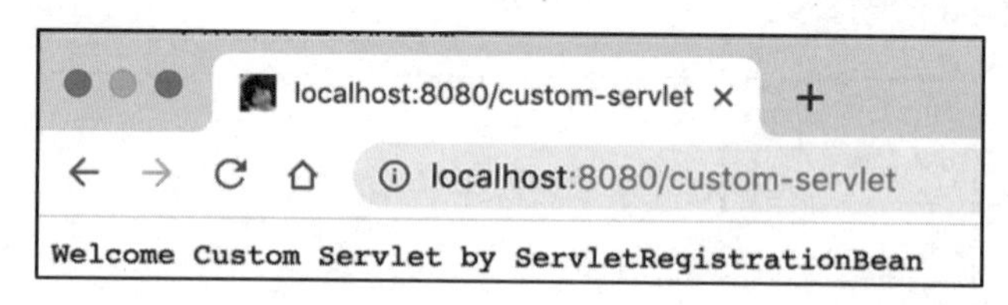

图 5-51

3. 使用 Servlet 注解注册 Bean

我们可以使用 Servlet 注解@WebServlet、@WebFilter 和@WebListener 来注册 Servlet、Filter 和 Listener 的 Bean，即通过@ServletComponentScan 将它们注册成 Bean。

Servlet：

```
@WebServlet({"/web-servlet"})
public class CustomServlet extends HttpServlet {
//……
}
```

Filter：

```
@WebFilter({"/*"})
public class CustomFilter extends HttpFilter {
//......
}
```

Listener：

```
@WebListener
public class CustomListener implements ServletContextListener {
//......
}
```

通过@ServletComponentScan 扫描注解注册：

```
@Configuration
@ServletComponentScan({"top.wisely.learningspringmvc.servlet"})
public class WebConfiguration implements WebMvcConfigurer {
    @Bean
    String msg(){
        return "Custom Servlet by Sevlet annotation";
    }
}
```

执行结果如图 5-52 所示。

图 5-52

4. ServletContextInitializer

在 Servlet 3.0 之前，Servlet 容器初始化是通过 web.xml 实现的；在 Servlet 3.0 中，Servlet 容器初始化是通过 javax.servlet.ServletContainerInitializer 接口的实现类实现的；在 Spring MVC 中，Servlet 容器初始化是通过实现 org.springframework.web.WebApplicationInitializer 接口完成的；在 Spring Boot 的内嵌式 Servlet 容器中，Servlet 容器初始化是通过定义 org.springframework.boot.web.servlet.ServletContextInitializer 接口的实现的 Bean 完成的。

通过覆写 ServletContextInitializer 的方法 onStart 即可定制初始化 Servlet 容器，而 ServletContextInitializer 是一个函数接口，所以定义 ServletContextInitializer 十分简捷。

```
@Bean
ServletContextInitializer servletContextInitializer(){
  return servletContext -> {
     servletContext.addServlet("new servlet", new CustomServlet("Custom Servlet by
ServletContextInitializer"))
            .addMapping("/new-servlet");
     servletContext.addFilter("new filter", new CustomFilter())
            .addMappingForUrlPatterns(EnumSet.of(DispatcherType.REQUEST), true, "/*");
     servletContext.addListener(new CustomListener());

  };
}
```

执行结果如图 5-53 所示。

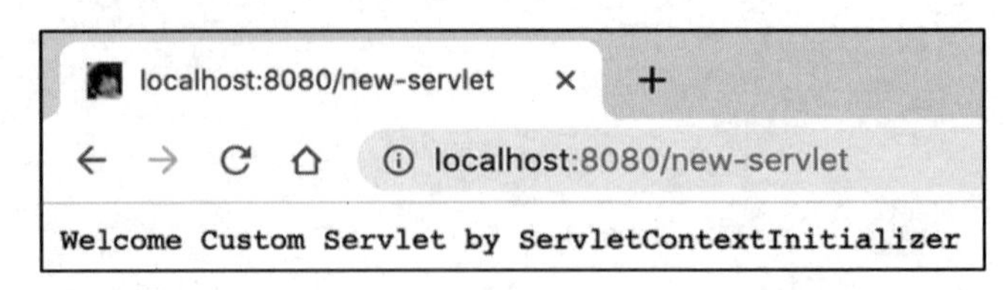

图 5-53

5.4.2 配置 Servlet 容器

1. 外部属性配置

◎ 网络配置：server.port、server.address 等。

◎ 用户会话配置：server.servlet.session.*。

◎ 错误配置：server.error.*。

◎ HTTP 压缩：server.compression.* ，支持 HTML、XML、CSS、JS、JSON 和 text。默认是关闭的，可用 server.compression.enabled: true 开启。

◎ SSL 配置：server.ssl.*。

◎ Tomcat 专有配置：server.tomcat.*。

◎ Jetty 专有配置：server.jetty.*。

◎ Undertow 专有配置：server.undertow.*。

◎ Servlet 相关配置：server.servlet.*。

更多关于 Servlet 容器外部属性配置的内容可参考：*Spring Boot Reference Documentation* 中的“common application properties”部分。

2. WebServerFactoryCustomizer

Spring Boot 提供了 WebServerFactoryCustomizer 来定制 Servlet 容器：

```
@Component
public class CustomWebServerFactoryCustomizer implements
WebServerFactoryCustomizer<ConfigurableServletWebServerFactory> {

   @Override
   public void customize(ConfigurableServletWebServerFactory server) {
      server.setPort(9000);
   }
}
```

3. ConfigurableServletWebServerFactory

在默认情况下，Spring Boot 通过 ServletWebServerApplicationContext 自动寻找配置 ServletWebServerFactory 的 Bean；Tomcat 的为 TomcatServletWebServerFactory，Jetty 的为 ServletWebServerFactory，Undertow 的为 UndertowServletWebServerFactory。

我们可以通过定义 ConfigurableServletWebServerFactory 的 Bean 来定制 Servlet 容器：

```
@Bean
ConfigurableServletWebServerFactory customWebServerFactory(){
    TomcatServletWebServerFactory server = new TomcatServletWebServerFactory();
    server.setPort(9000);
    return server;
}
```

4. 切换容器

Spring Boot 为不同的容器提供了不同的 starter，当切换容器时，只需在 build.gradle 中切换 starter 依赖即可。

◎ Tomcat：spring-boot-starter-tomcat，默认为 Tomcat。

◎ Jetty：spring-boot-starter-jetty。

◎ Undertow：spring-boot-starter-undertow。

切换为 Jetty：

```
implementation ('org.springframework.boot:spring-boot-starter-web'){
  exclude module: 'spring-boot-starter-tomcat'
}
implementation 'org.springframework.boot:spring-boot-starter-jetty'
```

切换为 Undertow：

```
implementation ('org.springframework.boot:spring-boot-starter-web'){
  exclude module: 'spring-boot-starter-tomcat'
}
implementation 'org.springframework.boot:spring-boot-starter-undertow'
```

5. 配置 SSL

SSL 的全称为 Secure Sockets Layer，它是建立 Web 服务器和客户端（浏览器）之间安全连接的标准安全技术，它保证了 Web 服务器和客户端之间传递数据的私有性和完整性。

为了创建 SSL 连接，Web 服务器需要一个 SSL 证书。一般情况下，需要购买证书，本例通过 Java 的工具 keytool 生成一个证书，如图 5-54 所示。

```
$ keytool -genkey -keyalg RSA -alias wisely -keystore keystore.jks -storepass
pass1234 -validity 4000 -keysize 2048
```

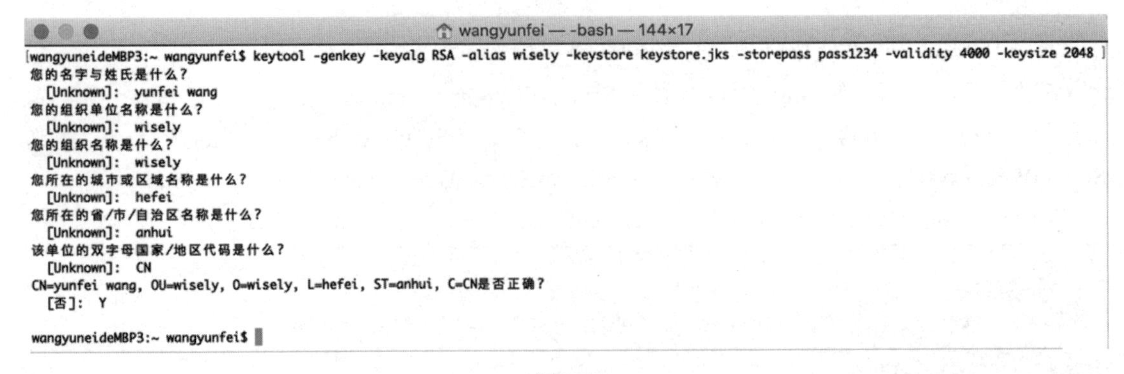

图 5-54

执行上面的命令后，在目录中会生成一个文件 keystore.jks。将此文件拷贝到 src/main/resources 目录下，如图 5-55 所示。

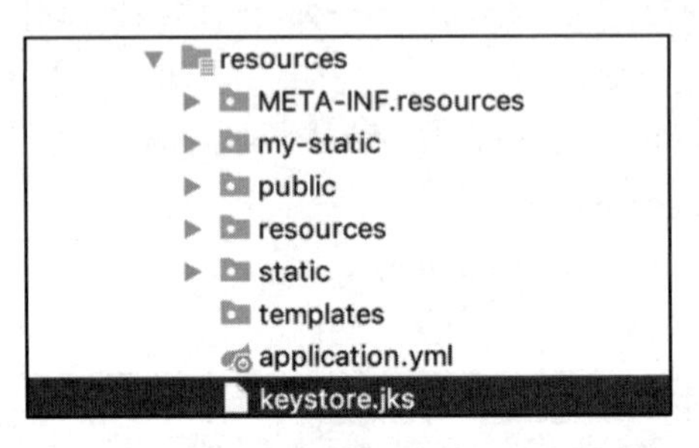

图 5-55

在外部文件 application.yml 中配置 SSL 连接。

```
server:
  ssl:
    key-store: classpath:keystore.jks
    key-store-password: pass1234
    key-store-type: JKS
    key-alias: wisely
    key-password: pass1234
```

启动应用，访问 https://localhost:8080/，如图 5-56 所示。

在开发中我们可以忽略警告，在 Chrome 浏览器地址栏输入 chrome://flags/#allow-insecure-localhost，设置为 Enabled 并重启浏览器，如图 5-57 所示。

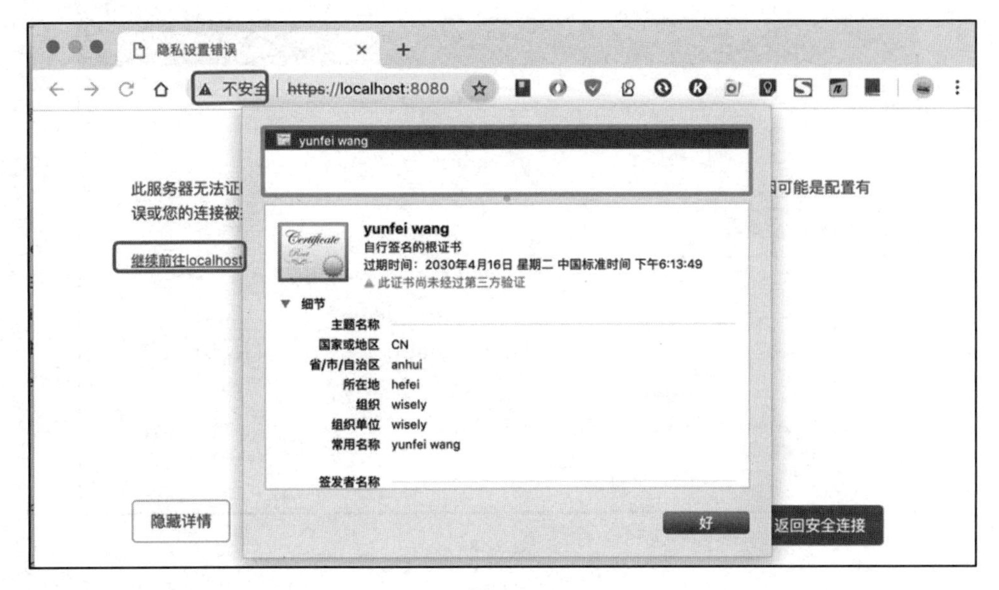

图 5-56

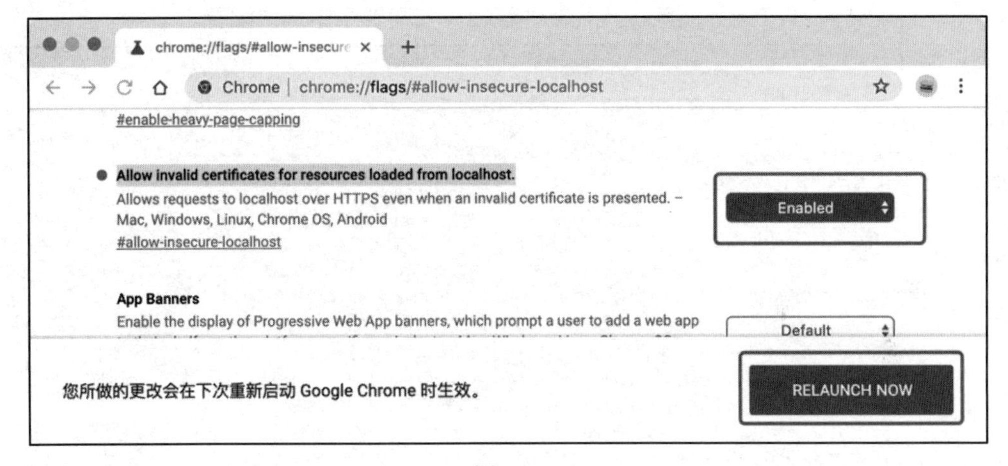

图 5-57

在 Postman 下，可以通过单击 Settings → General → SSL certificate verification 来关闭 SSL 连接，如图 5-58 所示。

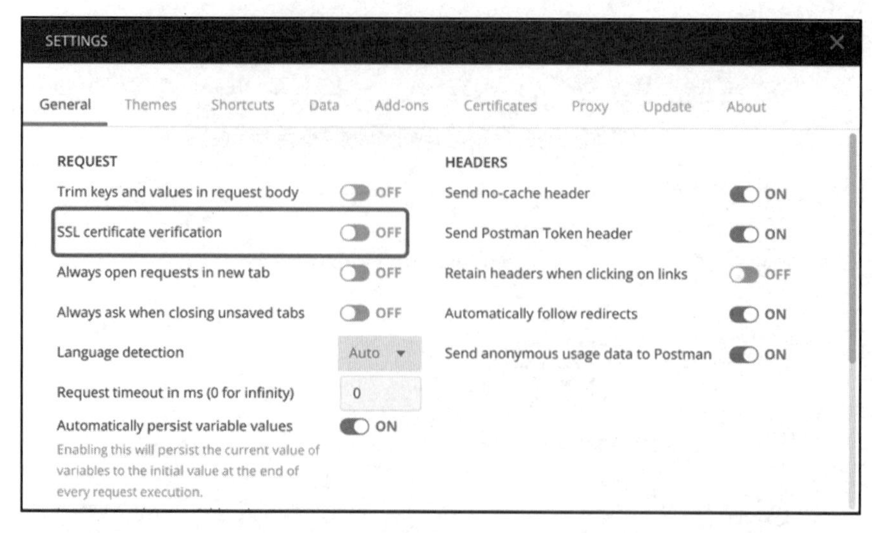

图 5-58

通过配置 ConfigurableServletWebServerFactory 和 Connector 的 Bean，可以帮助我们在访问 HTTP 时，自动转向 HTTPS。注意，此处要注释掉外部配置中的端口配置。

```
#server.port: 8080
```

```
@Bean
ConfigurableServletWebServerFactory customWebServerFactory(){
    TomcatServletWebServerFactory server = new TomcatServletWebServerFactory(){
        @Override
        protected void postProcessContext(Context context) {
            SecurityConstraint securityConstraint = new SecurityConstraint();
            securityConstraint.setUserConstraint("CONFIDENTIAL");
            SecurityCollection collection = new SecurityCollection();
            collection.addPattern("/*");
            securityConstraint.addCollection(collection);
            context.addConstraint(securityConstraint);
        }
    };
    server.setPort(8443);
    server.addAdditionalTomcatConnectors(redirectConnector());
    return server;
}

@Bean
Connector redirectConnector() {
    Connector connector = new Connector(TomcatServletWebServerFactory.DEFAULT_PROTOCOL);
    connector.setScheme("http");
    connector.setPort(8080);
    connector.setSecure(false);
```

```
    connector.setRedirectPort(8443);
    return connector;
}
```

当访问 http://localhost:8080 时，自动转向 https://localhost:8443/。

6. 配置 HTTP/2

HTTP/2 是新版本的网络协议，它提供了服务端推送资源到客户端的功能。

在开启 HTTP/2 时，必须要使用 SSL 连接，然后通过外部配置开启 HTTP/2。

```
server.http2.enabled: true
```

Servlet 4 容器支持 HTTP/2

◎ Tomcat 配置：
  - Tomcat 9.0.x +。
  - JDK 9 +。

◎ Jetty 配置：
  - Jetty 9.4.8 +。
  - org.eclipse.jetty:jetty-alpn-conscrypt-server 依赖。
  - org.eclipse.jetty.http2:http2-server 依赖。

◎ Undertow 配置：
  - Undertow 1.4.0 +。
  - JDK 8 +。

当前版本的 Postman 还不支持 HTTP/2，使用 Chrome 查看，效果如图 5-59 所示。

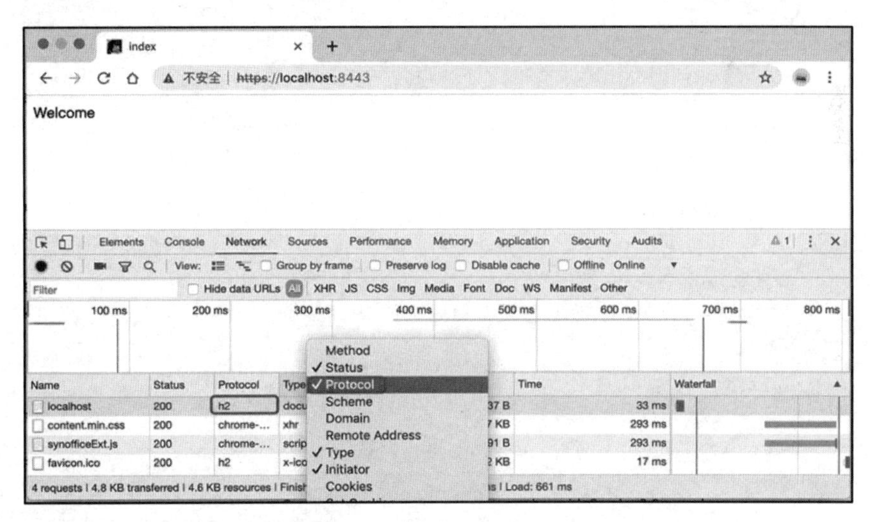

图 5-59

5.5 异步请求

当一个 Web 请求到达 Servlet 容器时，Servlet 线程将被阻碍，直至 Web 请求处理完成。这种阻碍称为同步。我们需要通过一些方式来降低 Servlet 容器的负载。

5.5.1 Servlet 3.0 异步返回

Spring MVC 支持 Servlet 3.0 提供的异步返回，它可将 Web 请求处理放在另外一个线程中（如线程 B），由线程 B 来处理耗时的任务。让 Servlet 线程去处理其他 Web 请求，而不是阻碍 Servlet 线程。此时的返回（response）是打开的。当处理完成后，Servlet 将处理结果返回给客户端。

我们无须与 Servlet API 进行交互，Spring MVC 支持 DeferredResult 和 Callable 作为控制器方法返回值，提供单个值的异步返回。

Callable：当异步运算结束后返回结果。

DeferredResult：由另外一个线程在稍后的异步运算后返回结果。

1. Callable

只需将控制器方法的返回值设置为 java.util.concurrent.Callable 即可，它会使用 Spring MVC 提供的 TaskExecutor（SimpleAsyncTaskExecutor，线程名以 MvcAsync 开头）来控制线程。

定义 TaskService，用来演示处理：

```
@Service
@Slf4j
public class TaskService {

   public String callableTask() throws InterruptedException{
      Thread.sleep(5000);
      log.info("+++++Callable 数据返回+++++");
      return "result from Callable";
   }
}
```

定义控制器，用来演示异步任务：

```
@RestController
@RequestMapping("/async")
@Slf4j
public class AsyncController {

   private TaskService taskService;

   public AsyncController(TaskService taskService) {
      this.taskService = taskService;
   }
```

```
@GetMapping("/callable")
public Callable<String> callable(){
   log.info("+++++servlet 线程已释放+++++"); // a
   return taskService::callableTask; // b
}
```

a. 控制器方法直接将 Servlet 线程释放。

b. Callable 是函数接口，不接收参数，只产生返回值，属于 Supplier。callableTask 方法符合 Callable 的定义需要。当然，返回值也可以用 Lambda 表达式表示。

```
return () -> {
   Thread.sleep(5000);
   log.info("+++++Callable 数据返回+++++");
   return "result from Callable";
};
```

在 Chrome 地址栏中输入 https://localhost:8443/async/callable，打开 Chrome 控制台，如图 5-60 所示。

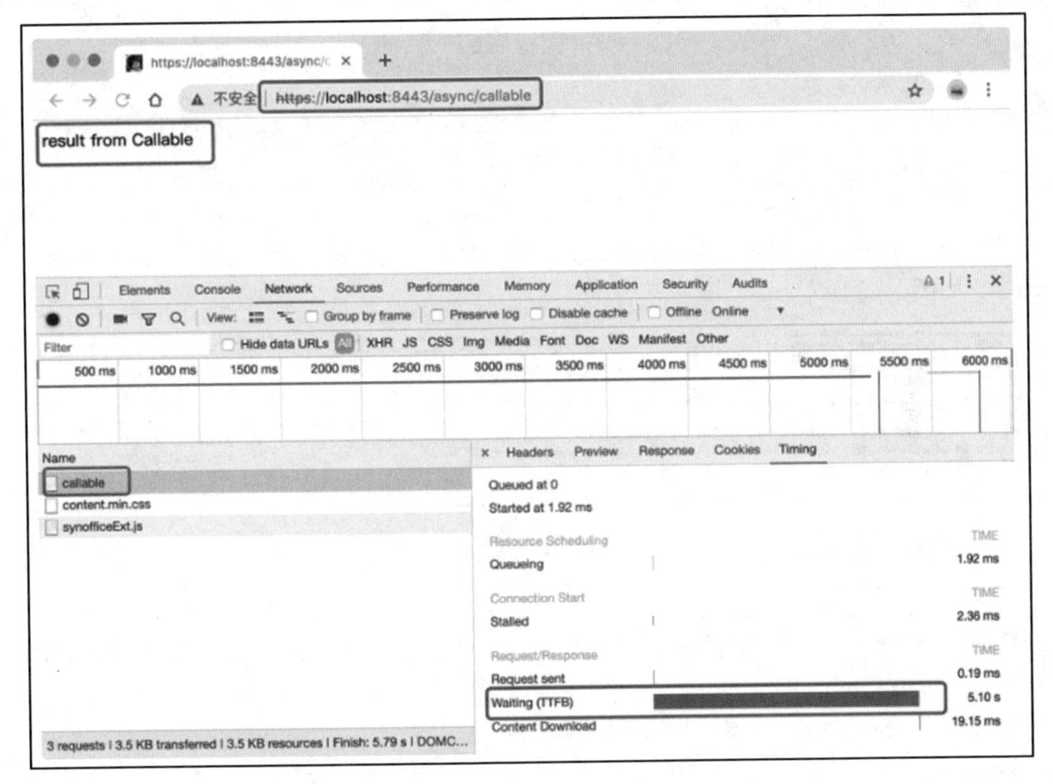

图 5-60

这时 IDE 控制台提示：需要使用定制线程执行器，如图 5-61 所示。

```
2019-05-05 18:46:34.265  INFO 61735 --- [nio-8443-exec-4] t.w.l.interceptor.CustomInterceptor      : preHandle处理中...
2019-05-05 18:46:34.266  INFO 61735 --- [nio-8443-exec-4] t.w.l.controller.AsyncController         : +++++servlet线程已释放+++++
2019-05-05 18:46:34.272  WARN 61735 --- [nio-8443-exec-4] o.s.w.c.request.async.WebAsyncManager    :
!!!
An Executor is required to handle java.util.concurrent.Callable return values.
Please, configure a TaskExecutor in the MVC config under "async support".
The SimpleAsyncTaskExecutor currently in use is not suitable under load.
-------------------------------
Request URI: '/async/callable'
!!!
2019-05-05 18:46:39.289  INFO 61735 --- [      MvcAsync1] t.w.l.service.TaskService                : +++++Callable数据返回+++++
2019-05-05 18:46:39.291  INFO 61735 --- [nio-8443-exec-6] t.w.l.interceptor.CustomInterceptor      : preHandle处理中...
2019-05-05 18:46:39.361  INFO 61735 --- [nio-8443-exec-6] t.w.l.interceptor.CustomInterceptor      : postHandle处理中...
```

图 5-61

通过覆写 WebMvcConfigurer 接口的 configureAsyncSupport 方法，利用方法注入的 AsyncSupportConfigurer 实例来配置 TaskExecutor。

```
@Override
public void configureAsyncSupport(AsyncSupportConfigurer configurer) {
      configurer.setTaskExecutor(callableTaskExecutor());
}

@Bean
public ThreadPoolTaskExecutor callableTaskExecutor(){
   ThreadPoolTaskExecutor taskExecutor = new ThreadPoolTaskExecutor();
   taskExecutor.setThreadNamePrefix("callable-task-");
   return taskExecutor;
}
```

再次访问 https://localhost:8443/async/callable，IDE 控制台的线程已经切换成我们定制的线程池了，如图 5-62 所示。

```
2019-05-05 18:55:25.679  INFO 61775 --- [nio-8443-exec-4] t.w.l.servlet.CustomFilter               : --CustomFilter/doFilter---
2019-05-05 18:55:29.242  INFO 61775 --- [nio-8443-exec-5] t.w.l.servlet.CustomFilter               : --CustomFilter/doFilter---
2019-05-05 18:55:29.243  INFO 61775 --- [nio-8443-exec-5] t.w.l.interceptor.CustomInterceptor      : preHandle处理中...
2019-05-05 18:55:29.245  INFO 61775 --- [nio-8443-exec-5] t.w.l.controller.AsyncController         : +++++servlet线程已释放+++++
2019-05-05 18:55:34.266  INFO 61775 --- [callable-task-1] t.w.l.service.TaskService                : +++++Callable数据返回+++++
2019-05-05 18:55:34.268  INFO 61775 --- [nio-8443-exec-7] t.w.l.interceptor.CustomInterceptor      : preHandle处理中...
2019-05-05 18:55:34.286  INFO 61775 --- [nio-8443-exec-7] t.w.l.interceptor.CustomInterceptor      : postHandle处理中...
2019-05-05 18:55:34.287  INFO 61775 --- [nio-8443-exec-7] t.w.l.interceptor.CustomInterceptor      : 请求处理时间为：18毫秒
```

图 5-62

2. DeferredResult

DeferredResult 的结果是由另外一个线程再稍后计算后返回给客户端的。DeferredResult 的线程执行器可由 CompletableFuture 来定制。

同样，在 TaskService 中编写我们的处理代码：

```
@Service
@Slf4j
public class TaskService {
```

```
    public String deferredTask() {
        log.info("+++++DeferredResult 数据返回+++++");
        return "result from DeferredResult";
    }
}
```

在 WebConfiguration 中定义 TaskExecutor：

```
@Bean
public ThreadPoolTaskExecutor deferredTaskExecutor(){
    ThreadPoolTaskExecutor taskExecutor = new ThreadPoolTaskExecutor();
    taskExecutor.setThreadNamePrefix("deferred-task-");
    return taskExecutor;
}
```

定义要演示的控制器：

```
@RestController
@RequestMapping("/async")
@Slf4j
public class AsyncController {

    private static Map<Long, DeferredResult<String>> deferredResultMap = new HashMap<>();
    @Autowired
    private TaskExecutor deferredTaskExecutor;

    private TaskService taskService;

    public AsyncController(TaskService taskService) {
        this.taskService = taskService;
    }

    @GetMapping("/{id}/deferred") // a
    public DeferredResult<String> deferred(@PathVariable Long id) throws
InterruptedException {
        log.info("+++++servlet 线程已释放+++++");
        DeferredResult<String> deferredResult = new DeferredResult();//b
        deferredResultMap.put(id, deferredResult); // c
        return deferredResult;
    }

    @GetMapping("/{id}/invoke-deferred") //d
    public void  invokeDeferred(@PathVariable Long id){
        DeferredResult<String> deferredResult = deferredResultMap.get(id); //e
        CompletableFuture.supplyAsync(taskService::deferredTask, deferredTaskExecutor)
                .whenCompleteAsync((result,throwable) ->
deferredResult.setResult(result));//f
        deferredResultMap.remove(id);
    }
}
```

a. 定义此控制器方法，使其可以接收 Web 请求。

b. 定义 DeferredResult<String>变量的直接返回，此时 Servlet 线程已释放。

c. 将当前的 DeferredResult<String>放入 Map 中，这个 Map 是类的静态变量，我们可以在其他地方取出 Map 中的内容，Map 中的数据的 key 为 id。

d. 定义此控制器方法，用来演示在稍后另外一个运算的线程。

e. 从 Map 中获取 DeferredResult<String>变量。

f. 使用 CompletableFuture 指定运算方法、taskExecutor，当计算完成后通过 deferredResult.setResult(result)返回结果。

首先用浏览器访问 https://localhost:8443/async/1/deferred，此时请求一直处于等待状态；然后在访问 https://localhost:8443/async/1/invoke-deferred 时，在稍后的另外的线程中计算并返回结果，该结果会显示在 https://localhost:8443/async/1/deferred 的页面上，如图 5-63 所示。

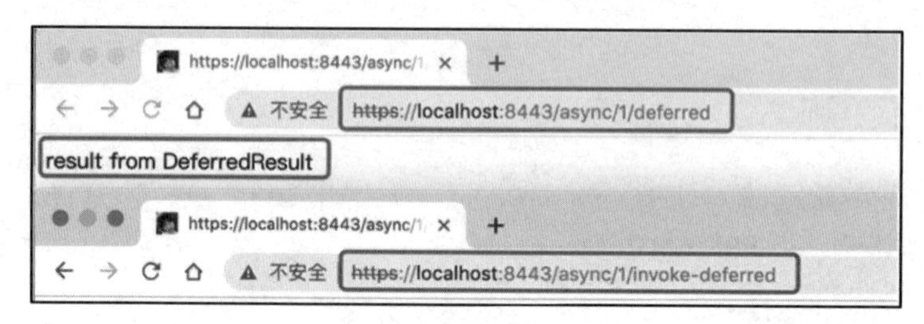

图 5-63

IDE 控制台显示如图 5-64 所示。

```
- [nio-8443-exec-7] t.w.l.interceptor.CustomInterceptor      : preHandle处理中...
- [nio-8443-exec-7] t.w.l.interceptor.CustomInterceptor      : postHandle处理中...
- [nio-8443-exec-7] t.w.l.interceptor.CustomInterceptor      : 请求处理时间为: 18毫秒
- [nio-8443-exec-8] t.w.l.servlet.CustomFilter               : --CustomFilter/doFilter---
- [nio-8443-exec-8] t.w.l.interceptor.CustomInterceptor      : preHandle处理中...
- [nio-8443-exec-8] t.w.l.controller.AsyncController         : +++++servlet线程已释放+++++
- [nio-8443-exec-2] t.w.l.servlet.CustomFilter               : --CustomFilter/doFilter---
- [nio-8443-exec-2] t.w.l.interceptor.CustomInterceptor      : preHandle处理中...
- [deferred-task-1] t.w.l.service.TaskService                : +++++DeferredResult数据返回+++++
- [nio-8443-exec-3] t.w.l.interceptor.CustomInterceptor      : preHandle处理中...
```

图 5-64

5.5.2 HTTP Streaming

DeferredResult 和 Callable 只能异步返回单个值；如果想要异步返回多个值，则可以用 HTTP Streaming 来实现。

HTTP Streaming 是一种推送形式的数据传输技术，它通过无限期开放的 HTTP 连接让 Web 服务器（Tomcat）能持续向客户端（浏览器）传送数据。

在 HTTP Streaming 下，Web 服务器会“握住”客户端的请求并让返回保持开放，这样服务器可以通过返回一直发送数据。这个“请求—返回”的通道会一直开启，除非显式地要求关闭。

1. ResponseBodyEmitter

可以使用 ResponseBodyEmitter 作为返回值返回数据对象的流。可以用 ResponseEntity 来包装 ResponseBodyEmitter，这样即可定制返回的状态和头信息。

定义本例所需的控制器方法，这个例子和 DeferredResult 类似。

```
@RestController
@RequestMapping("/async")
@Slf4j
public class AsyncController {

   private static Map<Long, ResponseBodyEmitter> responseBodyEmitterMap = new
HashMap<>();

   @GetMapping(value = "/{id}/rbe") //a
   public ResponseEntity<ResponseBodyEmitter> responseBodyEmitter(@PathVariable Long
id){
       ResponseBodyEmitter emitter = new ResponseBodyEmitter();
       responseBodyEmitterMap.put(id, emitter);
       return ResponseEntity
               .ok()
               .contentType(MediaType.TEXT_HTML)
               .body(emitter); //b

   }

   @GetMapping("/{id}/invoke-rbe") //c
   public void invokeResponseBodyEmitter(@PathVariable Long id) throws Exception {
      ResponseBodyEmitter emitter = responseBodyEmitterMap.get(id); //d
      emitter.send("Hello World", MediaType.TEXT_PLAIN); //e
      Thread.sleep(1000);
      emitter.send(new Person(1l, "wyf", 35), MediaType.APPLICATION_JSON);
      Thread.sleep(1000);
      emitter.send(new Person(2l,"foo", 40), MediaType.APPLICATION_XML);
      Thread.sleep(1000);
      emitter.send(new AnotherPerson(3l,"bar", 50), new
MediaType("application","another-person"));
   }

   @GetMapping("/{id}/close-rbe") //f
   public void closeResponseBodyEmitter(@PathVariable Long id) throws IOException {
      ResponseBodyEmitter emitter = responseBodyEmitterMap.get(id);
      emitter.complete(); //g
      responseBodyEmitterMap.remove(id);
   }
}
```

a. 定义此控制器接收 Web 请求。

b. 使用 ResponseEntity 包装 ResponseBodyEmitter 对象作为返回值，状态为 OK。头信息制定内容默认为 HTML，这样返回能直接显示在浏览器上。

c. 定义此控制器异步发送信息。

d. 从 Map 中获取 ResponseBodyEmitter 对象。

e. 通过 ResponseBodyEmitter 的 send(Object object,MediaType mediaType)方法发送数据；第一个参数是要发送的数据；第二个参数指定该条数据的数据类型，从而选用对应的 HttpMessageConverter 来转换数据。

f. 定义此控制器，关闭 ResponseBodyEmitter 对象。

g. 通过 ResponseBodyEmitter 的 complete()方法关闭 ResponseBodyEmitter。

用浏览器访问 https://localhost:8443/async/1/rbe，这时浏览器处于等待状态。当访问 https://localhost:8443/async/1/invoke-rbe 时，浏览器会每隔一秒输出相应的内容，刷新后会再次输出，如图 5-65 所示。

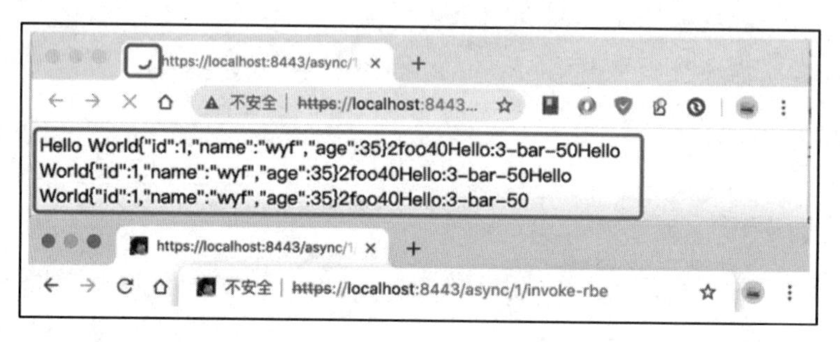

图 5-65

当访问 https://localhost:8443/async/1/close-rbe 时，浏览器关闭连接，https://localhost:8443/async/1/rbe 会结束等待状态；再次刷新则不再输出内容，如图 5-66 所示。

图 5-66

2. SSE

Spring MVC 中的 SseEmitter 可提供对 W3C 的 SSE 规范的支持。SSE 是 Server-Sent Events 的缩写。HTML5 API 提供了 EventSource 对象和服务端交互。当服务端返回数据时，头数据里的内容类型为 text/event-stream。

控制器的代码和 ResponseBodyEmitter 的代码相差不多。

```
@RestController
@RequestMapping("/async")
@Slf4j
public class AsyncController {

   private static Map<Long, SseEmitter> sseEmitterMap = new HashMap<>();

        //可使用 produces 属性执行返回的头信息中的内容类型
   @GetMapping(value = "/{id}/sse", produces = {MediaType.TEXT_EVENT_STREAM_VALUE})
   public SseEmitter sseEmitter(@PathVariable Long id){
      SseEmitter emitter = new SseEmitter();
      sseEmitterMap.put(id, emitter);
      return emitter;
   }

   @GetMapping("/{id}/invoke-sse")
   public void invokeSseEmitter(@PathVariable Long id) throws Exception {
      SseEmitter emitter = sseEmitterMap.get(id);
      emitter.send(new AnotherPerson(31,"bar", 50), new
MediaType("application","another-person"));
      Thread.sleep(1000);
      emitter.send(new Person(11, "wyf", 35), MediaType.APPLICATION_JSON);
      Thread.sleep(1000);
      emitter.send(new Person(21,"foo", 40), MediaType.APPLICATION_XML);
      Thread.sleep(1000);
      emitter.send("Hello World", MediaType.TEXT_PLAIN);
   }

   @GetMapping("/{id}/close-sse")
   public void closeSseEmitter(@PathVariable Long id) throws IOException {
      SseEmitter emitter = sseEmitterMap.get(id);
      emitter.complete();
      responseBodyEmitterMap.remove(id);
   }
}
```

下面使用 HTML5 的 EventSource 对象写一个极简单的静态页面：

```
<!DOCTYPE html>
<html xmlns="http://www.w3.org/1999/xhtml">
<head><title>SSE Demo</title></head>
<body>
```

```
<div id="msgFromSse"></div>
<script>
var source = new EventSource("/async/1/sse") //新建EventSource对象连接服务器的地址为
/async/1/sse
source.addEventListener('open', function (ev) { //监听连接成功的事件
   console.log("连接成功")
}, false);

source.addEventListener('message', function (ev) { { //监听数据发送的事件
   document.getElementById("msgFromSse").innerHTML = ev.data;
} });

source.addEventListener('error',function (ev) { //监听错误的事件
   if (ev.readyState == EventSource.CLOSED){
      console.log("连接已关闭");
   } else {
      console.log(e.readyState);
   }
}, false);
</script>
</body>
</html>
```

在 Chrome 中访问 https://localhost:8443/sse.html，Network 里有一个类型为 eventsource 的连接，且在连接的详细信息中有专门监控数据的 EventStream 标签页，如图 5-67 和图 5-68 所示。

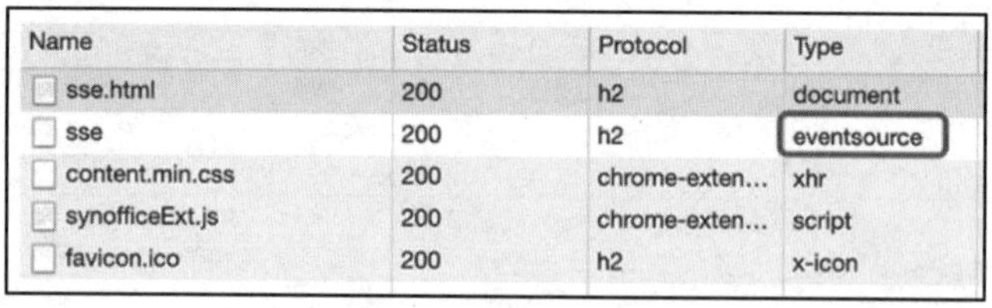

Name	Status	Protocol	Type
sse.html	200	h2	document
sse	200	h2	eventsource
content.min.css	200	chrome-exten...	xhr
synofficeExt.js	200	chrome-exten...	script
favicon.ico	200	h2	x-icon

图 5-67

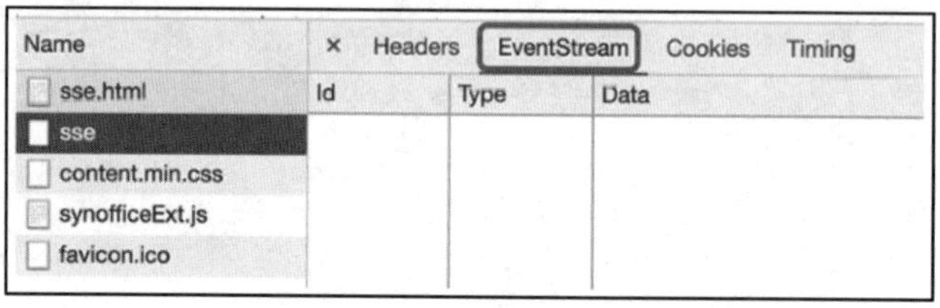

Name	× Headers EventStream Cookies Timing		
	Id	Type	Data
sse.html			
sse			
content.min.css			
synofficeExt.js			
favicon.ico			

图 5-68

访问 https://localhost:8443/async/1/invoke-sse，从另外的线程向 https://localhost:8443/sse.html 发送数据，这时页面会每隔一秒显示一次推送的消息，控制台也可以监控到数据的传送，如图 5-69 所示。

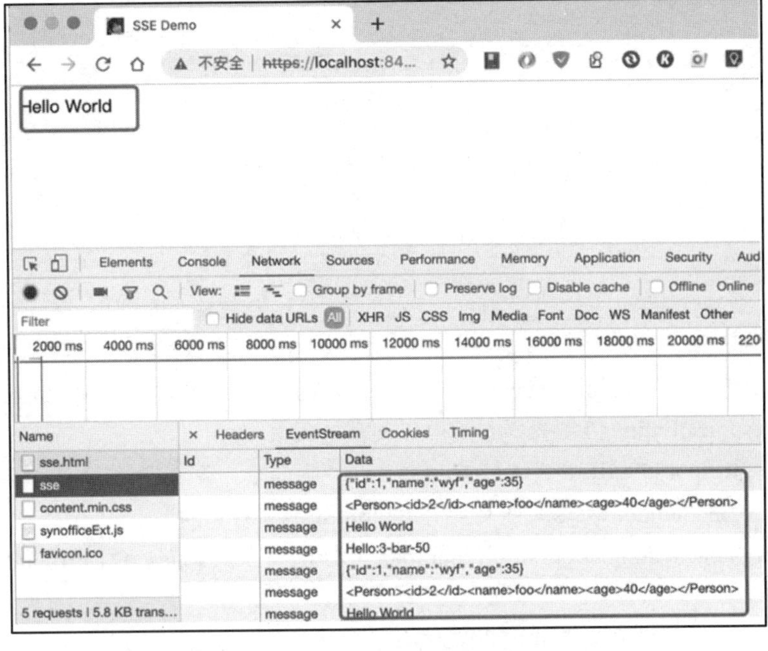

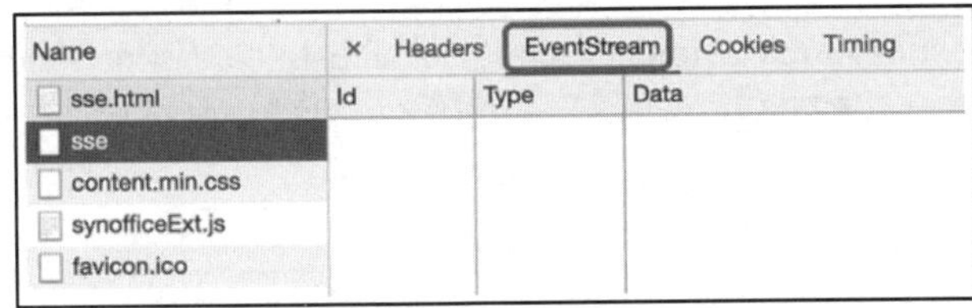

图 5-69

3. StreamingResponseBody

可以通过返回 StreamingResponseBody 来直接使用返回的 OutputStream 来自己控制数据返回。也可以使用 ResponseEntity 来定制状态和头的信息。

本例使用 StreamingResponseBody 返回一张图片，示例控制器如下：

```
@RestController
@RequestMapping("/async")
@Slf4j
public class AsyncController {
  @Value("classpath:wyn.jpg") //a
  private Resource image;

  @GetMapping("/img")
  public ResponseEntity<StreamingResponseBody> streamingResponseBody() {
    return ResponseEntity.ok()
          .contentType(MediaType.IMAGE_JPEG) //b
```

```
            .body(outputStream -> { //c
                IOUtils.copy(image.getInputStream(), outputStream); //d
            });
    }

    @GetMapping("/sync-img") //e
    public ResponseEntity<Resource> syncImage() {
        return ResponseEntity.ok()
                .contentType(MediaType.IMAGE_JPEG)
                .body(image);
    }

}
```

a. 将图片拷贝到 src/main/resouces 下，可以用@Value 注解读取图片到 Resource 对象中。

b. 使用 ResponseEntity 指定头信息的内容类型为图片。

c. StreamingResponseBody 是一个函数接口，接口内的唯一方法为 void writeTo(OutputStream outputStream)，使用 Lambda 表达式即可操作 OutputStream 对象。

d. 使用 org.apache.tomcat.util.http.fileupload.IOUtils 将 Resouce 中的 InputStream 复制到 OutputStream 中。

e. 此控制器用来同步返回图片，用作对比。

访问 https://localhost:8443/async/img 和 https://localhost:8443/async/sync-img，在同步情况下，阻碍式已经知道数据的大小和类型；而在异步情况下，阻碍式对数据的大小和类型是未知的，如图 5-70 所示。

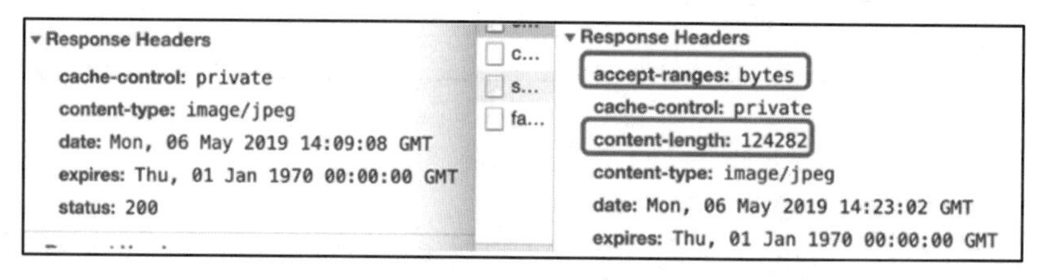

图 5-70

在图 5-70 中，左侧是异步的，右侧是同步的。

5.5.3 HTTP/2

前面已经配置了 SSL 并开启了 HTTP/2，现在只需在控制器方法的参数中使用 javax.servlet.http.PushBuilder 对象，即可使用 HTTP/2 的推送技术。PushBuilder 主动将资源推送到客户端（浏览器）。这个特性需要浏览器的支持，当然，现在绝大部分浏览器都支持这个功能。

HttpBuilder 属于 builder 模式，主要方法如下。

◎　path：中间操作，指定需推送的资源的地址。

◎　addHeader：中间操作，设定资源的内容类型。

◎　push：终结操作，执行推送动作。

HttpBuilder 推送的服务端资源是通过 ResourceHttpMessageConverter 来确定媒体类型和转换的，所以在 Spring MVC 下，是不需要显示指定媒体类型的。

在 static 目录下新建 push 目录，并加入如图 5-71 所示资源。

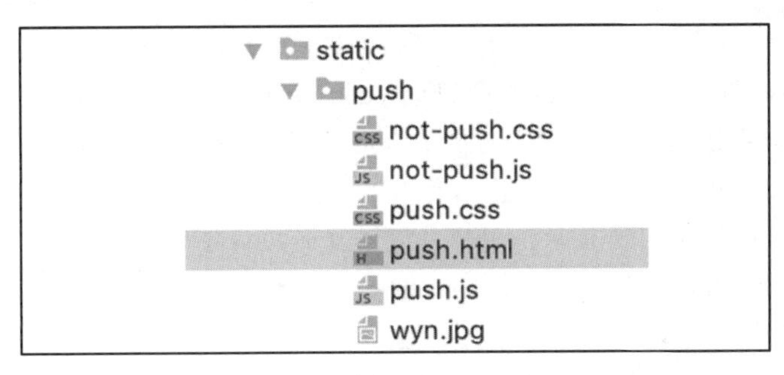

图 5-71

这里面最重要的是 push.html，它是服务端推送的终点。

```
<!DOCTYPE html>
<html xmlns="http://www.w3.org/1999/xhtml">
<head>
   <title>HTTP/2 Push Demo</title>
   <link rel="stylesheet" href="/push/push.css" />
   <link rel="stylesheet" href="/push/not-push.css" />
   <script type="application/javascript" src="/push/push.js"></script>
   <script type="application/javascript" src="/push/not-push.js"></script>
</head>
<body>
   <img src="/push/wyn.jpg"/>
</body>
</html>
```

当前页面的资源 push.css、push.js 和 wyn.jpg 是服务端推送的，not-push.css 和 not-push.js 没有做特别设置。

演示的控制器内容：

```
@Value("classpath:/static/push/push.html")
private Resource pushPage; //a

@GetMapping(value = "/http2-push")
public Resource http2Push(PushBuilder pushBuilder) { //b
   pushBuilder
         .path("push/wyn.jpg") //c
//          .addHeader("content-type", MediaType.IMAGE_JPEG_VALUE) //d
```

```
            .push(); //e
    pushBuilder
            .path("push/push.css") //c
//              .addHeader("content-type", "text/css") //d
            .push(); //e
    pushBuilder
            .path("push/push.js") //c
//              .addHeader("content-type", "application/javascript") //d
            .push(); //e
    return pushPage;
}
```

a. 本书不会引入页面模板引擎技术（Thymeleaf 等），所有的网页内容都是静态的。为了使用 REST 服务打开静态页面，方法返回值 Resource 通过 ResourceHttpMessageConverter 确定资源的媒体类型（text/html），并将载静态页面返回给浏览器。这个静态页面即我们要推送的目的地。若使用了页面模板引擎技术，则简单返回页面的路径即可。

b. 可以在控制器方法参数中直接注入 PushBuilder 对象。

c. 使用 path 方法指定需要推送资源的位置。

d. 使用 addHeadr 设定推送资源的内容类型。

e. 使用 push 方法推送。

使用 Chrome 访问 https://localhost:8443/async/http2-push，并打开 Chrome 的控制台，如图 5-72 所示。

Name	Status	Protocol	Type	Initiator
http2-push	200	h2	document	Other
push.css	200	h2	stylesheet	Push / http2-push
not-push.css	200	h2	stylesheet	http2-push
push.js	200	h2	script	Push / http2-push
not-push.js	200	h2	script	http2-push
wyn.jpg	200	h2	jpeg	Push / http2-push

图 5-72

5.6 小结

无论是使用 Spring 进行开发，还是使用 Spring Boot 进行开发，最重要的基础技术都是 Spring MVC。本章介绍了 Spring MVC 的用法和原理，以便读者更好地理解和掌握后续的内容。

第 6 章 数据访问

Spring Data 是一个伞型项目，包含主流数据库的访问技术。这些不同的数据访问项目均使用相同的编程模型，它们都是基于 Repository 的规范接口。本章介绍 Spring Data 的相关内容，包含领域驱动设计的基本知识、关系型数据库访问技术 Spring Data JPA 和 NoSQL 非关系型数据库访问技术 Spring Data Elasticsearch，以及数据缓存相关内容。

6.1 Spring Data Repository

6.1.1 DDD 与 Spring Data Repository

1. DDD

领域驱动设计（Domain-Driven Design，DDD）是解决复杂业务需求的一系列高级技术。在 DDD 的战术设计部分，有几个重要概念和 Spring Data 相关。

◎ 实体（Entity）：对一个实际的事物进行抽象建模，每个实体都有唯一标识，此唯一标识用来区分不同的实体。实体是可变的，会随着时间的推移而发生变化。

◎ 值对象（Value Object）：值对象等同于值，值是对一个不变的概念整体进行建模，它没有唯一标识，相等性通过比较值对象的属性来实现。值对象不能被修改，只能被替换。它和值一样都是用来描述、量化或衡量实体的。

◎ 聚合（Aggregate）：聚合是由一个或多个实体组合而成的，其中表示核心概念的实体叫作聚合根。在实际设计中，有个原则是“设计小聚合”，一般情况下，一个聚合应只包含一个实体类。

◎ 领域事件（Domain Event）：通过发布领域事件可进行聚合之间的通信，事件由聚合根发布。

◎ 库（Repository）：Respository 是用来存储聚合的，每个聚合都有一个 Respository，它们之间是一对一的关系。在库中，对聚合的存储操作类似于集合类。Respository 和 Set 一样，用来保障数据的唯一性。

下面通过代码来说明上述这些概念：

```
public class Person {
   private Long id;
   private String name;
   private Integer age;
   private Address address;
   private Collection<Child> children;
    //……
}
```

◎ Person 是实体类，因为有"设计小聚合"的原则，所以 Person 既是聚合，也是聚合根。

◎ id 属性是唯一标识。

◎ name 和 age 是值，用来描述实体 Person。

◎ 同样，address 和 children 也用来描述某个 Person，并且是对象形式的，所以它们是值对象。

2. Spring Data Repository

Spring Data Repository 为访问不同的数据库提供统一的抽象，它极大地减少了数据访问层的样板代码。Spring Data Repository 抽象的核心是 org.springframework.data.repository.Repository<T, ID>，T 代表它处理的实体的类型，ID 代表实体的唯一标识。它的主要子接口是 CrudRepository，定义了新增、查询、更新和删除的功能接口。PagingAndSortingRepository 是 CrudRepository 的子接口，定义了分页和排序的功能接口。

```
public interface CrudRepository<T, ID> extends Repository<T, ID> {
  <S extends T> S save(S entity); //保存一个实体
  <S extends T> Iterable<S> saveAll(Iterable<S> entities); //保存多个实体
  Optional<T> findById(ID id); // 按照 id 查询实体
  boolean existsById(ID id); // 按照 id 查询实体是否存在
  Iterable<T> findAll(); //查询所有实体
  Iterable<T> findAllById(Iterable<ID> ids); //按照多个 id 查询多个实体
  long count(); //计数实体
  void deleteById(ID id); //按照 id 删除实体
  void delete(T entity); //删除实体
  void deleteAll(Iterable<? extends T> entities);//删除多个实体
  void deleteAll(); //删除所有实体
}
public interface PagingAndSortingRepository<T, ID> extends CrudRepository<T, ID> {
  Iterable<T> findAll(Sort sort); //根据排序参数查询所有实体
```

```
  Page<T> findAll(Pageable pageable); //根据分页参数查询所有实体
}
```

针对不同的数据库，有特定的子接口抽象，如 JpaRepository、ElasticsearchRepository 等。

可以通过继承上面的接口来定义实体的 Repository 开启配置（@EnableJpaRepositories）注解后，实体的 Repository 会被 Spring Data 注册成一个 Bean，我们可以使用这个 Bean 进行数据访问操作。

定义 Repository：

```
public interface PersonRepository extends JpaRepository<Person, Long> {
}
```

使用 Repository：

```
@RestController
@RequestMapping("/people")
public class PersonController {

   private PersonRepository personRepository;

   public PersonController(PersonRepository personRepository) {
      this.personRepository = personRepository;
   }

   @GetMapping("/findByName")
   public List<Person> findByName(@RequestParam String name){
      return personRepository.findByName(name);
   }
}
```

用注解将相应数据库的领域模型标识为实体，如 JPA 使用的是@Entity，Elasticsearch 使用的是@Document。

JPA：

```
@Entity
public class Person {
   @Id
   private Long id;
   private String name;
   private Integer age;
   //……
}
Elasticsearch:
@Document(indexName = "person")
public class Person {}
```

6.1.2 查询方法

Spring Data 可根据方法名中的属性进行推导查询，关键词为 find...By、read...By 或 get…By，By 之后是查询条件（where 之后的条件）：

```
public interface PersonRepository extends JpaRepository<Person, Long> {
   List<Person> findByName(String name);
    List<Person> findDistinctPersonByName(String name);
}
```

Spring Data 可以通过 count...By 进行推导计数查询：

```
public interface PersonRepository extends JpaRepository<Person, Long> {
   long countByName(String name);
}
```

还可以通过 delete...By 进行推导删除查询：

```
public interface PersonRepository extends JpaRepository<Person, Long> {
   long deleteByName(String name);
   List<Person> removeByName(String name);
}
```

By 之前可以用 firtst 或 top 关键字来限制查询数量：

```
public interface PersonRepository extends JpaRepository<Person, Long> {
   List<Person> queryTop5ByName(String name);
   List<Person> getFirst10ByName(String name);
}
```

更多关于 Spring Data 的查询关键字会在后面详细讲解。

6.2 关系数据库——Spring Data JPA

6.2.1 JPA、Hibernate 和 Spring Data JPA

JPA（Java Persistence API Java 持久化 API）是规范，当用它对数据库进行各种操作时，实际操作的是 JPA 提供者（如 Hibernate）提供的实现。在使用规范开发时，只需和规范的 API 打交道即可，无须关心规范的实现。JPA 2.2 为 JSR-338 规范，属于 Jakarta EE 的一部分。Spring Data JPA 是在 JPA 之上所做的更高级别的抽象，使对数据库的操作更简单、更语义化。

6.2.2 环境准备

1. 安装 MySQL

使用 docker compose 安装 MySQL。

stack.yml：

```
version: '3.1'

services:

  db:
    image: mysql
    command: --default-authentication-plugin=mysql_native_password
    restart: always
    ports:
      - 3306:3306
    environment:
      MYSQL_DATABASE: first_db
      MYSQL_ROOT_PASSWORD: zzzzzz

  adminer:
    image: adminer # 全功能数据库管理工具
    restart: always
    ports:
      - 8081:8080
```

执行命令。

```
$ docker-compose -f stack.yml up -d
```

打开 http://localhost:8081/，访问 Adminer，如图 6-1 所示。

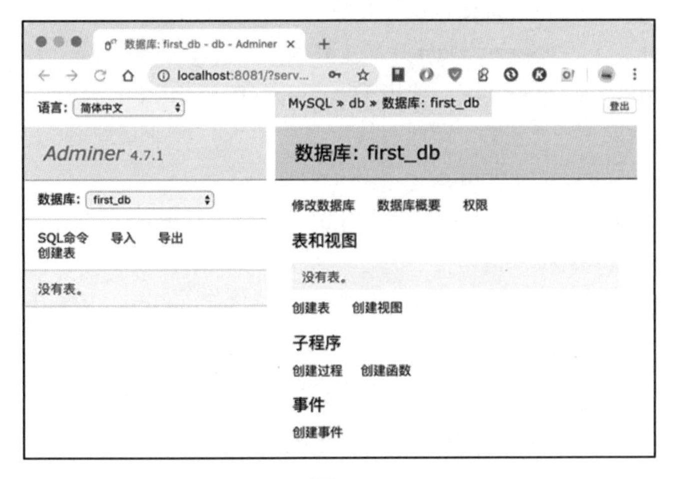

图 6-1

2. 新建应用

新建应用，信息如下。

Group：top.wisely。

Artifact：learning-spring-data-jpa。

Dependencies：Spring Web Starter、Spring Data JPA、MySQL Driver、Lombok。

build.gradle 文件中的依赖如下：

```
dependencies {
  implementation 'org.springframework.boot:spring-boot-starter-data-jpa'
  implementation 'org.springframework.boot:spring-boot-starter-web'
  runtimeOnly 'mysql:mysql-connector-java'
  compileOnly 'org.projectlombok:lombok'
     annotationProcessor 'org.projectlombok:lombok'
 ...
}
```

Spring Data JPA 的版本由 Spring Boot 来维护。

6.2.3 自动配置

在非 Spring Boot 下，需要具备下面的条件才能用 JPA 进行开发。

◎ DataSource：用来连接数据库。

◎ EntityManagerFactory：获取 EntityManager 和数据库进行交互。

◎ PlatformTransactionManager：用来管理事务。

◎ @EnableJpaRepositories：开启 JPA Repository 的配置。

◎ @EnableTransactionManagement：开启事务管理的配置。

在 Spring Boot 下，对于 Spring Data JPA 主要有以下 5 个自动配置。

（1）DataSourceAutoConfiguration：使用 DataSourceProperties 中的 spring.datasource.*来自动配置 DataSource。

（2）HibernateJpaAutoConfiguration：

◎ 导入 HibernateJpaConfiguration，使用 HibernateProperties 中的 spring.jpa.hibernate.*来自动配置 Hibernate。

◎ HibernateJpaConfiguration 的父类 JpaBaseConfiguration 使用 JpaProperties 中的 spring.jpa.*自动配置 JPA，包括 PlatformTransactionManager 和 EntityManagerFactory。

（3）JpaRepositoriesAutoConfiguration：使用@EnableJpaRepositories，开启 JPA 自动配置。

（4）TransactionAutoConfiguration：使用@EnableTransactionManagement，开启事务自动配置。

（5）SpringDataWebAutoConfiguration：使用@EnableSpringDataWebSupport，开启 Web 相关配置。

配置数据源，连接数据库：

```
spring:
 datasource:
   url: jdbc:mysql://localhost:3306/first_db?useSSL=false
```

```
  username: root
  password: zzzzzz
  driver-class-name: com.mysql.cj.jdbc.Driver
```

其他配置：

```
spring:
 jpa:
  show-sql: true # 控制台显示执行的 SQL 语句
  hibernate:
   ddl-auto: update # 启动更新 DDL（create、alter 和 drop 等）
```

6.2.4　定义聚合

1. 定义实体类

使用 javax.persistence.Entity 注解定义实体类：

```
@Entity
public class Person {}
```

一个实体类会映射数据库的一张表。这里启用了 spring.jpa.hibernate.ddl-auto: update，Spring Data 会自动创建一张首字母小写的表 Person。

可以通过@Table(name = "tb_person")来制定表名。

2. 定义唯一标识

使用 javax.persistence.Id 注解可定义唯一标识（Id）。唯一标识可以和数据库中的主键进行映射。使用 javax.persistence.GeneratedValue 注解可以指定唯一标识的产生策略：

```
@Entity
public class Person {
  @Id
  @GeneratedValue
  private Long id;
}
```

@GeneratedValue 默认使用自动选择唯一标识生成策略，在 MySQL 下，自动使用 Hibernate 生成的 hibernate_sequence 表。可以使用@GeneratedValue 的 generator 属性配合@SequenceGenerator 或@TableGenerator 来指定唯一标识生成器。

3. 定义属性值

使用 javax.persistence.Column 可定义一个属性值。属性值可以和数据库中的字段进行映射。当不使用属性值时，实体的值会自动映射为表的字段。命名规则是将驼峰式的属性名转换成下画线的字段名（如将 lastName 映射成 last_name）。

```
@Data
@AllArgsConstructor
```

```
@NoArgsConstructor
@Entity
public class Person {
   @Id
   @GeneratedValue
   private Long id;

   @Column(name = "name", length = 10)
   private String name;

   private Integer age;

   private Boolean active = true;
   }

}
```

4. 定义单个值对象

需要通过 javax.persistence.Embeddable 注解表明值对象是实体的一部分。在实体中，用 javax.persistence.Embedded 注解来标注单个值对象：

```
@Data
@AllArgsConstructor
@NoArgsConstructor
@Embeddable
public class Address {
   private String city;
   private String province;
}
```

Address 的属性会映射成 Person 表中的字段：

```
@Data
@AllArgsConstructor
@NoArgsConstructor
@Entity
public class Person {
   @Id
   @GeneratedValue
   private Long id;
   @Column
   private String name;

   private Integer age;

   @Embedded
   private Address address;
}
```

5．定义列表值对象

同样，需要通过 javax.persistence.Embeddable 注解表明列表值对象是实体的一部分。在实体中，通过 javax.persistence.ElementCollection 注解来标注列表值对象。

```
@Data
@AllArgsConstructor
@NoArgsConstructor
@Embeddable
public class Child {

   private String name;

   @Enumerated(EnumType.STRING)
   private Gender gender;

}
```

对于多个固定值选择的属性，建议使用枚举型表示：

```
public enum  Gender {
   MALE,
   FEMALE
}
@Data
@AllArgsConstructor
@NoArgsConstructor
@Entity
public class Person {
   @Id
   @GeneratedValue
   private Long id;

   @Column
   private String name;

   private Integer age;

   @Embedded
   private Address address;

   @ElementCollection(fetch = FetchType.EAGER)
   private Collection<Child> children;
}
```

生成一个 person_children 表，字段有 person_id、gender 和 name。

6．属性校验

实体的属性的校验也支持 JSR-303/349/380，可以用注解来校验，如@NotNull 和@Size 等：

```
@Data
@AllArgsConstructor
@NoArgsConstructor
@Entity
public class Person {
   @Id
   @GeneratedValue
   @NotNull //Id不能为空
   private Long id;

   @NotNull
   @Size(min = 3, max = 10)
   @Column(name = "name", length = 10)
   private String name;

   private Integer age;

   @Embedded
   private Address address;

   @ElementCollection
   private Collection<Child> children;
}
```

6.2.5 定义聚合 Repository

在定义聚合 Repository 时，只需确定三个问题。

◎ 当前数据访问技术：JpaRepository。

◎ 聚合根的实体类型：Person。

◎ 实体的唯一标识的类型：Long。

```
public interface PersonRepository extends JpaRepository<Person, Long> {
}
```

JpaRepository 的代理实现为 SimpleJpaRepository。Spring Data JPA 会将 PersonRepository 注册成一个 Bean，只需注入此 Bean 即可进行数据库操作。

1. 保存

直接使用继承自 CrudRepository 的 save*方法即可保存。

保存单个实体：

```
@SpringBootApplication
@Slf4j
public class LearningSpringDataJpaApplication {

  public static void main(String[] args) {
    SpringApplication.run(LearningSpringDataJpaApplication.class, args);
```

```
    }

    @Bean
    CommandLineRunner saveOne(PersonRepository personRepository){
        return args -> {
            Address address = new Address("hefei", "Anhui");
            Collection<Child> children = Arrays.asList(new Child("wyn", Gender.FEMALE), new
Child("wbe", Gender.MALE));
            Person person = new Person("wyf", 35, address, children);
            Person savedPerson = personRepository.save(person);
            log.info(person.toString());
        };
    }
}
```

保存多个实体：

```
@Bean
CommandLineRunner saveAll(PersonRepository personRepository){
    return args -> {
        Address address1 = new Address("beijing", "Beijing");
        Address address2 = new Address("shanghai", "Shanghai");
        Collection<Child> children1 = Arrays.asList(new Child("aaa", Gender.FEMALE), new
Child("bbb", Gender.MALE));
         Collection<Child> children2 = Arrays.asList(new Child("ccc", Gender.FEMALE), new
Child("ddd", Gender.MALE));
        Person person1 = new Person("foo", 36, address1, children1);
        Person person2 = new Person("bar", 34, address2, children1);
        List<Person> savedPeople = personRepository.saveAll(Arrays.asList(person1,
person2));
        savedPeople.forEach(p ->{
            log.info(p.toString());
        });
    };
}
```

保存结果如图 6-2 所示。

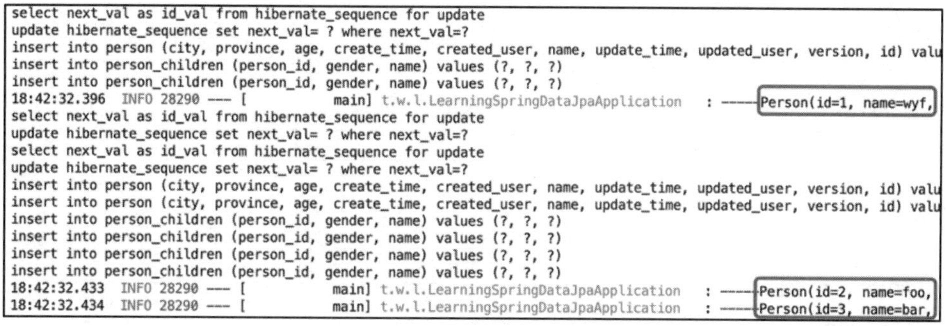

```
select next_val as id_val from hibernate_sequence for update
update hibernate_sequence set next_val= ? where next_val=?
insert into person (city, province, age, create_time, created_user, name, update_time, updated_user, version, id) valu
insert into person_children (person_id, gender, name) values (?, ?, ?)
insert into person_children (person_id, gender, name) values (?, ?, ?)
18:42:32.396  INFO 28290 --- [           main] t.w.l.LearningSpringDataJpaApplication    : -----Person(id=1, name=wyf,
select next_val as id_val from hibernate_sequence for update
update hibernate_sequence set next_val= ? where next_val=?
select next_val as id_val from hibernate_sequence for update
update hibernate_sequence set next_val= ? where next_val=?
insert into person (city, province, age, create_time, created_user, name, update_time, updated_user, version, id) valu
insert into person (city, province, age, create_time, created_user, name, update_time, updated_user, version, id) valu
insert into person_children (person_id, gender, name) values (?, ?, ?)
insert into person_children (person_id, gender, name) values (?, ?, ?)
insert into person_children (person_id, gender, name) values (?, ?, ?)
insert into person_children (person_id, gender, name) values (?, ?, ?)
18:42:32.433  INFO 28290 --- [           main] t.w.l.LearningSpringDataJpaApplication    : -----Person(id=2, name=foo,
18:42:32.434  INFO 28290 --- [           main] t.w.l.LearningSpringDataJpaApplication    : -----Person(id=3, name=bar,
```

图 6-2

2. 删除

使用继承自 CrudRepository 的 delete*方法来删除数据：

```
@Bean
CommandLineRunner delete(PersonRepository personRepository){
  return args -> {
    personRepository.deleteById(1l); //根据 id 删除数据
    log.info("-----剩余数量为" + personRepository.count() + "------");
    personRepository.delete(personRepository.getOne(2l)); //根据实体删除数据
    log.info("-----剩余数量为" + personRepository.count() + "------");
    personRepository.deleteAll(); // 删除所有数据
    log.info("-----剩余数量为" + personRepository.count() + "------");
  };
}
```

删除结果如图 6-3 所示。

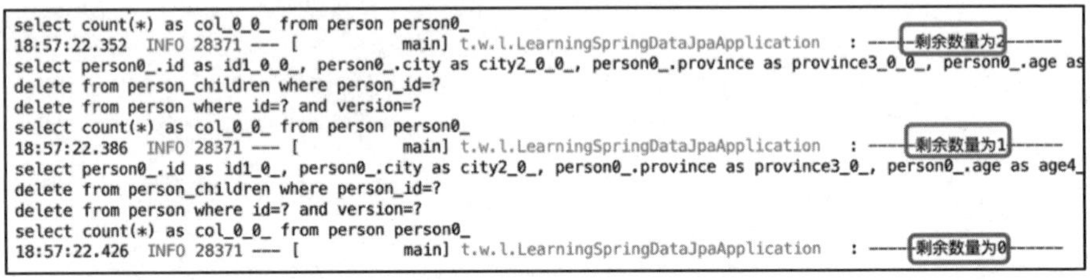

```
select count(*) as col_0_0_ from person person0_
18:57:22.352  INFO 28371 --- [           main] t.w.l.LearningSpringDataJpaApplication   : -----剩余数量为2------
select person0_.id as id1_0_0_, person0_.city as city2_0_0_, person0_.province as province3_0_0_, person0_.age as
delete from person_children where person_id=?
delete from person where id=? and version=?
select count(*) as col_0_0_ from person person0_
18:57:22.386  INFO 28371 --- [           main] t.w.l.LearningSpringDataJpaApplication   : -----剩余数量为1------
select person0_.id as id1_0_, person0_.city as city2_0_, person0_.province as province3_0_, person0_.age as age4_
delete from person_children where person_id=?
delete from person where id=? and version=?
select count(*) as col_0_0_ from person person0_
18:57:22.426  INFO 28371 --- [           main] t.w.l.LearningSpringDataJpaApplication   : -----剩余数量为0------
```

图 6-3

6.2.6 查询

1. 查询方法

（1）推导查询。

Spring Data JPA 支持根据方法名中的属性来推导查询语句：

```
public interface PersonRepository extends JpaRepository<Person, Long> {
  List<Person> findByNameAndAge(String name, Integer age);
}
```

把方法名 findByNameAndAge 翻译成查询 JPQL 语句：

```
select p from Person where p.name = ?1 and p.age = ?2
```

方法名中除属性外，关键字如表 6-1 所示。

表 6-1

关 键 字	示 例	JPQL 片段
And	findByNameAndAge	where p.name = ?1 and p.age = ?2

续表

关键字	示例	JPQL 片段
Or	findByNameOrAge	where p.name = ?1 or x.age = ?2
Is,Equals	findByName findByNameIs findByNameEquals	where p.name = ?1
Between	findByCreateTimeBetween	where p.createTime between ?1 and ?2
LessThan	findByAgeLessThan	where p.age < ?1
LessThanEqual	findByAgeLessThanEqual	where p.age <= ?1
GreaterThan	findByAgeGreaterThan	where p.age > ?1
GreaterThanEqual	findByAgeGreaterThanEqual	where p.age >= ?1
After	findByCreateTimeAfter	where p.createTime > ?1
Before	findByCreateTimeBefore	where p.createTime < ?1
IsNull	findByAgeIsNull	where p.age is null
IsNotNull,NotNull	findByAge(Is)NotNull	where p.age not null
Like	findByNameLike	where p.name like ?1
NotLike	findByNameNotLike	where p.name not like ?1
StartingWith	findByNameStartingWith	where p.name like ?1 (参数前附加%)
EndingWith	findByNameEndingWith	where p.name like ?1 (参数后附加%)
Containing	findByNameContaining	where p.name like ?1 (参数两边附加%)
OrderBy	findByAgeOrderByNameDesc	where p.age = ?1 order by p.name desc
Not	findByNameNot	where p.name <> ?1
In	findByAgeIn(Collection<Age> ages)	where p.age in ?1
NotIn	findByAgeNotIn(Collection<Age> ages)	where p.age not in ?1
True	findByActiveTrue()	where p.active = true
False	findByActiveFalse()	where p.active = false
IgnoreCase	findByFirstnameIgnoreCase	where UPPER(p.name) = UPPER(?1)

若以值对象的属性作为查询条件，则同样向后加属性名，中间可以用“_”隔开：

```
public interface PersonRepository extends JpaRepository<Person, Long> {
   List<Person> findByAddress_City(String city);
   List<Person> findByAddressCity(String city);
   List<Person> findByChildren_Name(String name);
   List<Person> findByChildrenName(String name);
}
```

使用 CommandLineRunner 验证：

```
@Bean
CommandLineRunner sortQuery(PersonRepository personRepository){
  return args -> {
    List<Person> people1 = personRepository.findByAgeLessThan(40, Sort.by("name"));
    List<Person> people2 = personRepository.findByAgeLessThanWithJqal(40,
```

```
JpaSort.by(Sort.Direction.DESC, "name"));
    List<Person> people3 = personRepository.findAll(Sort.by("address.city"));
    List<Person> people4 = personRepository.findAll(JpaSort.by(Sort.Direction.DESC,
"age"));
    Page<Person> people5 = personRepository.findByAgeLessThan(40, PageRequest.of(0, 2,
Sort.by("age")));
      Page<Person> people6 = personRepository.findAll(PageRequest.of(0, 2,
Sort.by("name")));

    people1.forEach(System.out::println);
    System.out.println("--------------");
    people2.forEach(System.out::println);
    System.out.println("--------------");
    people3.forEach(System.out::println);
    System.out.println("--------------");
    people4.forEach(System.out::println);
    System.out.println("--------------");
    people5.forEach(System.out::println);
    System.out.println("--------------");
    people6.forEach(System.out::println);
    };
}
```

```
Person(id=1, name=wyf, age=35, active=true, address=Address(city=hefei, province=Anhui), children=[Child(name=wyn, gender=
--------------
Person(id=2, name=foo, age=36, active=true, address=Address(city=beijing, province=Beijing), children=[Child(name=aaa, gen
--------------
Person(id=3, name=bar, age=34, active=true, address=Address(city=shanghai, province=Shanghai), children=[Child(name=ccc, g
--------------
Person(id=1, name=wyf, age=35, active=true, address=Address(city=hefei, province=Anhui), children=[Child(name=wyn, gender=
--------------
Person(id=1, name=wyf, age=35, active=true, address=Address(city=hefei, province=Anhui), children=[Child(name=wyn, gender=
--------------
Person(id=3, name=bar, age=34, active=true, address=Address(city=shanghai, province=Shanghai), children=[Child(name=ccc, g
--------------
Person(id=3, name=bar, age=34, active=true, address=Address(city=shanghai, province=Shanghai), children=[Child(name=ccc, g
--------------
```

（2）JPA 命名查询。

可以使用@NamedQuery 注解在实体上做命名查询：

```
@Entity
@NamedQuery(name = "Person.findByNameWyf",
  query = "select p from Person p where p.name = 'wyf'")
public class Person {
```

若命名的方法和推导查询的方法同名，则命令方法会覆盖推导查询的方法。

在 PersonRepository 中声明方法：

```
public interface PersonRepository extends JpaRepository<Person, Long> {
  List<Person> findByNameWyf();
}
```

用代码来检验结果，如图 6-4 所示。

```
@Bean
CommandLineRunner namedQuery(PersonRepository personRepository){
  return args -> {
    List<Person> people1 = personRepository.findByNameWyf();
    people1.forEach(System.out::println);
  };
}
```

```
Hibernate: select person0_.id as id1_0_, person0_.active as
Hibernate: select children0_.person_id as person_i1_1_0_, c
Person(id=1, name=wyf, age=35, active=true, address=Address
```

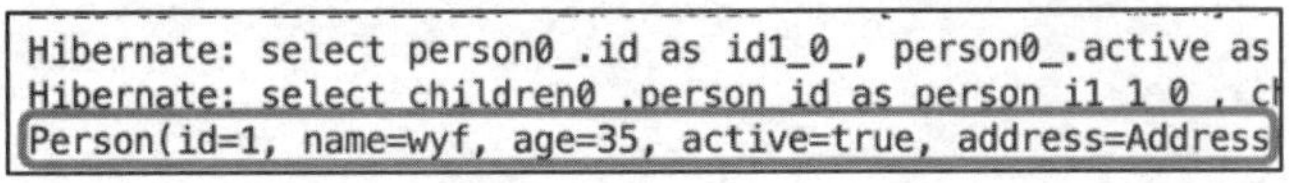

图 6-4

（3）JPQL 查询。

@NamedQuery 适合少量的查询，我们在 Repository 方法上可以使用@Query 注解，让方法执行定义的 JPQL：

```
public interface PersonRepository extends JpaRepository<Person, Long> {
   @Query("select p from Person p where p.name = ?1")
   List<Person> findByJpql(String name);
}
```

findByJpql 的 JPQL 和 findByName 的 JPQL 一致：

```
@Bean
CommandLineRunner jpqlQuery(PersonRepository personRepository){
  return args -> {
    List<Person> people1 = personRepository.findByJpql("wyf");
    people1.forEach(System.out::println);
  };
}
```

更多关于 JPQL 的用法可参考：JPQL Language Reference 一书。

（4）SQL 查询。

只要在@Query 注解上设置属性 nativeQuery = true，即可使用原生 SQL 查询：

```
public interface PersonRepository extends JpaRepository<Person, Long> {
   @Query(value = "select * from person where name = ?1", nativeQuery = true)
   List<Person> findBySql(String name);
}
```

执行检验代码：

```
@Bean
CommandLineRunner nativeQuery(PersonRepository personRepository){
  return args -> {
    List<Person> people1 = personRepository.findBySql("wyf");
    people1.forEach(System.out::println);
  };
}
```

（5）排序和分页。

只需在 Repository 的方法中使用 Sort 作为参数即可排序；方法还可接收 Pageable 参数，既可分页，也可排序：

```
public interface PersonRepository extends JpaRepository<Person, Long> {

   List<Person> findByAgeLessThan(Integer age, Sort sort);

   @Query("select p from Person p where p.age < ?1")
   List<Person> findByAgeLessThanWithJqal(Integer age, Sort sort);

   Page<Person> findByAgeLessThan(Integer age, Pageable pageable);
}
```

使用代码验证：

```
@Bean
CommandLineRunner sortQuery(PersonRepository personRepository){
  return args -> {
     List<Person> people1 = personRepository.findByAgeLessThan(40, Sort.by("name")); //a
     List<Person> people2 = personRepository.findByAgeLessThanWithJqal(40,
JpaSort.by(Sort.Direction.DESC, "name")); //b
     List<Person> people3 = personRepository.findAll(Sort.by("address.city")); //c
     List<Person> people4 = personRepository.findAll(JpaSort.by(Sort.Direction.DESC,
"age"));
     Page<Person> people5 = personRepository.findByAgeLessThan(40, PageRequest.of(1, 2,
Sort.by("age"))); //d
             Page<Person> people6 = personRepository.findAll(PageRequest.of(1, 2,
Sort.by("name")));

     people1.forEach(System.out::println);
     System.out.println("--------------");
     people2.forEach(System.out::println);
     System.out.println("--------------");
     people3.forEach(System.out::println);
     System.out.println("--------------");
     people4.forEach(System.out::println);
     System.out.println("--------------");
             people5.forEach(System.out::println);
             System.out.println("--------------");
             people6.forEach(System.out::println);
     };
}
```

a. 可以使用 Sort 的静态方法 by 构建 Sort 对象，默认为升序，按照 name 排序。

b. 可以使用 JpaSort 的静态方法 by 构建 Sort 对象，使用 Direction 来设定是升序，还是降序。

c. 可以通过 address.city 根据内嵌对象的属性进行排序。

d. PageRequest 是 Pageable 接口的实现类，可以用它的静态方法 of 来构造分页和排序。分页的起始索引是 0，每页数量设置为 2。

验证结果如图 6-5 所示。

```
Person(id=3, name=bar, age=34, active=true, address=Address(city=shanghai,
Person(id=2, name=foo, age=36, active=true, address=Address(city=beijing,
Person(id=1, name=wyf, age=35, active=true, address=Address(city=hefei, pr
--------------
Person(id=1, name=wyf, age=35, active=true, address=Address(city=hefei, pr
Person(id=2, name=foo, age=36, active=true, address=Address(city=beijing,
Person(id=3, name=bar, age=34, active=true, address=Address(city=shanghai,
--------------
Person(id=2, name=foo, age=36, active=true, address=Address(city=beijing,
Person(id=1, name=wyf, age=35, active=true, address=Address(city=hefei, pr
Person(id=3, name=bar, age=34, active=true, address=Address(city=shanghai,
--------------
Person(id=2, name=foo, age=36, active=true, address=Address(city=beijing,
Person(id=1, name=wyf, age=35, active=true, address=Address(city=hefei, pr
Person(id=3, name=bar, age=34, active=true, address=Address(city=shanghai,
--------------
Person(id=3, name=bar, age=34, active=true, address=Address(city=shanghai,
Person(id=1, name=wyf, age=35, active=true, address=Address(city=hefei, pr
--------------
Person(id=3, name=bar, age=34, active=true, address=Address(city=shanghai,
Person(id=2, name=foo, age=36, active=true, address=Address(city=beijing,
```

图 6-5

（6）命名参数。

前面的入参是通过参数位置来设置的，这是 Spring Data JPA 的默认行为。可以通过使用@Param 注解来绑定参数，在查询语句中使用“：参数名”：

```
public interface PersonRepository extends JpaRepository<Person, Long> {

   @Query("select p from Person p where p.name = :name")
   List<Person> findByJpqlWithNamedParameter(@Param("name") String name);

}
@Bean
CommandLineRunner namedParamQuery(PersonRepository personRepository){
  return args -> {
    List<Person> people = personRepository.findByJpqlWithNamedParameter("wyf");
    people.forEach(System.out::println);
  };
}
```

（7）修改查询。

当在@Query 中进行新增、更新或删除时，可以使用@Modifying 来注解这个语句是个修改查询语句：

```
public interface PersonRepository extends JpaRepository<Person, Long> {
```

```
    @Transactional
    @Modifying
    @Query("update Person p set p.name = ?1 where p.name =?2")
    int updatePersonName(String newName, String oldName);
}
```

因为 JpaRepository 的代理实现 SimpleJpaRepository 对全局的事务设置为@Transactional(readOnly = true)，所以此处要额外声明用@Transactional 来覆盖默认配置；

```
@Bean
CommandLineRunner modifyingQuery(PersonRepository personRepository){
  return args -> {
    int result = personRepository.updatePersonName("foooo", "foo");
    List<Person> people = personRepository.findByName("foooo");
    people.forEach(System.out::println);

  };
}
```

（8）Projection。

通常情况下，Spring Data JPA 的查询方法返回的是聚合根的一个多或多个实例。可以使用 Projection，通过聚合根的属性来定制查询返回值。

使用接口来定制返回值：

```
public interface PersonProjectionInterface {

  String getName(); //a

  Address getAddress(); //b

  @Value("#{target.name + ' s age is' + target.age}") //c
  String getAgeDesc();

  default String getCityDesc(){ //d
    return getName() + " lives in " + getAddress().getCity();
  }

  @Value("#{@personProjectionHelper.getInfo(target)}") //e
  String getInfo();

  @Value("#{args[0] + ' ' + target.name + ' !'}") //f
  String getHello(String greeting);
}
```

a. 可以在返回里直接设置聚合根的属性。

b. 在返回里设置聚合根的属性对象。

c. 可以使用@Value 注解计算新的值，target 代表聚合根的值。

d. 可以使用 default 方法来定义运算逻辑。

e. 可调用外部 Bean 的方法得到其运算结果。

```
@Component
public class PersonProjectionHelper {

   public String getInfo(Person person){
      return person.toString();
   }
}
```

f. 可获取方法参数来参与运算。

在 Repository 的返回值中使用上面的接口：

```
public interface PersonRepository extends JpaRepository<Person, Long> {

   List<PersonProjectionInterface> findByNameIs(String name);

}
```

代码验证：

```
@Bean
CommandLineRunner interfaceProjectionQuery(PersonRepository personRepository){
  return args -> {
     List<PersonProjectionInterface> people = personRepository.findByNameIs("wyf");
     people.forEach(person -> {
        System.out.println(person.getName());
        System.out.println(person.getAddress());
        System.out.println(person.getAgeDesc());
        System.out.println(person.getCityDesc());
        System.out.println(person.getInfo());
        System.out.println(person.getHello("Hello"));
     });
  };
}
```

通过类（DTO：Data Transfer Object ，数据传输对象）来定制返回：

```
public class PersonDto {
   private final String name;
   private final Integer age;

   public PersonDto(String name, Integer age) {
      this.name = name;
      this.age = age;
   }

   public String getName() {
      return name;
   }

   public Integer getAge() {
```

```
        return age;
    }
}
```

可以使用 Lombok 的 lombok.Value 注解来简化上面的声明：

```
@Value
public class PersonDto {
    String name;
    Integer age;
}
```

在 Repository 中声明方法：

```
public interface PersonRepository extends JpaRepository<Person, Long> {

    List<PersonDto> findByNameEquals(String name);

}
```

代码验证：

```
@Bean
CommandLineRunner classProjectionQuery(PersonRepository personRepository){
   return args -> {
      List<PersonDto> people = personRepository.findByNameEquals("wyf");
      people.forEach(System.out::println);
   };
}
```

还可以动态地选择返回的类型。在 Repository 中，可以用泛型来定义返回值，并在参数中指定返回的类型。

```
public interface PersonRepository extends JpaRepository<Person, Long> {

    <T> Collection<T> findByNameAndAge(String name, Integer age, Class<T> type);

}
```

代码验证：

```
@Bean
CommandLineRunner dynamicProjectionQuery(PersonRepository personRepository){
   return args -> {
      Collection<Person> people1 = personRepository.findByNameAndAge("wyf", 35,
Person.class);
      Collection<PersonProjectionInterface> people2 =
personRepository.findByNameAndAge("wyf", 35, PersonProjectionInterface.class);
      Collection<PersonDto> people3 = personRepository.findByNameAndAge("wyf", 35,
PersonDto.class);
      people1.forEach(person -> {
         System.out.println(person.getName());
         System.out.println(person.getAddress());
```

```
    });
    people2.forEach(personProjectionInterface -> {
      System.out.println(personProjectionInterface.getHello("Hello"));
      System.out.println(personProjectionInterface.getInfo());
    });
    people3.forEach(personDto -> {
      System.out.println(personDto.getName());
      System.out.println(personDto.getAge());
    });
  };
}
```

2. 存储过程

Spring Data JPA 支持使用@Procedure 注解调用数据库的存储过程。

先定义一个存储过程，入参为 name，出参为 prefix_name，在入参的 name 前加上 Mr./Mrs.:

```
DROP PROCEDURE IF EXISTS add_name_prefix;
DELIMITER $$
CREATE PROCEDURE add_name_prefix(IN name VARCHAR(255), OUT prefix_name VARCHAR(255))
BEGIN
   set prefix_name = CONCAT('Mr./Mrs. ', name);
END
```

在 Repository 中映射这个存储过程：

```
public interface PersonRepository extends JpaRepository<Person, Long> {

    @Procedure(procedureName = "add_name_prefix") //指定存储过程的名称
   String getPrefixName(String name);

}
```

验证结果，如图 6-6 所示。

```
@Bean
CommandLineRunner storedProcrdureQuery(PersonRepository personRepository){
  return args -> {
    System.out.println(personRepository.getPrefixName("yyy"));
  };
}
```

```
2019-05-12 15:49:34.264  INFO 34121 ---
2019-05-12 15:49:34.732  WARN 34121 ---
2019-05-12 15:49:34.840  INFO 34121 ---
2019-05-12 15:49:35.015  INFO 34121 ---
2019-05-12 15:49:35.020  INFO 34121 ---
Hibernate: {call add_name_prefix(?,?)}
Mr./Mrs. yyy
```

图 6-6

3. Specification

可以通过定义 Specification 来定制查询。需要用 Repository 继承 JpaSpecificationExecutor 接口，这样就可以在方法中使用 Specification 参数了。

Specification 是函数接口，只有一个方法：

```
Predicate toPredicate(Root<T> root, CriteriaQuery<?> query, CriteriaBuilder
criteriaBuilder);
```

◎ root：要查询的实体。

◎ query：用来进行高级别的查询，如 where、select 方法等。

◎ criteriaBuilder：用来构造查询，可使用 like、equal、lessThan 等，返回值为 Predicate。

◎ javax.persistence.criteria.Predicate：用来作为查询的条件，可用多个 predicate 组合查询。

下面用自定义类来演示 Specification：

```
public class CustomSpecs {
    public static Specification<Person> nameEqual(String name){
        return (root, query, criteriaBuilder) -> criteriaBuilder.equal(root.get("name"),
name); //a
    }

    public static Specification<Person> ageLessThanAndNameLike(Integer age, String name){
        return (root, query, criteriaBuilder) -> {
            Predicate ageLessThanPredicate = criteriaBuilder.lessThan(root.get("age"),
age); //b
            Predicate nameLikePredicate = criteriaBuilder.like(root.get("name"), "%" + name
+ "%"); //c
            query.where(ageLessThanPredicate, nameLikePredicate); //d
            return query.getRestriction(); //e
        };
    }
}
```

a. 可以用 Lambda 表达式作为实现，使用 CriteriaBuilder 的 equal 方法比较 root.get("name") 和 name 是否相等。

b. 使用 lessThan 方法比较 root.get("age")是否小于 age。

c. 使用 like 方法比较 root.get("name")与"%" + name + "%"。

d. 使用 CriteriaBuilder 的 where 组合两个限制的 Predicate。

e. 获得 query 的组合的限制 Predicate。

代码验证：

```
...
import static top.wisely.learningspringdatajpa.domain.specification.CustomSpecs.*;
...
@Bean
CommandLineRunner specificationQuery(PersonRepository personRepository){
```

```
  return args -> {
    List<Person> people1 = personRepository.findAll(nameEqual("wyf")); //a
    people1.forEach(System.out::println);
    List<Person> people2 = personRepository.findAll(ageLessThanAndNameLike(37,"o"));
    people2.forEach(System.out::println);
    List<Person> people3 =
personRepository.findAll(nameEqual("bar").or(ageLessThanAndNameLike(37, "o"))); //b
    people3.forEach(System.out::println);
  };
}
```

a. 继承 JpaSpecificationExecutor 接口，获得了 List<T> findAll(Specification<T> spec)方法。

b. 可以用 Specification 的静态方法 or 或 and 等来连接多个 Specification。

代码运行结果如图 6-7 所示。

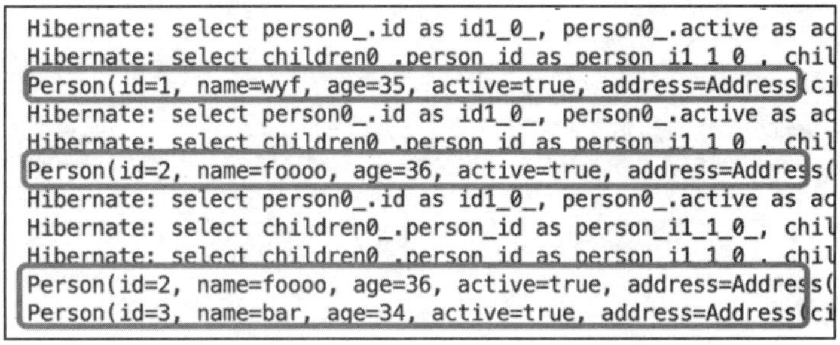

```
Hibernate: select person0_.id as id1_0_, person0_.active as ac
Hibernate: select children0_.person_id as person_i1_1_0_, chil
Person(id=1, name=wyf, age=35, active=true, address=Address(ci
Hibernate: select person0_.id as id1_0_, person0_.active as ac
Hibernate: select children0_.person_id as person_i1_1_0_, chil
Person(id=2, name=foooo, age=36, active=true, address=Address(
Hibernate: select person0_.id as id1_0_, person0_.active as ac
Hibernate: select children0_.person_id as person_i1_1_0_, chil
Hibernate: select children0_.person_id as person_i1_1_0_, chil
Person(id=2, name=foooo, age=36, active=true, address=Address(
Person(id=3, name=bar, age=34, active=true, address=Address(ci
```

图 6-7

4. Query by Example

Qurery by Example（简称 QBE）会根据部分属性已经设置的实体，动态地进行查询。QBE 主要分为 3 部分：

◎ Probe：设置属性的实体。

◎ ExampleMatcher：设置实体属性的匹配规则。

◎ Example：组合 Probe 和 ExampleMatcher 进行查询。

代码如下：

```
@Bean
CommandLineRunner queryByExample(PersonRepository personRepository){
  return args -> {
    Person person = new Person();
    person.setName("Y"); //a

    ExampleMatcher matcher = ExampleMatcher.matching() //b
         .withIgnoreCase("name") //c
         .withStringMatcher(StringMatcher.CONTAINING); //d

    Example<Person> example = Example.of(person, matcher); //e
```

```
        List<Person> people = personRepository.findAll(example); //f
        people.forEach(System.out::println);

    };
}
```

a. 构造 Probe 来设置 name 属性的值，此处设置为大写的“Y”。

b. 构造 ExampleMatcher 对象。

c. 在匹配 name 属性时可忽略大小写。

d. 字符类型匹配模式使用“包含”模式。

e. 使用 Probe 和 ExampleMatcher 来构造 Example。

f. 使用 Example 作为参数进行查询。

代码运行结果如图 6-8 所示。

```
Hibernate: select person0_.id as id1_0_, person0_.active as
Hibernate: select children0_.person_id as person_i1_1_0_, ch
Person(id=1, name=wyf, age=35, active=true, address=Address(
```

图 6-8

6.2.7 事件监听

在聚合根实体上，可以使用下面的注解来监听聚合操作的事件。

◎ @PrePersist：监听实体保存前事件。

◎ @PostPersist：监听实体保存后事件。

◎ @PreUpdate：监听实体更新前事件。

◎ @PostUpdate：监听实体更新后事件。

◎ @PreRemove：监听实体删除前事件。

◎ @PostRemove：监听实体删除后事件。

◎ @PostLoad：监听实体加载后事件。

```
@Data
@AllArgsConstructor
@NoArgsConstructor
@Entity
public class Person {
    //……
    @PrePersist
    public void prePersist(){
        System.out.println("prePersist:" + this);
    }
```

```
    @PostPersist
    public void postPersist(){
        System.out.println("postPersist:" + this);
    }

    @PreRemove
    public void preRemove(){
        System.out.println("preRemove:" + this);
    }

    @PostRemove
    public void postRemove(){
        System.out.println("postRemove:" + this);
    }

    @PreUpdate
    public void preUpdate(){
        System.out.println("preUpdate:" + this);
    }

    @PostUpdate
    public void postUpdate(){
        System.out.println("postUpdate:" + this);
    }

    @PostLoad
    public void postLoad(){
        System.out.println("postLoad:" + this);
    }

}
```

通过下面代码验证，结果如图 6-9 所示。

```
@Bean
CommandLineRunner entityListeners(PersonRepository personRepository){
    return args -> {
        Address address = new Address("wuhan","Hubei");
        Collection<Child> children = Arrays.asList(new Child("xxx", Gender.FEMALE));
        Person savedPerson = personRepository.save(new Person("www", 33, address, children));
        savedPerson.setAge(34);
        Person updatedPerson =  personRepository.save(savedPerson);
        personRepository.delete(updatedPerson);
    };
}
```

```
prePersist:Person(id=null, name=www, age=33, active=t
Hibernate: select next_val as id_val from hibernate_s
Hibernate: update hibernate_sequence set next_val= ?
Hibernate: insert into person (active, city, province
postPersist:Person(id=16, name=www, age=33, active=tr
Hibernate: insert into person_children (person_id, ge
Hibernate: select person0_.id as id1_0_0_, person0_.a
Hibernate: select children0_.person_id as person_i1_1
postLoad:Person(id=16, name=www, age=33, active=true,
preUpdate:Person(id=16, name=www, age=34, active=true
Hibernate: update person set active=?, city=?, provin
postUpdate:Person(id=16, name=www, age=34, active=tru
Hibernate: select person0_.id as id1_0_0_, person0_.a
Hibernate: select children0_.person_id as person_i1_1
postLoad:Person(id=16, name=www, age=34, active=true,
preRemove:Person(id=16, name=www, age=34, active=true
Hibernate: delete from person_children where person_i
Hibernate: delete from person where id=? and version=
postRemove:Person(id=16, name=www, age=34, active=tru
```

图 6-9

在实体类中写大量的代码显然不好，因此可以通过@EntityListeners({PersonListener.class})指定其他的类来处理。若需要让 Spring 对当前事件进行处理，则可以使用@Configurable 来注解该类，将依赖注入到非 Spring 容器管理，即将依赖注入到 Spring 容器之外的初始化的类中。

```
@Data
@AllArgsConstructor
@NoArgsConstructor
@Entity
@EntityListeners({PersonListener.class})
public class Person {}
@Configurable
public class PersonListener {
   private ListenerService listenerService;

   public PersonListener(ListenerService listenerService) {
      this.listenerService = listenerService;
   }

   @PrePersist
   public void prePersist(Person person){
      listenerService.process("prePersist:" + person);
   }

   @PostPersist
   public void postPersist(Person person){
      listenerService.process("postPersist:"+ person);
   }

   @PreRemove
   public void preRemove(Person person){
```

```
        listenerService.process("preRemove:" + person);
    }
    @PostRemove
    public void postRemove(Person person){
        listenerService.process("postRemove:" + person);
    }

    @PreUpdate
    public void preUpdate(Person person){
        listenerService.process("preUpdate:" + person);
    }

    @PostUpdate
    public void postUpdate(Person person){
        listenerService.process("postUpdate:" + person);
    }

    @PostLoad
    public void postLoad(Person person){
        listenerService.process("postLoad:" + person);
    }
}
```

被注入的依赖如下：

```
@Service
public class ListenerService {
    public void process(String msg){
        System.out.println("由 Spring 处理：" + msg);
    }
}
```

执行检验代码，结果如图 6-10 所示。

```
由Spring处理: prePersist:Person(id=null, name=www, age=33
Hibernate: select next_val as id_val from hibernate_sequ
Hibernate: update hibernate_sequence set next_val= ? whe
Hibernate: insert into person (active, city, province, a
由Spring处理: postPersist:Person(id=17, name=www, age=33,
Hibernate: insert into person_children (person_id, gende
Hibernate: select person0_.id as id1_0_0_, person0_.acti
Hibernate: select children0_.person_id as person_i1_1_0_
由Spring处理: postLoad:Person(id=17, name=www, age=33, ac
由Spring处理: preUpdate:Person(id=17, name=www, age=34, a
Hibernate: update person set active=?, city=?, province=
由Spring处理: postUpdate:Person(id=17, name=www, age=34,
Hibernate: select person0_.id as id1_0_0_, person0_.acti
Hibernate: select children0_.person_id as person_i1_1_0_
由Spring处理: postLoad:Person(id=17, name=www, age=34, ac
由Spring处理: preRemove:Person(id=17, name=www, age=34, a
Hibernate: delete from person_children where person_id=?
Hibernate: delete from person where id=? and version=?
由Spring处理: postRemove:Person(id=17, name=www, age=34,
```

图 6-10

6.2.8 领域事件

由于在 DDD 中采用了“设计小聚合”的原则，因此避免了领域模型的相互关联，从而避免了在应用演进中形成“大泥球”（Big Ball of Mud），也因为上述原因，本书不讲解@OneToMany、@ManyToMany 等关联注解。当聚合之间没有关联关系后，聚合之间的数据通信是通过领域事件来完成的，领域事件是由聚合根发出的。

Spring Data 对领域事件做了专门的支持，使用@DomainEvents 可注解注册领域事件；或者继承 AbstractAggregateRoot，使用它的 registerEvent 方法也可注册领域事件。

```
@Data
@AllArgsConstructor
@NoArgsConstructor
@Entity
public class Person {
   @DomainEvents // 使用集合类注册事件列表
   Collection<Object> domainEvents(){
      List<Object> events= new ArrayList<Object>();
      events.add(new PersonSaved(this.id, this.name, this.age));
      return events;
   }

   @AfterDomainEventPublication //在所有事件发布完成后调用，一般用来清空事件列表
   void callbackMethod() {
     domainEvents().clear();
   }
}
```

Repository 每一次调用 save 方法，领域事件都会被发布。

领域事件定义如下：

```
import lombok.Value;
@Value
public class PersonSaved {
  private Long id;
  private String name;
  private Integer age;
}
```

现在定义另外一个聚合根为雇员（Employee），它和 Person 是一一对应的关系，但是多了公司的信息。基于“设计小聚合”原则，并没有给它们配置@OneToOne，而是当 Person 保存成功后发布领域事件 PersonSaved，在事件监听的位置，于另外一个事务中新建对应的 Employee。“小聚合”的另外一个好处就是事务边界变小，从而有更快的速度和更好的性能。

新的雇员聚合：

```
@Data
@AllArgsConstructor
```

```
@NoArgsConstructor
@Entity
public class Employee {
    @Id
   private Long id;
   private String name;
   private Integer age;
   @Embedded
   private Company company;
}
@Data
@AllArgsConstructor
@NoArgsConstructor
@Embeddable
public class Company {
   private String name;
   private String city;
}
```

聚合的 Repository：

```
public interface EmployeeRepository extends JpaRepository<Employee, Long> {
  List<Employee> findByName(String name);
}
```

领域发布的是 Spring 事件，可以使用@EventListener 来接收。Spring Data 提供了专门的事务监听注解@TransactionalEventListener，它组合了@EventListener。

```
@Component
public class DomainEventListener {

   private EmployeeRepository employeeRepository;

   public DomainEventListener(EmployeeRepository employeeRepository) {
      this.employeeRepository = employeeRepository;
   }

   @Async //a
   @TransactionalEventListener
   public void handlePersonSavedEvent(PersonSaved personSaved){
      Company company = new Company("某某公司", "hefei");
      Optional<Employee> employeeOptional =
employeeRepository.findById(personSaved.getId());
      employeeOptional.ifPresent(employee -> { //b
         employee.setName(personSaved.getName());
         employee.setAge(personSaved.getAge());
        employeeRepository.save(employee);
         return;
      });
      employeeRepository.save(new Employee(personSaved.getId(), personSaved.getName(),
personSaved.getAge(), company));//c
```

```
    }
}
```

a. 使用@Async 注解在另外一个线程处理；需要在配置类启用异步支持@EnableAsync:

```
@SpringBootApplication
@EnableAsync
public class LearningSpringDataJpaApplication {}
```

b. 若存在，则更新雇员。

c. 若不存在，则保存新的雇员。

执行检验代码，结果如图 6-11 所示。

```
@Bean
CommandLineRunner domainEvents(PersonRepository personRepository,
                EmployeeRepository employeeRepository){
  return args -> {
    Address address = new Address("nanjing","Jiangsu");
    Collection<Child> children = Arrays.asList(new Child("xxxx", Gender.FEMALE));
    Person savedPerson = personRepository.save(new Person("wwww", 33, address,
children));
    Thread.sleep(100); //监听是在异步线程执行的，所以需要等待
    List<Employee> employees1 = employeeRepository.findByName("wwww");
    employees1.forEach(System.out::println);
    savedPerson.setName("wwwww");
    personRepository.save(savedPerson);
    Thread.sleep(100);
    List<Employee> employees2 = employeeRepository.findByName("wwwww");
    employees2.forEach(System.out::println);
  };
}
```

```
Hibernate: insert into employee (age, city, company_name, name, id) values (?, ?, ?,
Hibernate: select employee0_.id as id1_0_, employee0_.age as age2_0_, employee0_.city
Employee(id=23, name=wwww, age=33, company=Company(companyName=某某公司, city=hefei))
Hibernate: select person0_.id as id1_1_, person0_.active as active2_1_, person0_.
Hibernate: select children0_.person_id as person_i1_2_0_, children0_.gender as gender
由Spring处理: postLoad:Person(id=23, name=wwww, age=33, active=true, address=Address(c
由Spring处理: preUpdate:Person(id=23, name=wwwww, age=33, active=true, address=Address
Hibernate: update person set active=?, city=?, province=?, age=?, create_time=?, crea
由Spring处理: postUpdate:Person(id=23, name=wwwww, age=33, active=true, address=Addres
Hibernate: select employee0_.id as id1_0_0_, employee0_.age as age2_0_0_, employee0_.
Hibernate: select employee0_.id as id1_0_0_, employee0_.age as age2_0_0_, employee0_.
Hibernate: update employee set age=?, city=?, company_name=?, name=? where id=?
Hibernate: select employee0_.id as id1_0_0_, employee0_.age as age2_0_0_, employee0_.
Hibernate: select employee0_.id as id1_0_, employee0_.age as age2_0_, employee0_.city
Employee(id=23, name=wwwww, age=33, company=Company(companyName=某某公司, city=hefei)
```

图 6-11

6.2.9 审计功能

Spring Data JPA 提供了审计功能，注解如下。

◎ @CreatedDate：数据创建事件。

◎ @CreatedBy：数据创建人。

◎ @LastModifiedDate：数据最后修改时间。

◎ @LastModifiedBy：数据最后修改人。

需要在入口类上开启配置：

```
@SpringBootApplication
@EnableJpaAuditing
public class LearningSpringDataJpaApplication {}
```

在实体类上指定监听器：

```
@Data
@AllArgsConstructor
@NoArgsConstructor
@Entity
@EntityListeners({AuditingEntityListener.class})
public class Person {
   @CreatedDate
   private LocalDateTime createTime;

   @CreatedBy
   private String createdUser;

   @LastModifiedDate
   private LocalDateTime updateTime;

   @LastModifiedBy
   private String updatedUser;
}
```

另外，还需指定创建人和修改人，这里使用硬编码，后面将指定为系统用户：

```
@Bean
AuditorAware<String> auditorProvider(){
  return () -> Optional.of("wyf");
}
```

执行检验代码，结果如图 6-12 所示。

```
@Bean
CommandLineRunner audit(PersonRepository personRepository,
                EmployeeRepository employeeRepository){
  return args -> {
    Address address = new Address("zhengzhou","Henan");
    Collection<Child> children = Arrays.asList(new Child("ppp", Gender.FEMALE));
```

```
    Person savedPerson = personRepository.save(new Person("zzz", 33, address, children));
    Thread.sleep(1000);
    savedPerson.setName("zzzz");
    Person updatedPerson = personRepository.save(savedPerson);
    System.out.println("创建时间:" + updatedPerson.getCreateTime());
    System.out.println("创建人:" + updatedPerson.getCreatedUser());
    System.out.println("最后更新时间:" + updatedPerson.getUpdateTime());
    System.out.println("最后更新人" + updatedPerson.getUpdatedUser());
  };
}
```

```
创建时间:2019-05-13T20:10:55.364798
创建人:wyf
最后更新时间:2019-05-13T20:10:56.455834
最后更新人wyf
```

图 6-12

6.2.10 Web 支持

Spring Data 给 Web 开发提供了一定的支持，它通过@EnableSpringDataWebSupport 开启支持。EnableSpringDataWebSupport 导入了 SpringDataWebConfiguration 配置，通过 SpringDataWebConfiguration 配置注册了 DomainClassConverter 和 PageableHandlerMethodArgumentResolver。由于 SpringDataWebAutoConfiguration 可以自动配置，所以无须手动定义。

通过 DomainClassConverter，可以用控制器方法路径变量直接查询 Repository 中的实例，或者在请求参数中直接查询 Repository 中的实例。

```
@RestController
@RequestMapping("/people")
public class PersonController {

   @GetMapping("/{id}")
   public Person findOne(@PathVariable("id") Person person){
      return person;
   }

}
```

运行结果如图 6-13 所示。

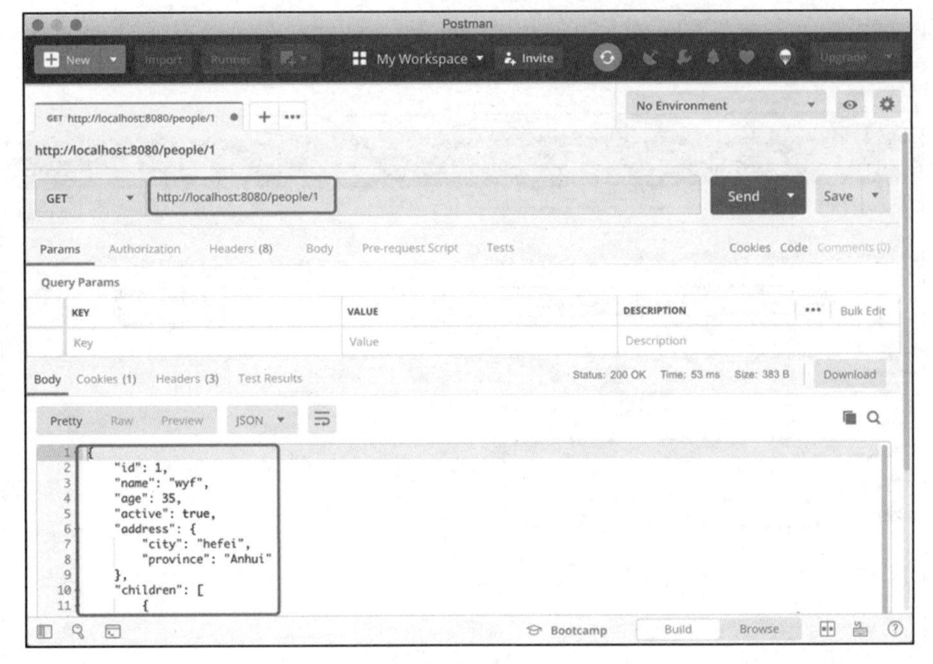

图 6-13

在 PageableHandlerMethodArgumentResolver 中，控制器方法可以通过 Pageable 获得客户端传递的分页和排序参数，参数如下。

◎ page：页数，默认为 0。

◎ size：每页数量，默认为 20。

◎ Sort：排序。

```
@RestController
@RequestMapping("/people")
public class PersonController {

   private PersonRepository personRepository;

   public PersonController(PersonRepository personRepository) {
      this.personRepository = personRepository;
   }

   @GetMapping
   public Page<Person> findAllByPage(Pageable pageable){
     return personRepository.findAll(pageable);
   }

}
```

访问 http://localhost:8080/people?page=0&size=2&sort=name,desc，通过参数即可完成分页和排序，如图 6-14 所示。

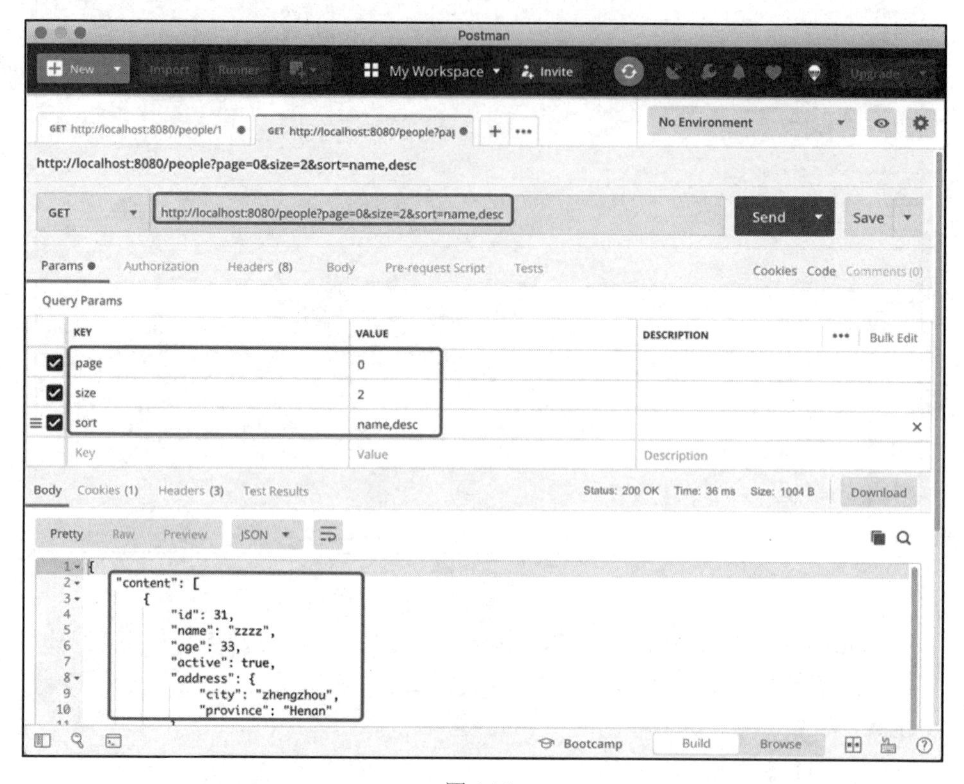

图 6-14

6.2.11 数据库初始化

1. 数据库结构初始化

Spring Boot 会根据实体类的配置自动生成 DDL（Data Definition Language，即数据定义语言，用来定义数据库结构）。DDL 可由下面两个外部配置属性开启。

```
spring.jpa.generate-ddl: true
```

是否在启动时初始化数据库结构。

```
spring.jpa.hibernate.ddl-auto: create
```

该配置一共有 5 个可选项。

◎ none：关闭 DDL 处理。

◎ validate：验证数据库结构，数据库不发生改变。

◎　update：更新数据库结构。

◎　create：创建数据库结构，并销毁以前的数据。

◎　create-drop：创建数据库结构，并在会话关闭后销毁数据库结构。

在使用内嵌数据库时，默认为 create-drop；其他情况下默认为 none。

2. 脚本初始化

DataSourceAutoConfiguration 提供了使用 SQL 脚本来初始化数据库的功能。Spring Boot 会自动处理根类路径位置的 schema.sql 和 data.sql，还可以根据 spring.datasource.platform 定义的 platform 自动处理 schema-${platform}.sql 和 data-${platform}.sql。platform 可使用的数据库有 HSQLDB、H2、Oracle、MySQL 和 PostgreSQL 等。

这里使用 schema.sql 来存放 DDL（CREATE、ALTER、DROP），使用 data.sql 来存放 DML（Data Manipulation Language，数据操作语言，包含 INSERT、UPDATE 和 DELETE）。

首先在外部属性做下面配置：

```
spring.datasource.initialization-mode: always
```

关闭 Hibernate，自动生成 DDL，让脚本来接管数据库的初始化。直接删除配置语句：

```
spring.jpa.hibernate.ddl-auto: none
```

建立 schema.sql，放置在 src/main/resources 目录下：

```
CREATE TABLE IF NOT EXISTS another_person(
  id bigint(20) NOT NULL AUTO_INCREMENT, //ID 使用自增
  name varchar(255) NOT NULL,
  age int(11) DEFAULT NULL,
  PRIMARY KEY (id)
);
```

建立 data.sql，把它放置在 src/main/resources 目录下：

```
INSERT INTO another_person (name, age) VALUES ('wyf', 35);
INSERT INTO another_person (name, age) VALUES ('foo', 34);
INSERT INTO another_person (name, age) VALUES ('bar', 33);
INSERT INTO another_person (name, age) VALUES ('www', 32);
```

对应的实体如下：

```
@Data
@AllArgsConstructor
@NoArgsConstructor
@Entity
@Table(name = "another_person")
public class AnotherPerson {
    @Id
    @GeneratedValue(strategy = GenerationType.IDENTITY) //使用 MySQL 的自增
    private Long id;
```

```
    private String name;
    private Integer age;
}
```

启动应用，如图 6-15 所示。

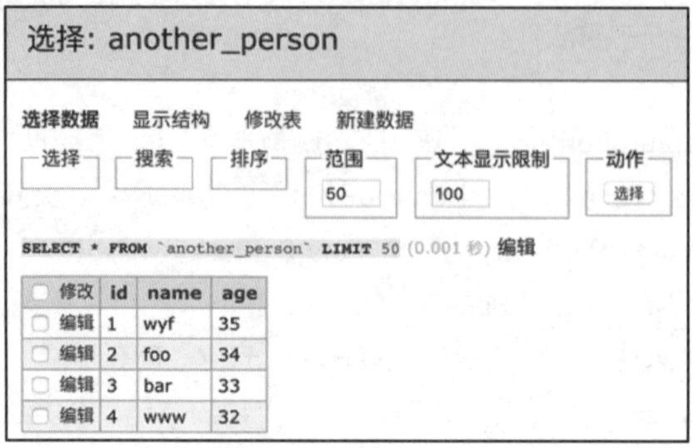

图 6-15

在第二次启动之前，关闭初始化，以防止在第二次启动时 data.sql 插入重复的数据，因为 ID 是自动生成的。

```
spring.datasource.initialization-mode: never
```

若在 data.sql 里指定了数据的 ID，则在第二次启动时会发生主键冲突错误。如果有应用因 SQL 脚本执行导致错误，则应做如下设置：

```
spring.datasource.continue-on-error: true
```

3. 数据库迁移工具

Spring Boot 支持的数据库迁移工具有 Flyway 和 Liquibase 两种。本节重点讲解 Flyway 的用法。

在 Spring Boot 中，FlywayAutoConfiguration 使用 FlywayProperties 对 Flyway 进行自动配置，外部配置通过 spring.flyway.*前缀进行。

在 build.gradle 中添加 Flyway 的依赖：

```
dependencies {
    //……
  runtimeOnly 'org.flywaydb:flyway-core'
  //……
}
```

默认自动执行在 classpath:db/migration 目录下的 V<VERSION>__<NAME>.sql（注意，中间的下画线是两个，如 V1__Initial_Setup.sql、V1_1__Some_Changes.sql 等，VERSION 内部使用下画线分隔）。

在 src/main/resources/db/migration 中建立 V1__Initial_Setup.sql：

```
CREATE TABLE IF NOT EXISTS third_person(
  id bigint(20) NOT NULL AUTO_INCREMENT,
  name varchar(255) NOT NULL,
  age int(11) DEFAULT NULL,
  PRIMARY KEY (id)
);

INSERT INTO third_person (name, age) VALUES ('wyf', 35);
INSERT INTO third_person (name, age) VALUES ('foo', 34);
INSERT INTO third_person (name, age) VALUES ('bar', 33);
INSERT INTO third_person (name, age) VALUES ('www', 32);
```

◎ 因为 Flyway 需要数据库处于空白状态，所以删除 first_db 中所有的表及存储过程。启动应用，数据库创建了两张表：flyway_schema_history 和 third_person，分别如图 6-16 和图 6-17 所示。

◎ flyway_schema_history 用来记录数据库的演进历史。

选择：flyway_schema_history

SELECT * FROM `flyway_schema_history` LIMIT 50 (0.001 秒) 编辑

修改	installed_rank	version	description	type	script
编辑	1	1	Initial Setup	SQL	V1__Initial_Setup.sql

图 6-16

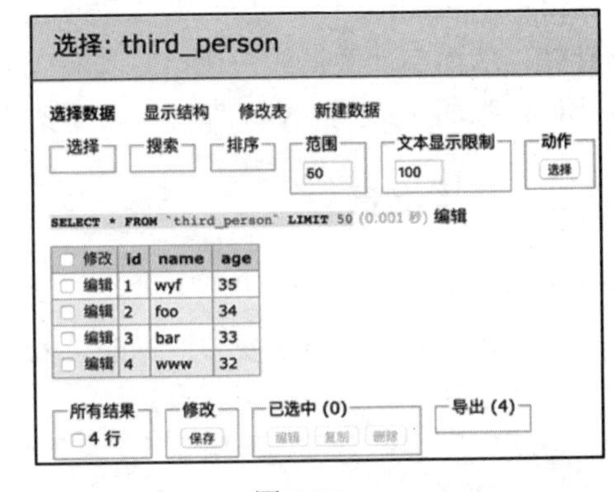

图 6-17

在 src/main/resources/db/migration 中建立 V1_1__Some_Changes.sql，对数据库进行以下修改：

```
ALTER TABLE third_person add COLUMN city varchar(255) NOT NULL;

UPDATE third_person set city = 'hefei';
```

◎ 再次启动应用，flyway_schema_history 和 third_person 分别如图 6-18 和图 6-19 所示。

修改	installed_rank	version	description	type	script
编辑	1	1	Initial Setup	SQL	V1__Initial_Setup.sql
编辑	2	1.1	Some Changes	SQL	V1_1__Some_Changes.sql

图 6-18

修改	id	name	age	city
编辑	1	wyf	35	hefei
编辑	2	foo	34	hefei
编辑	3	bar	33	hefei
编辑	4	www	32	hefei

图 6-19

第一个 V1__Initial_Setup.sql 版本是 1，在 Flyway 中称之为基线版本，默认也是 1，可以通过 spring.flyway.baseline-version 来设置。Flyway 默认连接当前数据库，如果要初始化到另外的数据库，则可使用 spring.flyway.schemas 来设置。

6.3 NoSQL——Spring Data Elasticsearch

6.3.1 Elasticsearch 简介

Elasticsearch 是一个分布式、RESTful 风格的搜索和数据分析引擎，可以把它作为集中的数据存储，从而实现数据的快速检索和高级分析。在使用 Spring Data Elasticsearch 之后，可以使用我们熟悉的 Repository 编程模型，而无须使用 Elasticsearch 的查询自定义语言。

6.3.2 环境准备

1．安装 Elasticsearch

使用 docker compose 安装 Elasticsearch。

stack.yml：

```
version: '3.1'

services:
  elasticsearch:
    image: elasticsearch:7.0.1
    environment:
      - cluster.name=docker-cluster
      - discovery.type=single-node #测试使用单节点的 Elascticsearch
    ports:
      - "9200:9200"
      - "9300:9300"
```

执行命令：

```
$ docker-compose -f stack.yml up -d
```

代码运行结果如图 6-20 所示。

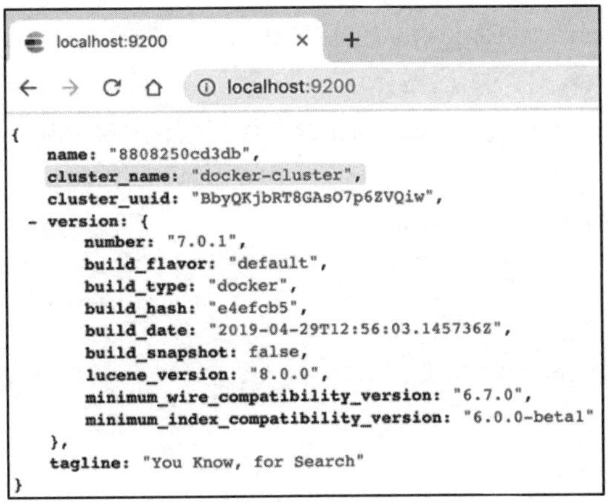

localhost:9200

localhost:9200


```
{
   name: "8808250cd3db",
   cluster_name: "docker-cluster",
   cluster_uuid: "BbyQKjbRT8GAsO7p6ZVQiw",
 - version: {
       number: "7.0.1",
       build_flavor: "default",
       build_type: "docker",
       build_hash: "e4efcb5",
       build_date: "2019-04-29T12:56:03.145736Z",
       build_snapshot: false,
       lucene_version: "8.0.0",
       minimum_wire_compatibility_version: "6.7.0",
       minimum_index_compatibility_version: "6.0.0-beta1"
   },
   tagline: "You Know, for Search"
}
```


图 6-20

2. 新建项目

新建项目，信息如下。

Group：top.wisely。

Artifact：learning-spring-data-elasticsearch。

Dependencies：Spring Web Starter、Spring Data Elasticsearch 和 Lombok。

build.gradle 文件中的依赖如下：

```
dependencies {
  implementation 'org.springframework.boot:spring-boot-starter-data-elasticsearch'
  implementation 'org.springframework.boot:spring-boot-starter-web'
  compileOnly 'org.projectlombok:lombok'
     annotationProcessor 'org.projectlombok:lombok'
  //……
}
```

6.3.3 自动配置

Spring Boot 提供的 Elasticsearch 的自动配置主要包括以下内容。

（1）RestClientAutoConfiguration：使用 RestClientProperties 通过 spring.elasticsearch.rest.*来自动配置 Elasticsearch REST 客户端。

- ◎ 在类路径中依赖 org.elasticsearch.client:elasticsearch-rest-client，可自动配置 RestClient。
- ◎ 在类路径中依赖 org.elasticsearch.client:elasticsearch-rest-high-level-client ，可自动配置 RestHighLevelClient。RestHighLevelClient 包装了已有的 RestClient 的 Bean，可复用 HTTP 配置。

（2）ElasticsearchRepositoriesAutoConfiguration：自动通过@EnableElasticsearchRepositories

开启对 Spring Data Elasticsearch Repository 的支持。

注意，Elasticsearch Jest 客户端已被弃用。

Elasticsearch TransportClient 在 Elasticsearch 7.0.0 版本中被弃用，将在 Elasticsearch 8.0 版本中被移除。

通过 REST 客户端连接 Elasticsearch：

```
spring:
  elasticsearch:
    rest:
      uris: http://localhost:9200
```

6.3.4 定义聚合

Spring Data 提供了下面的注解来定义聚合。

◎ @Document：注解在聚合根上，将此类与 Elasticsearch 的 index 进行映射，可设置 index 名称和 index 类型。

◎ @Id：聚合根的唯一标识，注解在 String 类型的 ID 上，Elasticsearch 会自动生成唯一标识。

◎ @Field：表示当前属性是值或者是值对象，可设置成与 Elasticsearch 的 Field 相关，如映射的名称，可省略。

◎ @Transient：注解的属性将不会持久化到 Elasticsearch 中。

聚合根定义如下：

```
@Data
@AllArgsConstructor
@NoArgsConstructor
@Document(indexName = "person")
public class Person {
    @Id
    private String id;
    private String name;
    private Integer age;
    private Address address;
    Collection<Child> children;

    public Person(String name, Integer age, Address address, Collection<Child> children)
{
        this.name = name;
        this.age = age;
        this.address = address;
        this.children = children;
    }
}
```

另外，两个值对象定义如下：

```
@Data
@AllArgsConstructor
@NoArgsConstructor
public class Address {
   private String city;
   private String province;
}
@Data
@AllArgsConstructor
@NoArgsConstructor
public class Child {
   private String name;
   private Gender gender;
}
```

6.3.5　定义聚合 Repository

定义聚合 Repository，需继承 ElasticsearchRepository 接口。

```
public interface PersonRepository extends ElasticsearchRepository<Person,String> {
}
```

ElasticsearchRepository 接口的代理实现是 SimpleElasticsearchRepository，它需要 ElasticsearchOperations 的 Bean，而 AbstractElasticsearchConfiguration 提供了配置 Elasticsearch 的能力，包括配置好的 ElasticsearchOperations 的 Bean（ElasticsearchRestTemplate）。

```
@Configuration
public class RestEsConfig extends AbstractElasticsearchConfiguration {
   private RestHighLevelClient restHighLevelClient;

   public RestEsConfig(RestHighLevelClient restHighLevelClient) {
      this.restHighLevelClient = restHighLevelClient;
   }

   @Override
   public RestHighLevelClient elasticsearchClient() {
      return this.restHighLevelClient;
   }
}
```

保存三条测试数据：

```
@Bean
CommandLineRunner saveAll(PersonRepository personRepository){
  return args -> {
    Address address = new Address("he fei", "Anhui");
    Collection<Child> children = Arrays.asList(new Child("wyn", Gender.FEMALE),
                                new Child("wbe", Gender.MALE));
```

```
        Person person = new Person("wyf", 35, address, children);

        Address address1 = new Address("bei jing", "Beijing");
        Collection<Child> children1 = Arrays.asList(new Child("aaa", Gender.FEMALE),
             new Child("bbb", Gender.MALE));
        Person person1 = new Person("foo", 34, address1, children1);

        Address address2 = new Address("shang hai", "Shanghai");
        Collection<Child> children2 = Arrays.asList(new Child("ccc", Gender.FEMALE),
             new Child("ddd", Gender.MALE));
        Person person2 = new Person("bar", 36, address2, children2);

        personRepository.saveAll(Arrays.asList(person, person1, person2));
    };
```

执行完成后，注释@Bean，避免重复插入。

6.3.6 查询

1. 查询方法

在 PersonRepository 中定义查询方法，这里的查询和 Spring Data JPA 一致：

```
public interface PersonRepository extends ElasticsearchRepository<Person,String> {
    List<Person> findByName(String name); //a

    List<Person> findByAddress_City(String city); //b

    List<Person> findByChildren_Name(String childName); //c

    @Query("{\"bool\" : {\"must\" : {\"range\" : {\"age\" : {\"gte\" : \"?0\", \"lte\" :
\"?1\"}}}}}")
    Page<Person> findByAgeRange(Integer startAge, Integer endAge, Pageable pageable); //d
}
```

a. 通过值 name 查询。

b. 通过值对象 Address 的 city 属性查询。

c. 通过列表值对象 children 的 name 查询。

d. 通过@Query 使用 Elasticsearch 查询自定义语言，查询 age 在 startAge 和 endAge 之间的 Person，并使用分页和排序功能。

通过代码验证：

```
@Bean
CommandLineRunner query(PersonRepository personRepository){
  return args -> {
    List<Person> people1 = personRepository.findByName("wyf");
    List<Person> people2 = personRepository.findByAddress_City("bei jing");
    List<Person> people3 = personRepository.findByChildren_Name("ccc");
    Page<Person> personPage = personRepository.findByAgeRange(30, 40,
```

```
        PageRequest.of(0,3, Sort.by(Sort.Direction.DESC,"age")));
    people1.forEach(System.out::println);
    people2.forEach(System.out::println);
    people3.forEach(System.out::println);
    System.out.println("总数为: " + personPage.getTotalElements()
        + " 总页数为: " + personPage.getTotalPages());
    personPage.forEach(System.out::println);
  };
}
```

运行结果如图 6-21 所示。

```
Person(id=8Le4umoBJcSue5oMyKLB, name=wyf, age=35,
Person(id=8be4umoBJcSue5oMyKLC, name=foo, age=34,
Person(id=8re4umoBJcSue5oMyKLC, name=bar, age=36,
总数为: 3 总页数为: 1
Person(id=8re4umoBJcSue5oMyKLC, name=bar, age=36,
Person(id=8Le4umoBJcSue5oMyKLB, name=wyf, age=35,
Person(id=8be4umoBJcSue5oMyKLC, name=foo, age=34,
```

图 6-21

2. search 查询

ElasticsearchRepository 除提供 CRUD 和分页排序功能外，还提供了支持 Elasticsearch API 的 search 方法。

```
@NoRepositoryBean
public interface ElasticsearchRepository<T, ID> extends ElasticsearchCrudRepository<T,
ID> {
    //……
  Iterable<T> search(QueryBuilder query);

  Page<T> search(QueryBuilder query, Pageable pageable);

  Page<T> search(SearchQuery searchQuery);
    //……
}
```

org.springframework.data.elasticsearch.core.query.SearchQuery 可使用 org.elasticsearch.index.query.QueryBuilder 构造查询条件，所以三个 search 方法是类似的，验证代码如下：

```
@Bean
CommandLineRunner queryBuilderAndSearchQuery(PersonRepository personRepository){
  return args -> {
    QueryBuilder queryBuilder = QueryBuilders.matchAllQuery(); //a

    SearchQuery searchQuery = new NativeSearchQueryBuilder() //b
        .withQuery(queryBuilder) // c
        .withPageable(PageRequest.of(0, 3, Sort.by("age"))) //d
        .build(); //e
```

```
    Page<Person> personPage1 = personRepository.search(queryBuilder, PageRequest.of(0,
3, Sort.by("age"))); //f
    Page<Person> personPage2 = personRepository.search(searchQuery); //g
            System.out.println("personPage1 总数为: " + personPage1.getTotalElements()
                    + " 总页数为: " + personPage1.getTotalPages());
            personPage1.forEach(System.out::println);
            System.out.println("personPage2 总数为: " + personPage2.getTotalElements()
                    + " 总页数为: " + personPage2.getTotalPages());
            personPage2.forEach(System.out::println);
  };
}
```

a. 在 QueryBuilders 中有大量的静态方法可用来构造 QueryBuilder。

◎ matchAllQuery()：查询所有的数据，返回值是 QueryBuilder 的子类 MatchAllQueryBuilder。

◎ matchQuery(String name, Object text)：匹配属性值，返回值是 QueryBuilder 的子类 MatchQueryBuilder。

◎ fuzzyQuery(String name, String value)：模糊匹配属性值，返回值是 QueryBuilder 的子类 FuzzyQueryBuilder。

◎ rangeQuery(String name)：范围查询，返回值是 QueryBuilder 的子类 RangeQueryBuilder。

b. 通过 NativeSearchQueryBuilder 构造 SearchQuery。

c. 使用 withQuery(QueryBuilder queryBuilder)，通过 QueryBuilder 设置查询。

d. 设置分页和排序信息。

e. 创建 SearchQuery。

f. 使用 Page<T> search(QueryBuilder query, Pageable pageable)方法查询。

g. 使用 Page<T> search(SearchQuery searchQuery)方法查询，两次查询的结果是一致的。

代码运行结果如图 6-22 所示。

```
personPage1总数为: 3 总页数为: 1
Person(id=8be4umoBJcSue5oMyKLC, name=foo, age=34,
Person(id=8Le4umoBJcSue5oMyKLB, name=wyf, age=35,
Person(id=8re4umoBJcSue5oMyKLC, name=bar, age=36,
personPage2总数为: 3 总页数为: 1
Person(id=8be4umoBJcSue5oMyKLC, name=foo, age=34,
Person(id=8Le4umoBJcSue5oMyKLB, name=wyf, age=35,
Person(id=8re4umoBJcSue5oMyKLC, name=bar, age=36,
```

图 6-22

使用 SearchQuery 不仅更友好，而且可读性更强，下面再演示几个查询的例子：

```
@Bean
CommandLineRunner search(PersonRepository personRepository){
```

```
    return args -> {
        SearchQuery queryByAgeRangeAndNameRegex = new NativeSearchQueryBuilder()
            .withQuery(rangeQuery("age").from(30).to(35)) // a
            .withFilter(matchQuery("name", "wyf")) // b
            .build();

        SearchQuery fuzzyQuery = new NativeSearchQueryBuilder()
            .withQuery(fuzzyQuery("address.city", "shang").fuzziness(Fuzziness.AUTO))
//c
            .build();

        SearchQuery anotherFuzzyQuery = new NativeSearchQueryBuilder()
            .withQuery(matchQuery("address.city", "jing")
                    .fuzziness(Fuzziness.AUTO))//d
            .build();

        Page<Person> personPage1 = personRepository.search(queryByAgeRangeAndNameRegex);
        Page<Person> personPage2 = personRepository.search(fuzzyQuery);
        Page<Person> personPage3 = personRepository.search(anotherFuzzyQuery);

        personPage1.forEach(System.out::println);
        System.out.println("--------------");
        personPage2.forEach(System.out::println);
        System.out.println("--------------");
        personPage3.forEach(System.out::println);
        System.out.println("--------------");

    };
}
```

a. 构造范围查询，form 和 to 是 RangeQueryBuilder 的方法。

b. withFilter(QueryBuilder filterBuilder)的入参为 QueryBuilder。

c. 构造模糊查询，fuzziness 是 FuzzyQueryBuilder 的方法。

d. 使用匹配查询构造模糊查询，fuzziness 是 MatchQueryBuilder 的方法。

代码运行结果如图 6-23 所示。

```
Person(id=Greiu2oBJcSue5oMgK05, name=wyf, age=35, address=Address(city=he fei, province=Anhui), chil
--------------
Person(id=HLeiu2oBJcSue5oMgK05, name=bar, age=36, address=Address(city=shang hai, province=Shanghai)
--------------
Person(id=G7eiu2oBJcSue5oMgK05, name=foo, age=34, address=Address(city=bei jing, province=Beijing), 
```

图 6-23

3. 查询统计

可以使用 SearchQuery 构造聚合查询统计，如总数、最大值、最小值、平均值、统计信息等。需要通过 ElasticsearchOperations（由 AbstractElasticsearchConfiguration 提供的 Bean）来执行查询统计。

```
@Bean
CommandLineRunner aggregateQuery(ElasticsearchOperations elasticsearchOperations){

  return args -> {

    SearchQuery aggregateQuery = new NativeSearchQueryBuilder().withIndices("person")
//a
        .addAggregation(AggregationBuilders.sum("sumAge").field("age")) //b
        .addAggregation(AggregationBuilders.max("maxAge").field("age"))
        .addAggregation(AggregationBuilders.min("minAge").field("age"))
        .addAggregation(AggregationBuilders.count("countAge").field("age"))
        .addAggregation(AggregationBuilders.avg("avgAge").field("age"))
        .addAggregation(AggregationBuilders.stats("ageInfo").field("age")) //c
        .addAggregation(AggregationBuilders.range("ageRange")
                                .field("age")
                                .addRange(30,35)) //d
        .addAggregation(AggregationBuilders.filter("leftPerson",
                               rangeQuery("age").from(30).to(35))

    .subAggregation(AggregationBuilders.sum("leftSum").field("age"))) //e
        .build();

    Aggregations aggs = elasticsearchOperations.query(aggregateQuery, searchResponse ->
{
      Aggregations aggregations = searchResponse.getAggregations();
      return aggregations;
    }); //f

    double sum = ((ParsedSum)aggs.get("sumAge")).getValue(); //g
    System.out.println("sum age is " + sum);
    double max = ((ParsedMax)aggs.get("maxAge")).getValue();
    System.out.println("max age is " + max);
    double min = ((ParsedMin)aggs.get("minAge")).getValue();
    System.out.println("min age is " + min);
    double count = ((ParsedValueCount)aggs.get("countAge")).getValue();
    System.out.println("count is " + count);
    double avg = ((ParsedAvg)aggs.get("avgAge")).getValue();
    System.out.println("avg age is " + avg);

    ParsedStats stats = ((ParsedStats)aggs.get("ageInfo")); //h
    System.out.println("stats sum age is " + stats.getSumAsString());
    System.out.println("stats max age is " + stats.getMaxAsString());
    System.out.println("stats min age is " + stats.getMinAsString());
    System.out.println("stats avg age is " + stats.getAvgAsString());
    System.out.println("stats count is " + stats.getCount());

    ((ParsedRange)aggs.get("ageRange")).getBuckets().forEach(bucket -> { //i
      System.out.println(bucket.getFromAsString()
              + "到" + bucket.getToAsString()
               + "数量为: " + bucket.getDocCount());
```

```
    });

            Aggregations filterAggs =
((ParsedFilter)aggs.get("leftPerson")).getAggregations();//j
            double leftSum = ((ParsedSum)filterAggs.get("leftSum")).getValue();
            System.out.println("left sum is " + leftSum);
  };
}
```

a. 通过 withIndices 方法指定索引名称，若不调用此方法，则统计整个 Elasticsearch 所有属性为 age 的索引。

b. 通过 addAggregation 添加聚合统计。使用 AggregationBuilders 的静态方法来构造聚合统计，如 sum()、max()、min()、avg()或 count()等。静态方法接收的参数是聚合统计的名称。通过 field 方法可设置需要统计的属性。

c. 构造统计查询，stats()方法包含的信息有 sum、max、min、avg 和 count。

d. 使用 range()构造范围统计。

e. 先用 filter 方法过滤人员，再通过 subAggregation 构造子聚合来统计剩下来的人员的年龄总和。

f. 通过 ElasticsearchOperations 的 query(SearchQuery query, ResultsExtractor<T> resultsExtractor)方法执行查询，聚合是通过 ResultsExtractor 函数接口获得的：

```
public interface ResultsExtractor<T> {
  T extract(SearchResponse response);
}
```

g. 不同的聚合有不同的类型，可以用 Aggregations 的 get("聚合名称")来获得聚合，并强制转换成 ParsedSum、ParsedMax、ParsedMin、ParsedValueCount 或 ParsedAvg 等，再使用 getValue()方法获取实际值。

h. ParsedStats 包含了 sum、max、min、avg 和 count。

i. ParsedRange 可通过 Bucket 的 getDocCount()方法获得符合范围的数量。

j. 可以从 ParsedFilter 聚合中获得它的子聚合 ParsedSum。

6.4 数据缓存

6.4.1 Spring Boot 与缓存

缓存服务数据能极大地提升应用的性能。在使用 Spring 缓存前，需开启@EnableCaching，说明如下。

◎ @Cacheable：注解方法可以被缓存。对于特定缓存 key，方法只执行一次，后续的请求将不执行方法，而是从缓存中获取数据。

◎ @CachePut：注解方法触发缓存添加操作。每次请求都会执行@CachePut 注解的方法。

◎ @CacheEvict：注解方法触发从缓存中移除旧数据的操作。

◎ @Caching：支持@Cacheable、@CachePut 和@CacheEvict 组合注解在一个方法上。

◎ @CacheConfig：类级别的共享配置。

一旦开启了缓存支持，Spring Boot 会按照下面的顺序查找缓存提供者。在使用缓存技术前，需要配置相关的 CacheManager 的 Bean。

◎ Generic：SimpleCacheManager。

◎ JCache（JSR-107）：JCacheCacheManager。

◎ EhCache 2.x：EhCacheCacheManager。

◎ Hazelcast：HazelcastCacheManager。

◎ Infinispan：SpringEmbeddedCacheManager。

◎ CouchBase：CouchbaseCacheManager。

◎ Redis：RedisCacheManager。

◎ Caffeine：CaffeineCacheManager。

◎ Simple：ConcurrentMapCacheManager。

Spring Boot 通过 CacheAutoConfiguration 实现对这些缓存提供技术的 CacheManager 的自动配置，可以通过使用 CacheProperties 提供的 spring.cache.*来定制一些配置。

```
spring.cache.type: redis #设置默认缓存提供者
```

Redis 是知名的内存数据库，也是优秀的缓存提供者。Spring Boot 通过 CacheAutoConfiguration 自动加载 RedisCacheConfiguration，而 RedisCacheConfiguration 自动配置了 RedisCacheManager 的 Bean。

当 Redis 相关库在类路径中时，Spring Boot 会自动使用 RedisAutoConfiguration，通过 RedisProperties 提供的 spring.redis.*来配置连接 Redis。

6.4.2 环境准备

1. 安装 Redis

使用 docker compose 安装 Redis。

stack.yml：

```
version: '3.1'

services:
  redis:
    image: 'bitnami/redis:5.0'
    environment:
```

```
    - REDIS_PASSWORD=zzzzzz
  ports:
    - '6379:6379'
```

执行命令：

```
$ docker-compose -f stack.yml up -d
```

2．新建项目

新建项目，信息如下。

Group：top.wisely。

Artifact：learning-caching。

Dependencies：Spring Cache abstraction、Spring Data Redis、Spring Web Starter、Spring Data JPA、MySQL Driver 和 Lombok。

build.gradle 文件中的依赖如下：

```
dependencies {
  implementation 'org.springframework.boot:spring-boot-starter-cache'
  implementation 'org.springframework.boot:spring-boot-starter-data-jpa'
  implementation 'org.springframework.boot:spring-boot-starter-data-redis'
  implementation 'org.springframework.boot:spring-boot-starter-web'
  compileOnly 'org.projectlombok:lombok'
  runtimeOnly 'mysql:mysql-connector-java'
  annotationProcessor 'org.projectlombok:lombok'
 //……
}
```

外部配置如下：

```
spring:
  cache:
    type: redis
  redis: # 连接Redis
    host: localhost
    password: zzzzzz
    port: 6379
  datasource: # 连接数据库
    url: jdbc:mysql://localhost:3306/first_db?useSSL=false
    username: root
    password: zzzzzz
    driver-class-name: com.mysql.cj.jdbc.Driver
    initialization-mode: always
    continue-on-error: true
  jpa:
    show-sql: true
    hibernate:
      ddl-auto: update
```

测试的实体类如下：

```
@Data
@AllArgsConstructor
@NoArgsConstructor
@Entity
public class Student implements Serializable {
   @Id
   @GeneratedValue
   private Long id;
   private String name;

   public Student(String name) {
      this.name = name;
   }
}
```

Repository 如下：

```
public interface StudentRepository extends JpaRepository<Student, Long> {
}
```

测试数据 data.sql：

```
insert into student values (1, "xxx");
```

6.4.3 使用缓存注解

下面使用 Service 来演示缓存注解：

```
@Service
@CacheConfig(cacheNames = "student") //a
public class StudentService {

   StudentRepository studentRepository;

   public StudentService(StudentRepository studentRepository) {
      this.studentRepository = studentRepository;
   }

   @CachePut(key = "#student.id") //b
   public Student saveOrUpdate(Student student){
      return studentRepository.save(student);
   }

   @CacheEvict //c
   public void remove(Long id){
      studentRepository.deleteById(id);
   }

   @Cacheable(unless = "#result == null") //d
```

```
    public Optional<Student> findOne(Long id){
        return studentRepository.findById(id);
    }

}
```

a. @CacheConfig 设置缓存名称作为整个类级别的共享配置，类中的方法注解可以不指定缓存名称。

b. 缓存注解默认使用参数作为缓存的 key，可以用 SPEL 从参数对象中获取值，如获得 student 的 id 可以使用 key = "#student.id"。@CachePut 将 key 为 student 的 id、值为返回值的数据存入缓存。

c. @CacheEvict 可以删除方法中参数（即 id）为 key 的值。

d. 当 key（方法参数的 id）第一次请求到此方法时，查询数据库，并将返回值放入缓存中。后续相同 key 的请求将直接从缓存中获取。unless 可做否决，当表达式为 true 时，不存入缓存；当表达式为 false 时，存入缓存。unless 在方法执行完成后进行评估，可以获得方法的返回值 result。当返回值为 Optional 时，result 仍然是其包装的 student。unless 中的表达式的含义是：如果返回值为空，则不放入缓存。

1．验证@Cacheable

```
@Bean
CommandLineRunner cacheableOperation(StudentService studentService){
    return args -> {
        System.out.println("第一次查询，需要查询数据，然后缓存数据，方法被调用，有查询 SQL");
        System.out.println(studentService.findOne(1l));
        System.out.println("第一次查询，从缓存中获取数据而不调用方法，下面不会有 SQL");
        System.out.println(studentService.findOne(1l));
    };
}
```

运行结果如图 6-24 所示。

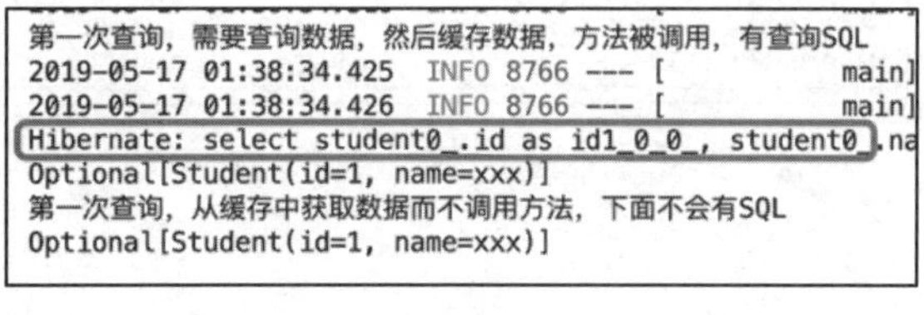

```
第一次查询，需要查询数据，然后缓存数据，方法被调用，有查询SQL
2019-05-17 01:38:34.425  INFO 8766 --- [           main]
2019-05-17 01:38:34.426  INFO 8766 --- [           main]
Hibernate: select student0_.id as id1_0_0_, student0_.na
Optional[Student(id=1, name=xxx)]
第一次查询，从缓存中获取数据而不调用方法，下面不会有SQL
Optional[Student(id=1, name=xxx)]
```

图 6-24

2．验证@CachePut

```
@Bean
CommandLineRunner cachePutOperation(StudentService studentService){
    return args -> {
```

```
        Student student = studentService.saveOrUpdate(new Student("wyf"));
        Long id = student.getId();
        System.out.println("因使用@CachePut已经将key为id的值存入了缓存,下面的查询将不会有SQL");
        System.out.println(studentService.findOne(id));

        student.setName("wangyunfei");
        studentService.saveOrUpdate(new Student("wyf"));
        System.out.println("@CachePut在修改时会被调用，缓存新的值，下面的查询也将不会有SQL");
        System.out.println(studentService.findOne(id));
    };
}
```

运行结果如图 6-25 所示。

```
因使用@CachePut已经将key为id的值存入了缓存，下面的查询将不会有SQL
Optional[Student(id=57, name=wyf)]
Hibernate: select next_val as id_val from hibernate_sequence
Hibernate: update hibernate_sequence set next_val= ? where n
Hibernate: insert into student (name, id) values (?, ?)
@CachePut在修改时会被调用，缓存新的值，下面的查询也将不会有SQL
Optional[Student(id=57, name=wyf)]
```

图 6-25

3．验证@CacheEvit

```
@Bean
CommandLineRunner cacheEvitOperation(StudentService studentService){
    return args -> {
        studentService.remove(1l);
        System.out.println("通过@CacheEvict删除了key为id的缓存，所以下面的查询会有SQL");
        studentService.findOne(1l);
    };
}
```

运行结果如图 6-26 所示。

```
2019-05-17 01:34:07.234  INFO 8748 --- [           main]
2019-05-17 01:34:07.235  INFO 8748 --- [           main]
通过@CacheEvict删除了key为id的缓存，所以下面的查询会有SQL
Hibernate: select student0_.id as id1_0_0_, student0_.na
```

图 6-26

6.5 小结

数据库操作是应用的核心功能，本章利用 Spring Data JPA 和 Spring Data Elasticsearch 分别演示了关系型数据库和 NoSQL 数据库的操作使用，它们使用了相同的 Spring Data Repository 编程模型。我们可以借助这个编程模型学习 Spring Data 支持的其他数据库技术。在本章的最后，介绍了通过数据缓存来提升项目速度的方法，通过简单的注解即可轻松完成数据缓存功能。

第 7 章 安全控制

在学习 Spring Security 前，需要了解认证和授权这两个重要的概念。

◎ 认证（Authentication）：用来确定谁在访问资源。在访问绝大部分系统时，都需要提供用户名和密码，由系统确定提供的用户名和密码与存储的是否一致。若认证通过，则可以访问受保护的资源；若认证无法通过，则不可以访问受保护的资源。

◎ 授权（Authoraztion）：用来确定当前访问者是否有权限访问指定的受保护资源。指定的受保护资源需要指定的权限才能访问，只有当前访问者拥有这个权限，才可以访问指定的受保护资源。

7.1 Spring Security 的应用

7.1.1 Spring Boot 的自动配置

Spring Boot 提供了 SecurityAutoConfiguration，它导入了三个自动配置，并通过 SecurityProperties 使用 spring.security.*来配置 Spring Security。

（1）SecurityAutoConfiguration。

◎ SpringBootWebSecurityConfiguration：提供默认的 Web 安全配置（DefaultConfigurerAdapter 继承了 WebSecurityConfigurerAdapter）。一般来说，只需定制自己需要的部分，其余部分保持默认即可。

◎ WebSecurityEnablerConfiguration：如果系统中有 WebSecurityConfigurerAdapter 的 Bean，则使用@EnableWebSecurity 可以自动开启 Web 安全配置。

◎ SecurityDataConfiguration：当 jar 包 org.springframework.security:spring-security-data 在类路径中时，允许使用 Spring Data 在查询时引用安全相关的表达式。

（2）UserDetailsServiceAutoConfiguration：自动配置一个内存中的用户，可通过spring.security.user.*来配置。如果自定义了认证的实现，则自动配置无效。

7.1.2 开启 Web 安全配置

在 Spring Boot 下，无须手动使用@EnableWebSecurity（组合了@Configuration 注解）来开启 Web 安全配置，但还是需要了解一下它做了什么。它导入的配置如下。

（1）WebSecurityConfiguration：注册 Bean 名为 springSecurityFilterChain 的 Servlet 过滤器（Filter），它创建了 Spring Security Filter Chain，负责应用中所有关于安全的内容。

（2）WebMvcSecurityConfiguration：实现了 Spring MVC 的 WebMvcConfigurer，并额外注册了 3 个 ArgumentResolvers，可以在控制器方法参数中直接获取。

◎ AuthenticationPrincipalArgumentResolver：使用@AuthenticationPrincipal 注解用户参数，获得用户信息。

◎ CurrentSecurityContextArgumentResolver：使用@CurrentSecurityContext 注解用户参数，获得 SecurityContext 中包含的信息。

◎ CsrfTokenArgumentResolver：获取参数中的 CsrfToken。

（3）OAuth2ClientConfiguration：OAuth 2.0 客户端支持的配置。

（4）EnableGlobalAuthentication：配置全局的 AuthenticationManagerBuilder。

新建应用，信息如下。

Group：top.wisely。

Artifact：learning-spring-security。

Dependencies：Spring Security、Spring Web Starter、Spring Data JPA、MySQL Driver 示中 Lombok。

build.gradle 文件中的依赖如下。

```
dependencies {
  implementation 'org.springframework.boot:spring-boot-starter-security'
  implementation 'org.springframework.boot:spring-boot-starter-data-jpa'
  implementation 'org.springframework.boot:spring-boot-starter-web'
  compileOnly 'org.projectlombok:lombok'
  runtimeOnly 'mysql:mysql-connector-java'
  annotationProcessor 'org.projectlombok:lombok'
 //……
}
```

7.1.3 定制 Web 安全配置

和 Spring MVC 类似，只需让一个配置类继承 WebSecurityConfigurerAdapter 类或实现

WebSecurityConfigurer 接口即可。

```
@Configuration
public class WebSecurityConfig extends WebSecurityConfigurerAdapter {
}
```

7.1.4 Authentication

Spring Security 提供了一个专门的 org.springframework.security.core.Authentication 接口来代表认证；它最常用的实现类是 UsernamePasswordAuthenticationToken。

一旦请求被认证后，Authentication 对象就会自动存储在由 SecurityContextHolder 管理的 SecurityContext 中。

认证的过程是通过加载下列类的处理顺序来进行的。

（1）FilterChainProxy：Servlet 过滤器（Filter）springSecurityFilterChain 的实际类型是 FilterChainProxy，它可能包含多个过滤器链（DefaultSecurityFilterChain），每个过滤器链包含多个过滤器。特定的过滤器会将请求中的认证信息（如用户名、密码）构造成 Authentication 对象，并交由 AuthenticationManager 的 authenticate 方法处理。主要过滤器如下。

◎ UsernamePasswordAuthenticationFilter：通过提交表单（用户名、密码）来进行信息认证，构造的 Authentication 对象类型为 UsernamePasswordAuthenticationToken，并调用 AuthenticationManager 的 authenticate 来进行认证操作。

◎ BasicAuthenticationFilter：使用 HTTP 请求的基础授权头提交认证信息。同样，构造的 Authentication 对象的类型为 UsernamePasswordAuthenticationToken，并调用 AuthenticationManager 的 authenticate 方法来进行认证操作。

◎ ExceptionTranslationFilter：处理过滤器链中的异常。

- AuthenticationException：认证异常，返回 401 状态码。
- AccessDeniedException：授权异常，返回 403 状态码。

◎ FilterSecurityInterceptor：它是 AbstractSecurityInterceptor 的子类，当认证成功后，再使用 AccessDecisionManager 对 Web 路径资源（Web URI）进行授权操作。

（2）AuthenticationManager：AuthenticationManager 接口的实现为 ProviderManager，可以使用 AuthenticationManagerBuilder 来定制构建 AuthenticationManager。

（3）ProviderManager：ProviderManager 通过它的 authenticate 方法将认证交给了一组顺序的 AuthenticationProvider 来完成认证。

（4）AuthenticationProvider：AuthenticationProvider 接口包含以下两个方法。

◎ supports：是否支持认证安全过滤器缓解构造的 Authentication。

◎ authenticate：对 Authentication 进行认证，若认证通过，则返回 Authentication；若认证未通过，则抛出异常。

（5）DaoAuthenticationProvider：DaoAuthenticationProvider 是 AuthenticationProvider 接口的实现，它支持认证的 Authentication 类型为 UsernamePasswordAuthenticationToken。在认证中，主要用到了下面 3 个部分。

◎ UserDetailsService：首先从指定的位置（如数据库）获得用户信息；然后把获得的用户信息和 Authentication（UsernamePasswordAuthenticationToken）中的用户名和密码信息进行对比，若完全一致，则认证通过；最后构建新的 Authentication（UsernamePasswordAuthenticationToken），包含用户的权限信息。

◎ PasswordEncoder：使用 PasswordEncoder 将请求传来的明文密码和存储的编码后的密码进行对比。

1．配置 AuthenticationManager

可以覆写 WebSecurityConfigurerAdapter 类的方法，使用 AuthenticationManagerBuilder 来配置 AuthenticationManager。

```
@Configuration
public class WebSecurityConfig extends WebSecurityConfigurerAdapter {

   @Override
   protected void configure(AuthenticationManagerBuilder auth) throws Exception {
      //……
   }

}
```

可以通过配置 UserDetailsService 或 AuthenticationProvider 来定制认证。

2．UserDetailsService

本例定制一个 UserDetailsService，通过 Spring Data JPA 从数据库中获取用户。

基本外部配置如下：

```
spring:
 datasource:
  url: jdbc:mysql://localhost:3306/first_db?useSSL=false
  username: root
  password: zzzzzz
  driver-class-name: com.mysql.cj.jdbc.Driver
 jpa:
  show-sql: true
  hibernate:
   ddl-auto: update
```

用户的实体：

```
@Data
@AllArgsConstructor
@NoArgsConstructor
@Entity
public class SysUser implements UserDetails {
   @Id
   @GeneratedValue(strategy = GenerationType.IDENTITY)
   private Long id;

   private String realName;

   @Column(unique = true)
   private String username;

   private String password;

   public SysUser(String realName, String username, String password) {
      this.realName = realName;
      this.username = username;
      this.password = password;
   }

   @Override
   public Collection<? extends GrantedAuthority> getAuthorities() { //a
      return null;
   }

   @Override
   public String getPassword() { //b
      return this.password;
   }

   @Override
   public String getUsername() { //c
      return this.username;
   }

   @Override
   public boolean isAccountNonExpired() { //d
      return true;
   }

   @Override
   public boolean isAccountNonLocked() { //e
      return true;
   }

   @Override
   public boolean isCredentialsNonExpired() { //f
      return true;
```

```
    }

    @Override
    public boolean isEnabled() { //g
        return true;
    }
}
```

实现 UserDetails 接口的用户，可通过接口的方法构建 Authentication 对象的用户信息。

a. getAuthorities 方法获得用户的权限信息，会在后面详细讲解。

b. getPassword 获得用户的密码，使用存储的密码。

c. getUsername 获得用户名，使用存储的用户名。

d. isAccountNonExpired 判断账号是否过期，设置为 true，即未过期。

e. isAccountNonLocked 判断账号是否被锁定，设置为 true，即未被锁定。

f. isCredentialsNonExpired 判断密码是否过期，设置为 true，即未过期。

g. isEnabled 判断用户是否弃用，设置为 true，即弃用。

用户的 Repository：

```
public interface SysUserRepository extends JpaRepository<SysUser, Long> {
  Optional<SysUser> findByUsername(String username);
}
```

自定义 UserDetailsService：

```
public class CusotmUserDetailsService implements UserDetailsService {

    SysUserRepository sysUserRepository;

    public CusotmUserDetailsService(SysUserRepository sysUserRepository) {
        this.sysUserRepository = sysUserRepository; //a
    }

    @Override
    public UserDetails loadUserByUsername(String username) throws
UsernameNotFoundException {
        Optional<SysUser> sysUserOptional = sysUserRepository.findByUsername(username);
//b
        return  sysUserOptional
                .orElseThrow(() -> new UsernameNotFoundException("Username not found"));
//c
    }
}
```

a. 注入 SysUserRepository，用来查询数据库用户信息。

b. 通过用户名从数据库中查询用户对象。

c. 如果存在，则返回用户对象；如果不存在，则抛出异常。本方法的返回值要求为 UserDetails 类型，用户对象也可通过 UserDetails 直接返回。

可以通过配置 AuthenticationManager 来注册 UserDetailsService：

```
@Configuration
public class WebSecurityConfig extends WebSecurityConfigurerAdapter {

   @Autowired
   SysUserRepository sysUserRepository; //a

   @Override
   protected void configure(AuthenticationManagerBuilder auth) throws Exception {
      auth.userDetailsService(new CusotmUserDetailsService(sysUserRepository)); //b
   }

   @Bean
   PasswordEncoder passwordEncoder(){ //c
      return new BCryptPasswordEncoder();
   }

}
```

a. 注入 SysUserRepository，给 CustomUserDetailsService 的构造使用。

b. 使用 AuthenticationManagerBuilder 的 userDetailsService 方法注册自定义的 UserDetailsService。

c. 使用 BCrypt 作为密码编码加密算法，给 DaoAuthenticationProvider 使用。

还可以直接通过声明 UserDetailsService 的 Bean 让 DaoAuthenticationProvider 使用自定义 UserDetailsService：

```
@Configuration
public class WebSecurityConfig extends WebSecurityConfigurerAdapter {
   @Bean
   UserDetailsService userDetailsService(SysUserRepository sysUserRepository){
      return new CusotmUserDetailsService(sysUserRepository);
   }

   @Bean
   PasswordEncoder passwordEncoder(){
      return new BCryptPasswordEncoder();
   }

}
```

建立一个控制器用来测试访问：

```
@RestController
public class IndexController {

   @GetMapping("/")
```

```
    public String hello(){
        return "Hello Spring Security";
    }
}
```

在应用启动时，向系统添加一个用户：

```
@Bean
CommandLineRunner createUser(SysUserRepository sysUserRepository, PasswordEncoder
passwordEncoder){
    return args -> {
        SysUser user = new SysUser("wangyunfei", "wyf", passwordEncoder.encode("111111"));
        sysUserRepository.save(user);
    };
}
```

Spring Security 默认使用表单登录，且自动提供一个表单。当访问 http://localhost:8080/时，会自动转向 http://localhost:8080/login。当用错误的账号或密码登录时，会提示“用户名或密码错误”；当用正确的账号和密码访问时，会显示测试的控制器并返回内容，如图 7-1 所示。

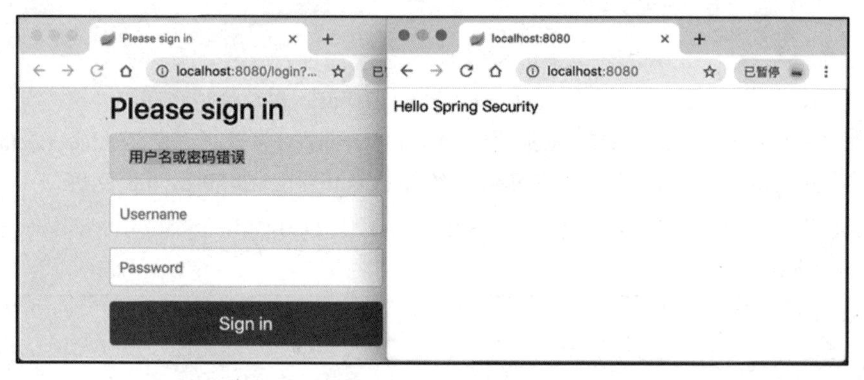

图 7-1

3. AuthenticationProvider

前面自定义的 UserDetailsService 实际上还是 AuthenticationProvider 的一部分，也可以通过 Bean 来使用自定义的 AuthenticationManager 来设置 AuthenticationProvider。

```
@Configuration
public class WebSecurityConfig extends WebSecurityConfigurerAdapter {

    @Autowired
    SysUserRepository sysUserRepository;

    @Override
    protected void configure(AuthenticationManagerBuilder auth) throws Exception {
        DaoAuthenticationProvider authProvider = new DaoAuthenticationProvider();
        authProvider.setUserDetailsService(new
```

```
CusotmUserDetailsService(sysUserRepository));
        authProvider.setPasswordEncoder(new BCryptPasswordEncoder());
        auth.authenticationProvider(authProvider);
    }
}
```

若自定义 AuthenticationProvider，则需要完全使用自己的验证逻辑：

```
public class CustomAuthenticationProvider implements AuthenticationProvider {

    SysUserRepository sysUserRepository;

    PasswordEncoder passwordEncoder;

    public CustomAuthenticationProvider(SysUserRepository sysUserRepository,
PasswordEncoder passwordEncoder) {
        this.sysUserRepository = sysUserRepository;
        this.passwordEncoder = passwordEncoder;
    }

    @Override
    public Authentication authenticate(Authentication authentication) throws
AuthenticationException {
        String usernameFromRequest = authentication.getName(); //a
        String passwordFromRequest = authentication.getCredentials().toString(); //b
        Optional<SysUser> sysUserOptional =
sysUserRepository.findByUsername(usernameFromRequest);
        SysUser sysUser = sysUserOptional
                .orElseThrow(() -> new UsernameNotFoundException("Username not found"));
//c
        if(passwordEncoder.matches(passwordFromRequest, sysUser.getPassword()) && //d
            sysUser.isAccountNonExpired() &&
            sysUser.isAccountNonLocked() &&
            sysUser.isCredentialsNonExpired() &&
            sysUser.isEnabled())
            return new UsernamePasswordAuthenticationToken(sysUser.getUsername(),
sysUser.getPassword(), sysUser.getAuthorities()); //e
        else
            throw new BadCredentialsException("Bad Credentials");
    }

    @Override
    public boolean supports(Class<?> authentication) { //f
        return (UsernamePasswordAuthenticationToken.class
                .isAssignableFrom(authentication));
    }
}
```

a. 获取请求传递的用户名。

b. 获取请求传递的密码。

c. 检查用户是否存在于数据库中。

d. 比较传递密码和存储的编码密码是否一致，包含用户有效性的对比。

e. 构建新的 Authentication：UsernamePasswordAuthenticationToken 对象。

f. 声明当前自定义 AuthenticationProvider 能处理 Authentication 类型为 UsernamePassword AuthenticationToken 的认证。

这个 AuthenticationProvider 同样可以通过 AuthenticationManager 来配置：

```
@Configuration
public class WebSecurityConfig extends WebSecurityConfigurerAdapter {

   @Autowired
   SysUserRepository sysUserRepository;

   @Bean
   PasswordEncoder passwordEncoder(){
      return new BCryptPasswordEncoder();
   }

   @Override
   protected void configure(AuthenticationManagerBuilder auth) throws Exception {
      auth.authenticationProvider(new CustomAuthenticationProvider(sysUserRepository,
passwordEncoder()));
   }
}
```

注册 AuthenticationProvider 的 Bean，效果与上面相同：

```
@Configuration
public class WebSecurityConfig extends WebSecurityConfigurerAdapter {

   @Bean
   PasswordEncoder passwordEncoder(){
      return new BCryptPasswordEncoder();
   }

   @Bean
   AuthenticationProvider authenticationProvider(SysUserRepository
sysUserRepository){
      return new CustomAuthenticationProvider(sysUserRepository, passwordEncoder());
   }
}
```

运行的结果和自定义 UserDetailsService 一致。一般来说，使用自定义 UserDetailsService 即可，只有在涉及复杂的认证逻辑时，才需要自定义 AuthenticationProvider。

4. HTTP 基础认证

HttpSecurity 用来针对不同的 HTTP 请求进行 Web 安全配置。前面的认证都是通过表单登录进行认证的，可以通过 HttpSecurity 配置在请求时将账号信息放置于头部进行登录。服务端和客户端（Web、App 等）分离的应用，一般都是通过类似的方式来请求服务端的接口的。

```
@Configuration
public class WebSecurityConfig extends WebSecurityConfigurerAdapter {
   @Override
   protected void configure(HttpSecurity http) throws Exception {
      http.authorizeRequests()
            .anyRequest().authenticated() //a
            .and()
            .httpBasic().authenticationEntryPoint(authenticationEntryPoint()); //b
   }

   @Bean
   AuthenticationEntryPoint authenticationEntryPoint(){ //c
      BasicAuthenticationEntryPoint authenticationEntryPoint = new
BasicAuthenticationEntryPoint();
      authenticationEntryPoint.setRealmName("wisely");
      return authenticationEntryPoint;
   }

   @Bean
   UserDetailsService userDetailsService(SysUserRepository sysUserRepository){
      return new CusotmUserDetailsService(sysUserRepository);
   }

   @Bean
   PasswordEncoder passwordEncoder(){
      return new BCryptPasswordEncoder();
   }

}
```

a. 设置所有的请求都需要在认证后才能访问。

b. 设置登录方式为 HTTP Basic，设置认证入口点为 BasicAuthenticationEntryPoint。

c. BasicAuthenticationEntryPoint 将认证信息放置于头部，当认证未通过时，返回 401 状态码。

Postman 支持 Basic Auth，如图 7-2 所示。

HTTP Basic 实际上是在请求头的 Authorization 中添加值为 Basic 账号:密码的 Base64 编码。wyf:111111 的 Base64 编码为 d3lmOjExMTExMQ==，所以当请求头为 Authorization: Basic d3lmOjExMTExMQ==时即可完成认证，如图 7-3 所示。

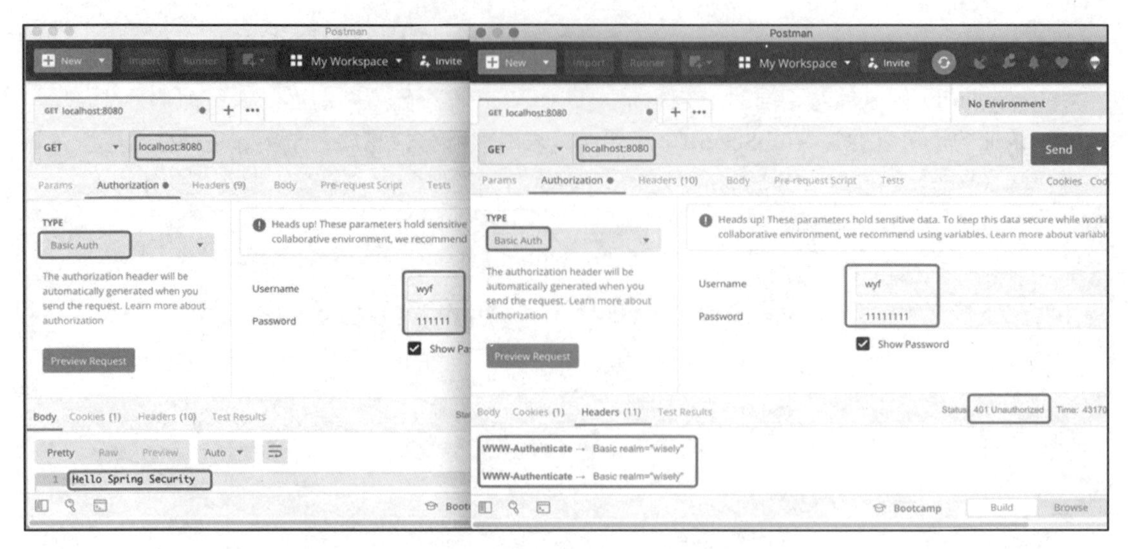
Postman
New
My Workspace
Invite
GET localhost:8080
GET
localhost:8080
Params
Authorization
Headers (9)
Body
Pre-request Script
Tests
TYPE
Basic Auth
The authorization header will be automatically generated when you send the request. Learn more about authorization
Preview Request
Heads up! These parameters hold sensitive data. To keep this data secure while working in a collaborative environment, we recommend using variables. Learn more about variables
Username
wyf
Password
111111
Show Password
Body
Cookies (1)
Headers (10)
Test Results
Pretty
Raw
Preview
Auto
1 Hello Spring Security
Bootcamp
No Environment
Send
Headers (10)
Cookies
11111111
Headers (11)
401 Unauthorized
WWW-Authenticate → Basic realm="wisely"
WWW-Authenticate → Basic realm="wisely"
Build
Browse

图 7-2

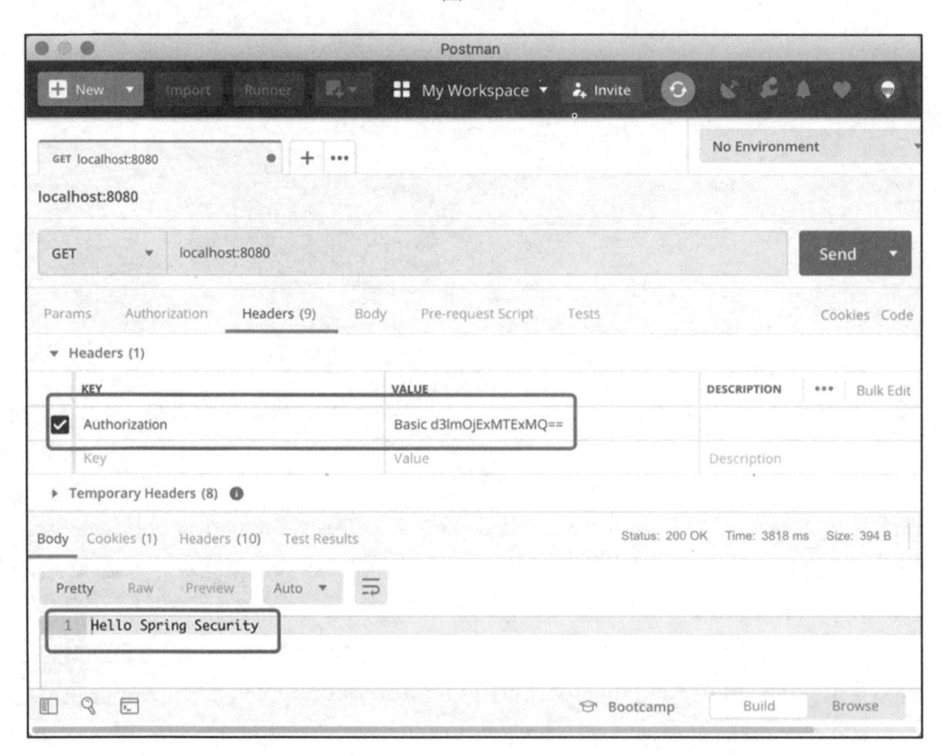
Postman
New
Import
Runner
My Workspace
Invite
GET localhost:8080
No Environment
localhost:8080
GET
localhost:8080
Send
Params
Authorization
Headers (9)
Body
Pre-request Script
Tests
Cookies
Code
Headers (1)
KEY
VALUE
DESCRIPTION
Bulk Edit
Authorization
Basic d3lmOjExMTExMQ==
Key
Value
Description
Temporary Headers (8)
Body
Cookies (1)
Headers (10)
Test Results
Status: 200 OK
Time: 3818 ms
Size: 394 B
Pretty
Raw
Preview
Auto
1 Hello Spring Security
Bootcamp
Build
Browse

图 7-3

5. 获取用户信息

可以从 SecurityContextHolder 中获取 SecurityContext，从而获得 Authentication：

```
SecurityContext context = SecurityContextHolder.getContext(); //获得
//SecurityContext
Authentication auth = context.getAuthentication(); //获得 Authentication
Object principal = auth.getPrincipal(); //获取用户信息
Object details = auth.getDetails(); //认证请求的更多信息
```

Spring Security 注册了 ArgumentResolvers，因此可以直接通过@CurrentSecurityContext 注解获取 SecurityContext。另外，@CurrentSecurityContext 注解还支持表达式获取 Authentication、principal 和 details。

除此之外，更简单的方式是使用@AuthenticationPrincipal 注解获取用户信息：

```
   @GetMapping("/user")
   public Map<String, Object> getUserInfo(@AuthenticationPrincipal SysUser sysUser, //a
                    @CurrentSecurityContext SecurityContext securityContext, //b
 @CurrentSecurityContext(expression = "authentication")  Authentication authentication,
//c
    @CurrentSecurityContext(expression = "authentication.principal") Object
principal,//d
     @CurrentSecurityContext(expression = "authentication.details") Object
details){ //e
      Map<String, Object> map = new HashMap<>();
      map.put("sysUser", sysUser);
      map.put("authentication", authentication);
      map.put("principal", principal);
        map.put("details", details);
      return map;
   }
```

a. @AuthenticationPrincipal 可获得系统用户对象。

b. @CurrentSecurityContext 可直接获得 SecurityContext 对象。

c. 使用表达式 expression = "authentication"，可获得 Authentication 对象。

d. 使用表达式 expression = "authentication.principal"，可获得用户信息。

e. 使用表达式 expression = "authentication.details"，可获得认证请求的额外信息。

在 Postman 中访问 http://localhost:8080/user，并使用 Basic Auth 认证，如图 7-4 所示。

```
{
    "principal": {
        "id": 1,
        "realName": "wangyunfei",
        "username": "wyf",
        "password": "$2a$10$aFSkksuBZ.YvU5TBJ6ZQeeGHUlqa4.KHN7sBrBoodHRhYdE7WyRHu",
        "enabled": true,
        "authorities": null,
        "accountNonExpired": true,
        "accountNonLocked": true,
        "credentialsNonExpired": true
    },
    "details": {
        "remoteAddress": "0:0:0:0:0:0:0:1",
        "sessionId": null
    },
    "sysUser": {
        "id": 1,
        "realName": "wangyunfei",
        "username": "wyf",
        "password": "$2a$10$aFSkksuBZ.YvU5TBJ6ZQeeGHUlqa4.KHN7sBrBoodHRhYdE7WyRHu",
        "enabled": true,
        "authorities": null,
        "accountNonExpired": true,
        "accountNonLocked": true,
        "credentialsNonExpired": true
    },
    "authentication": {
        "authorities": [],
        "details": {
            "remoteAddress": "0:0:0:0:0:0:0:1",
            "sessionId": null
        },
        "authenticated": true,
        "principal": {
            "id": 1,
            "realName": "wangyunfei",
            "username": "wyf",
            "password": "$2a$10$aFSkksuBZ.YvU5TBJ6ZQeeGHUlqa4.KHN7sBrBoodHRhYdE7WyRHu",
            "enabled": true,
            "authorities": null,
            "accountNonExpired": true,
            "accountNonLocked": true,
            "credentialsNonExpired": true
        },
        "credentials": null,
        "name": "wyf"
    }
}
```

图 7-4

把第 6 章“审计功能”中获得用户的代码修改如下。

```
@Bean
AuditorAware<String> auditorProvider(){
  return () ->
Optional.of(SecurityContextHolder.getContext().getAuthentication().getName());
}
```

6. 密码编码

在生产中，我们肯定会对密码进行编码，Spring Security 提供了 PassowrdEncoder 接口来编码密码和匹配密码。建议使用工业标准的 Bcrypt 的实现 BCryptPasswordEncoder 来进行编码。

```
public interface PasswordEncoder {
  String encode(CharSequence rawPassword); //对明文密码进行编码
```

```
  boolean matches(CharSequence rawPassword, String encodedPassword); //匹配明文密码和
//编码密码
  default boolean upgradeEncoding(String encodedPassword) { //更新密码编码机制
    return false;
  }
}
```

前面在应用中已经注册了 BCryptPasswordEncoder 的 Bean：

```
@Bean
PasswordEncoder passwordEncoder(){
  return new BCryptPasswordEncoder();
}
```

代码检验如下，结果如图 7-5 所示。

```
@Bean
CommandLineRunner passwordOperation(PasswordEncoder passwordEncoder){
  return args -> {
    String passwordPlain = "123456";
    String passwordEncoded = passwordEncoder.encode(passwordPlain);
    boolean isMatched = passwordEncoder.matches(passwordPlain, passwordEncoded);
    System.out.println("明文密码为: " + passwordPlain);
    System.out.println("编码密码为: " + passwordEncoded);
    System.out.println("密码是否匹配: " + isMatched);
  };
}
```

```
明文密码为: 123456
编码密码为: $2a$10$FEYOG/5PYPcCtHwDd68vEOQwq/PYVBP7XjNT2ehV0Ca2E.273qiA6
密码是否匹配: true
```

图 7-5

7.1.5 Authorization

当认证成功后，下一步要做的就是授权（访问控制）。只有通过授权（访问控制），用户才可以访问受保护的资源。

1. Web 路径安全

接前面认证的过程，从 FilterSecurityInterceptor 开始学习对 Web 路径的安全控制。

（1）FilterSecurityInterceptor：它是 AbstractSecurityInterceptor 的子类，当认证成功后，再使用 AccessDecisionManager 对 Web 路径资源（Web URI）进行授权操作。

（2）FilterInvocationSecurityMetadataSource：存储 Web 路径安全元数据并提供查询功能，用来存储系统中所有的 Web 路径安全配置，包括请求路径与对应的 ConfigAttribute 的 requestMap。当请求 Web 路径为/everyCanAccess 时，会从 requestMap 中获取它的 ConfigAttribute: permitAll()。

FilterInvocationSecurityMetadataSource 的功能由它的子类 ExpressionBasedFilterInvocationSecurityMetadataSource 实现。

（3）AccessDecisionManager：类似于 AuthenticationManager（ProviderManager）包含一组 AuthenticationProvider，AccessDecisionManager（默认实现为 AffirmativeBased）包含一组 AccessDecisionVoter。

```
public interface AccessDecisionManager {
  void decide(Authentication authentication, Object object,
      Collection<ConfigAttribute> configAttributes) throws AccessDeniedException,
      InsufficientAuthenticationException; //a
  boolean supports(ConfigAttribute attribute); //b
  boolean supports(Class<?> clazz); //c
}
```

a. decide 通过 3 个参数决定是否通过授权。

◎ Authentication authentication：Authentication 对象里包含了用户的权限信息，可通过 getAuthorities 获取。

◎ Object object：被保护的资源对象，在 Web 路径中，被保护的资源对象的类型为 FilterInvocation。

◎ Collection<ConfigAttribute> configAttributes：和被保护的资源对象一起的配置属性。

b. 见 AffirmativeBased 的 supports(ConfigAttribute attribute)。

c. 见 AffirmativeBased 的 supports(Class<?> clazz)。

（4）AffirmativeBased：包含一组 AccessDecisionVoter，只要有一个 AccessDecisionVoter 授权，整组就立即被授权，无论其他 AccessDecisionVoter 是否授权。

◎ decide：让一组 AccessDecisionVoter 中的每一个成员都进行投票。

◎ supports(ConfigAttribute attribute)：在当前 AccessDecisionManager 包含的一组 AccessDecisionVoter 中，若有支持该配置属性的，则为支持。

◎ supports(Class<?> clazz)：确认是否每一个 AccessDecisionVoter 都支持被保护资源对象的类型处理。

（5）AccessDecisionVoter：路径安全控制的主要实现为 WebExpressionVoter，通过计算表达式的结果来进行访问控制。

覆写 WebSecurityConfigurerAdapter 的方法 configure(HttpSecurity http)，通过 HttpSecurity 来配置 Web 路径的安全访问：

```
@Configuration
public class WebSecurityConfig extends WebSecurityConfigurerAdapter {
   @Override
   protected void configure(HttpSecurity http) throws Exception {
      http.authorizeRequests() //a
            .antMatchers("/everyCanAccess").permitAll() //b
```

```
            .antMatchers("/authenticatedCanAccess").authenticated() //c
            .antMatchers("/adminCanAccess").hasRole("ADMIN") //d
            .antMatchers("/userCanAccess").access("hasRole('USER') or
hasRole('ADMIN')") //e
           .antMatchers("/threeCanAccess").access("@webSecurity.checkUsernameLenEq3(
authentication)")//f
            .anyRequest().authenticated()//g
            .and()
            .httpBasic().authenticationEntryPoint(authenticationEntryPoint());
   }

   @Bean
   AuthenticationEntryPoint authenticationEntryPoint(){
     //……
   }

   @Bean
   UserDetailsService userDetailsService(SysUserRepository sysUserRepository){
      return new CusotmUserDetailsService(sysUserRepository);
   }

   @Bean
   PasswordEncoder passwordEncoder(){
      return new BCryptPasswordEncoder();
   }

}
```

a. authorizeRequests方法对所有的请求进行授权，它由底下的多个部分组成。

b. antMatchers方法中的路径/everyCanAccess允许任何人访问。

◎ antMatchers支持用HTTP方法来缩小匹配范围：antMatchers(HttpMethod.GET,"/everyCanAccess")。

◎ 若需要匹配路径下的所有路径，则可使用/**：antMatchers("/secure/**")。

c. antMatchers方法中的路径/authenticatedCanAccess，需要认证后才能访问。

d. antMatchers方法中的路径/adminCanAccess，需要用户有ROLE_ADMIN角色。

e. antMatchers方法中的路径/userCanAccess，需要用户有ROLE_ADMIN或ROLE_USER角色，access可使用其他表达式。

f. antMatchers方法中的路径/threeCanAccess，通过access中的“@”表达式调用Bean的方法进行访问控制。

WebSecurity：

```
@Service
public class WebSecurity {
   public boolean checkUsernameLenEq3(Authentication authentication){
     if ( authentication.getName().length() == 3)
```

```
        return true;
      return false;
   }
}
```

g. 其余路径都需要通过认证。

上面的访问控制方法都是配置属性（ConfigAttribute），下面列出了可作为配置属性的表达式，如表 7-1 所示。

表 7-1

表达式	描　述
hasRole([role])	当用户包含指定角色时可访问的 Web 路径，省略 ROLE_前缀
hasAnyRole([role1,role2])	当用户包含任意角色时可访问的 Web 路径，角色之间用“,”隔开，省略 ROLE_前缀
hasAuthority([authority])	当用户包含指定权限时，可访问的 Web 路径
hasAnyAuthority([authority1,authority2])	当用户包含任意权限时，可访问的 Web 路径，角色之间用“,”隔开
principal	用户信息
authentication	允许从 SecurityContext 中直接访问 Authentication 对象
permitAll()	允许所有访问
denyAll()	总是不可访问指定的 Web 路径
anonymous()	当用户为匿名用户时，可访问指定的 Web 路径
rememberMe()	当用户为 remember-me 用户时，可访问指定的 Web 路径
authenticated()	当前用户不是匿名用户，可访问指定的 Web 路径
fullyAuthenticated()	当前用户既不是匿名用户，也不是 remember-me 用户，可访问指定的 Web 路径
hasIpAddress	只有来自指定 IP 地址（或子网）的用户才可访问的 Web 路径
access	可接受上面的任意表达式，可调用 Bean 的方法

受保护的控制器定义：

```
@RestController
public class IndexController {
   @GetMapping("/everyCanAccess")
   public String everyCanAccess(){
      return "任何用户可访问";
   }

   @GetMapping("/authenticatedCanAccess")
   public String authenticatedCanAccess(){
      return "任何登录用户可访问";
   }

   @GetMapping("/userCanAccess")
```

```
    public String userCanAccess(){
        return "角色为 ROLE_USER 或 ROLE_ADMIN 的用户可访问";
    }

    @GetMapping("/adminCanAccess")
    public String adminCanAccess(){
        return "角色为 ROLE_ADMIN 的用户可访问";
    }

    @GetMapping("/threeCanAccess")
    public String threeCanAccess(){
        return "只有用户名字符串长度为 3 的用户可以访问";
    }
}
```

权限的获取是通过 SysUser 覆写 UserDetails 的 getAuthorities 方法获得的，可以定义新的 role 字段作为权限：

```
@Data
@AllArgsConstructor
@NoArgsConstructor
@Entity
public class SysUser implements UserDetails {
    @Id
    @GeneratedValue(strategy = GenerationType.IDENTITY)
    private Long id;

    private String realName;

    @Column(unique = true)
    private String username;

    private String password;

    private String role;

    public SysUser(String realName, String username, String password, String role) {
        this.realName = realName;
        this.username = username;
        this.password = password;
        this.role = role;
    }

    @Override
    public Collection<? extends GrantedAuthority> getAuthorities() {
        Collection<SimpleGrantedAuthority> authorities = new ArrayList<>();
        SimpleGrantedAuthority authority = new SimpleGrantedAuthority(this.role);
        authorities.add(authority);
        return authorities;
    }
```

```
    // ……
}
```

用户权限用 GrantedAuthority 接口代表，SimpleGrantedAuthority 是它的实现类，可接收字符串的权限值。删除数据库中 wyf 的用户记录数据，并重新初始化两条数据。

```
@Bean
CommandLineRunner createUser(SysUserRepository sysUserRepository, PasswordEncoder passwordEncoder){
  return args -> {
    SysUser user = new SysUser("wangyunfei", "wyf", passwordEncoder.encode("111111"), "ROLE_USER");
    SysUser admin = new SysUser("administrator", "admin", passwordEncoder.encode("admin"), "ROLE_ADMIN");
    sysUserRepository.save(user);
    sysUserRepository.save(admin);
  };
}
```

现在有两个新的用户：

◎ admin：角色为 ROLE_ADMIN，密码为 admin。

◎ wyf：角色为 ROLE_USER，密码为 111111。

验证 http://localhost:8080/everyCanAccess，任何人都可访问，无须登录，如图 7-6 所示。

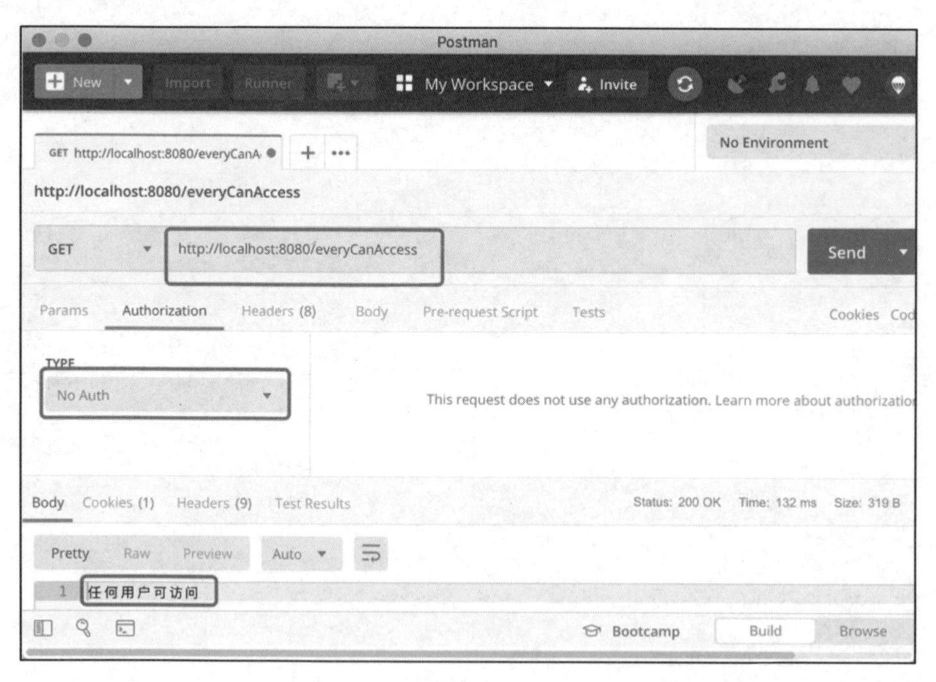

图 7-6

验证 http://localhost:8080/authenticatedCanAccess，任何认证用户都可访问，如图 7-7 所示。

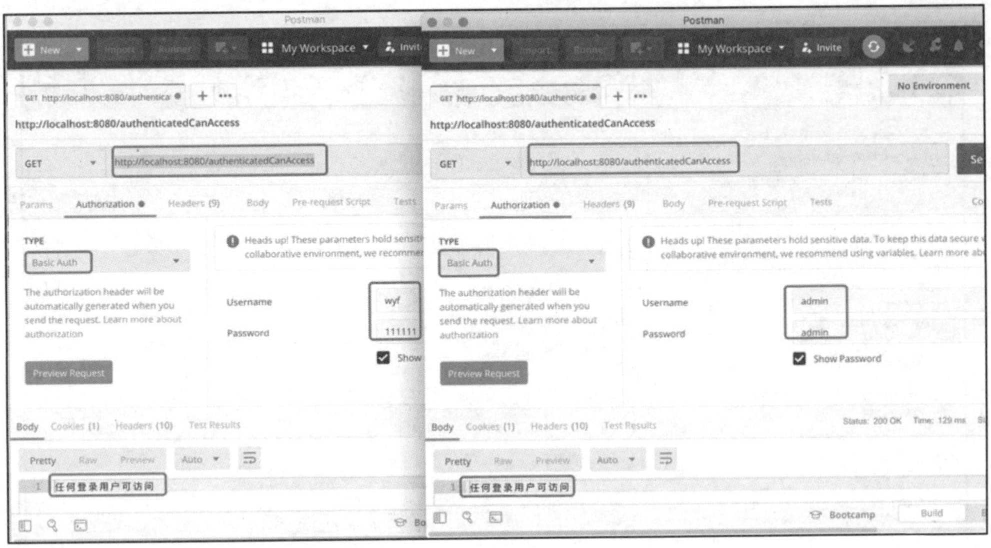

图 7-7

验证 http://localhost:8080/adminCanAccess，只有角色为 ROLE_ADMIN 的用户可访问，如图 7-8 所示。

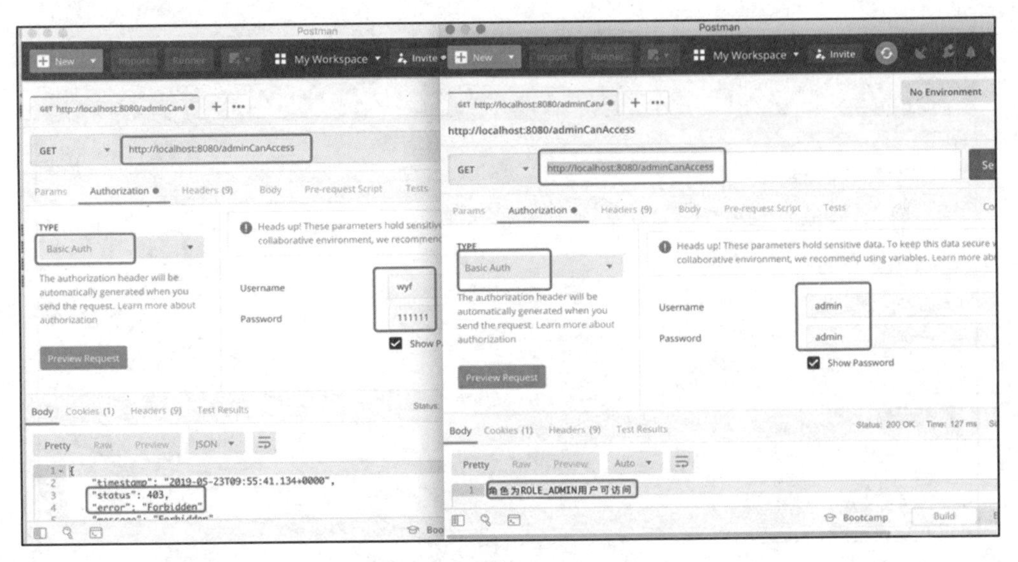

图 7-8

验证 http://localhost:8080/userCanAccess，角色为 ROLE_ADMIN 或 ROLE_USER 的用户可访问，如图 7-9 所示。

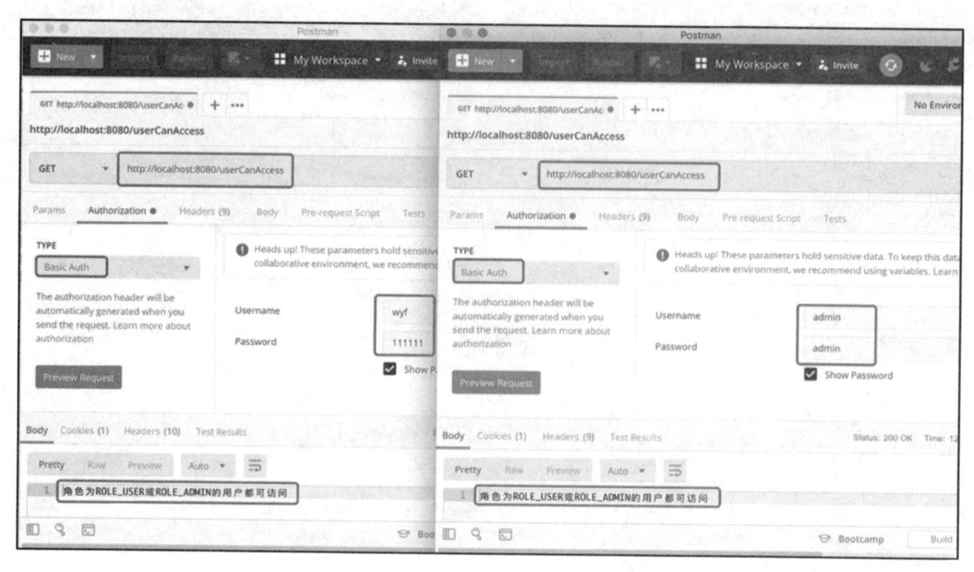

图 7-9

验证 http://localhost:8080/threeCanAccess，只有用户名字符串长度为 3 的用户才能访问，如图 7-10 所示。

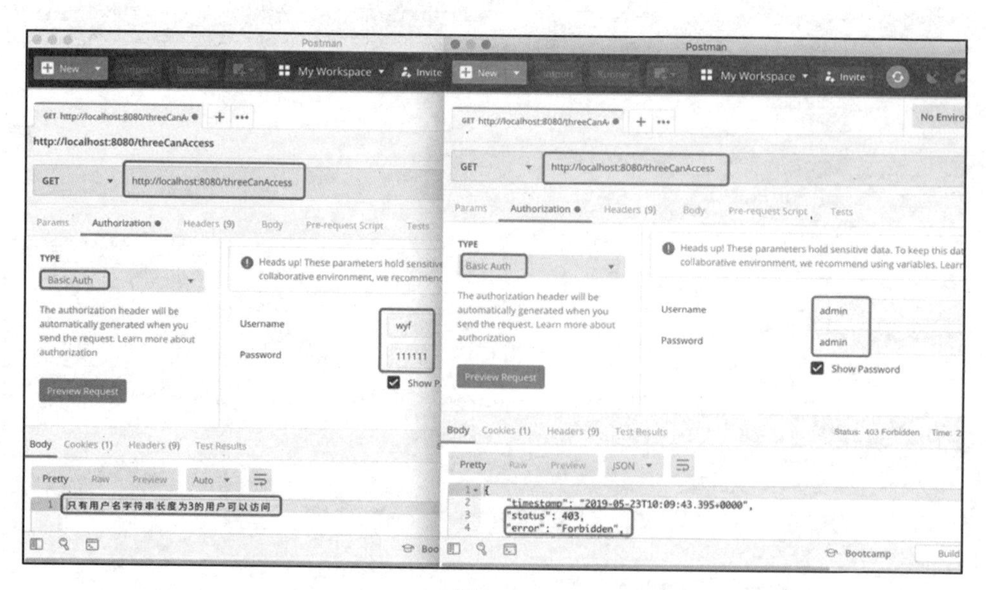

图 7-10

2．方法安全

Spring Security 在方法级别上进行权限控制，这样可以让权限控制更加灵活。使用注解@EnableGlobalMethodSecurity 开启对“方法安全”的支持（组合了@Configuration 注解）。

```
@EnableGlobalMethodSecurity(prePostEnabled = true)
public class WebSecurityConfig extends WebSecurityConfigurerAdapter {
  //……
}
```

在使用 prePostEnabled = true 之后，可支持下面注解进行方法安全控制，其他类型的注解本书不做讨论。

◎ @PreAuthorize：在方法调用前，通过访问控制表达式决定方法是否允许被调用。

◎ @PreFilter：定义规则过滤注解方法的参数（集合），filterObject 代表集合或数据里的当前对象。

◎ @PostAuthorize：当方法被调用后，访问控制表达式将被检查。若在检查后返回值，则表达式可使用 returnObject。

◎ @PostFilter：定义规则过滤方法返回值（集合或数组），filterObject 代表集合或数据里的当前对象。

同样，需要学习一下方法安全的原理。Web 路径安全是从 FilterSecurityInterceptor 开始的，而方法在安装时是通过 MethodSecurityInterceptor 开始的，它们都是 AbstractSecurityInterceptor 的子类。

◎ MethodSecurityInterceptor：对方法安全进行授权操作；由@EnableGlobalMethodSecurity 导入的配置 GlobalMethodSecurityConfiguration 进行初始化。MethodSecurityInterceptor 还实现了用 AOP 的 MethodInterceptor 接口来拦截注解方法；拦截方法调用 AccessDecisionManager 对注解方法进行授权操作。

  ➢ MethodSecurityInterceptor：对使用了特定注解（如@PreAuthorize）的方法的元数据进行提取、存储和查询，它是由 PrePostAnnotationSecurityMetadataSource 实现类提供功能的。

◎ AccessDecisionManager：decide 方法中被保护的资源对象类型为 MethodInvocation。

◎ AffirmativeBased：由一组 AccessDecisionVoter 进行投票来授权。

◎ AccessDecisionVoter：方法安全控制的主要实现是 PreInvocationAuthorizationAdviceVoter，它使用 ExpressionBasedPreInvocationAdvice 对表达式进行计算，从而达到对方法的访问控制。

演示控制器内容如下：

```
@RestController
public class IndexController {
   @GetMapping("/methodAdmin")
   @PreAuthorize("hasRole('ADMIN')") //a
   public String methodAdmin(){
      return "只有角色为 ROLE_ADMIN 的用户可访问";
   }

   @GetMapping("/methodDiffName")
   @PreAuthorize("#user.username != authentication.name") //b
   public String methodDiffName(@RequestBody  SysUser user){
      return "传输的用户名和当前用户名不相同的可访问";
   }

   @GetMapping("/methodNameThree")
   @PreAuthorize("@webSecurity.checkUsernameLenEq3(authentication)") //c
   public String methodNameThree(){
      return "只有用户名字符串长度为 3 的用户可以访问";
   }

   @GetMapping("/methodAnotherName3")
   @PostAuthorize("returnObject.length() == 5") //d
   public String anotherTree(@AuthenticationPrincipal SysUser sysUser){
      return "Hi" + sysUser.getUsername();
   }

   @GetMapping("/methodFilterIn")
   @PreAuthorize("hasRole('USER')")
   @PreFilter("filterObject%2 == 0") //e
   public List<Integer>  methodFilterIn (@RequestParam List<Integer> numbers){
      return numbers;
   }

   @GetMapping("/methodFilterOut")
   @PostFilter("hasRole('USER') and filterObject%2 == 0") //f
   public Integer[] methodFilterIn (){
      Integer[] numbers = {1,2,3,4,5,6,7,8,9};
      return numbers;
   }

}
```

a. 在@PreAuthorize 中，表达式的使用方式和“Web 路径安全”一致。只有 ROLE_ADMIN 角色可访问 http://localhost:8080/methodAdmin，如图 7-11 所示。

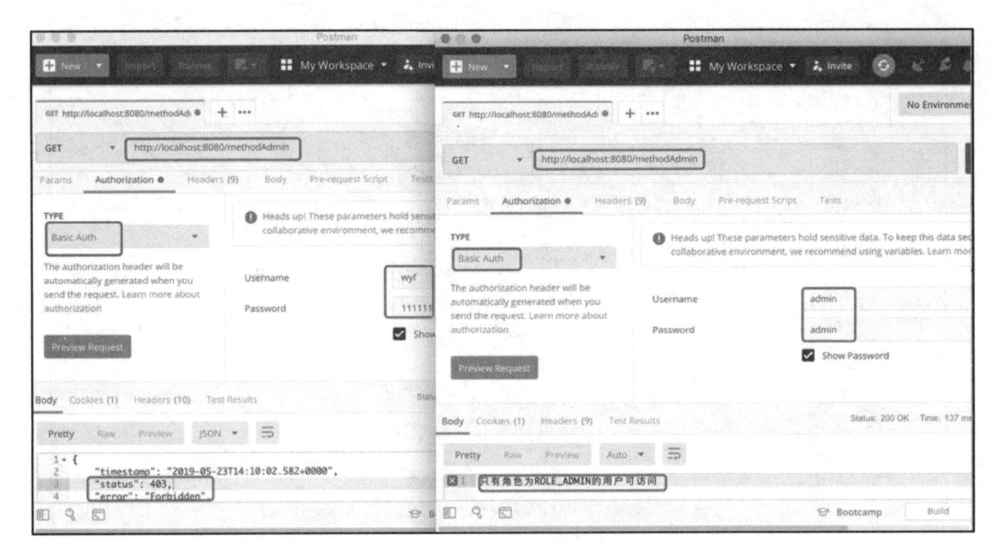

图 7-11

b. 可以使用#在表达式中获取参数中的数据。访问 http://localhost:8080/methodDiffName，在请求体中传递用户名，只有传递用户名和当前用户的用户名不一致时才可访问，如图 7-12 所示。

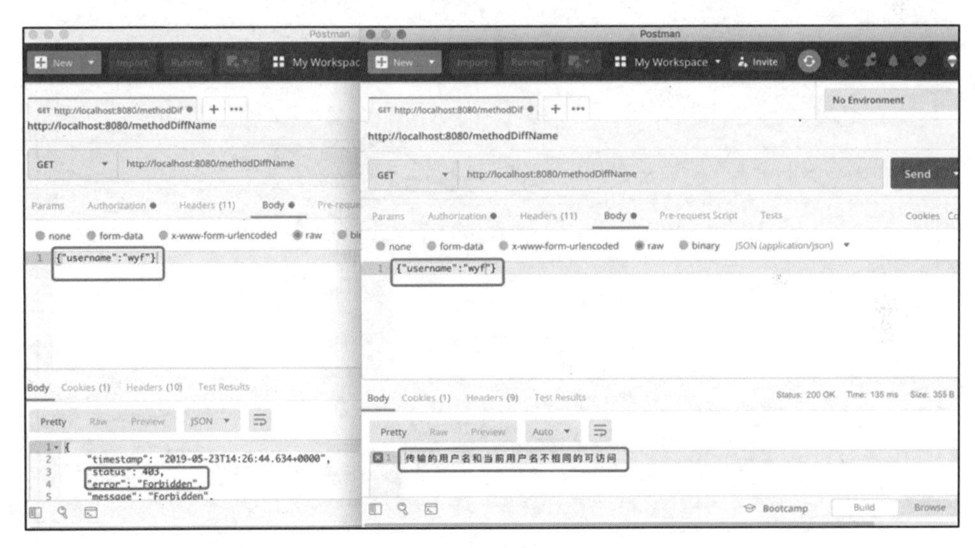

图 7-12

c. 在表达式中使用@调用 Bean 的方法访问 http://localhost:8080/methodNameThree，只有用户名字符串长度为 3 的用户才可以访问，如图 7-13 所示。

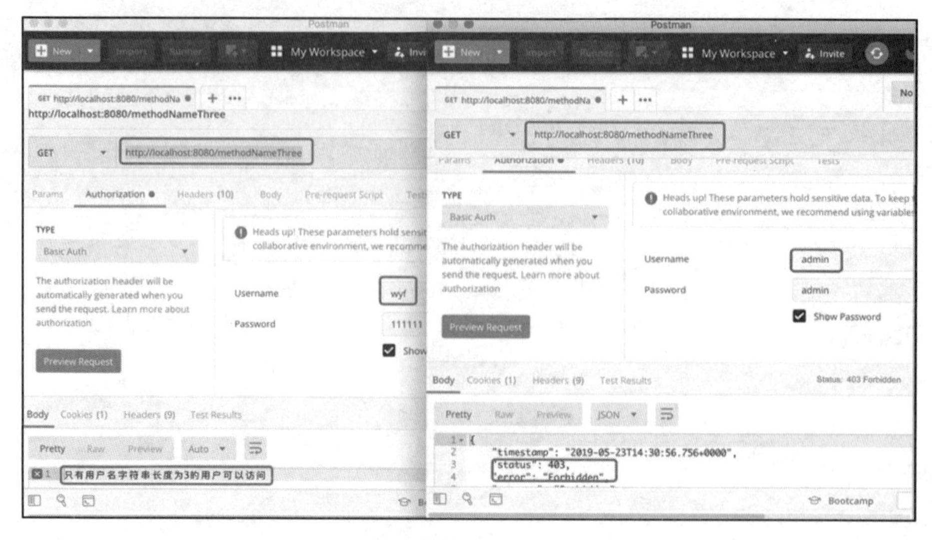

图 7-13

d. 通过@PostAuthorize 对返回值进行检验。访问 http://localhost:8080/methodAnotherName3，只有返回值字符串长度为 5 的用户才可访问，如图 7-14 所示。

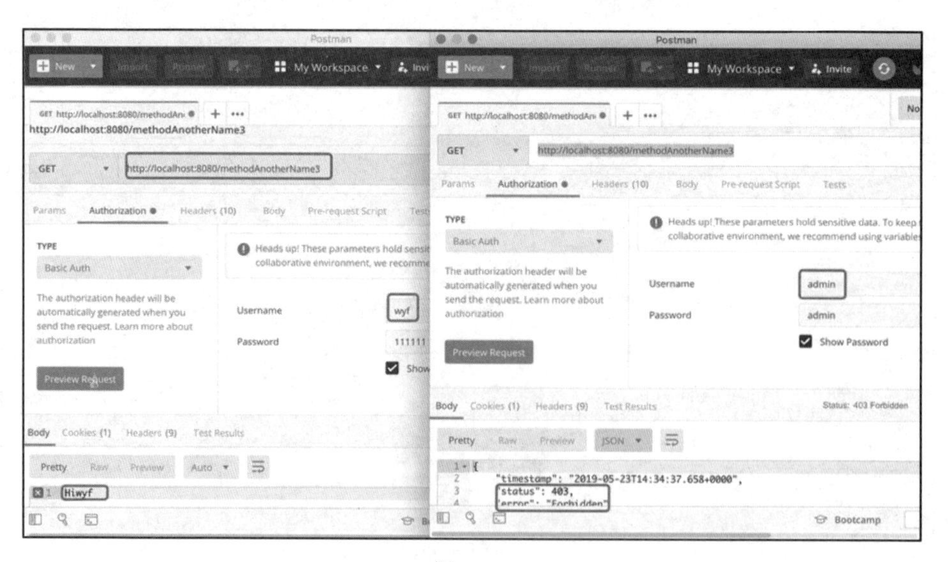

图 7-14

e. 使用@PreFilter 过滤参数的数据，参数只支持集合类，不支持数组。访问 http://localhost:8080/methodFilterIn?numbers=1,2,3,4,5,6,7,8,9,10，在参数列表中只保留偶数，如图 7-15 所示。

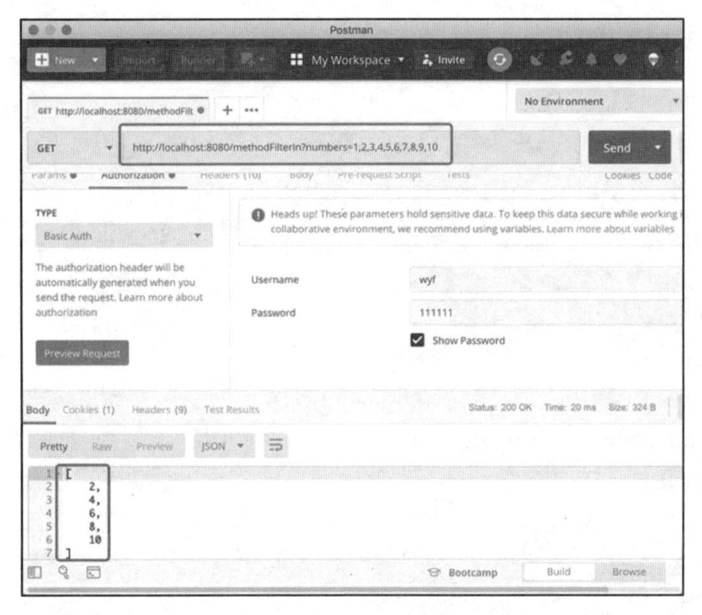

图 7-15

f. 使用@PostFilter 过滤返回值中的数据。访问 http://localhost:8080/methodFilterOut，在返回值列表中只包含偶数，如图 7-16 所示。

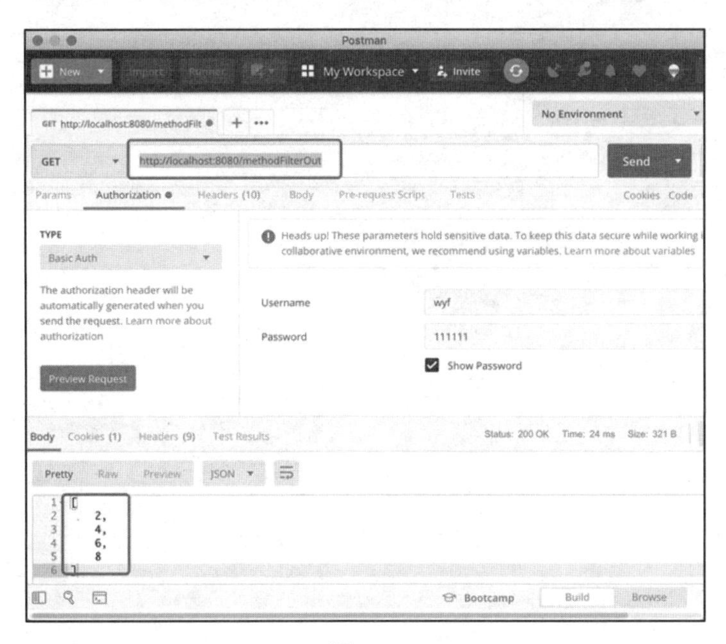

图 7-16

7.1.6 Spring Data 集成

应用依赖增加：

```
implementation 'org.springframework.security:spring-security-data'
```

Spring Boot 允许在使用 Spring Data 进行查询时引用 Spring Security 的安全表达式：

```
public interface SysUserRepository extends JpaRepository<SysUser, Long> {
  @Query("select u from SysUser u where u.id = ?#{principal?.id} and 
true=?#{hasRole('ROLE_ADMIN')}")
  Optional<SysUser> findRoleAdminMyself();
}
```

可以在@Query 查询中使用安全表达式。当表达式为空时，使用“?”可以不抛出异常（BeanExpressionException），而是返回 null，如图 7-17 所示。本例是查询当前用户信息，且当前用户的类型是 ROLE_ADMIN。

通过控制器方法进行验证：

```
@Autowired
SysUserRepository sysUserRepository;

@GetMapping("/findRoleAdminMyself")
public Optional<SysUser> findMyself(){
   return sysUserRepository.findRoleAdminMyself();
}
```

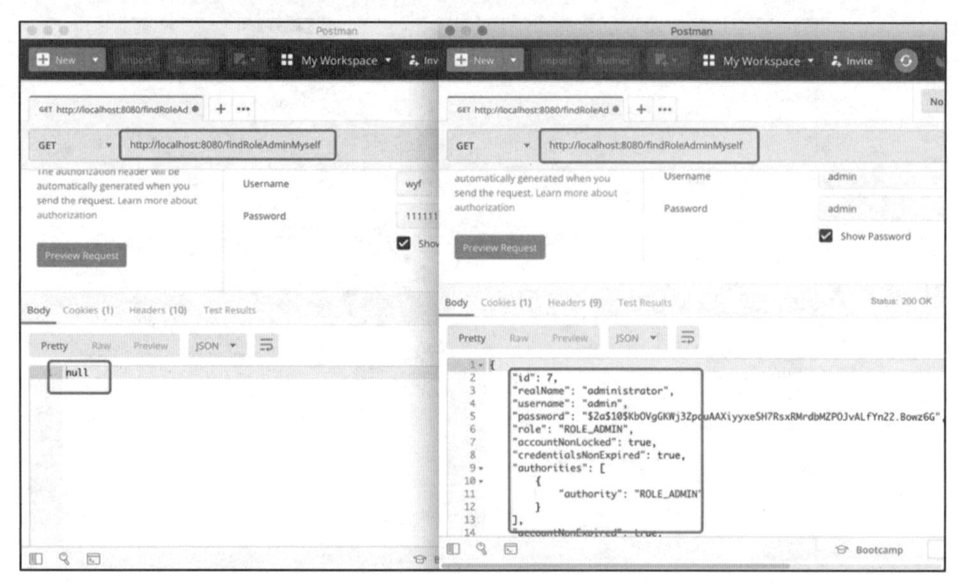

图 7-17

7.2　Spring Security 实战

本节演示一个基于用户、角色和权限的例子。一个用户有一个或多个角色，每个角色有一个或多个权限。

新建应用，信息如下。

Group：top.wisely。

Artifact：learning-spring-security-in-battle。

Dependencies：Spring Security、Spring Web Starter、Spring Data JPA、MySQL Driver 和 Lombok。

build.gradle 文件中的依赖如下：

```
dependencies {
  implementation 'org.springframework.boot:spring-boot-starter-data-jpa'
  implementation 'org.springframework.boot:spring-boot-starter-security'
  implementation 'org.springframework.boot:spring-boot-starter-web'
  compileOnly 'org.projectlombok:lombok'
  runtimeOnly 'mysql:mysql-connector-java'
  annotationProcessor 'org.projectlombok:lombok'
  //……
}
```

删除上一个应用的数据库 SYS_USER 表，在 application.yml 中连接数据库：

```
spring:
 datasource:
  url: jdbc:mysql://localhost:3306/first_db?useSSL=false
  username: root
  password: zzzzzz
  driver-class-name: com.mysql.cj.jdbc.Driver
 jpa:
  show-sql: true
  hibernate:
   ddl-auto: update
```

用户、角色和权限的实体分别是 SysUser、SysRole 和 SysAuthority。

权限的实体：

```
@Data
@AllArgsConstructor
@NoArgsConstructor
@Entity
public class SysAuthority {
   @Id
   @GeneratedValue(strategy = GenerationType.IDENTITY)
   private Long id;

   private String name; //权限名称
```

```java
    private String value; //权限值

    public SysAuthority(String name, String value) {
        this.name = name;
        this.value = value;
    }
}
```

角色的实体：

```java
@Data
@AllArgsConstructor
@NoArgsConstructor
@Entity
public class SysRole {
    @Id
    @GeneratedValue(strategy = GenerationType.IDENTITY)
    private Long id;

    private String name;

    @ManyToMany(targetEntity = SysAuthority.class)
    private Set<SysAuthority> authorities; // 角色和权限是多对多的关系

    public SysRole(String name, Set<SysAuthority> authorities) {
        this.name = name;
        this.authorities = authorities;
    }
}
```

用户的实体：

```java
@Data
@AllArgsConstructor
@NoArgsConstructor
@Entity
public class SysUser implements UserDetails {
    @Id
    @GeneratedValue(strategy = GenerationType.IDENTITY)
    private Long id;

    @Column(unique = true)
    @NotNull
    private String username;

    private String password;

    private String realName;

    private Boolean enable;
```

```
    @ManyToMany(targetEntity = SysRole.class, fetch = FetchType.EAGER)
    private Set<SysRole> roles; //用户和角色是多对多的关系

    @Override
    public Collection<? extends GrantedAuthority> getAuthorities() {
        Collection<GrantedAuthority> authorities = new HashSet<>();
        roles.forEach(role -> {
            role.getAuthorities().forEach(authority -> {
                authorities.add(new SimpleGrantedAuthority(authority.getValue()));
            });
        });
        return authorities; //获取当前用户的角色集合，通过角色集合获得当前用户的权限集合
    }

    public SysUser(String username, String password, String realName, Boolean enable,
Set<SysRole> roles) {
        this.username = username;
        this.password = password;
        this.realName = realName;
        this.enable = enable;
        this.roles = roles;
    }
     // 省略其他接口方法
}
```

三个 Repository 分别如下：

```
public interface SysUserRepository extends JpaRepository<SysUser, Long> {
    Optional<SysUser> findByUsername(String username);
}
public interface SysRoleRepository extends JpaRepository<SysRole, Long> {
}
public interface SysAuthorityRepository extends JpaRepository<SysAuthority, Long> {
}
```

安全配置：

```
@EnableGlobalMethodSecurity(prePostEnabled = true)
public class WebSecurityConfig extends WebSecurityConfigurerAdapter {

    @Override
    protected void configure(HttpSecurity http) throws Exception {
        http.authorizeRequests()
                .anyRequest().authenticated()
                .and()
                .httpBasic().authenticationEntryPoint(authenticationEntryPoint());
    }

    @Bean
    AuthenticationEntryPoint authenticationEntryPoint(){
```

```
        BasicAuthenticationEntryPoint authenticationEntryPoint = new
BasicAuthenticationEntryPoint();
        authenticationEntryPoint.setRealmName("wisely");
        return authenticationEntryPoint;
    }

    @Bean
    UserDetailsService userDetailsService(SysUserRepository sysUserRepository){
        return new CusotmUserDetailsService(sysUserRepository);
    }

    @Bean
    PasswordEncoder passwordEncoder(){
        return new BCryptPasswordEncoder();
    }
}
```

演示权限的控制器方法：

```
    @GetMapping("/userCan1")
    @PreAuthorize("hasAuthority('userCan1')") //当前用户拥有的Authorities是否包含
//userCan1
    public Map<String, Object> userCan1(@AuthenticationPrincipal SysUser sysUser){
        return getReturnMap(sysUser,"userCan1");
    }

    @GetMapping("/userCan2")
    @PreAuthorize("hasAuthority('userCan2')")//当前用户拥有的Authorities是否包含
//userCan2
    public Map<String, Object> userCan2(@AuthenticationPrincipal SysUser sysUser){
        return getReturnMap(sysUser,"userCan2");

    }

    @GetMapping("/adminCan1")
    @PreAuthorize("hasAuthority('adminCan1')")//当前用户拥有的Authorities是否包含
//adminCan1
    public Map<String, Object> adminCan1(@AuthenticationPrincipal SysUser sysUser){
        return getReturnMap(sysUser,"adminCan1");
    }

    @GetMapping("/adminCan2")
    @PreAuthorize("hasAuthority('adminCan2')")//当前用户拥有的Authorities是否包含
//adminCan2
    public Map<String, Object> adminCan2(@AuthenticationPrincipal SysUser sysUser){
        return getReturnMap(sysUser,"adminCan2");
    }

    private Map<String, Object> getReturnMap(SysUser sysUser, String currentAuthority)
{
```

```
        Map<String, Object> map = new HashMap<>();
        map.put("user-authorities" , sysUser.getAuthorities()); //
        map.put("current-authority", "userCan1");
        return map;
    }

}
```

通过 CommandLineRunner 初始化系统的所有权限、角色和用户。注意，当第二次运行系统时，要注释掉这些代码。

```
@SpringBootApplication
public class LearningSpringSecurityInBattleApplication {

  public static void main(String[] args) {
    SpringApplication.run(LearningSpringSecurityInBattleApplication.class, args);
  }

  @Bean
  CommandLineRunner init(SysUserRepository sysUserRepository,
              SysRoleRepository sysRoleRepository,
              SysAuthorityRepository sysAuthorityRepository,
              PasswordEncoder passwordEncoder){
    return args -> {
      SysAuthority authority1 = sysAuthorityRepository.save(new
SysAuthority("userCan1","userCan1"));
      SysAuthority authority2 = sysAuthorityRepository.save(new
SysAuthority("userCan2","userCan2"));
      SysAuthority authority3 = sysAuthorityRepository.save(new
SysAuthority("adminCan1","adminCan1"));
      SysAuthority authority4 = sysAuthorityRepository.save(new
SysAuthority("adminCan2","adminCan2"));

      SysRole role1 = sysRoleRepository.save(new SysRole("普通用户",
          Stream.of(authority1,authority2).collect(Collectors.toSet())));
      SysRole role2 = sysRoleRepository.save(new SysRole("管理员",

Stream.of(authority1,authority2,authority3,authority4).collect(Collectors.toSet())));

      SysUser user1 = sysUserRepository.save(new SysUser("wyf",
          passwordEncoder.encode("111111"),
          "wangyunfei",
          true,
          Stream.of(role1).collect(Collectors.toSet())));

      SysUser user2 = sysUserRepository.save(new SysUser("admin",
          passwordEncoder.encode("admin"),
          "administrator",
          true,
          Stream.of(role2).collect(Collectors.toSet())));
```

```
        };
    }

}
```

普通用户角色的用户 wyf，可以访问 userCan1 和 userCan2；管理员角色的用户 admin，可以访问 userCan1、userCan2、adminCan1 和 adminCan2。

两个用户均能访问 http://localhost:8080/userCan1，如图 7-18 所示。

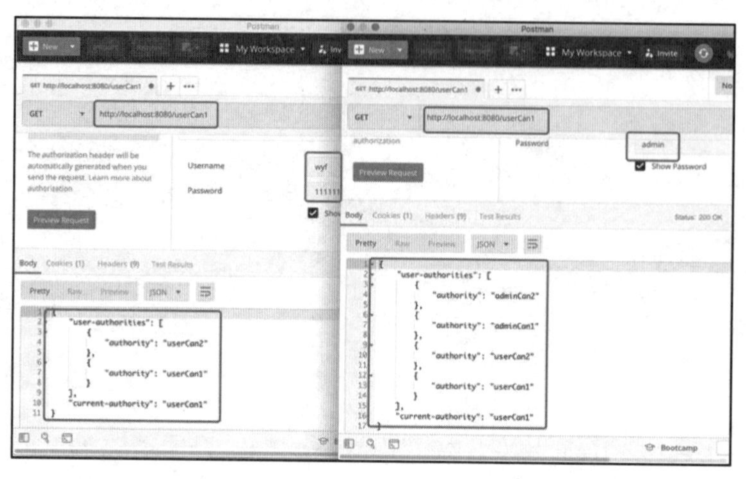

图 7-18

只有管理员角色的用户才能访问 http://localhost:8080/adminCan1，管理员角色拥有 adminCan1 的权限，如图 7-19 所示。

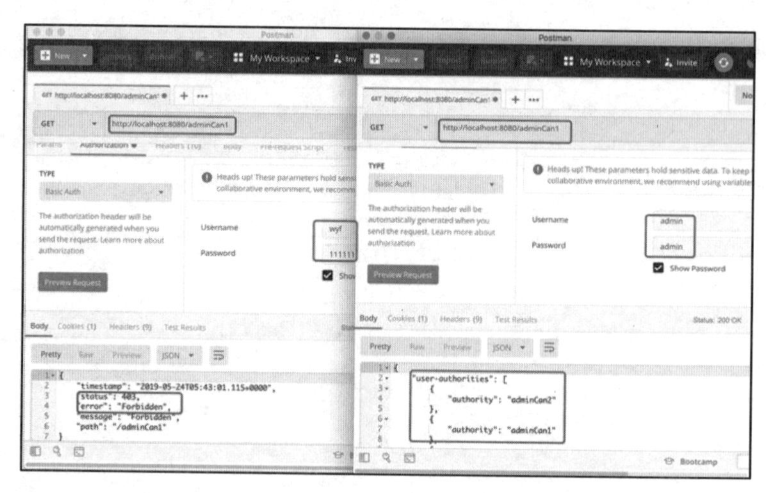

图 7-19

7.3 OAuth 2.0

OAuth 2.0 是安全授权的工业标准协议，在了解它之前，需要理解下面的专用术语。

◎ 交互参与方。

- Client：客户端需要访问终端用户（Resource Sever）受保护资源的应用。
- Resource Owner：终端用户，客户端通过终端用户进行不同类型的授权（Grant Type）。
- Authorization Server：授权服务，提供访问授权的应用。客户端使用某种 Grant Type 向授权服务获取客户端。
- Resource Sever：资源服务，包含受保护资源的应用，客户端使用客户端访问终端用户的受保护资源。

◎ 授权类型（Grant Type）。

- Authorization Code：当用户访问客户端页面时，页面导向 Authorization Server 的登录页面；登录后，显示授权访问页面。在授权成功后，客户端即可获得访问令牌，进而访问资源服务。
- Password：通过提供用户名和密码获得访问令牌，一般是给应用服务的客户端使用（iOS、Android、Web App）的。
- Client Credentials：客户端通过 Client Id 和 Client Secret 直接向授权服务请求访问令牌。Client Credentials 主要用于非用户参与的应用，如后台服务。

◎ Token。

- Access Token：访问令牌，用来访问受保护资源的唯一令牌。
- Refresh Token：当访问令牌失效时，可以使用 Refresh Token 来获取一个新的访问令牌，它的时效性要远远大于访问令牌。
- JWT：JSON Web Token，它代表双方之间安全传输的信息；使用数字签名，传输的信息可以被验证和信任。

7.3.1 OAuth 2.0 Authorization Server

新建应用，信息如下。

Group：top.wisely。

Artifact：auth-server。

Dependencies：Spring Security、Spring Web Starter、Spring Data JPA、MySQL Driver 和 Lombok。

build.gradle 文件中的依赖如下：

```
dependencies {
  implementation 'org.springframework.boot:spring-boot-starter-data-jpa'
  implementation 'org.springframework.boot:spring-boot-starter-security'
  implementation 'org.springframework.boot:spring-boot-starter-web'
  compileOnly 'org.projectlombok:lombok'
  compileOnly 'org.projectlombok:lombok'
  runtimeOnly 'mysql:mysql-connector-java'
  annotationProcessor 'org.projectlombok:lombok'
 //……
}
```

Spring Security 不提供 OAuth 2.0 授权服务的功能，使用 spring-security-oauth2-autoconfigure 可以实现 OAuth 2.0 授权服务的功能和配置。Spring Boot 不提供 spring-security-oauth2-autoconfigure 版本的维护，所以需要指定它的版本。

```
implementation
'org.springframework.security.oauth.boot:spring-security-oauth2-autoconfigure:2.2.X.
RELEASE'
```

1. spring-security-oauth2-autoconfigure 提供的自动配置

spring-security-oauth2-autoconfigure 使用自动配置类 OAuth2AutoConfiguration 导入授权服务的配置 OAuth2AuthorizationServerConfiguration，通过 AuthorizationServerProperties 用 security.oauth2.authorization.*来进行定制配置。OAuth2AuthorizationServerConfiguration 做了下面的配置。

（1）AuthorizationServerConfigurerAdapter：继承此类，通过覆写其方法进行配置。

◎ configure(ClientDetailsServiceConfigurer clients)：配置客户端的信息，可通过 security.oauth2.client.*来配置，只能配置一个客户端的信息。

◎ configure(AuthorizationServerEndpointsConfigurer endpoints)：配置授权服务端点的非安全特性。配置了 TokenStore 和 AccessTokenConverter。若 Grant Type 为 password，则需要设置 AuthenticationManager。

◎ configure(AuthorizationServerSecurityConfigurer security)：配置授权服务的安全性。

- tokenKeyAccess：配置端点/oauth/token_key 的访问安全，提供 JWT 编码的 Token；可通过 security.oauth2.authorization.tokenKeyAccess 来配置。
- checkTokenAccess：配置端点/oauth/check_token 的访问安全，解码 Access Token 用来检查和确认生成的 token；可通过 security.oauth2. authorization. checkTokenAccess 来配置。

（2）导入 AuthorizationServerTokenServicesConfiguration：配置 JWT。

- TokenStore：用来生成和获取 token；自动配置 JwtTokenStore 的 Bean.

➢ AccessTokenConverter：将 access token 转换成不同的格式；自动配置 JwtAccessTokenConverter，将其转换成 JWT 格式。

2. 获取用户配置

授权服务需要获取终端用户来获取用户相关信息。沿用前面的 SysAuthority、SysRole、SysUser、SysAuthorityRepository、SysRoleRepository、SysUserRepository 和 CusotmUserDetailsService，配置如下。

```
@Configuration
public class WebSecurityConfig extends WebSecurityConfigurerAdapter {
   @Autowired
   SysUserRepository sysUserRepository;
   @Bean
   protected UserDetailsService userDetailsService() {
      return new CusotmUserDetailsService(sysUserRepository);
   }

   @Bean
   PasswordEncoder passwordEncoder(){
      return new BCryptPasswordEncoder();
   }

   @Bean
   @Override //暴露 authenticationManager Bean
   protected AuthenticationManager authenticationManager() throws Exception {
      return super.authenticationManager();
   }

   @Override
   protected void configure(HttpSecurity http) throws Exception {
      http.authorizeRequests()
            .anyRequest().authenticated();
   }
}
```

3. 配置 JWT

在配置 JWT 时，需要添加 spring-security-jwt 的依赖：

```
implementation 'org.springframework.security:spring-security-jwt'
```

因为 AuthorizationServerTokenServicesConfiguration 做了自动配置，所以只需在 application.yml 中即可配置 JWT。当签名生成 JWT Token 时，需要使用自签名的证书，当前证书使用的是第 5 章中生成的证书：

```
security:
  oauth2:
    authorization:
```

```
    jwt:
      key-store: classpath:keystore.jks
      key-store-password: pass1234
      key-alias: wisely
      key-password: pass1234
```

4. 配置授权服务

OAuth2AuthorizationServerConfiguration 自动配置的客户端只有一个不符合笔者要求，需要使用@EnableAuthorizationServer 和继承 AuthorizationServerConfigurerAdapter 完全定制自己的 Authorization Server 配置：

```
@EnableAuthorizationServer
@Configuration
public class AuthServerConfig extends AuthorizationServerConfigurerAdapter {

   private final PasswordEncoder passwordEncoder;
   private final DataSource dataSource;
   private final AuthenticationManager authenticationManager;
   private final UserDetailsService userDetailsService;
   private final TokenStore tokenStore;
   private final AccessTokenConverter accessTokenConverter;
   private final AuthorizationServerProperties properties;

   public AuthServerConfig(PasswordEncoder passwordEncoder,
                     DataSource dataSource,
                     AuthenticationManager authenticationManager,
                     UserDetailsService userDetailsService,
                     TokenStore tokenStore,
                     AccessTokenConverter accessTokenConverter,
                     AuthorizationServerProperties properties) {
      this.passwordEncoder = passwordEncoder;
      this.dataSource = dataSource;
      this.authenticationManager = authenticationManager;
      this.userDetailsService = userDetailsService;
      this.tokenStore = tokenStore;
      this.accessTokenConverter = accessTokenConverter;
      this.properties = properties;
   }
}
```

上面注入的 Bean 都是下面配置需要的。

（1）配置客户端：

```
@Override
public void configure(ClientDetailsServiceConfigurer clients) throws Exception {
   clients
      .jdbc(this.dataSource)
      .passwordEncoder(passwordEncoder);
}
```

ClientDetailsServiceConfigurer 指定了客户端来源于数据库，数据源为 dataSource，且默认了存储客户端的数据库表结构。

新建 schema.sql：

```
create table if not exists oauth_client_details (
  client_id VARCHAR(255) PRIMARY KEY,
  resource_ids VARCHAR(255),
  client_secret VARCHAR(255),
  scope VARCHAR(255),
  authorized_grant_types VARCHAR(255),
  web_server_redirect_uri VARCHAR(255),
  authorities VARCHAR(255),
  access_token_validity INTEGER,
  refresh_token_validity INTEGER,
  additional_information VARCHAR(4096),
  autoapprove varchar(255)
);
```

新建 data.sql，新增两个 Client：

```
INSERT INTO oauth_client_details (client_id, client_secret, scope,
authorized_grant_types, autoapprove)
values ('postman', # a
       '$2a$10$KCroi.THbmXdKXOgBud1zOzdmHrfpNxSytd/o5ZQLhCTzFXib1p66', #b
       'any'
       'password,client_credentials,refresh_token', # c
       true);
INSERT INTO oauth_client_details (client_id, client_secret, authorized_grant_types,
autoapprove)
values ('app',
       '$2a$10$1vUvCP9fzPeXIRzoOaPROuIRvq2nrh7iauWLIa371qZSmaP0p6ave',
       'any'
       'password,client_credentials,refresh_token',
       true);
```

a. Client Id。

b. Client Secret；postman 的密码是 postman，app 的密码是 app；可使用 passwordEncoder.encode("postman")获得。

c. 该客户端支持的 Grant Type。

当需要指定客户端来源时，可通过 ClientDetailsService 接口来定制。与 UserDetailsService 类似，可以通过覆写其 loadClientByClientId 来定制。这里通过下面代码来配置：

```
@Override
public void configure(ClientDetailsServiceConfigurer clients) throws Exception {
    clients.withClientDetails(new SomeCustomClientDetailsService());
}
```

上面的代码：

```
@Override
public void configure(ClientDetailsServiceConfigurer clients) throws Exception {
   clients
       .jdbc(this.dataSource)
       .passwordEncoder(passwordEncoder);
}
```

等同于代码：

```
@Override
public void configure(ClientDetailsServiceConfigurer clients) throws Exception {
   JdbcClientDetailsService detailsService = new
JdbcClientDetailsService(this.dataSource);
   detailsService.setPasswordEncoder(passwordEncoder);
   clients.withClientDetails(detailsService);
}
```

一般使用内存来注册客户端：

```
public void configure(ClientDetailsServiceConfigurer clients) throws Exception {
   clients
       .inMemory()
          .withClient("postman")
          .secret(passwordEncoder.encode("postman"))
          .scopes("any")
          .authorizedGrantTypes("password", "authorization_code", "refresh_token")
       .and()
          .withClient("app")
          .secret(passwordEncoder.encode("app"))
          .scopes("any")
          .authorizedGrantTypes("password", "authorization_code", "refresh_token");
}
```

（2）配置端点非安全特性。

```
@Override
public void configure(AuthorizationServerEndpointsConfigurer endpoints) throws
Exception {
      endpoints.accessTokenConverter(this.accessTokenConverter); //a
      endpoints.tokenStore(this.tokenStore); //b
      endpoints.authenticationManager(this.authenticationManager); //c
      endpoints.userDetailsService(userDetailsService); //d
}
```

a. 设置自动配置的 accessTokenConverter，实际设置为 JwtAccessTokenConverter 的 Bean。

b. 设置自动配置的 tokenStore，实际设置为 JwtTokenStore 的 Bean。

c. 为了让 Grant Type 支持 password，需设置 authenticationManager。

d. 当 Grant Type 为 refresh_token 时，需设置 userDetailsService。

（3）配置 Authorization Server 安全:

```
@Override
public void configure(AuthorizationServerSecurityConfigurer security) throws Exception
{
      security.checkTokenAccess(this.properties.getCheckTokenAccess());
      security.tokenKeyAccess(this.properties.getTokenKeyAccess());
      security.realm(this.properties.getRealm());
}
```

上面的配置可通过 application.yml 来配置：

```
security:
  oauth2:
    authorization:
      jwt:
        key-store: classpath:keystore.jks
        key-store-password: pass1234
        key-alias: wisely
        key-password: pass1234
      tokenKeyAccess: permitAll()
      checkTokenAccess: isAuthenticated()
      realm: wisely
```

5. Authorization Server 的端点

- ◎ TokenEndpoint：路径为/oauth/token，按照 OAuth 2.0 规范请求 Token。
- ◎ TokenKeyEndpoint：路径为/oauth/token_key，提供 JWT 编码的 Token。
- ◎ CheckTokenEndpoint：路径为/oauth/check_token，解码客户端的访问令牌用来检查和确认生成的 Token。
- ◎ AuthorizationEndpoint：路径为/oauth/authorize，遵循 OAuth 2.0 规范的授权实现。

6. 获取 Token

同样，客户端也是独立的，可以采用获取 Token 的方式来访问服务。

（1）获取访问令牌。

postman 为一个 Client，使用 POST 访问 http://localhost:8080/oauth/token，把 Client Id 和 Client Secret 填入 Basic Auth 中，如图 7-20 所示。

请求参数如下。

- ◎ grant_type：password。
- ◎ username：wyf。
- ◎ password：111111。

JWT 是一种编码形式，由 Header、PayLoad 和 Signature 三部分组成，通过 keystore 的公钥（Public Key）即可验证 JWT Token，如图 7-21 所示。

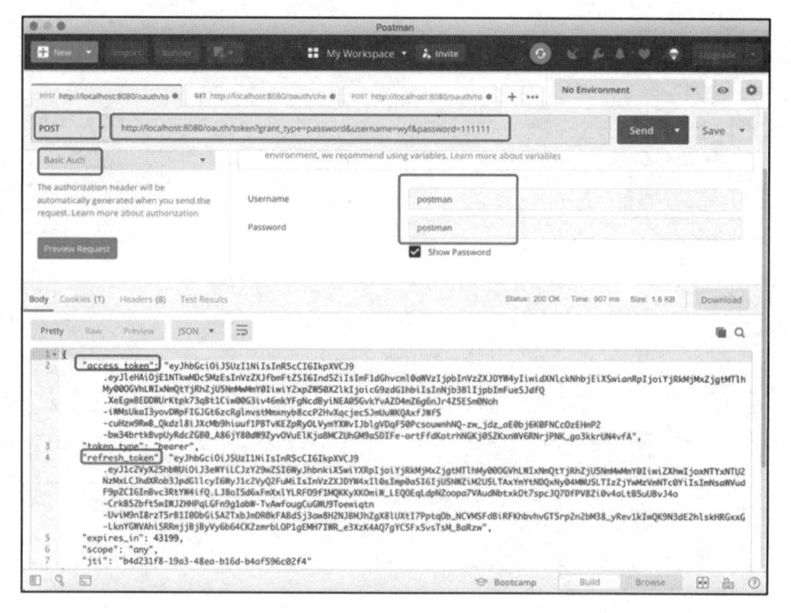

图 7-20

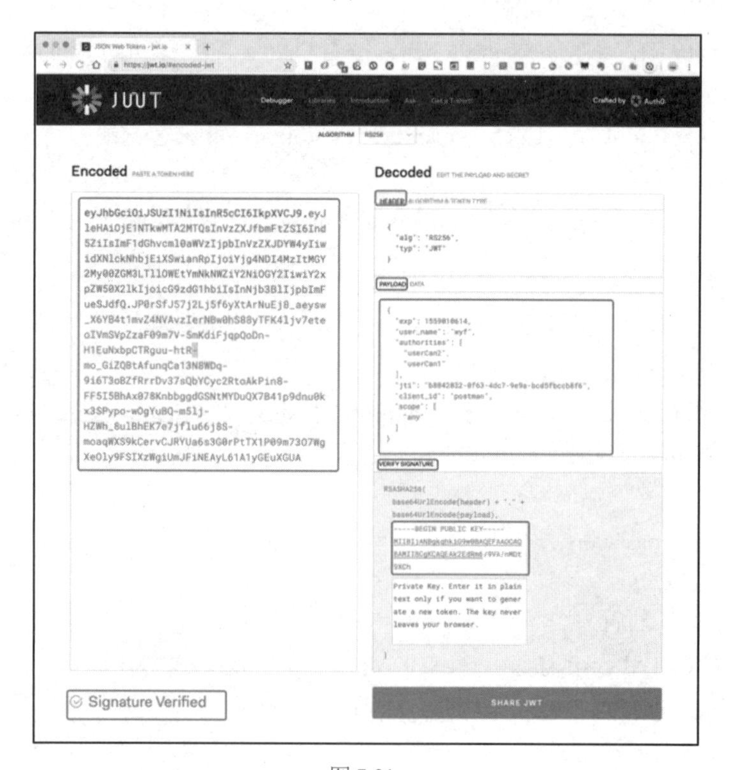

图 7-21

（2）检查 Token。

用 GET 访问 http://localhost:8080/oauth/check_token，把 Client Id 和 Client Secret 填入 Basic Auth 中，如图 7-22 所示。

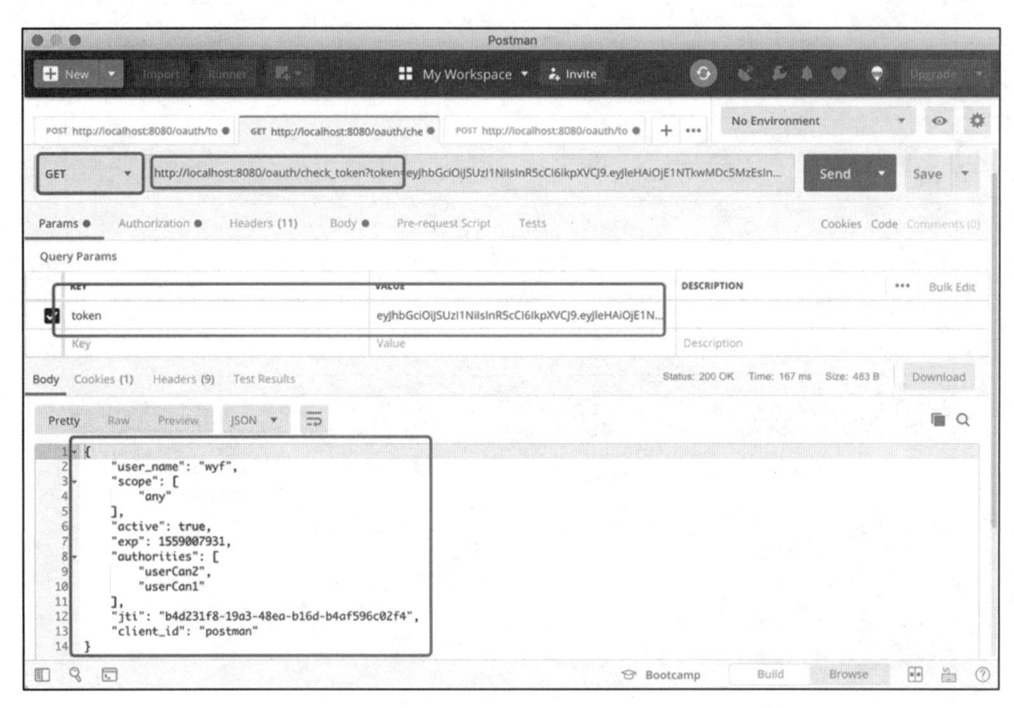

图 7-22

（3）刷新 Token。

用 POST 访问 7-24http://localhost:8080/oauth/token，如图 7-23 和图 7-24 所示，请求参数如下。

◎ grant_type：refresh_token。

◎ username：wyf。

◎ password：111111。

◎ refresh_token：需从请求返回中复制。

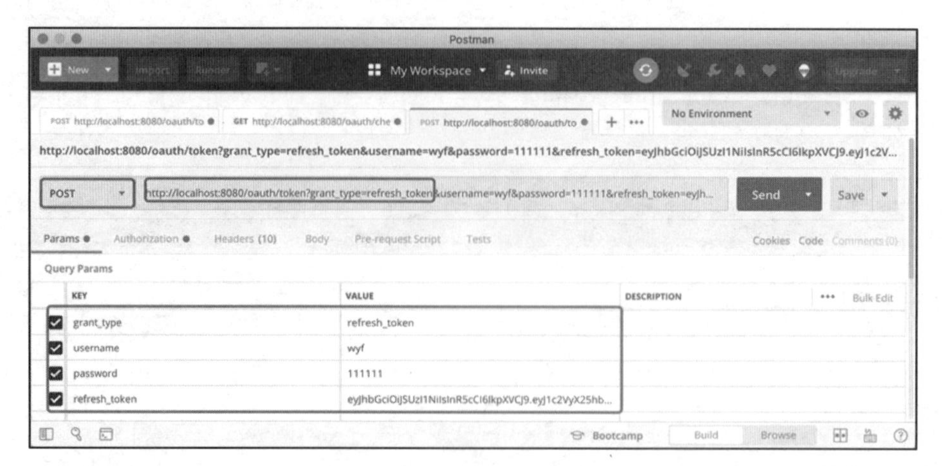

图 7-23

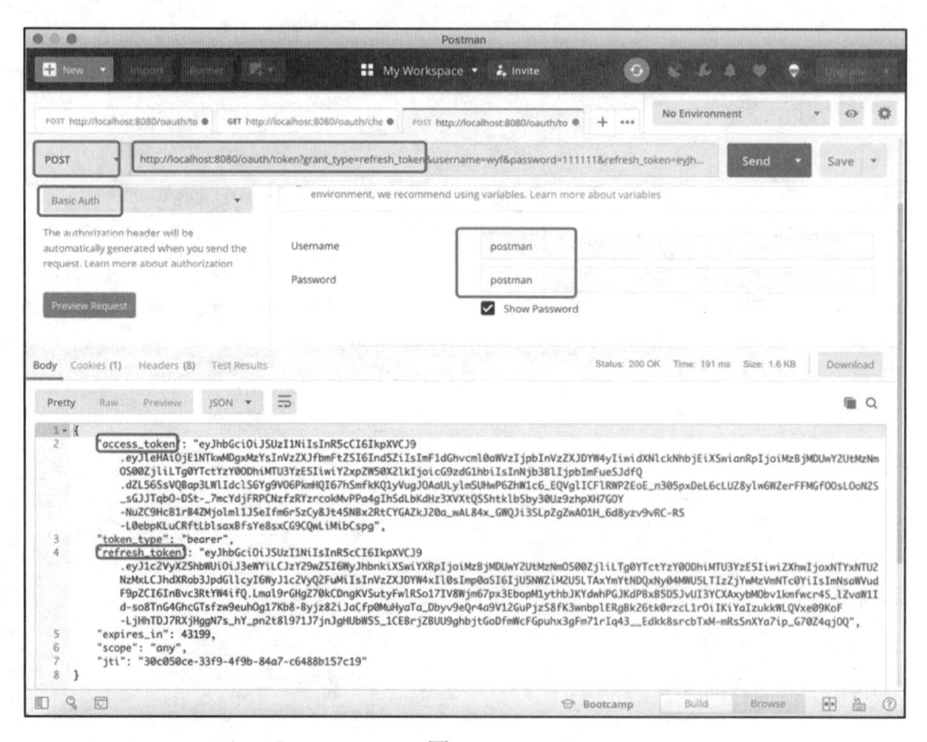

图 7-24

7.3.2 OAuth 2.0 Resource Server

新建应用，信息如下。

Group：top.wisely。

Artifact：resource-server。

Dependencies：Spring Security、OAuth2 Resource Server、Spring Web Starter 和 Lombok。

build.gradle 文件中的依赖如下：

```
dependencies {
    implementation
'org.springframework.boot:spring-boot-starter-oauth2-resource-server'
    implementation 'org.springframework.boot:spring-boot-starter-security'
    implementation 'org.springframework.boot:spring-boot-starter-web'
    compileOnly 'org.projectlombok:lombok'
    annotationProcessor 'org.projectlombok:lombok'
  //……
}
```

1. Spring Boot 的自动配置

Spring Boot 使用 OAuth2ResourceServerJwtConfigurationResource Server 做自动配置，它使用 OAuth2ResourceServerProperties 通过 spring.security.oauth2.resourceserver.* 进行配置。OAuth2ResourceServerJwtConfiguration 导入了下面两个配置。

（1）OAuth2ResourceServerJwtConfiguration：配置 JWT Token 解码的 JwtDecoder。

◎ 使用 Authorization Server 的 JWK Set URI 端点进行解码，使用 spring.security.oauth2.resourceserver.jwt.jwk-set-uri 配置。

◎ 使用 Authorization Server 的 keyStore 的公钥进行解码，使用 spring.security.oauth2.resourceserver.jwt.public-key-location 配置。

（2）OAuth2ResourceServerWebSecurityConfiguration：常规的 Spring Security 配置，使用 HttpSecurity 配置 Resource Server 和 JWT：

```
@Configuration(proxyBeanMethods = false)
@ConditionalOnBean(JwtDecoder.class)
static class OAuth2WebSecurityConfigurerAdapter extends WebSecurityConfigurerAdapter {

   @Override
   protected void configure(HttpSecurity http) throws Exception {
      http.authorizeRequests().anyRequest().authenticated().and()
            .oauth2ResourceServer().jwt();
   }

}
```

方式 1：JWK Set 路径配置。

通过在资源服务中设置 JWK Set 路径来解码 JWT Token。

（1）Authorization Server 的修改。

对 JWT 的操作依赖于 nimbus 包，需添加如下依赖：

```
implementation 'com.nimbusds:nimbus-jose-jwt:7.0.1'
```

资源服务端已通过 spring-boot-starter-oauth2-resource-server 自动依赖。

想要使用 JWK Set URI，则需授权服务提供支持，在授权服务上添加端点：

```
@FrameworkEndpoint //a
public class JwkSetEndpoint {

   KeyPair keyPair; //b

   public JwkSetEndpoint(KeyPair keyPair) {
      this.keyPair = keyPair;
   }

   @GetMapping("/.well-known/jwks.json") //c
   @ResponseBody
   public Map<String, Object> getKey(Principal principal) {
      RSAPublicKey publicKey = (RSAPublicKey) this.keyPair.getPublic(); //d
      RSAKey key = new RSAKey.Builder(publicKey).build();
      return new JWKSet(key).toJSONObject(); //e
   }
}
```

a. @FrameworkEndpoint 和@Controller 功能相同，只用于框架提供的端点。

b. 注入密钥对 keyPair。

```
@Bean //通过读取 key Store 的配置构造
public KeyPair keyPair(AuthorizationServerProperties properties, ApplicationContext
context){
   Resource keyStore = context
         .getResource(properties.getJwt().getKeyStore());
   char[] keyStorePassword = properties.getJwt().getKeyStorePassword()
         .toCharArray();
   KeyStoreKeyFactory keyStoreKeyFactory = new KeyStoreKeyFactory(keyStore,
         keyStorePassword);
   String keyAlias = properties.getJwt().getKeyAlias();
   char[] keyPassword = Optional
         .ofNullable(properties.getJwt().getKeyPassword())
         .map(String::toCharArray).orElse(keyStorePassword);
   return keyStoreKeyFactory.getKeyPair(keyAlias, keyPassword);
}
```

c. 构造端点地址为/.well-known/jwks.json。

d. 返回 JWK set 的 JSON 对象。

将/.well-known/jwks.json 配置为允许任意访问:

```
@Configuration
public class WebSecurityConfig extends WebSecurityConfigurerAdapter {
      //……
   @Override
```

```
    protected void configure(HttpSecurity http) throws Exception {
        http.authorizeRequests()
                .mvcMatchers("/.well-known/jwks.json").permitAll()
                .anyRequest().authenticated();
    }
    //……
}
```

（2）配置资源服务。

资源服务的配置很简单，具体如下：

```
spring:
  security:
    oauth2:
      resourceserver:
        jwt:
          jwk-set-uri: http://localhost:8080/.well-known/jwks.json
server:
  port: 8082
```

方式 2：公钥配置。

可以在资源服务中配置 keyStore 的公钥来解码 JWT Token。

可以借助 keytool 和 openssl 输出公钥，命令如下：

```
$ keytool -list -rfc --keystore keystore.jks | openssl x509 -inform pem -pubkey
-noout
```


```
-----BEGIN PUBLIC KEY-----
MIIBIjANBgkqhkiG9w0BAQEFAAOCAQ8AMIIBCgKCAQEAk2EdRm6/9VA/nMDt9XCh
96dTOv3wmZyh4LgG5pccsLgd5ZCSm5oENz+X6/m3yN7e+QMuP+2zSG/kdEH9vqGK
xKRd0/DV54s77OzRG+KHxxfmB10i36GNprfHN50pcyXoAbeIbztVIFWPIBedZrpZ
S8aufnK9PLOgH6C8cEoPf6Y/t2+Vxn/kjjMRU4oj0gl7j6tFTT79/g8qp0R5eBiJ
2KT/sdqMhRC0u5+i0ijcQycX52L+mfv4iyCyu7Z0g9bpQQDv069iPC9URyPuKyY0
HSlXUT/lXJ4o7wEYu7wrcmnkNk+4ot9uDvBUl2cdZ8BFlVyuNxJwYmRFyx2zgnAT
VwIDAQAB
-----END PUBLIC KEY-----
```


还可以通过 keyPair 来获取，在上例的授权服务中，可以通过下面的代码将图 7-25 中的内容复制到 src/main/resources/public.txt 中：

```
@Bean
CommandLineRunner publicKey(KeyPair keyPair){
    return args -> {
        System.out.println(Base64.encodeBase64String(keyPair.getPublic().getEncoded()));
    };
}
```

```
2019-05-28 15:22:19.569  INFO 7114 --- [           main] o.s.s.c
2019-05-28 15:22:19.830  INFO 7114 --- [           main] o.s.s.w
2019-05-28 15:22:19.836  INFO 7114 --- [           main] o.s.s.w
2019-05-28 15:22:19.914  INFO 7114 --- [           main] o.s.b.w
2019-05-28 15:22:19.918  INFO 7114 --- [           main] t.w.aut
MIIBIjANBgkqhkiG9w0BAQEFAAOCAQ8AMIIBCgKCAQEAk2EdRm6/9VA/nMD19XCh
```

图 7-25

可以通过 application.yml 指定公钥：

```
spring:
  security:
    oauth2:
      resourceserver:
        jwt:
          public-key-location: classpath:public.txt
server:
  port: 8082
```

2．获取资源服务权限

Spring Boot 已经做好了所有的配置，至此，资源服务的配置已完成。在默认情况下，权限是从客户端的 Scope 中获取的，而权限存在于 authorities 中，所以需要替代 OAuth2ResourceServerWebSecurityConfiguration 的配置：

```
@EnableGlobalMethodSecurity(prePostEnabled = true) //a
@EnableWebSecurity
public class WebSecurityConfig extends WebSecurityConfigurerAdapter {
    @Override
    protected void configure(HttpSecurity http) throws Exception {
        http.authorizeRequests()
                .anyRequest().authenticated()
                .and()
                .oauth2ResourceServer().jwt().jwtAuthenticationConverter(jwt -> { //b
                    Collection<SimpleGrantedAuthority> authorities =
                            ((Collection<String>) jwt.getClaims()
                                    .get("authorities")).stream() //c
                                                    .map(SimpleGrantedAuthority::new)
                                                    .collect(Collectors.toSet());
            return new JwtAuthenticationToken(jwt, authorities);
        });
    }

}
```

a. 弃用方法安全。

b. 通过自定义 Converter 来指定权限，Converter 是函数接口。当前上下文参数为 JWT 对象。

c. 获取 JWT 中的 authorities。

3．测试控制器

借用前面定义的控制器的权限声明：

```
@RestController
public class SecurityController {

   @GetMapping("/userCan1")
   @PreAuthorize("hasAuthority('userCan1')")
   public Jwt userCan1(@AuthenticationPrincipal Jwt jwt){ //获得用户信息
      return jwt;
   }

   @GetMapping("/userCan2")
   @PreAuthorize("hasAuthority('userCan2')")
   public Jwt userCan2(@AuthenticationPrincipal Jwt jwt){
      return jwt;
   }

   @GetMapping("/adminCan1")
   @PreAuthorize("hasAuthority('adminCan1')")
   public Jwt adminCan1(@AuthenticationPrincipal Jwt jwt){
      return jwt;
   }

   @GetMapping("/adminCan2")
   @PreAuthorize("hasAuthority('adminCan2')")
   public Jwt adminCan2(@AuthenticationPrincipal Jwt jwt){
      return jwt;
   }

}
```

4．验证

分别用 wyf 用户和 admin 用户请求 Token，如图 7-26 所示。

资源服务的认证请求是通过 BearerTokenAuthenticationFilter 来传递的，需要使用 Bearer Token 构造头部数据来传递 Token，格式为 Authorization:Bearer Token。Postman 支持直接使用 Bearer Token。

分别复制 wyf 和 admin 的 access_token 请求 http://localhost:8082/adminCan1，这个地址只有 admin 用户才可以访问。wyf 用户使用 Postman 的 Bearer Token，admin 用户可手动构造头部信息，如图 7-27 所示。

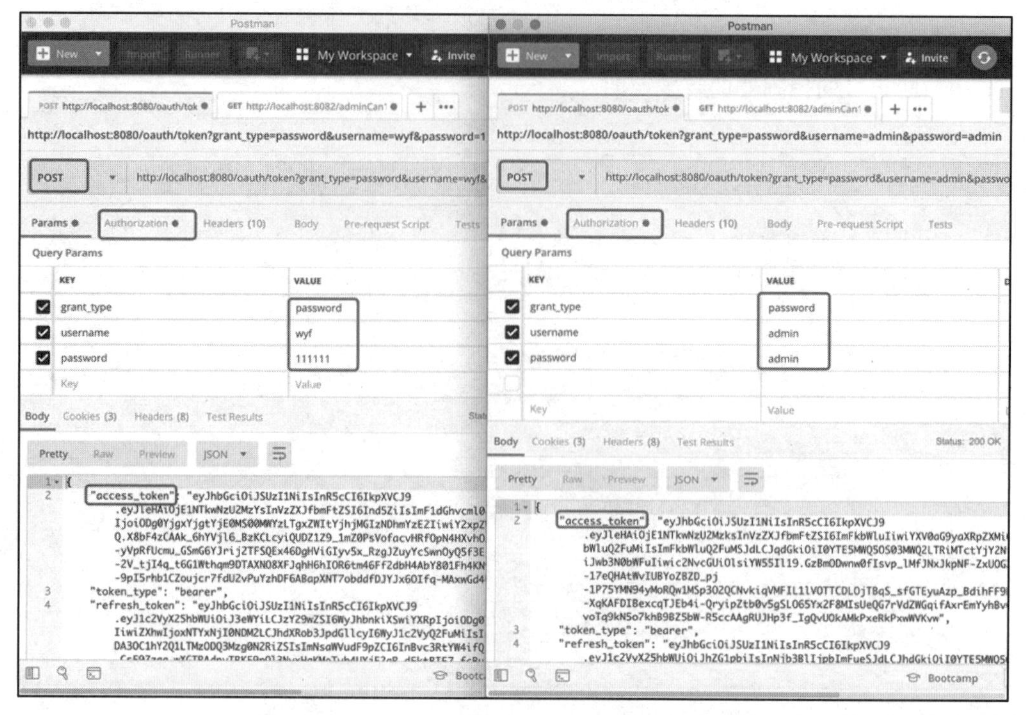

图 7-26

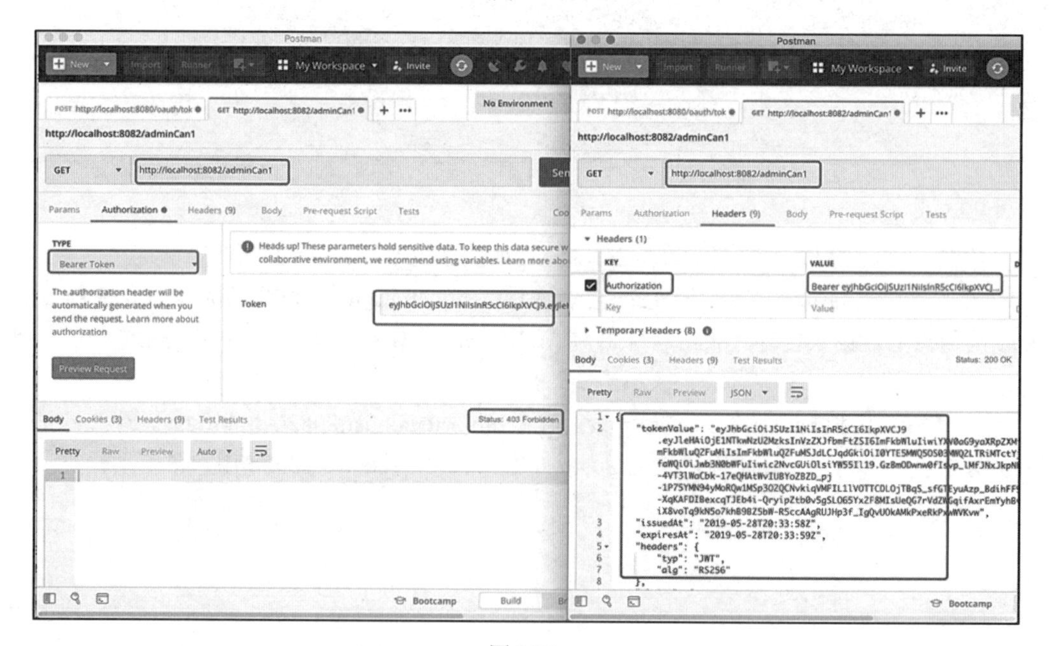

图 7-27

7.3.3　OAuth 2.0 Client

前面使用资源服务和授权服务演示了 Grant Type 为 password 的场景；下面结合客户端演示 Grant Type 为 Authorization Code 的场景。

新建应用，信息如下。

Group：top.wisely。

Artifact：client。

Dependencies：Spring Security、OAuth2 Client、Spring Web Starter 和 Lombok。

build.gradle 文件中的依赖如下：

```
dependencies {
    implementation 'org.springframework.boot:spring-boot-starter-oauth2-client'
    implementation 'org.springframework.boot:spring-boot-starter-security'
    implementation 'org.springframework.boot:spring-boot-starter-web'
    compileOnly 'org.projectlombok:lombok'
    annotationProcessor 'org.projectlombok:lombok'
  //……
}
```

1．Spring Boot 的自动配置

Spring Boot 通过 OAuth2ClientAutoConfiguration 来进行自动配置，通过 Auth2ClientProperties 使用 spring.security.oauth2.client.*来配置客户端。它导入了两个配置。

（1）OAuth2ClientRegistrationRepositoryConfiguration：读取外部配置 ClientRegistration（客户端注册）集合，将其添加到 InMemoryClientRegistrationRepository 中（在内存中存储 ClientClientRegistration 的库），并且将 InMemoryClientRegistrationRepository 注册成 Bean。在注册客户端时，需要指定客户端的 Registration，以及 Registration 对应客户端的 Provider（Authorization Server）信息。

- ◎ Registration：通过配置 spring.security.oauth2.client.registration.[registrationId].*来实现。
- ◎ Provider：通过配置 spring.security.oauth2.client.provider.[providerId].*来实现。Provider 和 Client 是对应关系，providerId 和 registrationId 相同。

（2）OAuth2WebSecurityConfiguration：注册了两个 Bean，并做了相关的 Spring Security 配置。

- ◎ OAuth2AuthorizedClientService：使用其实现类 InMemoryOAuth2AuthorizedClientService 注册 Bean，用来管理被授权的客户端（OAuth2AuthorizedClient，可获得用户的 Access Token）。
- ◎ OAuth2AuthorizedClientRepository：使用其实现类 AuthenticatedPrincipalOAuth2AuthorizedClientRepository 注册 Bean，用来持久化在请求间被授权的 Client。

◎ 使用 HttpSecurity 配置弃用 OAuth 2.0 的登录（oauth2Login()）和客户端（oauth2Client()）设置：

```
@Configuration(proxyBeanMethods = false)
@ConditionalOnMissingBean(WebSecurityConfigurerAdapter.class)
static class OAuth2WebSecurityConfigurerAdapter extends WebSecurityConfigurerAdapter {

   @Override
   protected void configure(HttpSecurity http) throws Exception {
      http.authorizeRequests().anyRequest().authenticated().and().oauth2Login()
            .and().oauth2Client();
   }

}
```

2. 配置 Client 和 Provider

```
spring:
  security:
    oauth2:
      client:
        registration:
          my-client: # a
            client-id: app # b
            client-secret: app # c
            authorization-grant-type: authorization_code # d
            redirect-uri: http://127.0.0.1:8083/login/oauth2/code/my-client # e
        provider:
          my-client: # f
            authorization-uri: http://localhost:8080/oauth/authorize # g
            token-uri: http://localhost:8080/oauth/token # h
            user-info-uri: http://localhost:8080/userInfo # i
            userNameAttribute: username # j
server:
  port: 8083
```

a. my-client 为 registrationId。

b. 设置在 Authorization Server 中配置的 Client Id。

c. 设置在 Authorization Server 中配置的 Client Secret。

d. 设置 Grant Type 为 authorization_code。

e. 设置转向的路径，默认规则为{baseUrl}/login/oauth2/code/{registrationId}。此处需使用 127.0.0.1，而不是使用 localhost（相同的域名会导致会话 cookie 被覆盖，产生 [authorization_request_not_found]错误）。另外，此处还需将数据库表 oauth_client_details 中 app 数据下的 web_server_redirect_uri 值修改为 http://127.0.0.1:8083/login/oauth2/code/my-client。如图 7-28 所示。

select client_id, web_server_redirect_uri from oauth_client_details

client_id	web_server_redirect_uri
app	http://127.0.0.1:8083/login/oauth2/code/my-client
postman	NULL

图 7-28

f. my-client 为 providerId.

g. Authorization Server 的授权路径。

h. Authorization Server 获取访问令牌的路径。

i. 获取用户信息的路径。

j. 从用户信息路径的返回值中获取用户名的属性值。

3. Authorization Server 的修改

因为在客户端的 Provider 中设置了获取用户信息的路径，所以需要在授权服务上提供用户信息的控制器：

```
@RestController
public class UserController {
    @RequestMapping("/userInfo")
    public Map<String, String> userInfo(@AuthenticationPrincipal Jwt jwt){
        Map<String,String> map = new HashMap<>();
        map.put("username", jwt.getClaimAsString("user_name"));
        return map;
    }
}
```

上面的 Provider 在设置 userNameAttribute 时指定了获取用户名的 key 为 username，如果需要在外部访问这个路径，则意味着授权服务需要由资源服务来提供用户信息的服务，因此对授权服务做如下修改。

（1）添加资源服务依赖：

```
implementation 'org.springframework.boot:spring-boot-starter-oauth2-resource-server'
```

（2）公钥配置。

◎　将 public.txt 放置在 src/main/resources 路径中。

◎　配置公钥：

```
spring:
  security:
    oauth2:
      resourceserver:
        jwt:
          public-key-location: classpath:public.txt
```

（3）Spring Security 配置资源服务支持并开启表单登录：

```
@Configuration
public class WebSecurityConfig extends WebSecurityConfigurerAdapter {
        //……
   @Override
   protected void configure(HttpSecurity http) throws Exception {
      http.authorizeRequests()
            .mvcMatchers("/.well-known/jwks.json").permitAll()
            .anyRequest().authenticated()
            .and()
            .formLogin().permitAll() //表单登录
            .and()
            .oauth2ResourceServer().jwt().jwtAuthenticationConverter(jwt -> {
         Collection<SimpleGrantedAuthority> authorities = ((Collection<String>)
jwt.getClaims().get("authorities")).stream()
               .map(SimpleGrantedAuthority::new)
               .collect(Collectors.toSet());
         return new JwtAuthenticationToken(jwt, authorities);
      });

   }
        //……
}
```

4．受保护控制器

下面在客户端中调用资源服务的 adminCan1 服务。当调用资源服务时，需要使用 RestTemplate。当调用资源服务时，需要使用 Bearer Token 在头部传递 Access Token。Spring Boot 的 RestTemplateAutoConfiguration 已经自动配置了 RestTemplateBuilder，它可以用来配置构造 RestTemplate，所以需要通过 RestTemplateCustomizer 对 RestTemplate 进行定制。

```
@Configuration
public class WebConfig {

   @Bean
   RestTemplateCustomizer restTemplateCustomizer(OAuth2AuthorizedClientService
clientService){
      return restTemplate -> { //a
         List<ClientHttpRequestInterceptor> interceptors =
               restTemplate.getInterceptors();
         if (CollectionUtils.isEmpty(interceptors)) {
            interceptors = new ArrayList<>();
         }
         interceptors.add((request, body, execution) -> { //b
            OAuth2AuthenticationToken auth = (OAuth2AuthenticationToken)
                  SecurityContextHolder.getContext().getAuthentication();
            String clientRegistrationId = auth.getAuthorizedClientRegistrationId();
            String principalName = auth.getName();
```

```
            OAuth2AuthorizedClient client =
                    clientService.loadAuthorizedClient(clientRegistrationId,
principalName); //c
            String accessToken = client.getAccessToken().getTokenValue(); //d
            request.getHeaders().add("Authorization", "Bearer " + accessToken); //e
            return execution.execute(request, body);
         });
         restTemplate.setInterceptors(interceptors);
      };
   }

}
```

a. RestTemplateCustomizer 是函数接口，入参是 RestTemplate。

b. 通过增加 RestTemplate 拦截器，让每次请求添加 Bearer Token（Access Token）。ClientHttpRequestInterceptor 是函数接口，可以用 Lambda 表达式来实现。

c. OAuth2AuthorizedClientService 可获得用户的 OAuth2AuthorizedClient。

d. OAuth2AuthorizedClient 可获得用户的 Access Token。

e. Access Token 通过头部的 Bearer Token 访问 Resource Server。

控制器内容如下：

```
@RestController
public class SecurityController {
   private RestTemplate restTemplate;

   public SecurityController(RestTemplateBuilder restTemplateBuilder) {
      this.restTemplate = restTemplateBuilder.build();
   }

   @GetMapping("/forAdminCan1")
   public Object forAdminCan1(){
      return
restTemplate.getForObject("http://localhost:8082/adminCan1",Object.class);
   }
}
```

5. 验证

此时 Authorization Server、Resource Server 和 Client 都启动了。在 Chrome 浏览器中访问 http://127.0.0.1:8083/forAdminCan1，这是一个受保护的资源，因此会自动导向授权服务的登录页面 http://localhost:8080/login，如图 7-29 所示。

输入 admin 的账号和密码，单击“Sign in”按钮，这时页面导向 http://localhost:8080/oauth/authorize?responsetype=code&clientid=app&state=RVTd7i3-bbjgofPUOUMZ82DqqzjwT3D--Voe-oP16YI%3D&redirect_uri=http://127.0.0.1:8083/login/oauth2/code/my-client，可以在此页面进行授权，如图 7-30 所示。

图 7-29

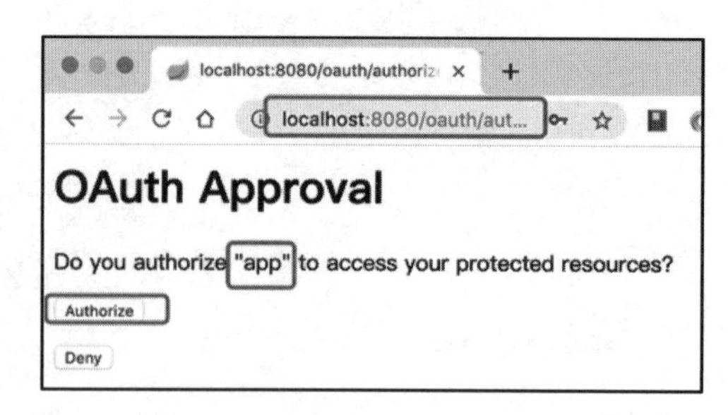

图 7-30

单击“Authorize”按钮，地址会自动导向刚开始访问的 http://127.0.0.1:8083/forAdminCan1。在授权服务上进行登录授权后，就可以访问资源服务上的服务了，如图 7-31 所示。

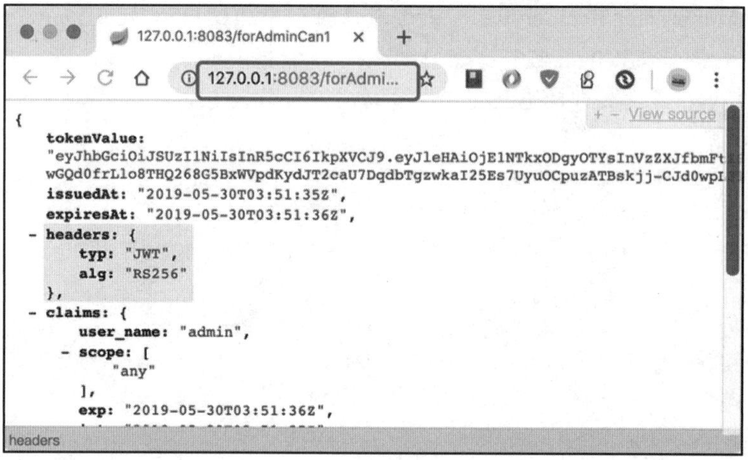

图 7-31

7.4 小结

本章学习了 Spring Security 的认证和授权这两大特性的原理和应用，并提供了一个简单的使用认证和授权进行安全管理的示例。在本章的最后，介绍了 OAuth 2.0 的相关知识，包括 Authorization Server、Resource Server 和 Client 的协作使用。

第 8 章 响应式编程

Spring 5.x 的最大更新是将响应式编程作为头等支持，响应式开发极大地提升了应用系统的性能。响应式应用是完全异步和非阻碍的，即需要应用的每个环节都是异步和非阻碍的。Spring 5.x 对响应式编程进行了全方位的支持。

◎ Web：Spring WebFlux 和 WebFlux.fn。

◎ 数据库：Reactive Spring Data。

◎ 安全：Reactive Spring Security。

◎ 客户端：WebClient。

Spring 的响应式编程是以 Project Reactor 为基石的，它用来构建基于 JVM Reactive Streams 规范的非阻碍式应用。

8.1 Project Reactor

使用 Intellij IDEA 新建 Gradle 应用。

GroupId：top.wisely。

ArtifactId：learning-reactor。

添加依赖：

```
dependencies {
   compile 'io.projectreactor:reactor-core:3.2.9.RELEASE'
   testCompile group: 'junit', name: 'junit', version: '4.12'
}
```

8.1.1 Reactive Streams 的基础接口

Reactor 基于 Reactive Streams 规范，下面介绍 Reactive Streams 的主要接口。

（1）Publisher：顾名思义为发布者，是一系列数据的提供者；它根据订阅者（Subscriber）的要求将数据推送给订阅者。一个发布者可以服务多个订阅者。

```
public interface Publisher<T> {
   public void subscribe(Subscriber<? super T> s);
}
```

订阅者通过 subscribe 方法订阅发布者，请求发布者传递数据。它可以被调用多次，每一次调用都会开始一个新的订阅（Subscription）；每一个订阅只为一个订阅者服务。

（2）Subscriber：订阅者。

```
public interface Subscriber<T> {

   public void onSubscribe(Subscription s); //a

   public void onNext(T t); //b

   public void onError(Throwable t); //c

   public void onComplete(); //d
}
```

a. 在 Publisher.subscribe(Subscriber)执行时被调用。此时没有数据传送，直至 Subscription.request(long)被调用。

b. 当请求 Subscription.request(long)时，由发布者发送数据通知。方法会被一次或多次调用，这依赖于 Subscription.request(long)中定义的最大数量。

c. 失败，结束状态。

d. 成功，结束状态。

（3）Subscription：订阅，代表一次订阅者订阅发布者的生命周期。它只能被一个订阅者使用，既可以用来通知传送数据，也可以取消传送数据。

```
public interface Subscription {

   public void request(long n); //a

   public void cancel(); //b
}
```

a. 若不调用此方法，则发布者不会发送任何数据；参数 n 的最大取值为 Long.MAX_VALUE。若取值为 Long.MAX_VALUE，则意味着发布者的数据是无限的。

b. 请求发布者停止发送数据并清理资源。

（4）Processor：处理器代表一个处理阶段，它既是订阅者，也是发布者。

```
public interface Processor<T, R> extends Subscriber<T>, Publisher<R> {
}
```

<T>：订阅者接收的数据类型；<R>：发布者发送数据的类型。

Reactor 提供了 Provider 接口的两个实现：Flux 和 Mono。

8.1.2　Flux 和 Mono

Flux 可以发送 0 到 n 个数据，而 Mono 可以发送 0 到 1 个数据。

1. 构建发布者

如何构建 Flux 或 Mono 的发布者（提供者）呢？答案是可以通过它们的静态方法来构建。

（1）构建 Flux 发布者：

```
Flux<String> flux1 = Flux.just("a", "b", "b");
Flux<String> flux2 = Flux.fromArray(new String[]{"c", "d", "e"});
Flux<String> flux3 = Flux.fromStream(Stream.of("e", "f", "g"));
Flux<Integer> flux4 = Flux.range(2019, 3);
```

（2）构建 Mono 发布者：

```
Mono<String> mono1 = Mono.just("a");
Mono<String> mono2 = Mono.from(Flux.just("b"));
```

2. 订阅发布者

若不主动订阅发布者，则不会有任何数据被发送。订阅者既可以是 Consumer 函数接口，也可以是 Subscriber 接口的实现。

（1）订阅 Flux，如图 8-1 所示。

```
System.out.println("-------------");
flux1.subscribe(System.out::print);
System.out.println();
flux2.subscribe(System.out::print);
System.out.println();
flux3.subscribe(System.out::print);
System.out.println();
flux4.subscribe(new Subscriber<Integer>() {
   volatile Subscription subscription;
   @Override
   public void onSubscribe(Subscription s) {
      subscription = s;
      System.out.println("初始化请求一个数据");
      subscription.request(1);
   }

   @Override
   public void onNext(Integer integer) {
      System.out.println("当前数据:" + integer);
      System.out.println("再请求一个数据");
```

```
        subscription.request(1);
    }

    @Override
    public void onError(Throwable t) {
        System.out.println(t.getMessage());
    }

    @Override
    public void onComplete() {
        System.out.println("处理完成");
    }
});
```

```
-------------
abb
cde
efg
初始化请求一个数据
当前数据:2019
再请求一个数据
当前数据:2020
再请求一个数据
当前数据:2021
再请求一个数据
处理完成
```

图 8-1

（2）订阅 Mono，如图 8-2 所示。

```
mono1.subscribe(System.out::println);
mono2.subscribe(new Subscriber<String>() {
    volatile Subscription subscription;
    @Override
    public void onSubscribe(Subscription s) {
        subscription = s;
        System.out.println("初始化请求一个数据");
        subscription.request(1);
    }

    @Override
    public void onNext(String s) {
        System.out.println("当前数据:" + s);
        System.out.println("再请求一个数据");
        subscription.request(1);
    }

    @Override
    public void onError(Throwable t) {
        System.out.println(t.getMessage());
    }

    @Override
```

```
    public void onComplete() {
        System.out.println("处理完成");
    }
});
```

3. 处理操作

Flux 和 Mono 可以像 Stream 一样对其中的数据进行处理操作，如图 8-3 所示。

```
Flux<Integer> flux1 = Flux.just(1, 6, 4, 3, 5, 2);
flux1
  .map(i -> i * 3) //a
  .filter(i -> i % 2 == 0) //b
  .sort(Comparator.comparingInt(Integer::intValue)) //c
  .subscribe(System.out::println);
```

a. 将发布者中的所有数据都乘以 3。

b. 通过过滤取出数据中的偶数。

c. 对数据进行排序。

```
-------------
a
初始化请求一个数据
当前数据:b
再请求一个数据
处理完成
```

图 8-2

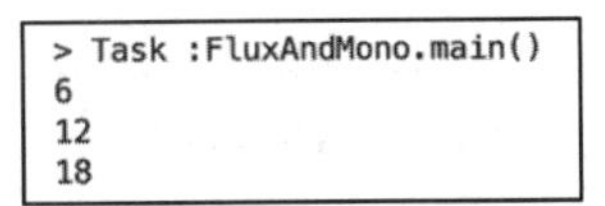

图 8-3

8.2 Spring WebFlux

新建应用，信息如下。

Group：top.wisely。

Artifact：learning-webflux。

Dependencies：Spring Reactive Web 和 Lombok。

build.gradle 文件中的依赖如下：

```
dependencies {
  implementation 'org.springframework.boot:spring-boot-starter-webflux'
  compileOnly 'org.projectlombok:lombok'
  annotationProcessor 'org.projectlombok:lombok'
    //……
}
```

8.2.1 Spring WebFlux 基础

Spring WebFlux 不依赖于 Servlet API，它可以运行在非 Servlet 容器之上，例如 Netty、Undertow 和任何 Servlet 3.1+的 Servlet 容器（Tomcat、Jetty）。虽然 Servlet 3.1 提供了非阻碍 I/O

的 API，但仍有很多其他的 API 依然是同步或阻碍式的，这使得 Spring 需要重新构建完全基于异步和非阻碍式的运行环境。

Spring WebFlux 的底层组件是 HttpHandler，它可用来适配不同的服务器引擎（Netty、Undertow、Tomcat、Jetty）。

在通过 HttpHandler 消除了服务器引擎的异构行后，Spring WebFlux 的 API 设计与 Spring MVC 是高度一致的。

Spring WebFlux 与 Spring MVC 在概念上有对应的关系，如表 8-1 所示。

表 8-1

Spring WebFlux	Spring Web MVC
DispatcherHandler	DispatcherServlet
WebFilter	Filter
HttpMessageWriter HttpMessageReader	HttpMessageConverter
HandlerMapping	HandlerMapping
HandlerAdapter	HandlerAdapter
ServerHttpRequest ServerHttpResponse	ServletRequest ServletResponse

Spring WebFlux 支持下面两种编程模型。

（1）注解控制器：和 Spring MVC 使用的注解式一致。

◎ RequestMappingHandlerMapping：关联请求与标注了@RequestMapping 的类和方法之间的映射。

◎ RequestMappingHandlerAdapter：调用标注了@RequestMapping 注解的方法。

Sping MVC 的两个类与这两个类名称一致，在不同的包里，有不同的实现，但功能一致。

（2）函数式端点：基于 Lambada 表达式、轻量级的函数式编程模型。

◎ RouterFunctionMapping：用来支持 RouterFunction。

◎ HandlerFunctionAdapter：用来支持 HandlerFunctions。

在 Spring 5.2.x 和 Spring 2.2.x 之后，Spring MVC 也支持这种编程模型。

8.2.2 Spring Boot 的自动配置

Spring Boot 提供的自动配置主要如下。

◎ CodecsAutoConfiguration ： Spring WebFlux 使用 HttpMessageReader 和 HttpMessageWriter 接口来转换 HTTP 的请求和返回。本配置类注册了 CodecCustomizer 的 Bean，默认使用 Jackson2JsonEncoder 和 Jackson2JsonDecoder。

◎ ReactiveWebServerFactoryAutoConfiguration：为响应式 Web 服务器进行自动配置。

◎ WebFluxAutoConfiguration：使用等同于@EnableWebFlux 的配置开启对 WebFlux 的支持。可通过 WebFluxProperties 使用 spring.webflux.*来对 WebFlux 进行配置。

```
spring:
  webflux:
    date-format: yyyy-MM-dd # 日期格式
    static-path-pattern: /resouces/static/** # 静态资源目录
```

◎ WebClientAutoConfiguration：为 WebClient 进行自动配置。

8.2.3 注解控制器

1．示例

下面演示一个简单的例子，先定义简单的领域模型：

```
@Data
@AllArgsConstructor
@NoArgsConstructor
public class Person {
    private Long id;
    private String name;
    private Integer age;
}
@RestController
@RequestMapping("/people")
public class PersonController {

    PersonRepository personRepository; //a

    public PersonController(PersonRepository personRepository) {
        this.personRepository = personRepository;
    }

    @PostMapping
    @ResponseStatus(HttpStatus.CREATED)
    public Mono<Person> add(@RequestBody Person person){ //b
        return Mono.just(personRepository.save(person));
    }

    @GetMapping("/{id}")
    public ResponseEntity<Mono<Person>> getById(@PathVariable Long id){ //c
        return ResponseEntity.ok()
              .body(Mono.just(personRepository.findOne(id)));
    }
    @GetMapping
    public Flux<Person> list(){ //d
        return Flux.fromIterable(personRepository.list());
    }

    @DeleteMapping("/{id}")
```

```
    @ResponseStatus(HttpStatus.NO_CONTENT)
    public Mono<Void> delete(@PathVariable("id") Long id) { //e
        personRepository.delete(id);
        return Mono.empty();
    }
}
```

a. 沿用第 5 章使用的 PersonRepository（包含 CommonRepository、MyBeanUtils），把它注入到控制器中。

b. 返回一个 Person，使用 Mono<Person>。

c. 使用 ResponseEntity 来构造返回，返回体为 Mono<Person>。

d. 返回多个 Person，使用 Flux<Person>。PersonRepository（包含 CommonRepository）没有 list 方法，代码如下：

```
@Override
public Set<Person> list() {
    return people;
}
```

e. Mono<Void> 是响应式下的 void。

2. WebClient 调用

WebClient 是 Spring WebFlux 提供的响应式客户端，Spring Boot 通过 WebClientAutoConfiguration 已自动配置。为方便起见，本例在当前应用中直接编写客户端代码。

```
@Component
public class ControllerClient {

    WebClient webClient;

    public ControllerClient(WebClient.Builder builder) {
        this.webClient = builder.build(); //a
    }

    public void add1(){
        System.out.println("添加第一条数据");
        Mono<Person> mono = webClient
                .post() //b
                .uri("http://localhost:8080/people") //c
                .body(Mono.just(new Person(1l, "wyf", 35)), Person.class) //d
                .retrieve() //e
                .bodyToMono(Person.class); //f
        mono.subscribe(System.out::println); //g
    }

    public void findOne(){
        System.out.println("查询一条数据");
        Mono<Person> mono = webClient
                .get()
```

```
            .uri("http://localhost:8080/people/{id}", 1l)//h
            .retrieve()
            .bodyToMono(Person.class);
    mono.subscribe(System.out::println);
}

public void add2(){
    System.out.println("添加第二条数据");
    Mono<Person> mono = webClient
            .post()
            .uri("http://localhost:8080/people")
            .body(Mono.just(new Person(21, "foo", 34)), Person.class)
            .retrieve()
            .bodyToMono(Person.class);
    mono.subscribe(System.out::println);
}

public void list(){
    System.out.println("获取列表数据");
    Flux<Person> flux = webClient
            .get()
            .uri("http://localhost:8080/people")
            .retrieve()
            .bodyToFlux(Person.class); //i
    flux.subscribe(System.out::println);
}

public void delete(){
    System.out.println("删除一条数据");
    Mono<Void> mono = webClient
            .delete()
            .uri("http://localhost:8080/people/{id}", 1l)
            .retrieve()
            .bodyToMono(Void.class);
    mono.subscribe(System.out::println);
}

}
```

a. 使用 Spring Boot 为自动配置的 WebClient.Builder Bean 构建 WebClient。

b. 针对 RESTful 服务，WebClient 提供了 post()、get()、put()、head()、put()和 patch()方法来指定 HTTP 方法。

c. 通过 uri()指定服务端的路径。

d. 可以用 body()方法向后台传送请求体，当前的请求体是响应式的；也可以通过 header()方法向后台传送请求体。

```
.header("headerName","headerValue")
```

e. 使用 retrieve()方法执行请求，并获取返回体。

f. 将请求体转换为响应式类型 Mono。

g. 通过 subscribe()订阅 Mono，只有订阅才能发送数据。

h. 可以通过 URI 模板设置 URI 参数，如：

```
.uri("/peopple?name={name}&age={age}", "bar", "33")
```

i. 将请求体转换为响应式类型 Flux。

使用 CommandLineRunner 执行上述请求：

```
@Bean
CommandLineRunner webClientClr(ControllerClient controllerClient){
  return args -> {
    controllerClient.add1();
    Thread.sleep(1000);
    controllerClient.findOne();
    Thread.sleep(1000);
    controllerClient.add2();
    Thread.sleep(1000);
    controllerClient.list();
    Thread.sleep(1000);
    controllerClient.delete();
    Thread.sleep(1000);
    controllerClient.list();
  };
}
```

因为每一个请求都是异步请求的，不能确保哪个请求先执行完，所以每一个请求后都睡眠了 1s，执行结果如图 8-4 所示。

```
添加第一条数据
Person(id=1, name=wyf, age=35)
查询一条数据
Person(id=1, name=wyf, age=35)
添加第二条数据
Person(id=2, name=foo, age=34)
获取列表数据
Person(id=1, name=wyf, age=35)
Person(id=2, name=foo, age=34)
删除一条数据
获取列表数据
Person(id=2, name=foo, age=34)
```

图 8-4

8.2.4 函数式端点

使用 RouteFunction 进行请求和方法映射，使用 HandlerFunction 接口来定义控制器方法。

HandlerFunction 是一个函数接口：

```
@FunctionalInterface
public interface HandlerFunction<T extends ServerResponse> {
  Mono<T> handle(ServerRequest request);
```

```
}
```

这意味着控制器方法的参数是ServerRequest，返回值是Mono<T>，其中，T是ServerResponse的实现类。当符合这个定义时，即为一个 HandlerFunction。首先定义控制器方法：

```
@Component
public class PersonHandler {
   PersonRepository personRepository;

   public PersonHandler(PersonRepository personRepository) {
      this.personRepository = personRepository;
   }

   public Mono<ServerResponse> add(ServerRequest request){
    Mono<Person> personMono = request.bodyToMono(Person.class); //a
    return personMono.flatMap(person -> ServerResponse
          .status(HttpStatus.CREATED)
          .body(Mono.just(personRepository.save(person)),Person.class)); //b
   }

   public Mono<ServerResponse> getById(ServerRequest request){
      Long id = Long.valueOf(request.pathVariable("id")); //c
      return ServerResponse
            .ok()
            .body(Mono.just(personRepository.findOne(id)), Person.class); //d

   }

   public Mono<ServerResponse> delete(ServerRequest request) {
      Long id = Long.valueOf(request.pathVariable("id"));
      personRepository.delete(id);
      return ServerResponse
            .noContent()
            .build(Mono.empty());
   }
}
```

a. 通过 bodyToMono 方法可以从 request 对象中将体取出并转换为 Mono。

b. 可以通过 ServerResponse 的静态方法来构造返回。因为 Repository 是非响应式的，所以还需要通过 Mono.just 转换一下。

c. 可以通过 request 对象的 pathVariable()方法获取路径变量。更多可从 request 中获取的信息请参照 ServerRequest 的 API。

下面在配置类中对 RouterFunction 进行配置。

```
//……
import static org.springframework.web.reactive.function.server.RequestPredicates.*;
import static org.springframework.web.reactive.function.server.RouterFunctions.route;

@Configuration
```

```
public class RoutingConfiguration {
   @Bean
   public RouterFunction<ServerResponse> personRouterFunction(PersonHandler
personHandler,
                                          PersonRepository personRepository){
      return route(POST("/people"), personHandler::add) //a
            .andRoute(GET("/people/{id}"), personHandler::getById)
            .andRoute(GET("/people"), serverRequest -> ServerResponse
                  .ok()
                  .body(Flux.fromIterable(personRepository.list()), Person.class))
            .andRoute(DELETE("/people/{id}"), personHandler::delete);

   }
}
```

a. 使用 RouterFunctions 静态方法 route()和 andRoute()来构建 RouterFunction，该方法接收两个参数。

◎ RequestPredicate：可通过 RequestPredicates 中的静态方法来构造请求。

◎ HandlerFunction：符合 HandlerFunction 接口声明的方法均可作为入参。可以使用方法引用和 Lambda 表达式作为参数。

注释掉 PersonController 上的@RestController 注解，使用 WebClient 检验函数式端点的响应式接口，效果等同于使用注解控制器。

8.2.5 Spring WebFlux 的配置

若需要配置 Spring WebFlux，则只需让配置实现接口 WebFluxConfigurer 即可，这样既能保留 Spring Boot 对 WebFlux 进行配置，又能添加定制配置。如果想完全控制 WebFlux，则需要在配置类中添加注解@EnableWebFlux：

```
@Configuration
public class WebFluxConfig implements WebFluxConfigurer {
}
```

配置方式和 Spring MVC 类似，本节不做具体演示。

8.3 Reactive NoSQL

前面的 Repository 是非响应式的，而响应式编程要求全栈技术都是响应式的。本节讨论响应式的 Spring Data。目前，Spring Data 支持的响应式的 NoSQL 如下。

◎ MongoDB：使用 spring-boot-starter-data-mongodb-reactive 依赖。

◎ Redis：使用 spring-boot-starter-data-redis-reactive 依赖。

◎ Cassandra：使用 spring-boot-starter-data-cassandra-reactive 依赖。

◎ Couchbase：使用 spring-boot-starter-data-couchbase-reactive 依赖。

◎ Elasticsearch：使用 spring-boot-starter-data-elasticsearch 依赖。

8.3.1 响应式 Elasticsearch

本节使用 Elasticsearch 作为演示。

新建应用，信息如下。

Group：top.wisely。

Artifact：learning-reactive-nosql。

Dependencies：Spring Reactive Web、Spring Data Elasticsearch 和 Lombok。

build.gradle 文件中的依赖如下：

```
dependencies {
  implementation 'org.springframework.boot:spring-boot-starter-data-elasticsearch'
  implementation 'org.springframework.boot:spring-boot-starter-webflux'
  compileOnly 'org.projectlombok:lombok'
  annotationProcessor 'org.projectlombok:lombok'
    //……
}
```

1. Spring Boot 的自动配置

Spring Boot 自 2.2.0 版本开始支持响应式 Elasticsearch 的自动配置，自动配置文件如下。

◎ ReactiveElasticsearchRepositoriesAutoConfiguration：通过@EnableReactiveElasticsearchRepositories 开启对响应式 Elasticsearch Repository 的支持。

◎ ElasticsearchDataAutoConfiguration：通过导入 lasticsearchDataConfiguration.ReactiveRestClientConfiguration 来配置响应式操作模板 ReactiveElasticsearchTemplate 的 Bean。

◎ ReactiveRestClientAutoConfiguration：通过 ReactiveRestClientProperties，以 spring.data.elasticsearch.client.reactive.* 开头来配置响应式 Elasticsearch 客户端 ReactiveElasticsearchClient 的 Bean。

2. 示例

配置连接 Elasticsearch：

```
spring:
  data:
    elasticsearch:
      client:
        reactive:
          endpoints: localhost:9200
```

下面的工作和 Spring Data 的其他项目类似，首先定义领域模型：

```
//……
import org.springframework.data.elasticsearch.annotations.Document;

@Data
@AllArgsConstructor
@NoArgsConstructor
@Document(indexName = "person")
public class Person {
   @Id
   private String id;
   private String name;
   private Integer age;

   public Person(String name, Integer age) {
      this.name = name;
      this.age = age;
   }
}
```

然后定义 Repository，此处继承的是 ReactiveElasticsearchRepository：

```
//……
import
org.springframework.data.elasticsearch.repository.ReactiveElasticsearchRepository;

public interface PersonRepository extends ReactiveElasticsearchRepository<Person,
String> {
}
```

ReactiveElasticsearchRepository 的父接口的定义如下：

```
@NoRepositoryBean
public interface ReactiveCrudRepository<T, ID> extends Repository<T, ID> {
  <S extends T> Mono<S> save(S entity);
  <S extends T> Flux<S> saveAll(Iterable<S> entities);
  <S extends T> Flux<S> saveAll(Publisher<S> entityStream);
  Mono<T> findById(ID id);
  Mono<T> findById(Publisher<ID> id);
  Mono<Boolean> existsById(ID id);
  Mono<Boolean> existsById(Publisher<ID> id);
  Flux<T> findAll();
  Flux<T> findAllById(Iterable<ID> ids);
  Flux<T> findAllById(Publisher<ID> idStream);
  Mono<Long> count();
  Mono<Void> deleteById(ID id);
  Mono<Void> deleteById(Publisher<ID> id);
  Mono<Void> delete(T entity);
  Mono<Void> deleteAll(Iterable<? extends T> entities);
  Mono<Void> deleteAll(Publisher<? extends T> entityStream);
  Mono<Void> deleteAll();
}
```

从父级 Repository 接口的方法声明中可以看出，所有的返回值都是响应式类型的，同时支持响应式的参数。

下面用响应式控制器来调用 Repository，代码逻辑和上例相同：

```
@RestController
@RequestMapping("/people")
public class PersonController {

   PersonRepository personRepository;

   public PersonController(PersonRepository personRepository) {
      this.personRepository = personRepository;
   }

   @PostMapping
   @ResponseStatus(HttpStatus.CREATED)
   public Mono<Person> add(@RequestBody Person person){
      return personRepository.save(person);
   }

   @GetMapping("/{id}")
   public Mono<Person> getById(@PathVariable String id) {
      return personRepository.findById(id);
   }
   @GetMapping
   public Flux<Person> list(){
      return personRepository.findAll();
   }

   @DeleteMapping("/{id}")
   @ResponseStatus(HttpStatus.NO_CONTENT)
   public Mono<Void> delete(@PathVariable("id") String id) {
      return personRepository.deleteById(id);
   }
}
```

下面使用上例的 WebClient 代码验证开发的结果，对 ControllerClient 的代码做一些修改：

```
public String add1(){
   System.out.println("添加第一条数据");
   Mono<Person> mono = webClient
         .post()
         .uri("http://localhost:8080/people")
         .body(Mono.just(new Person("wyf", 35)), Person.class) //使用 2 个参数的构造器
         .retrieve()
         .bodyToMono(Person.class);
   Person person = mono.block(); //使用 block()方法订阅 Mono 并将其变为同步
   System.out.println(person);
   return person.getId(); //获取 Id，后面在删除时使用
```

```
}
  public void delete(String id){
     System.out.println("删除一条数据");
     Mono<Void> mono = webClient
           .delete()
           .uri("http://localhost:8080/people/{id}", id) //使用外部传递的id来删除
           .retrieve()
           .bodyToMono(Void.class);
     mono.subscribe(System.out::println);
  }
```

CommandLineRunner 的代码修改如下：

```
@Bean
CommandLineRunner webClientClr(ControllerClient controllerClient){
  return args -> {
    String id = controllerClient.add1();
    controllerClient.add2();
    Thread.sleep(1000);
    controllerClient.list();
    Thread.sleep(1000);
    controllerClient.delete(id);
    Thread.sleep(1000);
    controllerClient.list();
  };
}
```

先删除第 6 章中生成的 Elasticsearch 的 person 索引，在 Postman 中用 DELETE 方法调用 http://localhost:9200/person，如图 8-5 所示。

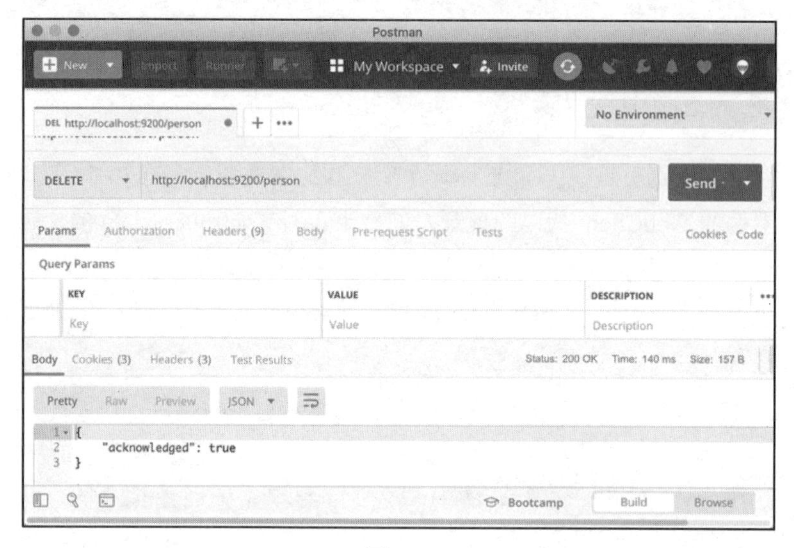

图 8-5

运行应用，结果如图 8-6 所示。

```
添加第一条数据
Person(id=nIjYKmsBeaqT2lxQ6-ry, name=wyf, age=35)
添加第二条数据
Person(id=nYjYKmsBeaqT2lxQ70pe, name=foo, age=34)
获取列表数据
Person(id=nIjYKmsBeaqT2lxQ6-ry, name=wyf, age=35)
Person(id=nYjYKmsBeaqT2lxQ70pe, name=foo, age=34)
删除一条数据
获取列表数据
Person(id=nYjYKmsBeaqT2lxQ70pe, name=foo, age=34)
```

图 8-6

8.3.2　响应式 MongoDB

本节演示响应式 MongoDB 的应用。

1. 安装 MongoDB

使用 docker compose 安装 MongoDB。

stack.yml：

```
version: '3.1'

services:

  mongo:
    image: mongo
    restart: always
    ports:
      - 27017:27017
    environment:
      MONGO_INITDB_ROOT_USERNAME: root
      MONGO_INITDB_ROOT_PASSWORD: example
```

执行如下命令：

```
$ docker-compose -f stack.yml up -d
```

使用 Navicat for MongoDB 连接 MongoDB，如图 8-7 所示。

建立数据库 first_db 及其对应的账号 wisely，密码为 zzzzzz，如图 8-8 所示。

编辑用户的角色为 first_db.dbOwner，如图 8-9 所示。

图 8-7

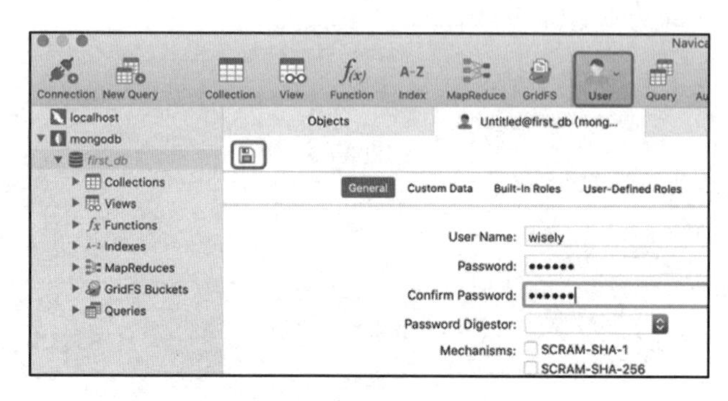

图 8-8

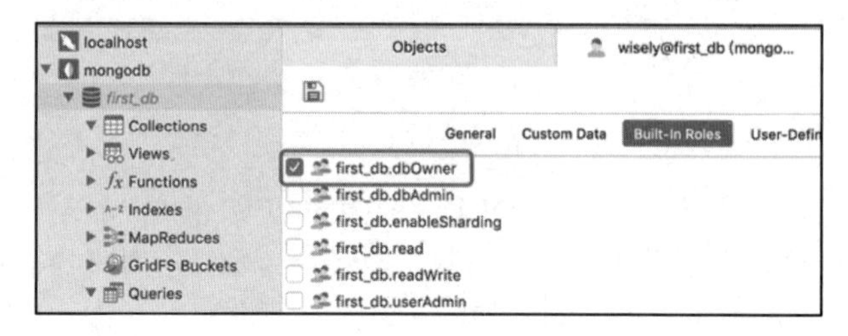

图 8-9

2. Spring Boot 的自动配置

Spring Boot 为响应式 MongoDB 提供的自动配置如下。

◎ MongoReactiveAutoConfiguration：自动配置响应式 MongoDB 客户端的 Bean（reactiveStreamsMongoClient）。

◎ MongoReactiveDataAutoConfiguration：自动配置 Spring Data 响应式 MongoDB 所需的 ReactiveMongoTemplate 的 Bean。

◎ MongoReactiveRepositoriesAutoConfiguration：自动配置 Spring Data 响应式 MongoDB Repository。

外部配置使用 MongoProperties，通过 spring.data.mongodb.*来配置。

3．示例

直接使用 learning-reactive-nosql 进行修改。

◎ 修改如下依赖：

implementation 'org.springframework.boot:spring-boot-starter-data-elasticsearch'

将其修改为：

implementation 'org.springframework.boot: spring-boot-starter-data-mongodb-reactive '

◎ 在 application.yml 中配置连接数据库：

```
spring:
  data:
    mongodb:
      host: localhost
      port: 27017
      username: wisely
      password: zzzzzz
      database: first_db
```

◎ 去除 ReactiveEsConfig 配置，Spring Boot 已为 MongoDB 做好了所有的自动配置。

◎ 领域模型：

```
//......
import org.springframework.data.mongodb.core.mapping.Document;

@Data
@AllArgsConstructor
@NoArgsConstructor
@Document(collection = "person")
public class Person {
    @Id
    private String id;
    private String name;
    private Integer age;

    public Person(String name, Integer age) {
        this.name = name;
        this.age = age;
    }
}
```

◎ Repository，此处使用 ReactiveMongoRepository：

```
//……
import org.springframework.data.mongodb.repository.ReactiveMongoRepository;

public interface PersonRepository extends ReactiveMongoRepository<Person, String> {
}
```

其余代码保持不变，执行结果如图 8-10 所示。

```
添加第一条数据
2019-06-06 13:28:06.113  INFO 12915 --- [ntLoopGroup-2-2]
Person(id=5cf8a465b16aa66d12d5290e, name=wyf, age=35)
添加第二条数据
Person(id=5cf8a466b16aa66d12d5290f, name=foo, age=34)
获取列表数据
Person(id=5cf8a465b16aa66d12d5290e, name=wyf, age=35)
Person(id=5cf8a466b16aa66d12d5290f, name=foo, age=34)
删除一条数据
获取列表数据
Person(id=5cf8a466b16aa66d12d5290f, name=foo, age=34)
```

图 8-10

8.4 Reactive 关系型数据库：R2DBC

与关系型数据库进行交互的 JDBC 不具备与数据库异步交互的能力，R2DBC（Reactive Relational Database Connectivity，Reactive 关系型数据库）将响应式编程 API 带给关系型 SQL 数据库。

Spring Data R2DBC 为此提供支持，同样，Spring Boot 也提供了 starter（spring-boot-starter-data-r2dbc）和自动配置（spring-boot-actuator-autoconfigure-r2dbc）。

在编写本书时，R2DBC 仍处在试验阶段，第一个正式版还没有发布。在正式版发布之后，Group（org.springframework.boot.experimental）和版本信息应该会有变化。

新建应用，信息如下。

Group：top.wisely。

Artifact：learning-reactive-sql。

Dependencies：Spring Reactive Web 和 Lombok。

添加 R2DBC 相关依赖，具体如下：

```
repositories {
    mavenCentral()
    maven { url 'https://repo.spring.io/snapshot' }
    maven { url 'https://repo.spring.io/milestone' }
}
dependencyManagement {
    imports {
```

```
    mavenBom("org.springframework.boot.experimental:spring-boot-dependencies-r2dbc:0
.1.0.BUILD-SNAPSHOT") //Group 和版本会有变化
    }
}
dependencies {
  implementation 'org.springframework.boot:spring-boot-starter-web'
  implementation
'org.springframework.boot.experimental:spring-boot-starter-data-r2dbc'
  implementation
'org.springframework.boot.experimental:spring-boot-actuator-autoconfigure-r2dbc'
  implementation 'io.r2dbc:r2dbc-postgresql' //PostgreSQL 的 R2DBC 驱动
  implementation 'io.r2dbc:r2dbc-pool' // 数据库连接池
     implementation 'io.r2dbc:r2dbc-client' // R2DBC 客户端
  compileOnly 'org.projectlombok:lombok'
  //……
}
```

本节使用 PostgreSQL 数据库作为演示。

8.4.1　安装 PostgreSQL

使用 docker compose 安装 PostgreSQL。

stack.yml：

```
version: '3.1'

services:
  db:
    image: postgres
    restart: always
    ports:
      - 5432:5432
    environment:
      POSTGRES_DB: first_db
      POSTGRES_USER: wisely
      POSTGRES_PASSWORD: zzzzzz
```

执行如下命令：

```
$ docker-compose -f stack.yml up -d
```

8.4.2　Spring Boot 的自动配置

Spring Boot 提供的自动配置在 spring-boot-actuator-autoconfigure-r2dbc 包中，主要如下。

◎ ConnectionFactoryAutoConfiguration：基于连接池的 connectionFactory Bean 的配置；使用 R2dbcProperties 通过 spring.r2dbc.*进行配置。

  ➢ 数据库连接配置：spring.r2dbc.url。

  - ➢ 连接池配置：spring.r2dbc.pool.*。
- ◎ R2dbcDataAutoConfiguration：自动配置数据库客户端 DatabaseClient 的 Bean，类似于 JdbcTemplate，可以直接进行查询。
- ◎ R2dbcRepositoriesAutoConfiguration：使用@EnableR2dbcRepositories 开启对 Spring Data R2DBC Repository 的支持。
- ◎ R2dbcTransactionManagerAutoConfiguration：自动配置事务管理器 R2dbcTransaction Manager 的 Bean，这样就可以使用注解@Transactional 进行声明式事务处理了。
- ◎ ReactiveTransactionAutoConfiguration：注册 TransactionalOperator 的 Bean 来简化处理编程式事务处理。

8.4.3 示例

Spring Data R2DBC 的用法和其他 Spring Data 项目几乎相同，下面一起来学习。

1. 连接数据库

在 application.yml 中配置连接数据库：

```
spring:
  r2dbc:
    url: r2dbc:postgresql://localhost:5432/first_db
    username: wisely
    password: zzzzzz
```

2. 定义领域模型

```
//……
import org.springframework.data.annotation.Id;
import org.springframework.data.relational.core.mapping.Table;

@Data
@AllArgsConstructor
@NoArgsConstructor
@Table("person") //a
public class Person {
   @Id//b
   private Long id;
   private String name;
   private Integer age;

   public Person(String name, Integer age) {
      this.name = name;
      this.age = age;
   }
}
```

a. 使用@Table 注解映射数据库的表，注意@Table 所在的包。

b. 使用@Id 注解映射数据库的主键。

定义 Repository：

```
public interface PersonRepository extends R2dbcRepository<Person, Long> {
   @Query("select id, name, age  from person")
   Flux<Person> findAll();
}
```

这里继承的接口是 R2dbcRepository；和其他 Spring Data 项目一样，这里演示用自定义的方法覆盖默认的 findAll()方法。

4. 定义控制器和 WebClient 客户端

这里的控制器 PersonController 和 ControllerClient 与前面的例子一致，把 ControllerClient 代码中的 id 修改为 Long 类型。

```
@Component
public class ControllerClient {
  //……
   public Long add1(){//返回值为 Long 类型
     //……
   }
  //……
   public void delete(Long id){ //入参为 Long 类型
     //……
   }

}
```

5. 初始化数据库

使用 schema.sql 建立 Person 对应的表：

```
create table if not exists person (
 id serial primary key,
 name varchar(255) not null,
 age int4 not null
);
```

目前，Spring Data R2DBC 不支持自动读取 schema.sql，下面编写数据库初始化代码，用来读取并执行 schema.sql 中的语句：

```
@Component
public class ReactiveDatabaseInitializer {

   private final DatabaseClient client;

   private final String sql;

   public ReactiveDatabaseInitializer(DatabaseClient client, //a
```

```
                                      @Value("classpath:schema.sql") Resource resource) throws
Exception{
        this.client = client;
        try (Reader in = new InputStreamReader(resource.getInputStream())) {
            this.sql = FileCopyUtils.copyToString(in); //b
        }
    }

    public Mono<Void> initialize () {
        return client.execute().sql(this.sql).then() //c
                .onErrorResume(throwable -> Mono.empty());
    }

}
```

a. 自动配置已经配置了 DatabaseClient 的 Bean，因而可以直接注入。

b. 读取 schema.sql 中的 SQL 语句。

c. 使用 DatabaseClient 执行 SQL 语句。

6. 启动验证

在 CommandLineRunner 中启动初始化数据库，并执行 WebClient：

```
@Bean
CommandLineRunner webClientClr(ControllerClient controllerClient,
                        ReactiveDatabaseInitializer initializer){
   return args -> {
      initializer.initialize().block();
      Long id = controllerClient.add1();
      controllerClient.add2();
      Thread.sleep(1000);
      controllerClient.list();
      Thread.sleep(1000);
      controllerClient.delete(id);
      Thread.sleep(1000);
      controllerClient.list();
   };
}
```

结果如图 8-11 所示。

```
添加第一条数据
Person(id=1, name=wyf, age=35)
添加第二条数据
Person(id=2, name=foo, age=34)
获取列表数据
Person(id=1, name=wyf, age=35)
Person(id=2, name=foo, age=34)
删除一条数据
获取列表数据
Person(id=2, name=foo, age=34)
```

图 8-11

8.5　Reactive Spring Security

8.5.1　Reactive Spring Security 原理

Spring MVC 的 Security 是通过 Servlet 的 Filter 实现的；而 WebFlux 的响应式 Security 是基于 WebFilter 实现的，是由一系列的 WebFilter 组成的过滤器链。

（1）认证。

Spring WebFlux 下的响应式安全和 Spring MVC 下的安全认证机制在概念上是对应的，如表 8-2 所示。

表 8-2

Spring WebFlux	Spring MVC
Spring WebFlux Security	Spring Web MVC Security
AuthenticationWebFilter	UsernamePasswordAuthenticationFilter
ReactiveAuthenticationManager	AuthenticationManager
UserDetailsRepositoryReactiveAuthenticationManager	ProviderManager AuthenticationManager
ReactiveUserDetailsService	UserDetailsService

（2）授权。

授权在概念上也是对应的，如表 8-3 所示。

表 8-3

Spring WebFlux	Spring MVC
Spring WebFlux Security	Spring WebFlux Security
AuthorizationWebFilter	FilterSecurityInterceptor
ReactiveAuthorizationManager	AccessDecisionManager

其他可对应的如表 8-4 所示。

表 8-4

Spring WebFlux	Spring MVC
Spring WebFlux Security	Spring WebFlux Security
@EnableWebFluxSecurity	@EnableWebSecurity
@EnableReactiveMethodSecurity	@EnableGlobalMethodSecurity
ReactiveSecurityContextHolder	SecurityContextHolder

8.5.2　Spring Boot 的自动配置

Spring Boot 为响应式安全提供了下面两个配置。

◎ ReactiveSecurityAutoConfiguration：使用 @EnableWebFluxSecurity 开启对 Spring WebFlux 安全的支持。

◎ ReactiveUserDetailsServiceAutoConfiguration：配置一个 reactiveUserDetailsService 的 Bean。

下面用一个例子来学习响应式的安全机制，它和前面介绍的 Spring MVC 下的安全用法十分相似。

8.5.3 示例

1. 新建应用

新建应用，信息如下。

Group：top.wisely。

Artifact：learning-reactive-security。

Dependencies：Spring Security、Spring Reactive Web、Spring Data Reactive MongoDB 和 Lombok。

build.gradle 文件中的依赖如下：

```
dependencies {
    implementation 'org.springframework.boot:spring-boot-starter-security'
    implementation
'org.springframework.boot:spring-boot-starter-data-mongodb-reactive'
    implementation 'org.springframework.boot:spring-boot-starter-webflux'
    compileOnly 'org.projectlombok:lombok'
    annotationProcessor 'org.projectlombok:lombok'
    //……
}
```

Spring WebFlux Security 和 Spring MVC Security 都在 Spring Security 中。

2. 连接 MongoDB

在 application.yml 中配置连接 MongoDB：

```
spring:
  data:
    mongodb:
      host: localhost
      port: 27017
      username: wisely
      password: zzzzzz
      database: first_db
```

3. 定义领域模型

（1）安全相关模型：

```
@Data
@AllArgsConstructor
@NoArgsConstructor
@Document(collection = "user")
public class SysUser implements UserDetails {
   @Id
   private String id;

   private String username;

   private String password;

   private Set<SysAuthority> sysAuthorities;

   private Boolean enable = true;

   public SysUser(String username, String password, Set<SysAuthority> sysAuthorities)
{
      this.username = username;
      this.password = password;
      this.sysAuthorities = sysAuthorities;
   }

   @Override
   public Collection<? extends GrantedAuthority> getAuthorities() {
      return this.sysAuthorities
            .stream()
            .map(authority -> new SimpleGrantedAuthority(authority.getName()))
            .collect(Collectors.toCollection(HashSet::new));
   }
  //省略其他 UserDetails 接口方法
}
@Data
@AllArgsConstructor
@NoArgsConstructor
public class SysAuthority {
   private String name;
}
```

（2）业务相关模型：

```
@Data
@AllArgsConstructor
@NoArgsConstructor
@Document(collection = "people")
public class Person {
   @Id
   private String id;
   private String name;
   private Integer age;
```

```
    public Person(String name, Integer age) {
        this.name = name;
        this.age = age;
    }
}
```

4. Repository

```
public interface SysUserRepository extends ReactiveMongoRepository<SysUser, String> {
    Mono<UserDetails> findByUsername(String username);
}
public interface PersonRepository extends ReactiveMongoRepository<Person, String> {
}
```

5. 获取用户

使用 ReactiveUserDetailsService 实现类来获取用户：

```
public class CustomReactiveUserDetailsService implements ReactiveUserDetailsService {

    @Autowired
    SysUserRepository sysUserRepository;

    @Override
    public Mono<UserDetails> findByUsername(String username) {
        return sysUserRepository.findByUsername(username)
                .switchIfEmpty(Mono.error(new UsernameNotFoundException("User Not Found")))
                .map(UserDetails.class::cast);
    }
}
```

6. 安全配置

```
@Configuration
@EnableReactiveMethodSecurity//a
public class ReactiveSecurityConfig {

    @Bean
    SecurityWebFilterChain springSecurityFilterChain(ServerHttpSecurity http){//b
        return http.authorizeExchange()
                .anyExchange().authenticated() //c
              .and()
                .csrf().disable() //d
                .httpBasic() //e
              .and()
                .build();
    }

    @Bean
    CustomReactiveUserDetailsService reactiveUserDetailsService(){ //f
```

```
        return new CustomReactiveUserDetailsService();
    }

    @Bean
    PasswordEncoder passwordEncoder(){ //g
        return new BCryptPasswordEncoder();
    }
}
```

a. 使用@EnableReactiveMethodSecurity 开启对响应式方法安全的支持。

b. 定义 SecurityWebFilterChain 对安全进行控制，使用 ServerHttpSecurity 构造过滤器链。

c. 所有请求都需要通过认证。

d. 关闭 CSRF（Cross-site request forgery）跨站请求伪造，本例的客户端是自己。

e. 使用 HTTP Basic 方式登录。

f. 注册 CusotmReactiveUserDetailsService。

g. 注册 PasswordEncoder。

7. 控制器

```
@RestController
public class UserController {
    @GetMapping("/user") //使用@AuthenticationPrincipal 获得用户信息，登录即可
    public Mono<String> user(@AuthenticationPrincipal Mono<SysUser> userMono){
        return userMono.map(user -> user.getUsername());
    }
}
@RestController
@RequestMapping("/people")
public class PersonController {

    PersonRepository personRepository;

    public PersonController(PersonRepository personRepository) {
        this.personRepository = personRepository;
    }
    @PostMapping
    @PreAuthorize("hasAuthority('personAdd')") //使用方法安全，定义控制器权限为 personAdd
    public Mono<Person> add(@RequestBody Person person){
        return personRepository.save(person);
    }

    @GetMapping
    @PreAuthorize("hasAuthority('personList')")//使用方法安全，定义控制器权限为 personList
    public Flux<Person> list(){
        return personRepository.findAll();
    }
}
```

8. WebClient

```
@Component
public class ControllerClient {

    WebClient webClient;

    public ControllerClient(WebClient.Builder builder) {
        this.webClient = builder.build();
    }

    public void addPersonByUser(){
        System.out.println("普通用户添加 Person，不具备权限");
        Mono<Person> mono = webClient
                .post()
                .uri("http://localhost:8080/people")
                .header(HttpHeaders.AUTHORIZATION,
                        "Basic " +
Base64Utils.encodeToString("wyf:111111".getBytes(Charset.defaultCharset())))
                .body(Mono.just(new Person("aaa", 33)), Person.class)
                .retrieve()
                .bodyToMono(Person.class);
        mono.subscribe(System.out::println);
    }

    public void addPersonByAdmin(){
        System.out.println("管理员用户添加 Person，具备权限");
                        //……
            .body(Mono.just(new Person("bbb", 34)), Person.class)
            .header(HttpHeaders.AUTHORIZATION,
                    "Basic " +
Base64Utils.encodeToString("admin:admin".getBytes(Charset.defaultCharset())))
                        //……
    }

    public void listPersonByUser(){
        System.out.println("普通用户查看 Person 列表，具备权限");
        Flux<Person> flux = webClient
                .get()
                .uri("http://localhost:8080/people")
                .header(HttpHeaders.AUTHORIZATION,
                        "Basic " +
Base64Utils.encodeToString("wyf:111111".getBytes(Charset.defaultCharset())))
                .retrieve()
                .bodyToFlux(Person.class);
        flux.subscribe(System.out::println);
    }

    public void listPersonByAdmin(){
```

```
        System.out.println("管理员用户查看 Person 列表，具备权限");
                                //……
                .header(HttpHeaders.AUTHORIZATION,
                        "Basic " +
Base64Utils.encodeToString("admin:admin".getBytes(Charset.defaultCharset())))
                                //……
    }

    public void getUserInfo(){
        System.out.println("普通用户查看用户信息，登录即可");
        Mono<String> mono = webClient
                .get()
                .uri("http://localhost:8080/user")
                .header(HttpHeaders.AUTHORIZATION,
                        "Basic " +
Base64Utils.encodeToString("wyf:111111".getBytes(Charset.defaultCharset())))
                .retrieve()
                .bodyToMono(String.class);
        mono.subscribe(System.out::println);
    }

    public void getAdminInfo(){
        System.out.println("管理员用户查看用户信息，登录即可");
                                //……
                .header(HttpHeaders.AUTHORIZATION,
                        "Basic " +
Base64Utils.encodeToString("admin:admin".getBytes(Charset.defaultCharset())))
                                //……
    }
}
```

值得注意的是，这里使用了 header()方法进行 HTTP Basic 登录：

```
.header(HttpHeaders.AUTHORIZATION,
                "Basic " +
Base64Utils.encodeToString("wyf:111111".getBytes(Charset.defaultCharset())))
```

9. 验证

（1）初始化用户：

```
@Bean
CommandLineRunner initUsersClr(PasswordEncoder passwordEncoder,
                SysUserRepository sysUserRepository){
    return args -> {
        SysUser user = new SysUser("wyf",
                passwordEncoder.encode("111111"),
                Stream.of(new SysAuthority("personList")).collect(Collectors.toSet()));
        SysUser admin = new SysUser("admin",
                passwordEncoder.encode("admin"),
```

```
        Stream.of(new SysAuthority("personList"), new SysAuthority("personAdd"))
            .collect(Collectors.toSet()));
    sysUserRepository.save(user).subscribe();
    sysUserRepository.save(admin).subscribe();
  };
}
```

wyf 用户具备 personList 权限；admin 用户具备 personList 和 personAdd 权限。

（2）调用接口：

```
@Bean
CommandLineRunner webClientClr(ControllerClient controllerClient){
  return args -> {
    controllerClient.addPersonByUser();
    Thread.sleep(1000);
    controllerClient.addPersonByAdmin();
    Thread.sleep(1000);
    controllerClient.listPersonByUser();
    Thread.sleep(1000);
    controllerClient.listPersonByAdmin();
    Thread.sleep(1000);
    controllerClient.getUserInfo();
    Thread.sleep(1000);
    controllerClient.getAdminInfo();
  };
}
```

（3）输出结果，如图 8-12 所示。

```
Error has been observed at the following site(s):
    |_ checkpoint - 403 FORBIDDEN from POST http://localhost:8080/people

管理员用户添加Peron，具备权限
Person(id=5cfbbcfd006889280647e180, name=bbb, age=34)
普通用户查看Person列表，具备权限
Person(id=5cfbbcfd006889280647e180, name=bbb, age=34)
管理员用户查看Person列表，具备权限
Person(id=5cfbbcfd006889280647e180, name=bbb, age=34)
普通用户查看用户信息，登录即可
wyf
管理员用户查看用户信息，登录即可
admin
```

图 8-12

8.6 小结

本章介绍了响应式 Spring 编程的全栈技术。通过 Spring WebFlux、Spring Data Reactive，以及 Reactive Spring Security，完全可以用响应式编程替代现有的编程方式。

第 9 章 事件驱动

事件驱动开发是一种异步的、用来开发高扩展应用的、分布式的架构。它涉及事件的产生、处理和消费。在学习事件驱动开发之前，需要了解下面 4 个概念。

◎ Message/Event：消息或事件，需要产生、处理和消费的数据。

◎ Publisher/Producer：消息的发布者或生产者。

◎ Subscriber/Consumer：消息的订阅者或消费者。

◎ Message Broker：消息代理或消息中间件，是除发布者和订阅者外的第三方。当通过消息代理来转发消息时，发布者和订阅者无须知道彼此的存在。

9.1 JMS

Spring 支持 ActiveMQ 和 Artemis 作为 JMS 消息中间件。Artemis 是 ActiveMQ 下一代的重新实现，本节以 Artemis 作为示例。

9.1.1 安装 Apache ActiveMQ Artemis

使用 docker compose 安装 Artemis。

stack.yml：

```
version: '3.1'

services:
  artemis:
    image: vromero/activemq-artemis
    restart: always
    ports:
      - 8161:8161 # 管理控制台端口
      - 61616:61616 # 连接端口
```

```
    environment:
      ARTEMIS_USERNAME: wisely
      ARTEMIS_PASSWORD: zzzzzz
```

执行命令：

```
$ docker-compose -f stack.yml up -d
```

9.1.2 新建应用

新建应用，信息如下。

Group：top.wisely。

Artifact：learning-jms。

Dependencies：Spring for Apache ActiveMQ Artemis 和 Lombok。

build.gradle 文件中的依赖如下：

```
dependencies {
  implementation 'org.springframework.boot:spring-boot-starter-artemis'
  compileOnly 'org.projectlombok:lombok'
  annotationProcessor 'org.projectlombok:lombok'
    //……
}
```

9.1.3 Spring Boot 的自动配置

Spring Boot 对 JMS 提供的自动配置如下。

◎ ArtemisAutoConfiguration：通过 ArtemisProperties，使用 spring.artemis.*来配置连接 Artemis 的 jmsConnectionFactory。

◎ JmsAutoConfiguration：

  ➢ 使用 jmsConnectionFactory 的 Bean 配置 jmsTemplate 的 Bean。

  ➢ 通过@EnableJms 开启对 JMS 的支持。

9.1.4 示例

（1）连接 JMS 中间件 Artemis：

```
spring:
  artemis:
    host: localhost
    port: 61616
    user: wisely
    password: zzzzzz
```

（2）消息定义：

```
@Data
@AllArgsConstructor
@NoArgsConstructor
public class MessageEvent implements Serializable{
   private String id;
   private String name;
}
```

（3）开启计划任务：

```
@SpringBootApplication
@EnableScheduling
public class LearningJmsApplication {}
```

（4）发送消息：

```
@Component
public class MessageProducer {

   JmsTemplate jmsTemplate;

   public MessageProducer(JmsTemplate jmsTemplate) {
      this.jmsTemplate = jmsTemplate;
   }

   @Scheduled(fixedRate = 2000) //a
   public void send(){
      jmsTemplate.convertAndSend("my-dest",
            new MessageEvent(UUID.randomUUID().toString(), "wyf"));  //b
   }

}
```

a. 使用@Scheduled 每隔两秒钟发送一次消息。

b. 通过 JmsTemplate 的 convertAndSend(String destinationName, final Object message)方法向指定的终点 my-dest 发送消息。

（5）接收消息并应答。

一般来说，消息的消费者和生产者是位于不同的应用中的，本例为演示简单，把消息的消费者和生产者放置在了一起。

```
@Component
public class MessageConsumer {

   @JmsListener(destination = "my-dest") //a
   @SendTo("confirm-dest") //b
   public String consume(MessageEvent event){
      System.out.println("在 consume 方法中处理事件" + event);
      return "接收到了:" + event.toString(); //c
```

```
    }
}
```

a. 使用@JmsListener 从指定终点中获取消息。

b. 在处理完成后，可通过@SendTo 向指定终点触发另外的处理。

c. 将返回值的内容发送到@SendTo 的指定终点。

（6）处理应答：

```
@Component
public class ConfirmConsumer {
    @JmsListener(destination = "confirm-dest")
    public void confirmReceived(String confirm){
        System.out.println(confirm);
    }
}
```

（7）执行结果如图 9-1 所示。

```
在consume方法中处理事件MessageEvent(id=e29b4d07-b1dc-4eb4-8dff-9761136b252f, name=wyf)
接收到了:MessageEvent(id=e29b4d07-b1dc-4eb4-8dff-9761136b252f, name=wyf)
在consume方法中处理事件MessageEvent(id=4d3fa3b8-b59b-42b2-b4ac-62bcf7b47deb, name=wyf)
接收到了:MessageEvent(id=4d3fa3b8-b59b-42b2-b4ac-62bcf7b47deb, name=wyf)
在consume方法中处理事件MessageEvent(id=b04968b3-5c60-4ab3-9cf2-157be1bb242a, name=wyf)
接收到了:MessageEvent(id=b04968b3-5c60-4ab3-9cf2-157be1bb242a, name=wyf)
在consume方法中处理事件MessageEvent(id=407c472d-02fa-4288-bbf8-a85e0d5ae95e, name=wyf)
接收到了:MessageEvent(id=407c472d-02fa-4288-bbf8-a85e0d5ae95e, name=wyf)
在consume方法中处理事件MessageEvent(id=ce2f8f39-c7db-4e0e-a563-a7a3d855e02a, name=wyf)
接收到了:MessageEvent(id=ce2f8f39-c7db-4e0e-a563-a7a3d855e02a, name=wyf)
在consume方法中处理事件MessageEvent(id=bea713a7-4c00-41a3-9d05-4a8a8c1aa033, name=wyf)
接收到了:MessageEvent(id=bea713a7-4c00-41a3-9d05-4a8a8c1aa033, name=wyf)
在consume方法中处理事件MessageEvent(id=1a3c0a78-7e04-4b7e-98b0-1490a2caeb53, name=wyf)
接收到了:MessageEvent(id=1a3c0a78-7e04-4b7e-98b0-1490a2caeb53, name=wyf)
在consume方法中处理事件MessageEvent(id=4f436252-e762-4d8d-94f6-e39ace92875f, name=wyf)
接收到了:MessageEvent(id=4f436252-e762-4d8d-94f6-e39ace92875f, name=wyf)
在consume方法中处理事件MessageEvent(id=1266d730-9974-41c8-b35f-20978c14694a, name=wyf)
接收到了:MessageEvent(id=1266d730-9974-41c8-b35f-20978c14694a, name=wyf)
在consume方法中处理事件MessageEvent(id=fa7295e5-0503-4387-a4fe-d60256a0297d, name=wyf)
接收到了:MessageEvent(id=fa7295e5-0503-4387-a4fe-d60256a0297d, name=wyf)
在consume方法中处理事件MessageEvent(id=40161c42-f15b-4d55-9309-7ee8d6af0f92, name=wyf)
接收到了:MessageEvent(id=40161c42-f15b-4d55-9309-7ee8d6af0f92, name=wyf)
在consume方法中处理事件MessageEvent(id=5549eb58-3b09-407a-a6c8-133fdcc31773, name=wyf)
接收到了:MessageEvent(id=5549eb58-3b09-407a-a6c8-133fdcc31773, name=wyf)
```

图 9-1

9.1.5 Topic 和 Queue

在前面的例子中只有一个消费者，如果有多个消费者，则含义如下。

◎ Queue：所有消费者会对消息进行负载轮流处理。

◎ Topic：所有消费者都会收到消息并进行处理。

在下面代码中：

```
jmsTemplate.convertAndSend("my-dest",
            new MessageEvent(UUID.randomUUID().toString(), "wyf"));
```

my-dest 是 Destination 的地址，Destination 的子接口是 Queue 和 Topic，默认自动建立地址为 my-dest 的 Queue：

```
@Component
public class MessageConsumer {

    @JmsListener(destination = "my-dest")
    @SendTo("confirm-desc")
    public String consume(MessageEvent event){
        System.out.println("在 consume 方法中处理事件" + event);
        return "接收到了:" + event.toString();
    }

    @JmsListener(destination = "my-dest")
    public void consume2(MessageEvent event){
        System.out.println("在 consume2 方法中处理事件" + event);
    }
}
```

启动应用，如图 9-2 所示。

```
在consume方法中处理事件MessageEvent(id=c24f5889-3de6-4d05-b385-061a1aca8826, name=wyf)
接收到了:MessageEvent(id=c24f5889-3de6-4d05-b385-061a1aca8826, name=wyf)
在consume2方法中处理事件MessageEvent(id=4e9a7492-0f61-4f63-b394-63bd335f4041, name=wyf)
在consume方法中处理事件MessageEvent(id=452424b8-cdd8-46ef-8182-9f36d6164506, name=wyf)
接收到了:MessageEvent(id=452424b8-cdd8-46ef-8182-9f36d6164506, name=wyf)
在consume2方法中处理事件MessageEvent(id=f2c3ce7f-ab34-46bf-b0cf-a522b2abbb94, name=wyf)
在consume方法中处理事件MessageEvent(id=00d2a525-88a0-4205-887c-514b8c9ee05e, name=wyf)
接收到了:MessageEvent(id=00d2a525-88a0-4205-887c-514b8c9ee05e, name=wyf)
在consume2方法中处理事件MessageEvent(id=d60f5083-930d-4cf8-be14-065e994fd243, name=wyf)
在consume方法中处理事件MessageEvent(id=7aafb022-a05b-436e-a74a-c8990806a2e2, name=wyf)
接收到了:MessageEvent(id=7aafb022-a05b-436e-a74a-c8990806a2e2, name=wyf)
在consume2方法中处理事件MessageEvent(id=2e643c14-5afe-4424-8a71-0f273c281e01, name=wyf)
在consume方法中处理事件MessageEvent(id=cc5b7537-1d59-4b15-8ef9-82fc6cbe6daf, name=wyf)
接收到了:MessageEvent(id=cc5b7537-1d59-4b15-8ef9-82fc6cbe6daf, name=wyf)
在consume2方法中处理事件MessageEvent(id=1f299995-14ac-4eef-a86f-0d2ff59f7217, name=wyf)
在consume方法中处理事件MessageEvent(id=f6167e62-7996-452d-a4a9-7b1d152851c3, name=wyf)
接收到了:MessageEvent(id=f6167e62-7996-452d-a4a9-7b1d152851c3, name=wyf)
在consume2方法中处理事件MessageEvent(id=94ed78d1-32fa-4a44-b89b-42ce8dd8bdb5, name=wyf)
```

图 9-2

当在 application.yml 中设置了 spring.jms.pub-sub-domain 为 true 时，默认会建立 Topic，这时所有的消费者都能收到单独的数据，如图 9-3 所示。

```
spring:
  artemis:
    host: localhost
    port: 61616
    user: wisely
    password: zzzzzz
  jms:
    pub-sub-domain: true
```

```
在consume方法中处理事件MessageEvent(id=15d86f31-bcf2-43d2-9ed3-7dddaeefa878, name=wyf)
在consume2方法中处理事件MessageEvent(id=15d86f31-bcf2-43d2-9ed3-7dddaeefa878, name=wyf)
接收到了:MessageEvent(id=15d86f31-bcf2-43d2-9ed3-7dddaeefa878, name=wyf)
在consume方法中处理事件MessageEvent(id=6680502e-4980-4e8e-8b82-9a268eb490f6, name=wyf)
在consume2方法中处理事件MessageEvent(id=6680502e-4980-4e8e-8b82-9a268eb490f6, name=wyf)
接收到了:MessageEvent(id=6680502e-4980-4e8e-8b82-9a268eb490f6, name=wyf)
在consume方法中处理事件MessageEvent(id=11a7038c-7975-48b8-98f8-911d5464d084, name=wyf)
在consume2方法中处理事件MessageEvent(id=11a7038c-7975-48b8-98f8-911d5464d084, name=wyf)
接收到了:MessageEvent(id=11a7038c-7975-48b8-98f8-911d5464d084, name=wyf)
在consume方法中处理事件MessageEvent(id=c5b05687-17bd-4dfa-8e29-637b3d328b4e, name=wyf)
在consume2方法中处理事件MessageEvent(id=c5b05687-17bd-4dfa-8e29-637b3d328b4e, name=wyf)
接收到了:MessageEvent(id=c5b05687-17bd-4dfa-8e29-637b3d328b4e, name=wyf)
在consume方法中处理事件MessageEvent(id=0830ea62-395a-49eb-9f43-9af1a4c56ef4, name=wyf)
在consume2方法中处理事件MessageEvent(id=0830ea62-395a-49eb-9f43-9af1a4c56ef4, name=wyf)
接收到了:MessageEvent(id=0830ea62-395a-49eb-9f43-9af1a4c56ef4, name=wyf)
在consume方法中处理事件MessageEvent(id=567b9425-e80f-4b89-aa64-e8a096f1c278, name=wyf)
在consume2方法中处理事件MessageEvent(id=567b9425-e80f-4b89-aa64-e8a096f1c278, name=wyf)
接收到了:MessageEvent(id=567b9425-e80f-4b89-aa64-e8a096f1c278, name=wyf)
在consume方法中处理事件MessageEvent(id=a9d2cff2-29fa-4d67-9a23-ea944e432a5f, name=wyf)
在consume2方法中处理事件MessageEvent(id=a9d2cff2-29fa-4d67-9a23-ea944e432a5f, name=wyf)
接收到了:MessageEvent(id=a9d2cff2-29fa-4d67-9a23-ea944e432a5f, name=wyf)
```

图 9-3

9.2 RabbitMQ

本节演示 RabbitMQ 支持的 AMQP（协议），在学习 AMQP 之前，需要了解下面 3 个概念。当发布者发送的消息和消息的 routing key 到 RabbitMQ 中间件时：

◎ exchange：基于消息的 routing key，将消息路由到一个或多个 queue。

◎ binding： binding 是设置 queue 绑定到 exchange 的连接。

◎ queue：消息队列。

Exchange 的主要类型如下。

◎ Default：将消息路由到名称为 routing key 的 queue，所有的 queue 都会自动绑定 default exchange。

◎ Direct：将消息路由到 binding key 与消息的 routing key 一致的 queue。

◎ Fanout：将消息路由到所有绑定的 queue，不考虑 binding key 和 routing key。

◎ Topic：将消息路由到 binding key，匹配 routing key 的一个或多个 queue，匹配可包含通配符。

◎ Headers：类似于 topic，但路由基于消息头，而不是 routing key。

1. 安装 RabbitMQ

使用 docker compose 安装 RabbitMQ。

stack.yml：

```
version: '3.1'

services:
  rabbitmq:
    image: rabbitmq:management
```

```
  restart: always
  ports:
    - "5672:5672" # 连接端口
    - "15672:15672" # 管理控制台端口
  environment:
    RABBITMQ_DEFAULT_USER: wisely
    RABBITMQ_DEFAULT_PASS: zzzzzz
```

执行命令：

```
$ docker-compose -f stack.yml up -d
```

2．新建应用

新建应用，信息如下。

Group：top.wisely。

Artifact：learning-amqp。

Dependencies：Spring for RabbitMQ 和 Lombok。

build.gradle 文件中的依赖如下：

```
dependencies {
    implementation 'org.springframework.boot:spring-boot-starter-amqp'
  compileOnly 'org.projectlombok:lombok'
   annotationProcessor 'org.projectlombok:lombok'
    //……
}
```

3．Spring Boot 的自动配置

Spring Boot 的自动配置如下。

（1）RabbitAutoConfiguration：

◎　配置连接 RabbitMQ 的 rabbitConnectionFactory 的 Bean。

◎　配置用来操作消息的 rabbitTemplate 的 Bean。

◎　通过@EnableRabbit 注解开启对 RabbitMQ 的支持。

通过 RabbitProperties 使用 spring.rabbitmq.*对 RabbitMQ 进行配置。

4．示例

（1）连接 RabbitMQ：

```
spring:
  rabbitmq:
    host: localhost
    port: 5672
    username: wisely
    password: zzzzzz
```

（2）消息定义：

```
@Data
@AllArgsConstructor
@NoArgsConstructor
public class MessageEvent implements Serializable {
    private String id;
    private String name;
}
```

（3）配置：

```
@SpringBootApplication
@EnableScheduling //a
public class LearningAmqpApplication {

  public static void main(String[] args) {
    SpringApplication.run(LearningAmqpApplication.class, args);
  }

  @Bean
  Queue myDest(){ //b
    return new Queue("my-dest");
  }

  @Bean
  Queue confirmDest(){ //c
    return new Queue("confirm-dest");
  }
}
```

a. 开启对计划任务的支持。

b. 新建用于接收消息的 queue：my-dest。

c. 新建用于确认消息的 queue：confirm-dest。

（4）发送消息：

```
@Component
public class MessageProducer {

    RabbitTemplate rabbitTemplate;

    public MessageProducer(RabbitTemplate rabbitTemplate) {
        this.rabbitTemplate = rabbitTemplate;
    }

    @Scheduled(fixedRate = 2000)//a
    public void send(){
        rabbitTemplate.convertAndSend("my-dest",
                new MessageEvent(UUID.randomUUID().toString(), "wyf"));//b
```

```
    }
}
```

a. 使用@Scheduled 每隔两秒钟发送一次消息。

b. 通过 RabbitTemplate 的 convertAndSend(String routingKey, final Object object)发送消息，此处的第一个参数是 routing Key my-dest，此时的 exchange 是一个 default exchange；消息会发到名称为 my-dest 的 queue。

（5）接收消息并应答：

```
@Component
public class MessageConsumer {

    @RabbitListener(queuesToDeclare = @Queue(name = "my-direct"))//a
    @SendTo("confirm-dest")
    public String consume(MessageEvent event){
        System.out.println("在 consume 方法中处理事件" + event);
        return "接收到了:" + event.toString();
    }
}
```

a. 使用@RabbitListener 从指定队列获取消息，使用 queuesToDeclare 可以在 RabbitMQ 上声明 queue。

也可以自己声明 Queue，使用@RabbitListener 的 queues 属性来指定 queue：

```
@Bean
Queue myDest(){
   return new Queue("my-dest");
}
@RabbitListener(queues = "my-dest")
@SendTo("confirm-dest")//b
public String consume(MessageEvent event){
    System.out.println("在 consume 方法中处理事件" + event);
    return "接收到了:" + event.toString();
}
```

b. 在处理完成后，可通过@SendTo 向指定 routing key 触发另外的处理。这里的 confirm-dest 同样为 routing key。如果既需要指定 exchange，又需要指定 routing key，则可以使用@SendTo("some-exchange/confirm-dest")。

（6）处理应答：

```
@Component
public class ConfirmConsumer {
    @RabbitListener(queuesToDeclare = {@Queue(name = "confirm-dest")})
    public void confirmReceived(String confirm){
        System.out.println(confirm);
    }
}
```

（7）启动应用，如图 9-4 所示。

```
在consume方法中处理事件MessageEvent(id=cfd485e4-ed66-4863-b08b-14fa44196d47, name=wyf)
接收到了:MessageEvent(id=cfd485e4-ed66-4863-b08b-14fa44196d47, name=wyf)
在consume方法中处理事件MessageEvent(id=dc308688-9869-4257-9c6f-ef69d814ee4b, name=wyf)
接收到了:MessageEvent(id=dc308688-9869-4257-9c6f-ef69d814ee4b, name=wyf)
在consume方法中处理事件MessageEvent(id=2027ec8e-b683-4719-8a65-80293e639e87, name=wyf)
接收到了:MessageEvent(id=2027ec8e-b683-4719-8a65-80293e639e87, name=wyf)
在consume方法中处理事件MessageEvent(id=00a85af1-3d7b-47a6-8240-3031998b7460, name=wyf)
接收到了:MessageEvent(id=00a85af1-3d7b-47a6-8240-3031998b7460, name=wyf)
在consume方法中处理事件MessageEvent(id=201745b3-256d-49e9-88db-d318b27c68cf, name=wyf)
接收到了:MessageEvent(id=201745b3-256d-49e9-88db-d318b27c68cf, name=wyf)
在consume方法中处理事件MessageEvent(id=2b673ec7-99f4-40c6-b89b-8febb0c714e5, name=wyf)
接收到了:MessageEvent(id=2b673ec7-99f4-40c6-b89b-8febb0c714e5, name=wyf)
在consume方法中处理事件MessageEvent(id=d516729c-4fda-49c7-86fb-e43da0dc35e1, name=wyf)
接收到了:MessageEvent(id=d516729c-4fda-49c7-86fb-e43da0dc35e1, name=wyf)
在consume方法中处理事件MessageEvent(id=040c7dee-085d-4474-baa6-3a7ad8e9c519, name=wyf)
接收到了:MessageEvent(id=040c7dee-085d-4474-baa6-3a7ad8e9c519, name=wyf)
在consume方法中处理事件MessageEvent(id=538ea2c6-74ae-4ab7-b286-cd9f40fec938, name=wyf)
接收到了:MessageEvent(id=538ea2c6-74ae-4ab7-b286-cd9f40fec938, name=wyf)
```

图 9-4

5. 其他类型的 exchange

（1）Direct Exchange。

发送消息：

```
@Component
public class MessageProducer {

   RabbitTemplate rabbitTemplate;

   public MessageProducer(RabbitTemplate rabbitTemplate) {
      this.rabbitTemplate = rabbitTemplate;
   }

   @Scheduled(fixedRate = 2000)
   public void sendDirect(){
      rabbitTemplate.convertAndSend("direct-exchange", "some-key",
            new MessageEvent(UUID.randomUUID().toString(), "wyf"));
   }
}
```

第一个参数 direct-exchange 为 exchange 的名称，some-key 为 routing key。

接收消息：

```
@Component
public class MessageConsumer {

   @RabbitListener(bindings = @QueueBinding(
         exchange = @Exchange(name = "direct-exchange", type = ExchangeTypes.DIRECT),
         value = @Queue(name = "direct-queue"),
         key = "some-key"
   ))
   public void consumeDirect(MessageEvent event){
```

```
        System.out.println("在consumeDirect方法中处理事件" + event);
    }

    @RabbitListener(bindings = @QueueBinding(
            exchange = @Exchange(name = "direct-exchange", type = ExchangeTypes.DIRECT),
            value = @Queue(name = "direct-queue2"),
            key = "some-key"
    ))
    public void consumeDirect2(MessageEvent event){
        System.out.println("在consumeDirect2方法中处理事件" + event);
    }
}
```

◎ 可以使用@RabbitListener 的 bindings 属性，通过@QueueBinding 来声明 binding。@QueueBinding 的属性如下。

◎ exchange：通过@Exchange 声明名称为 direct-exchange 的 exchange，默认类型是 Direct，可省略。

◎ value：通过@Queue 声明名称为 direct-queue 的 queue。

◎ key：此处为 binding key。

在上面的例子中，使用了两个 binding，分别绑定了 exchange 和不同的两个 queue，它们的 binding key 和 routing key 相同，两个 queue 都能收到消息。

启动应用，如图 9-5 所示。

```
在consumeDirect方法中处理事件MessageEvent(id=718d75a3-639d-451b-99f7-a1bd40c8b96f, name=wyf)
在consumeDirect2方法中处理事件MessageEvent(id=718d75a3-639d-451b-99f7-a1bd40c8b96f, name=wyf)
在consumeDirect方法中处理事件MessageEvent(id=2391796f-e058-492f-851c-9c51cd0cbcc7, name=wyf)
在consumeDirect2方法中处理事件MessageEvent(id=2391796f-e058-492f-851c-9c51cd0cbcc7, name=wyf)
在consumeDirect方法中处理事件MessageEvent(id=09b2bdf2-7a25-452c-9bbd-ef04dda97f92, name=wyf)
在consumeDirect2方法中处理事件MessageEvent(id=09b2bdf2-7a25-452c-9bbd-ef04dda97f92, name=wyf)
在consumeDirect方法中处理事件MessageEvent(id=723c3f31-1907-4357-8629-5e529967f8a1, name=wyf)
在consumeDirect2方法中处理事件MessageEvent(id=723c3f31-1907-4357-8629-5e529967f8a1, name=wyf)
在consumeDirect方法中处理事件MessageEvent(id=4d568295-723c-42ba-8973-801ed11154f1, name=wyf)
在consumeDirect2方法中处理事件MessageEvent(id=4d568295-723c-42ba-8973-801ed11154f1, name=wyf)
在consumeDirect方法中处理事件MessageEvent(id=f83ad9ef-e796-4670-9653-abcad25cc6e4, name=wyf)
在consumeDirect2方法中处理事件MessageEvent(id=f83ad9ef-e796-4670-9653-abcad25cc6e4, name=wyf)
在consumeDirect2方法中处理事件MessageEvent(id=0ddf71ad-ed0b-4c7e-991e-69f352399c57, name=wyf)
在consumeDirect方法中处理事件MessageEvent(id=0ddf71ad-ed0b-4c7e-991e-69f352399c57, name=wyf)
在consumeDirect方法中处理事件MessageEvent(id=c5c589ae-1193-47ca-959b-7c86048a0ba6, name=wyf)
在consumeDirect2方法中处理事件MessageEvent(id=c5c589ae-1193-47ca-959b-7c86048a0ba6, name=wyf)
```

图 9-5

当然，也可以自己声明 binding、exchange 和 queue：

```
    @Bean
    Queue directQueue(){
        return new Queue("direct-queue");
    }

    @Bean
    DirectExchange directExchange(){
```

```
        return new DirectExchange("direct-exchange");
    }

    @Bean
    Binding directBinding(){
        return
BindingBuilder.bind(directQueue()).to(directExchange()).with("some-key");
    }
@RabbitListener(queues = "direct-queue")
public void consumeDirect(MessageEvent event){
   System.out.println("在consumeDirect方法中处理事件" + event);
}
```

（2）Topic Exchange。

发送消息：

```
@Component
public class MessageProducer {

   RabbitTemplate rabbitTemplate;

   public MessageProducer(RabbitTemplate rabbitTemplate) {
      this.rabbitTemplate = rabbitTemplate;
   }

   @Scheduled(fixedRate = 2000)
   public void sendTopic(){
      rabbitTemplate.convertAndSend("topic-exchange", "some.key.topic",
            new MessageEvent(UUID.randomUUID().toString(), "wyf"));
   }
}
```

topic-exchange 为 exchange 的名称，some.key 为 routing key。发送给 topic exchange 的 routing key 不是任意的，词之间必须使用“.”隔开。

接收消息：

```
@Component
public class MessageConsumer {
   @RabbitListener(bindings = @QueueBinding(
         exchange = @Exchange(name = "topic-exchange", type = ExchangeTypes.TOPIC),
         value = @Queue(name = "topic-queue"),
         key = "some.*.topic"
   ))
   public void consumeTopic(MessageEvent event){
      System.out.println("在consumeTopic方法中处理事件" + event);
   }

   @RabbitListener(bindings = @QueueBinding(
         exchange = @Exchange(name = "topic-exchange", type = ExchangeTypes.TOPIC),
         value = @Queue(name = "topic-queue2"),
```

```
        key = "#.topic"
    ))
    public void consumeTopic2(MessageEvent event){
        System.out.println("在 consumeTopic 方法中处理事件" + event);
    }
}
```

这里的 binding key 有两个特殊的匹配字符。

◎ *：可替代一个词。

◎ #：可替代一个或多个词。

上面两种情况都可匹配。启动应用，如图 9-6 所示。

```
在consumeTopic方法中处理事件MessageEvent(id=74576e63-3433-439d-a8c6-b606716e1cb1, name=wyf)
在consumeTopic2方法中处理事件MessageEvent(id=74576e63-3433-439d-a8c6-b606716e1cb1, name=wyf)
在consumeTopic方法中处理事件MessageEvent(id=e3c11891-cfde-4850-aa22-54491106c73d, name=wyf)
在consumeTopic2方法中处理事件MessageEvent(id=e3c11891-cfde-4850-aa22-54491106c73d, name=wyf)
在consumeTopic2方法中处理事件MessageEvent(id=c99d96d8-b9d8-4d1d-82ec-694254b77a4a, name=wyf)
在consumeTopic方法中处理事件MessageEvent(id=c99d96d8-b9d8-4d1d-82ec-694254b77a4a, name=wyf)
在consumeTopic方法中处理事件MessageEvent(id=22af4586-d1a5-45ad-ad50-988e771faf9c, name=wyf)
在consumeTopic2方法中处理事件MessageEvent(id=22af4586-d1a5-45ad-ad50-988e771faf9c, name=wyf)
在consumeTopic方法中处理事件MessageEvent(id=ee48ae6c-39aa-4afc-8854-00c07d40ec14, name=wyf)
在consumeTopic2方法中处理事件MessageEvent(id=ee48ae6c-39aa-4afc-8854-00c07d40ec14, name=wyf)
在consumeTopic方法中处理事件MessageEvent(id=4a2c64ce-93c9-4f59-a014-f48b64336ca9, name=wyf)
在consumeTopic2方法中处理事件MessageEvent(id=4a2c64ce-93c9-4f59-a014-f48b64336ca9, name=wyf)
在consumeTopic2方法中处理事件MessageEvent(id=5c170a7b-1900-4cb8-89b3-de9ed8024995, name=wyf)
在consumeTopic方法中处理事件MessageEvent(id=5c170a7b-1900-4cb8-89b3-de9ed8024995, name=wyf)
在consumeTopic方法中处理事件MessageEvent(id=260de62f-b786-45e5-ace5-078c60221e94, name=wyf)
在consumeTopic2方法中处理事件MessageEvent(id=260de62f-b786-45e5-ace5-078c60221e94, name=wyf)
在consumeTopic方法中处理事件MessageEvent(id=61ba0b56-11ff-4998-9314-a46340351f8f, name=wyf)
在consumeTopic2方法中处理事件MessageEvent(id=61ba0b56-11ff-4998-9314-a46340351f8f, name=wyf)
在consumeTopic方法中处理事件MessageEvent(id=191f23ee-da76-46af-ae81-c4b5ac79f3e7, name=wyf)
在consumeTopic2方法中处理事件MessageEvent(id=191f23ee-da76-46af-ae81-c4b5ac79f3e7, name=wyf)
```

图 9-6

9.3 Kafka

Kafka 以 Topic 为导向，提供消息中间件的功能。一个类型的数据称之为一个 Topic。

1. 安装 Kafka

使用 docker compose 安装 Apache Kafka。

stack.yml：

```
version: '3.1'

services:
  zookeeper:
    image: wurstmeister/zookeeper
    restart: always

  kafka:
    image:  wurstmeister/kafka
    ports:
```

```
    - "9092:9092"
  environment:
    KAFKA_ADVERTISED_HOST_NAME: localhost
    KAFKA_ZOOKEEPER_CONNECT: zookeeper:2181
```

执行命令：

```
$ docker-compose -f stack.yml up -d
```

2．新建应用

新建应用，信息如下。

Group：top.wisely。

Artifact：learning-kafka。

Dependencies：Spring for Kafka 和 Lombok。

build.gradle 文件中的依赖如下：

```
dependencies {
    implementation 'org.springframework.boot:spring-boot-starter'
    implementation 'org.springframework.kafka:spring-kafka'
    implementation 'com.fasterxml.jackson.core:jackson-databind' //消息序列化、
//反序列化使用
    compileOnly 'org.projectlombok:lombok'
    annotationProcessor 'org.projectlombok:lombok'
  //……
}
```

3．Spring Boot 的自动配置

Spring Boot 提供的自动配置为 KafkaAutoConfiguration，它所做的自动配置如下。

◎ 注册 Kafka 消息操作的 Bean（kafkaTemplate）。

◎ 注册 kafkaAdmin 的 Bean 用来在应用中新建 Topic。

◎ 使用@EnableKafka 注解开启对 Kafka 的支持。

◎ 对 Kafka Stream 注解驱动配置的支持。

通过 KafkaProperties 使用 spring.kafka.*来配置 Kafka。

4．示例

（1）连接 Kafka 并配置：

```
spring:
 kafka:
   bootstrap-servers: #a
     - localhost:9092
   producer: # b
#     key-serializer: org.springframework.kafka.support.serializer.JsonSerializer #c
     value-serializer: org.springframework.kafka.support.serializer.JsonSerializer #d
```

```
    consumer: #e
#     key-deserializer: org.springframework.kafka.support.serializer.JsonDeserializer
#f
      value-deserializer:
org.springframework.kafka.support.serializer.JsonDeserializer #g
      group-id: consumers #h
      properties:
        spring:
          json:
            trusted:
              packages: top.wisely.learningkafka.messaging #h
```

a. 配置连接Kafka服务器，若是Kafka集群，则可连接多个服务器。

b. 配置Kafka消息提供者。

c. 消息提供者对Key序列化的方式，默认为StringSerializer。本例的Key为String类型，可省略。

d. 消息提供者对消息序列化的方式。

e. 配置Kafka消息消费者。

f. 消息消费者对Key反序列化的方式，默认为StringDeserializer，符合默认，可省略。

g. 消息消费者对消息反序列化的方式。

h. 配置消费者的group-id，若多个应用的消费者的group-id相同，则可负载消费消息。

i. 将top.wisely.learningkafka.messaging加到消费者信任的反序列化包中。

（2）消息定义：

```
@Data
@AllArgsConstructor
@NoArgsConstructor
public class MessageEvent implements Serializable {
    private String id;
    private String name;
}
```

（3）配置：

```
@SpringBootApplication
@EnableScheduling
public class LearningKafkaApplication {

   public static void main(String[] args) {
      SpringApplication.run(LearningKafkaApplication.class, args);
   }

   @Bean
   NewTopic topic(){ //使用NewTopic定义确认所需要的Topic
      return new NewTopic("confirm-topic",1 ,(short) 1);
   }
}
```

（4）发送消息：

```
@Component
public class MessageProducer {

   KafkaTemplate<String, MessageEvent> kafkaTemplate;

   public MessageProducer(KafkaTemplate kafkaTemplate) { //a
      this.kafkaTemplate = kafkaTemplate;
   }

   @Scheduled(fixedRate = 2000)
   public void send(){
      kafkaTemplate.send("my-topic","name",
                  new MessageEvent(UUID.randomUUID().toString(), "wyf")); //b
   }
}
```

a. 注入 kafkaTemplate 的 Bean，String 为 Key 的类型（决定消息在集群中存储的位置），MessageEvent 是 Value 类型（消息）；

b. 使用 kafkaTemplate 的 send(String topic, K key, @Nullable V data)发送消息；第一个参数为 Topic 的名称（此时的 Topic 会自动创建），第二个参数为 Key，第三个参数是 Value，即消息。

（5）接收并应答:

```
@Component
public class MessageConsumer {

   @KafkaListener(topics = {"my-topic"}) //a
   @SendTo("confirm-topic") //b
   public String consume(MessageEvent event){
      System.out.println("在consume方法中处理事件" + event);
      return "接收到了:" + event.toString();
   }
}
```

a. 使用@KafkaListener 的 topics 属性来设置接收指定的 Topic 消息。

b. 当处理完成后，可通过@SendTo 向指定 Topic 触发另外的处理。

（6）处理应答：

```
@Component
public class ConfirmConsumer {
   @KafkaListener(topics = {"confirm-topic"})
   public void confirmReceived(String confirm){
      System.out.println(confirm);
   }
}
```

（7）启动应用，如图 9-7 所示。

```
在consume方法中处理事件MessageEvent(id=2e55f7a0-8a4e-473c-a9f3-5374682be5ee, name=wyf)
接收到了:MessageEvent(id=2e55f7a0-8a4e-473c-a9f3-5374682be5ee, name=wyf)
在consume方法中处理事件MessageEvent(id=e1599c36-8ac7-436c-9877-1151a756c5af, name=wyf)
接收到了:MessageEvent(id=e1599c36-8ac7-436c-9877-1151a756c5af, name=wyf)
在consume方法中处理事件MessageEvent(id=da0b0043-28ec-4584-8e46-b810fa7b28c5, name=wyf)
接收到了:MessageEvent(id=da0b0043-28ec-4584-8e46-b810fa7b28c5, name=wyf)
在consume方法中处理事件MessageEvent(id=eccdbcd6-7cb6-4198-ae7f-8724b87bad15, name=wyf)
接收到了:MessageEvent(id=eccdbcd6-7cb6-4198-ae7f-8724b87bad15, name=wyf)
在consume方法中处理事件MessageEvent(id=c18609ae-7f40-42b2-8d4e-2f89bc5dc3cf, name=wyf)
接收到了:MessageEvent(id=c18609ae-7f40-42b2-8d4e-2f89bc5dc3cf, name=wyf)
在consume方法中处理事件MessageEvent(id=9404c879-a28d-4329-a3b6-2c81a8a9285a, name=wyf)
接收到了:MessageEvent(id=9404c879-a28d-4329-a3b6-2c81a8a9285a, name=wyf)
在consume方法中处理事件MessageEvent(id=0fd14263-e837-4f95-88c7-c3c6d94a903a, name=wyf)
接收到了:MessageEvent(id=0fd14263-e837-4f95-88c7-c3c6d94a903a, name=wyf)
在consume方法中处理事件MessageEvent(id=85eb6ebd-63aa-46e5-9572-39c36d174604, name=wyf)
接收到了:MessageEvent(id=85eb6ebd-63aa-46e5-9572-39c36d174604, name=wyf)
在consume方法中处理事件MessageEvent(id=ccc60d71-bafc-49be-a848-a8d09e838592, name=wyf)
接收到了:MessageEvent(id=ccc60d71-bafc-49be-a848-a8d09e838592, name=wyf)
```

图 9-7

5．Kafka Streams

Kafka Streams 是 Kafka 提供的用来构建高效实时的流式应用和微服务的客户端库，应用的输入和输出数据都存储在 Kafka 集群中。

Kafka Streams 可以从 Kafka Topic 中消费、分析或转换数据，还可以将处理后的数据发送给另外的 Topic。

（1）添加 Kafka Streams 依赖：

```
dependencies {
  implementation 'org.springframework.boot:spring-boot-starter'
  implementation 'org.springframework.kafka:spring-kafka'
  implementation 'org.apache.kafka:kafka-streams' // Kafka Streams 的依赖包
  implementation 'com.fasterxml.jackson.core:jackson-databind'
  compileOnly 'org.projectlombok:lombok'
  annotationProcessor 'org.projectlombok:lombok'
  //……
}
```

（2）外部配置：

```
spring:
  kafka:
    bootstrap-servers:
      - localhost:9092
    producer:
      value-serializer: org.springframework.kafka.support.serializer.JsonSerializer
    consumer:
      value-deserializer:
org.springframework.kafka.support.serializer.JsonDeserializer
      group-id: consumers
      properties:
        spring:
```

```
      json:
        trusted:
          packages: top.wisely.learningkafka.messaging
  streams: #a
    application-id: kafka-streams-demo #b
    properties:
      default:
        key:
          serde: org.apache.kafka.common.serialization.Serdes$StringSerde #c
        value:
          serde: org.springframework.kafka.support.serializer.JsonSerde #d
      spring:
        json:
          trusted:
            packages: top.wisely.learningkafka.messaging #e
```

a. 通过 spring.kafka.stream.*配置 Kafka Streams。

b. 必须配置 Kafka Streams 应用的 id，可使用 spring.application.name 替代。

c. Kafka Streams 涉及输入和输入，serde 包含了 serializer 和 deserializer。本句是默认的输入和输出 Key 的序列化和反序列化。

d. 本句是默认的输入和输出 Value 的序列化和反序列化。

e. 将包设置为受信任的包，不能和 consumer 的受信任包共用，需单独设置。

（3）示例：Kafka Stream 任务：

```
@SpringBootApplication
@EnableScheduling
@EnableKafkaStreams //a
public class LearningKafkaApplication {

  public static void main(String[] args) {
    SpringApplication.run(LearningKafkaApplication.class, args);
  }

  @Bean
  NewTopic topic(){
    return new NewTopic("confirm-topic",1 ,(short) 1);
  }

  @Bean
  public KStream<String, MessageEvent> kStream(StreamsBuilder streamsBuilder){ //b
    KStream<String, MessageEvent> stream = streamsBuilder.stream("my-topic");//c
    stream.map((key, value) -> {
      value.setName(value.getName().toUpperCase());
      return new KeyValue<>(key,value); //d
    }).to("another-topic"); //e
    return stream;
  }
}
```

a. 通过@EnableKafkaStreams 开启对 Kafka Streams 的支持，配置了 defaultKafkaStreamsBuilder 的 Bean。

b. 可直接注入 streamsBuilder 的 Bean。

c. 使用 streamsBuilder.stream("my-topic")，从 my-topic 中获得 KStream 输入。KStream 代表着一系列的 Key Value 对。

d. 这里的处理很简单，将 MessageEvent 里的 name 转换成大写。

e. 将 KStream 结果输出到 another-topic。

（4）处理 another-topic 中的数据。

发送数据的代码不变，下面是接收数据的代码：

```
@Component
public class MessageConsumer {

    @KafkaListener(topics = {"my-topic"})
    @SendTo("confirm-topic")
    public String consume(MessageEvent event){
        System.out.println("在 consume 方法中处理事件" + event);
        return "接收到了:" + event.toString();
    }

    @KafkaListener(topics = {"another-topic"})
    public void consumeAnother(MessageEvent event){
        System.out.println("在 consumeAnother 方法中处理 another-topic 中的事件" + event);
    }
}
```

（5）启动应用，如图 9-8 所示。

```
在consume方法中处理事件MessageEvent(id=29cc69db-bf61-4f7d-b1d8-4c112e948839, name=wyf)
接收到了:MessageEvent(id=29cc69db-bf61-4f7d-b1d8-4c112e948839, name=wyf)
在consumeAnother方法中处理another-topic中的事件MessageEvent(id=29cc69db-bf61-4f7d-b1d8-4c112e948839  name=WYF)
在consume方法中处理事件MessageEvent(id=f85b43a1-3ca1-4425-b8e3-46a758a79bd8, name=wyf)
接收到了:MessageEvent(id=f85b43a1-3ca1-4425-b8e3-46a758a79bd8, name=wyf)
在consumeAnother方法中处理another-topic中的事件MessageEvent(id=f85b43a1-3ca1-4425-b8e3-46a758a79bd8  name=WYF)
在consume方法中处理事件MessageEvent(id=9a2808cf-3ab2-4a2c-817a-6317a9a7135e, name=wyf)
接收到了:MessageEvent(id=9a2808cf-3ab2-4a2c-817a-6317a9a7135e, name=wyf)
在consumeAnother方法中处理another-topic中的事件MessageEvent(id=9a2808cf-3ab2-4a2c-817a-6317a9a7135e  name=WYF)
在consume方法中处理事件MessageEvent(id=d72d696b-563b-41b9-b067-283b9051dd20, name=wyf)
接收到了:MessageEvent(id=d72d696b-563b-41b9-b067-283b9051dd20, name=wyf)
在consumeAnother方法中处理another-topic中的事件MessageEvent(id=d72d696b-563b-41b9-b067-283b9051dd20  name=WYF)
在consume方法中处理事件MessageEvent(id=2e3e606e-34cb-4abb-96c0-7b01700ac99e, name=wyf)
接收到了:MessageEvent(id=2e3e606e-34cb-4abb-96c0-7b01700ac99e, name=wyf)
在consumeAnother方法中处理another-topic中的事件MessageEvent(id=2e3e606e-34cb-4abb-96c0-7b01700ac99e  name=WYF)
在consume方法中处理事件MessageEvent(id=c6cf7b31-14a1-4f43-b175-c0a9da602502, name=wyf)
接收到了:MessageEvent(id=c6cf7b31-14a1-4f43-b175-c0a9da602502, name=wyf)
在consumeAnother方法中处理another-topic中的事件MessageEvent(id=c6cf7b31-14a1-4f43-b175-c0a9da602502  name=WYF)
在consume方法中处理事件MessageEvent(id=a6a11495-7a5b-48c8-aecc-d408e3f416ff, name=wyf)
接收到了:MessageEvent(id=a6a11495-7a5b-48c8-aecc-d408e3f416ff, name=wyf)
在consumeAnother方法中处理another-topic中的事件MessageEvent(id=a6a11495-7a5b-48c8-aecc-d408e3f416ff  name=WYF)
在consume方法中处理事件MessageEvent(id=84916134-4f43-43c2-af6c-68a4884a1517, name=wyf)
接收到了:MessageEvent(id=84916134-4f43-43c2-af6c-68a4884a1517, name=wyf)
在consumeAnother方法中处理another-topic中的事件MessageEvent(id=84916134-4f43-43c2-af6c-68a4884a1517  name=WYF)
在consume方法中处理事件MessageEvent(id=2b2ffea6-963f-4b73-b2b2-520f470e8a94, name=wyf)
接收到了:MessageEvent(id=2b2ffea6-963f-4b73-b2b2-520f470e8a94, name=wyf)
在consumeAnother方法中处理another-topic中的事件MessageEvent(id=2b2ffea6-963f-4b73-b2b2-520f470e8a94  name=WYF)
```

图 9-8

9.4 Websocket

在 HTTP 下，可以通过 Websocket 对服务端和客户端进行全双工通信，即客户端和服务端都可通过通道直接向彼此发送数据。

9.4.1 STOMP Websocket

可以在 Websocket 之上使用 STOMP（Simple/Streaming Text Oriented Message Protocol）进行交互。当使用 STOMP 时，Spring Boot Websocket 应用的其中一个角色会作为所有连接客户端的消息代理，在生产中，一般使用支持 STOMP 的第三方消息代理（RabbitMQ）作为专门的消息代理。

1．新建应用

新建应用，信息如下：

Group：top.wisely。

Artifact：learning-websocket。

Dependencies：Websocket 、Spring Security 和 Lombok。

build.gradle 文件中的依赖如下：

```
dependencies {
    implementation 'org.springframework.boot:spring-boot-starter-websocket'
    implementation 'org.springframework.boot:spring-boot-starter-security'
    compileOnly 'org.projectlombok:lombok'
    annotationProcessor 'org.projectlombok:lombok'
  //……
}
```

2．示例

（1）配置 STOMP Websocket：

```
@Configuration
@EnableWebSocketMessageBroker //a
public class WebsocketConfig implements WebSocketMessageBrokerConfigurer { //b
   @Override
   public void configureMessageBroker(MessageBrokerRegistry registry) { //c
      registry.enableSimpleBroker("/topic"); //d
      registry.setApplicationDestinationPrefixes("/app"); //e
   }

   @Override
   public void registerStompEndpoints(StompEndpointRegistry registry) {
      registry.addEndpoint("/endpoint").withSockJS(); //f
   }

}
```

a. @EnableWebSocketMessageBroker 开启 Websocket 消息代理的支持，包含配置了消息发送模板 SimpMessagingTemplate 的 Bean。

b. 通过实现 WebSocketMessageBrokerConfigurer 接口并覆写其方法配置 Websocket 消息代理。

c. 通过 configureMessageBroker 方法配置 Websocket 消息代理。

d. 配置消息代理的终点，客户端可订阅监听终点获取信息。

e. 配置消息处理器（@MessageMapping 注解的方法）的前缀。

f. WebSocket 的端点地址，提供 SockJS 后备支持。

（2）配置安全:

```
@Configuration
public class SecurityConfig extends WebSecurityConfigurerAdapter {

   @Bean
   PasswordEncoder passwordEncoder(){
      return new BCryptPasswordEncoder();
   }
   @Override
   protected void configure(AuthenticationManagerBuilder auth) throws Exception {//a
      auth.inMemoryAuthentication()
               .withUser("wyf")
               .password(passwordEncoder().encode("111111"))
               .roles("USER")
            .and()
               .withUser("admin")
               .password(passwordEncoder().encode("admin"))
               .roles("ADMIN");
   }

   @Override
   protected void configure(HttpSecurity http) throws Exception {//b
      http.authorizeRequests()
            .anyRequest().authenticated()
            .and()
         .formLogin();
   }
}
```

a. 配置内存中的用户 wyf 和 admin。

b. 配置页面登录。

（3）消息处理:

```
@Controller
public class ChatController {

   SimpMessagingTemplate simpMessagingTemplate; //a
```

```
    public ChatController(SimpMessagingTemplate simpMessagingTemplate) {
        this.simpMessagingTemplate = simpMessagingTemplate;
    }

    @MessageMapping("/chat/{to}") //b
    @SendTo("/topic/status") //c
    public String chat(Principal principal,
                    @RequestBody String msg,
                    @DestinationVariable  String to){ //d
        simpMessagingTemplate.convertAndSendToUser(to,"/topic/response", msg); //e
        return principal.getName() + " 发送的 " + msg + " 消息已送达给" + to;
    }

    @RequestMapping("/user")
    @ResponseBody
    public String user(Principal principal){ //f
        return principal.getName();
    }
}
```

a. 注入发送消息的 simpMessagingTemplate 的 Bean。

b. Spring Messaging 的注解，用来映射终点和处理方法，类似于@RequestMapping；客户端可以向此路径发送消息。

c. 这里通过@SendTo 把消息发送到终点/topic/status，所有监听该终点的客户端都可订阅接收。和前面一样，发送的消息内容为方法返回值。

d. 通过@DestinationVariable 获取路径变量。

e. 通过 convertAndSendToUser 向指定用户、指定终点发送消息，被发送用户订阅/user/topic/response 即可获得消息；若不指定用户，则可使用 convertAndSend 方法发送。

f. 可通过注入 Principal 对象获得用户信息。

（4）客户端页面:

```
<!DOCTYPE html>
<html lang="en">
<head>
    <meta charset="UTF-8">
    <title>聊天室</title>
    <script src="https://cdn.bootcss.com/jquery/3.4.1/jquery.min.js"></script>
    <script src="https://cdn.bootcss.com/stomp.js/2.3.3/stomp.min.js"></script>
    <script src="https://cdn.bootcss.com/sockjs-client/1.3.0/sockjs.min.js"></script>
</head>
<body>
<div id="connect">
    <button onclick="connect()">连接</button>
    <button onclick="disconnect()">关闭</button>
</div>
```

```
<div id="input">
   <input id="msg" type="text"/>
   <button onclick="send()">发送</button>
</div>
<div id="response"></div>
<div id="status"></div>

</body>

<script>
   var stompClient = null;
   var from = null;
   var to = null;
   getUser();
   function getUser(){ //获取当前用户和聊天对象
      $.get('/user',function (data) {
         from = data;
         if (data == 'wyf'){
            to = 'admin';
         } else {
            to = 'wyf'
         }
         alert('当前用户为:' + from + ',' +"聊天对象为: " + to);
      })
   }

   function connect() { //连接
      var socket = new SockJS('/endpoint'); //a
      stompClient = Stomp.over(socket);
      stompClient.connect({}, function(frame) {
         alert("连接成功");
         stompClient.subscribe('/user/topic/response', function(messageOutput) { //b
            $("#response").append(messageOutput.body + '<br>');
         });

         stompClient.subscribe('/topic/status', function(messageOutput) { //c
           $("#status").html(messageOutput.body);
         });

      });
   }

   function send() {
      msg = $("#msg").val()
      stompClient.send("/app/chat/" + to,{},msg); //d
   }
   function disconnect() {
      stompClient.disconnect();
      alert("关闭成功");
   }
```

```
</script>
</html>
```

a. 此处填写的地址为服务端设置的断点。

```
registry.addEndpoint("/endpoint").withSockJS()
```

b. 订阅用户消息，和服务端对应。

```
simpMessagingTemplate.convertAndSendToUser(to,"/topic/response", msg)
```

c. 订阅广播消息，对应服务端。

```
@SendTo("/topic/status")。
```

d. 发送消息到消息处理器。

前缀配置：

```
registry.setApplicationDestinationPrefixes("/app")
```

映射配置：

```
@MessageMapping("/chat/{to}")
```

（5）启动应用。

用 Chrome 打开 http://localhost:8080/chat.html；再在 Chrome 浏览器上“以访客身份打开一个窗口”，同时访问 http://localhost:8080/chat.html。此时都要求登录系统，两边分别以 wyf 和 admin 用户登录成功后如图 9-9 所示。

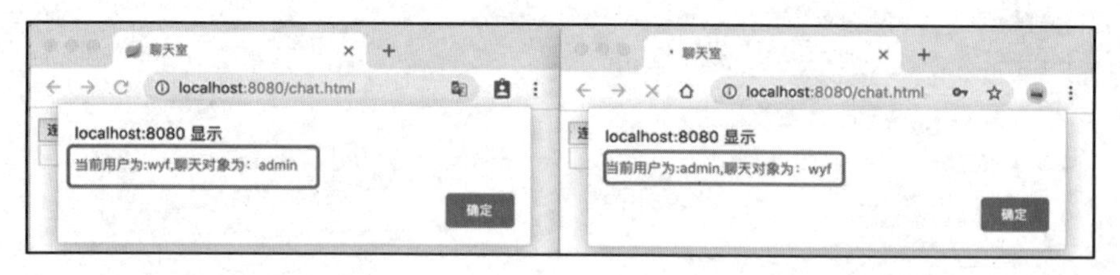

图 9-9

两边分别单击“确定”按钮，提示连接成功后，在左边窗口输入“你好”，如图 9-10 所示。

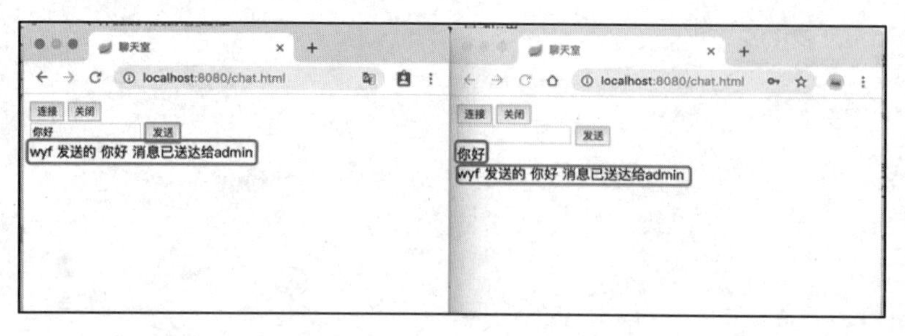

图 9-10

左边的信息是所有用户都能收到的信息；右边窗口的上面是左边用户发送的信息，下面是都能收到的信息。在右边发送“你也好”，如图 9-11 所示。

图 9-11

9.4.2 Reactive Websocket

本节对响应式的 Websocket 进行简单的演示。

1．新建应用

新建应用，信息如下。

Group：top.wisely。

Artifact：learning-websocket-reactive。

Dependencies：Spring Reactive Web 和 Lombok。

build.gradle 文件中的依赖如下：

```
dependencies {
    implementation 'org.springframework.boot:spring-boot-starter-webflux'
    implementation 'javax.websocket:javax.websocket-api' //添加 Websocket API
    compileOnly 'org.projectlombok:lombok'
    annotationProcessor 'org.projectlombok:lombok'
  //……
}
```

2．示例

（1）配置响应式 Websocket：

```
@Configuration
public class WebsocketConfig {
   @Bean
   public HandlerMapping handlerMapping() {//a
      Map<String, WebSocketHandler> map = new HashMap<>();
      map.put("/hello", new HelloHandler());

      SimpleUrlHandlerMapping mapping = new SimpleUrlHandlerMapping();
      mapping.setUrlMap(map);
      mapping.setOrder(-1); // before annotated controllers
```

```
        return mapping;
    }

    @Bean
    public HandlerAdapter handlerAdapter() {
        return new WebSocketHandlerAdapter();//b
    }
}
```

与 Spring MVC 和 WebFlux 的处理一样，配置主要由两个类组成。

a. HandlerMapping：负责地址和处理方法之间的映射；这里使用的是 SimpleUrlHandlerMapping，将地址/hello 与处理类 HelloHandler 对应。

b. HandlerAdapter：负责实际的方法调用处理；这里使用 WebSocketHandlerAdapter，可以把 WebSocketHandler 的实现类作为处理方法。

（2）处理方法：

```
public class HelloHandler implements WebSocketHandler {
    @Override
    public Mono<Void> handle(WebSocketSession session) { //a
        return session.send(session.receive()
                .map(msg -> "Hello:" + msg.getPayloadAsText()) //b
                .map(session::textMessage)
        );
    }
}
```

a. WebSocketSession 对象使用 send 方法发送消息，使用 receive 方法接收消息。

b. 给客户端传来的消息添加 Hello:并发回给客户端。

（3）客户端页面：

```
<!DOCTYPE html>
<html lang="en">
<head>
    <meta charset="UTF-8">
    <title>Hello</title>
    <script src="https://cdn.bootcss.com/jquery/3.4.1/jquery.min.js"></script>
    <script src="https://cdn.bootcss.com/stomp.js/2.3.3/stomp.min.js"></script>
</head>
<body>
<div>
    <button onclick="connect()">连接</button>
    <button onclick="disconnect()">关闭</button>

</div>
<div>
<input type="text" id="msg"/>
<button onclick="send()">发送</button>
```

```
</div>
<div id="response"></div>
</body>
<script>
   var ws = null;
   var url = "ws://localhost:8080/hello";
   function connect() {
      ws = new WebSocket(url);
      ws.onopen = function() {
         alert("连接成功")
      };
      ws.onmessage = function(event) { //监听服务端返回
         $("#response").append(event.data + '<br>');
      };

      ws.onclose = function(event) {
        alert("关闭成功")
      };
   }

   function disconnect() {
         ws.close();
   }

   function send() { //发送消息到服务端
      if (ws != null) {
         ws.send($("#msg").val());
      } else {
         alert('连接已关闭，请重连');
      }
   }

</script>
</html>
```

（4）启动应用。

访问 http://localhost:8080/hello.html，单击“连接”按钮后，在输入框输入消息，如图 9-12 所示。

图 9-12

9.5 RSocket

RSocket 是一个使用在字节流传输（TCP/Websocket）之上的二进制点对点通信协议；它主要用在分布式应用上，用来替换如 HTTP 这种通信协议。

RSocket 提供 4 种交互模型。

- ◎ request/response：请求返回一条流数据（Mono<T>）。
- ◎ request/stream：请求返回多条流数据（Flux<T>）。
- ◎ fire-and-forget：请求不返回数据（Mono<Void>）。
- ◎ channel：双向流数据通信（Flux<R> method(Publisher<T> input)）。

9.5.1 新建应用

1. Server

（1）新建应用，信息如下。

Group：top.wisely。

Artifact：rsocket-server。

Dependencies：RSocket 、Spring Data Reactive MongoDB 和 Lombok。

build.gradle 文件中的依赖如下：

```
dependencies {
    implementation 'org.springframework.boot:spring-boot-starter-rsocket'
    implementation
'org.springframework.boot:spring-boot-starter-data-mongodb-reactive'
    compileOnly 'org.projectlombok:lombok'
    annotationProcessor 'org.projectlombok:lombok'
  //……
    }
}
```

2. Client

新建应用，信息如下。

Group：top.wisely。

Artifact：rsocket-client。

Dependencies：RSocket 、Spring Reactive Web 的 Lombok。

build.gradle 文件中的依赖如下：

```
dependencies {
    implementation 'org.springframework.boot:spring-boot-starter-webflux'
    implementation 'org.springframework.boot:spring-boot-starter-rsocket'
    compileOnly 'org.projectlombok:lombok'
    annotationProcessor 'org.projectlombok:lombok'
```

```
//……
    }
}
```

9.5.2　Spring Boot 的自动配置

（1）服务端的配置如下。

◎ RSocketServerAutoConfiguration：通过 RSocketProperties，使用 spring.rsocket.server.* 配置 RSocket Server。

◎ RSocketStrategiesAutoConfiguration：使用 CBOR 和 Jackson 配置数据交互的编码和解码。

◎ RSocketMessagingAutoConfiguration：在 Spring Messaging 中配置对 Spring RSocket 的支持。

（2）客户端配置。

RSocketRequesterAutoConfiguration：配置一个 RSocketRequester.Builder 的 Bean，用来定义 RSocketRequester。

9.5.3　示例

1．Server

（1）配置 RSocket Server，连接 MongoDB：

```
spring:
 rsocket:
  server:
   address: localhost
   port: 9898
   transport: tcp #使用 TCP 监听 localhost 的 9898 端口
 data:
  mongodb:
   host: localhost
   port: 27017
   username: wisely
   password: zzzzzz
   database: first_db
```

（2）领域模型和 Repository：

```
@Data
@AllArgsConstructor
@NoArgsConstructor
@Document(collection = "people")
public class Person {
   @Id
```

```
    private String id;
    private String name;
    private Integer age;
    public Person(String name, Integer age) {
        this.name = name;
        this.age = age;
    }

}
public interface PersonRepository extends ReactiveMongoRepository<Person, String> {
}
```

（3）RSocket Server。

RSocket 是由 Spring Messaging 集成的，同样使用@MessageMapping 定义消息的终点：

```
@Controller
public class PersonController {

    PersonRepository personRepository;

    public PersonController(PersonRepository personRepository) {
        this.personRepository = personRepository;
    }

    @MessageMapping("people.findById")
    Mono<Person> getOne(Person person){ //a
        return personRepository.findById(person.getId());
    }

    @MessageMapping("people.findAll")
    Flux<Person> all(Person person){ //b
        return personRepository.findAll();
    }

    @MessageMapping("people.deleteById")
    Mono<Void> delete(Person person){ //c
        return personRepository.deleteById(person.getId());
    }

    @MessageMapping("people.save")
    Flux<Person> save(Publisher<Person> people){ //d
        return personRepository.saveAll(people);
    }

}
```

a. 演示 request/response。

b. 演示 request/stream。

c. 演示 fire-and-forget。

d. 演示 channel，双向流数据。

（4）添加演示数据：

```
@Bean
CommandLineRunner initPersonData(PersonRepository personRepository){
  return args -> {
    personRepository.deleteAll().subscribe();
    personRepository.save(new Person("wyf", 35)).subscribe();
    personRepository.save(new Person("foo", 34)).subscribe();
    personRepository.save(new Person("bar", 36)).subscribe();
  };
}
```

2. Client

在客户端，使用 RSocketRequester 作为客户端调用 RSocket Server，类似于 Spring WebFlux 的 WebClient 或 Spring MVC 的 RestTemplate。

（1）配置 RSocketRequester 的 Bean：

```
@Bean
RSocketRequester rSocketRequester(RSocketRequester.Builder builder){
  return builder
       .connectTcp("localhost", 9898) //连接服务端地址和端口
       .block();
}
```

（2）和服务端共享数据模型：

```
@Data
@AllArgsConstructor
@NoArgsConstructor
public class Person {
   private String id;
   private String name;
   private Integer age;

   public Person(String name, Integer age) {
      this.name = name;
      this.age = age;
   }

   public Person(String id) {
      this.id = id;
   }
}
```

（3）使用 RSocketRequester。

在 Spring WebFlux 中，使用 RSocketRequester 调用 RSocket Server：

```
@RestController
@RequestMapping("/people")
public class ClientPersonController {
   RSocketRequester rSocketRequester;

   public ClientPersonController(RSocketRequester rSocketRequester) {
      this.rSocketRequester = rSocketRequester;
   }
   @GetMapping("/{id}")
   public Mono<Person> getOne(@PathVariable String id){
      return this.rSocketRequester
             .route("people.findById") //a
             .data(new Person(id)) //b
             .retrieveMono(Person.class); //c
   }
   @GetMapping
   public Flux<Person> getAll(){
      return this.rSocketRequester
             .route("people.findAll")
             .data(new Person())
             .retrieveFlux(Person.class);//d
   }

   @DeleteMapping("/{id}")
   public Mono<Void> delete(@PathVariable String id){
      return this.rSocketRequester
             .route("people.deleteById")
             .data(new Person(id))
             .send(); //e
   }

   @PostMapping
   public Flux<Person> save(@RequestBody Flux<Person> personFlux){
      return this.rSocketRequester
             .route("people.save")
             .data(personFlux, Person.class)
             .retrieveFlux(Person.class);
   }
}
```

a. 通过 route()方法指定服务端的终点。

b. 通过 data()方法向服务端传送数据。

c. 通过 retrieveMono()方法获取返回值是 Mono。

d. 通过 retrieveFlux 方法获取返回值是 Flux。

e. 通过 send()方法发送数据而不关心返回。

（4）使用 WebClient 调用 WebFlux：

```
@Component
public class ControllerClient {
    WebClient webClient;

    public ControllerClient(WebClient.Builder builder) {
        this.webClient = builder.build();
    }

    public void getOne(){
        System.out.println("查询一条数据");
        Mono<Person> mono = webClient
                .get()
                    //此处id由查询MongoDB得到
                .uri("http://localhost:8080/people/{id}", "5d03608320802b1b10458227")
                .retrieve()
                .bodyToMono(Person.class);
        mono.subscribe(System.out::println);
    }

    public void getAll(){
        System.out.println("查询所有");
        Flux<Person> flux = webClient
                .get()
                .uri("http://localhost:8080/people")
                .retrieve()
                .bodyToFlux(Person.class);
        flux.subscribe(System.out::println);
    }

    public void delete(){
        System.out.println("删除一条数据");
        Mono<Void> mono = webClient
                .delete()
                .uri("http://localhost:8080/people/{id}", "5d03608320802b1b10458227")
                .retrieve()
                .bodyToMono(Void.class);
        mono.subscribe();
    }

    public void save(){
        System.out.println("新增多个");
        List<Person> people = Arrays.asList(new Person("aaa", 36),
                                    new Person("bbb", 37),
                                    new Person("ccc", 38));
        Flux<Person> flux = webClient
                .post()
                .uri("http://localhost:8080/people")
                .body(Flux.fromIterable(people), Person.class)
                .retrieve()
```

```
            .bodyToFlux(Person.class);
        flux.subscribe(System.out::println);
    }

}
```

（5）启动调用：

```
@Bean
CommandLineRunner webclient(ControllerClient client){
    return args -> {
        client.getOne();
        Thread.sleep(1000);
        client.getAll();
        Thread.sleep(1000);
        client.delete();
        Thread.sleep(1000);
        client.save();
        Thread.sleep(1000);
        client.getAll();
    };
}
```

3. 验证

启动 rsocket-server，自动初始化三条 Person 数据到 MongoDB。再次启动 rsocket-client，如图 9-13 所示。

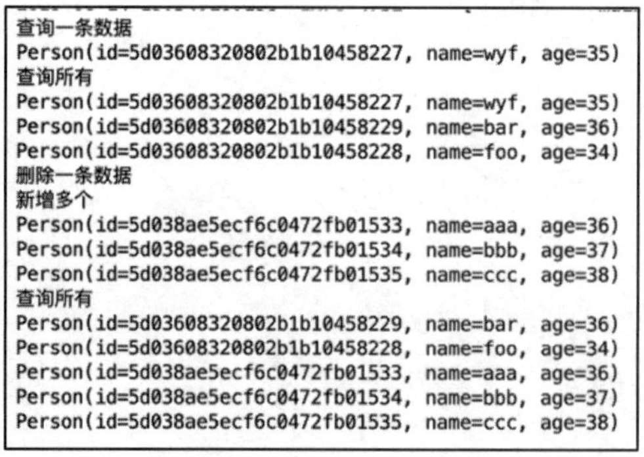

```
查询一条数据
Person(id=5d03608320802b1b10458227, name=wyf, age=35)
查询所有
Person(id=5d03608320802b1b10458227, name=wyf, age=35)
Person(id=5d03608320802b1b10458229, name=bar, age=36)
Person(id=5d03608320802b1b10458228, name=foo, age=34)
删除一条数据
新增多个
Person(id=5d038ae5ecf6c0472fb01533, name=aaa, age=36)
Person(id=5d038ae5ecf6c0472fb01534, name=bbb, age=37)
Person(id=5d038ae5ecf6c0472fb01535, name=ccc, age=38)
查询所有
Person(id=5d03608320802b1b10458229, name=bar, age=36)
Person(id=5d03608320802b1b10458228, name=foo, age=34)
Person(id=5d038ae5ecf6c0472fb01533, name=aaa, age=36)
Person(id=5d038ae5ecf6c0472fb01534, name=bbb, age=37)
Person(id=5d038ae5ecf6c0472fb01535, name=ccc, age=38)
```

图 9-13

9.6 小结

本章介绍了使用 Spring 支持的消息机制进行事件驱动开发，内容包括 JMS、RabbitMQ、Kafka、Websocket 和 RSocket，通过消息代理（不含响应式 Websocket、RSocket）进行事件驱动开发降低了分布式系统的耦合度，并让应用的扩展性大大提高。

第 10 章 系统集成与批处理

10.1 Spring Integration

Spring Integration 使用 Spring 编程模型来支持**企业集成模式**（Enterprise Integration Patterns）。企业集成模式一般用来解决在开发中遇到的系统集成问题，集成模式主要有 4 种。

◎ 文件传输：应用输出文件，其他应用读取文件，从而达到数据共享。同样，应用也可读取其他应用输出的文件。

◎ 共享数据库：存储需要共享的数据的数据库。

◎ 远程过程调用：应用将自己的功能暴露出来，这样其他应用即可远程调用它并交换数据。

◎ 消息机制：使用消息系统将应用连接起来，应用之间通过消息系统来交换数据和调用行为。

10.1.1 Spring Integration 基础

在理解 Spring Integration 之前，需要理解 4 个概念，分别为 Integration Flow、Message、Message Channel 和 Message Endpoint。

（1）Integration Flow：集成流程，使用 Message、Message Channel 和 Message Endpoint 进行消息处理。

（2）Message：消息，包含消息头（Header）和消息体（Payload）。

```
package org.springframework.messaging;

public interface Message<T> {
  T getPayload();
```

```
  MessageHeaders getHeaders();
}
```

（3）Message Channel：消息通道，通过消息通道将消息从发布者（提供者）传输给订阅者（订阅者）。发布者发布消息到消息通道，订阅者从消息通道中接收消息。

```
package org.springframework.messaging;

@FunctionalInterface
public interface MessageChannel {
  long INDEFINITE_TIMEOUT = -1;
  default boolean send(Message<?> message) {
    return send(message, INDEFINITE_TIMEOUT);
  }
  boolean send(Message<?> message, long timeout);
}
```

MessageChannel 是函数接口，只有一个发送消息的方法 send()。若消息发送成功，则返回值为 true；若消息发送失败，则返回值为 false。

MessageChannel 分为两类：PollableChannel 和 SubscribableChannel。

PollableChannel：通过轮询获取消息。PollableChannel 接口的主要实现有以下 3 种。

◎ QueueChannel：消息都放置在队列（先进先出）的通道中。

◎ PriorityChannel：它是 QueueChannel 的子类，消息的优先级基于 Comparator，而不是先进先出，默认的比较器为消息头中的 priority 头。

◎ RendezvousChannel：它也是 QueueChannel 的子类，但是这个队列的容量是 0；发布者会阻碍消息直至订阅者接收到消息。

SubscribableChannel：消息通道维护注册订阅者，并通知它们通过通道处理消息；SubscribableChannel 接口的主要实现有以下 3 种。

◎ PublishSubscribeChannel：通过此通道将消息发送给多个订阅者；

◎ DirectChannel：通过此通道将消息发送给单个订阅者；消息发布者和订阅者使用相同的线程。

◎ ExecutorChannel：和 DirectChannel 类似，但消息的分发是由 Executor 负责的，发布者和订阅者不在同一个线程。

Spring Integration 还提供了响应式的消息通道 FluxMessageChannel。在使用 Spring Integration Java DSL 进行开发时，输入通道是自动创建的，且默认为 DirectChannel。

（4）Message EndPoint：消息端点是从消息通道中获取消息的组件；端点需指定输入通道。若需要将消息发送给另外的端点，则需要配置输出通道。Spring Integration 提供的主要端点有以下 8 种。

◎ Filter：基于条件过滤消息并传向输出通道。

- ◎ Transformer：将从输入通道获得的消息转换成其他格式，然后发送给输出通道。
- ◎ Router：将从输入通道获得的消息路由给一个或多个不同的输出通道。路由规则一般是基于消息头来定的。
- ◎ Splitter：将从输入通道获得的消息拆分成多个消息，然后把它们分别发送给不同的输出通道。
- ◎ Aggregator：将从多个输入通道传来的消息合并成一个消息，然后把这个合并后的消息发送给输出通道。
- ◎ Service Activator：将从输入通道获得的消息发送给 Java 方法进行处理，然后将消息发送给输出通道。
- ◎ Channel Adapter：将通道连接到外部的系统或协议，既可输入消息，也可输出消息，它是集成流程的入口或出口。
  - ➢ Inbound Channel Adapter（集成流程入口）：从外部系统获取消息，并将它放置在通道。
  - ➢ Outbound Channel Adapter（集成流程出口）：从通道获取消息，并将它发送给外部系统。

Channel Adapter 是由 Spring Integration 的端点模块提供的，包含 AMQP、Spring Application Event、Feed、File、FTP、Gemfire、HTTP、JDBC、JMS、JMX、JPA、Mail、MongoDB、MQTT、MQTT、Redis、Resource、RMI、SFTP、STOMP、Stream、Syslog、TCP/UDP、Web Services、Web Sockets、XMPP 和 ZooKeeper，它们的依赖一般都为 spring-integration-名称。例如，File 的依赖为 spring-integration-file，Spring Application Event 的依赖为 spring-integration-event。特别地，TCP/UDP 的依赖为 spring-integration-ip。另外，每个模块还提供了工厂类来构造 Inbound Channel Adapter 和 Outbound Channel Adapter，如 Files.inboundAdapter()和 Files.outboundAdapter，其余的模块类似。

- ◎ Gateway：代码可分为两部分：业务代码和集成流程（Integration Flow）代码。业务代码通过定义的接口（Gateway）来调用集成流程代码。Gateway 通过指定请求通道（request channel）向集成流程发送消息，通过应答通道（reply channel）获取结果。

10.1.2 Spring Integration Java DSL

本节使用 Spring Integration Java DSL 来配置集成流程。通过注册 IntegrationFlow 的 Bean 来定义一个集成流程。一个集成流程通常从 IntegrationFlows.from()方法开始，它可以获得 IntegrationFlowBuilder 来构造 IntegrationFlow。from()方法支持多种形式的开始，例如，可从指定消息通道、Inbound Channel Adapter 或 Gateway 等开始。

Java DSL 提供了与 Spring Integration 的管道、端点对应的方法，具体如下。

◎ transform()：Transformer。

◎ filter()：Filter。

◎ handle()：Service Activator 或 Outbound Channel Adapter。

◎ split()：Splitter。

◎ aggregate()：Aggregator。

◎ route()：Router。

Spring Boot 的自动配置

Spring Boot 对 Spring Integration 的自动配置为 IntegrationAutoConfiguration，它的主要工作是使用@EnableIntegration 开启对 Spring Integration 的支持。

10.1.3 示例

1. 新建应用

新建应用，信息如下。

Group：top.wisely。

Artifact：learning-spring-integration。

Dependencies：Spring Integration 。

build.gradle 文件中的依赖如下：

```
dependencies {
    implementation 'org.springframework.boot:spring-boot-starter-integration'
    implementation 'org.springframework.integration:spring-integration-file' //file 模块
    implementation
('org.springframework.integration:spring-integration-mail'){ //mail 模块
        exclude module: 'jakarta.activation-api' //在 JDK9+下集成 Java Mail
    }
    implementation 'com.sun.activation:jakarta.activation:1.2.1' //在 JDK9+下集成//Java Mail
    implementation 'com.sun.mail:javax.mail:1.6.2' //JDK9+下集成 Java Mail
    //……
}
```

在本例中，通过 Gateway 向集成流程发送消息。这些消息一部分输出到文件，另外一部分通过邮件发送至指定邮箱。

2. Gateway 定义

```
public interface SendingGateway {
   void send(@Header(FileHeaders.FILENAME) String filename, String content);
}
```

Gateway 是一个接口，Spring Integration 会在运行时自动生成接口的实现，send()方法的数据会传送到集成流程中。第一个参数以通过@Header 注解来指定文件名，第二个参数为消息本身。

3. 主集成流程定义

```
@Configuration
public class IntegrationConfig {
   @Bean
   public IntegrationFlow flowFormGateway() { //a
      return IntegrationFlows.from(SendingGateway.class) //b
            .<String> filter(payload -> payload.startsWith("h"))//c
            .<String,String> transform(payload -> payload + "!") //d
            .<String, Boolean> route(payload -> payload.startsWith("hello"), //e
                  mapping -> mapping.channelMapping(true, "fileChannel") //f
                              .channelMapping(false, "emailChannel")) //g
            .get(); //h
   }
         //……
}
```

a. 通过定义 IntegrationFlow 的 Bean 定义集成流程。

b. 通过 from()方法，从 Gateway 开始集成流程。通过已经获得的 IntegrationFlowBuilder 的方法来构建 IntegrationFlow。

c. 通过 filter()方法过滤掉非 h 开头的字符，<String>指的是当前消息的类型。当前端点到下一端点 Transformer 之间的通道可省略。

d. 通过 transform()来转换消息，在所有的消息后加!。<String,String>：第一个 String 是原消息类型，第二个 String 是转换后的消息类型。

e. 通过 route()方法将消息路由到不通的通道，第一个参数为路由条件。

f. 把以 hello 开头的消息发送到 fileChannel 通道。

g. 把不以 hello 开头的消息发送到 emailChannel 通道。

h. 通过 get()方法获得 IntegrationFlow。

4. 写文件流程定义

```
@Bean
IntegrationFlow flowFromChannelToFile(){
  return IntegrationFlows.from("fileChannel") //a
    .handle(Files.outboundAdapter(new File("/Users/wangyunfei/file")) //b
          .fileExistsMode(FileExistsMode.APPEND) //c
          .appendNewLine(true)) //d
    .get();
}
```

a. 写文件集成流程从 fileChannel 通道读取消息。

b. 通过 handle()方法调用 Outbound Channel Adapter。通过 Files.outboundAdapter()的构造，将消息输出到文件，构造的参数为指定文件输出的位置。

c. 如果文件存在，则消息会附加到文件尾部。

d. 设置附加到文件的消息会另起一行。

5. 发邮件流程定义

```
@Bean
IntegrationFlow flowFromChannelToMail(){
   return IntegrationFlows.from("emailChannel") //a
         .enrichHeaders(Mail.headers() //b
               .subject("来自 localhost 的信息")
               .from("wisely-man@126.com")
               .to("wisely-man@126.com"))
         .handle(Mail.outboundAdapter("smtp.126.com")//c
               .port(25)
               .protocol("smtp")
               .credentials("wisely-man@126.com", "*******"))
         .get();
}
```

a. 发邮件集成流程从 emailChannel 通道读取消息。

b. 可以通过 enrichHeaders()方法丰富头信息，通过 Mail.headers()来设置头信息。subject()用来指定邮件主题，from()用来指定邮件发送人，to()用来指定接收人。

c. 通过 handle()方法调用 Outbound Channel Adapter。此处设置通过 SMTP 来发送邮件，通过 Mail.outboundAdapter()来构造，参数为服务器的地址。port()指定服务器端口，protocol()指定协议，credentials()指定发送者的账号与密码，读者可设置自己的邮箱来测试。

6. 拷贝文件流程定义

```
@Bean IntegrationFlow copyFileFlow(){ //a
   return IntegrationFlows.from(Files.inboundAdapter(new
File("/Users/wangyunfei/file")).patternFilter("*.txt"),
                              c -> c.poller(Pollers.fixedRate(5, TimeUnit.SECONDS)))
//b
         .enrichHeaders(h -> {
            h.header(FileHeaders.FILENAME, "copy.txt", true); //c
         })
         .handle(Files.outboundAdapter(new File("/Users/wangyunfei/output"))
                  .fileExistsMode(FileExistsMode.REPLACE_IF_MODIFIED)) //d
         .get();
}
```

a. 这是一个新的流程，和上面的流程没有关系，这个流程将/Users/wangyunfei/file 文件夹中的 txt 文件拷贝到/Users/wangyunfei/output。

b. from()方法从 Inbound Channel Adapter 开始。Files.inboundAdapter()方法指定需读取的文件，匹配模式为所有 txt 文件。from()方法的第二个参数为每隔 5 秒轮询这个文件夹。

c. 通过 enrichHeaders()来丰富头信息，通过 FileHeaders.FILENAME 修改文件名。

d. 通过 handle()方法调用 Files.outboundAdapter()，将文件输出到/Users/wangyunfei/output 文件夹。若文件存在且文件已修改，则替换文件。

下面使用 CommandLineRunner 来调用 Gateway：

```
@SpringBootApplication
public class LearningSpringIntegrationApplication {

    public static void main(String[] args) {
        SpringApplication.run(LearningSpringIntegrationApplication.class, args);
    }

    @Bean
    CommandLineRunner commandLineRunner(SendingGateway gateway){
        return args -> {
            gateway.send("greeting.txt","hello world");
            gateway.send("greeting.txt", "hello wyf");
            gateway.send("greeting.txt", "good morning");
            gateway.send("greeting.txt", "hi world");
            gateway.send("greeting.txt", "hi wyf");
        };

    }
}
```

7. 检验结果

（1）运行应用，查看/Users/wangyunfei/file 文件夹，如图 10-1 所示。

（2）登录并查看邮箱，如图 10-2 所示。

（3）查看/Users/wangyunfei/output 文件夹，如图 10-3 所示。

图 10-1

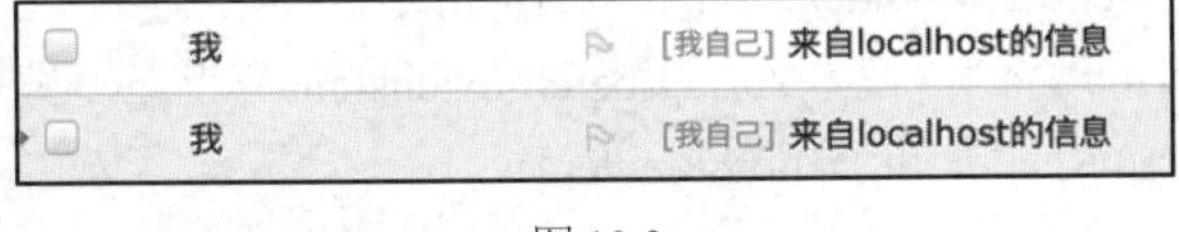

图 10-2

图 10-3

10.2　Spring Batch

Spring Batch 是 Spring 生态下批处理的解决方案。一个典型的 Spring Batch 应用是从一个来源读取大量的数据，经过一定处理后将数据写入另外一个位置。

10.2.1 Spring Batch 的流程

一个 Spring Batch 的流程如下。

任务启动器（JobLauncher）启动任务（Job），让其按照步骤（Step）执行。每个步骤都包含读取数据（ItemReader）、数据预处理（ItemProcessor）和数据写入（ItemWriter）三步。所有关于任务的执行情况都由 JobRepository 存储在数据库中。

在 Spring Batch 中，数据读取使用 ItemReader 接口、数据处理使用 ItemProcessor 接口，数据写入使用 ItemWriter 接口。Spring Batch 的读取、写入的主要实现如下。

◎ JSON：JsonItemReader、JacksonJsonObjectReader、GsonJsonObjectReader 和 JsonFileItemWrite。

◎ JMS：JmsItemReader 和 JmsItemWriter。

◎ AMQP：AmqpItemReader 和 AmqpItemWriter。

◎ Kafka：KafkaItemReader 和 KafkaItemWriter。

◎ Gemfire：GemfireItemWriter。

◎ MongoDB：MongoItemReader 和 MongoItemWriter。

◎ Neo4j：Neo4jItemReader 和 Neo4jItemWriter。

◎ Repository：RepositoryItemReader 和 RepositoryItemWriter。

◎ JDBC：JdbcCursorItemReader、JdbcPagingItemReader、JdbcBatchItemWriter。

◎ Hibernate：HibernateCursorItemReader、HibernatePagingItemReader 和 HibernateItemWriter。

◎ JPA：JpaPagingItemReader 和 JpaItemWriter。

◎ File：FlatFileItemReader 和 FlatFileItemWriter。

◎ Resource：ResourcesItemReader、MultiResourceItemReader 和 MultiResourceItemWriter。

◎ LDIF：LdifReader 和 MappingLdifReader。

◎ Mail：SimpleMailMessageItemWriter 和 MimeMessageItemWriter。

◎ XML：StaxEventItemReader、StaxEventItemWriter。

GitHub（访问 GiHub 官网，搜索 spring-batch-extensions）上还提供了以下 2 种实现。

◎ Excel：PoiItemReader 和 JxlItemReader。

◎ Elasticsearch：ElasticsearchItemReader 和 ElasticsearchItemWriter。

10.2.2 Spring Boot 的自动配置

Spring Boot 为 Spring Batch 提供的自动配置类为 BatchAutoConfiguration，它主要做的工作如下。

◎ 配置初始化存储 Spring Batch 任务执行情况的数据库，默认使用。

◎ 配置启动应用，自动执行 Spring Batch 任务。

Spring Boot 通过 BatchProperties 使用 spring.batch.*配置 Spring Batch。

10.2.3 示例

本节的示例是将 CSV 文件里的数据导入 MySQL 数据库中。

1. 新建应用

新建应用，信息如下。

Group：top.wisely。

Artifact：learning-spring-batch。

Dependencies：Spring Batch、MySQL Driver 和 Lombok。

build.gradle 文件中的依赖如下：

```
dependencies {
    implementation 'org.springframework.boot:spring-boot-starter-batch'
    runtimeOnly 'mysql:mysql-connector-java'
  compileOnly 'org.projectlombok:lombok'
    annotationProcessor 'org.projectlombok:lombok'
  //……
}
```

2. 配置 Spring Batch

（1）application.yml 配置如下：

```
spring:
  datasource:
    url: jdbc:mysql://localhost:3306/first_db?useSSL=false
    username: root
    password: zzzzzz
    driver-class-name: com.mysql.cj.jdbc.Driver #a
  batch:
    initialize-schema: always # b
    job:
      enabled: false #c
```

a. 配置 DataSource。这里的 DataSource 有两个作用：用作 Spring Batch 配置数据库和即将被导入的数据库。

b. 当设置应用启动时总是初始化 Spring Batch 数据库，默认只有嵌入式数据库才会初始化数据库。

c. 关闭任务自动执行，默认为自动执行。

（2）Spring Batch 配置类：

```
@Configuration
@EnableBatchProcessing
public class BatchConfig {
}
```

使用@EnableBatchProcessing 开启对 Spring Batch 的支持。

（3）CSV 数据如下：

```
name, gender, age
aaa, M, 20
bbb, F, 21
ccc, M, 22
ddd, F, 23
eee, M, 24
fff, F, 25
ggg, M, 26
hhh, F, 27
iii, M, 28
jjj, F, 29
```

文件位置为 src/main/resources/people.csv。

（3）配置数据读取

在 BatchConfig 配置类里：

```
@Bean
@StepScope //a
FlatFileItemReader<CsvPerson> itemReader(@Value("#{jobParameters['input.file.name']}")
Resource resource){ //1
   return new FlatFileItemReaderBuilder<CsvPerson>() //b
          .name("从 CSV 中读取数据") //c
          .resource(resource) //d
          .linesToSkip(1) //e
          .targetType(CsvPerson.class)//f
          .delimited().names("name", "gender", "age")//g
          .build();
}
```

a. 使用启动任务时传入的参数作为 CSV 文件的路径，必须用@StepScope 注解 Bean 才能获得值。

b. 使用 FlatFileItemReaderBuilder 来构造 FlatFileItemReader。在 Spring Batch 中，几乎所有的 ItemReader 和 ItemWriter 都有对应的 Builder，可以用 Builder 构建 ItemReader 和 ItemWriter。

c. 设置当前 ItemReader 的名称。

d. 指定读取的文件。

e. 忽略的行，CSV 中的第一行不是数据。

f. 用指定的类型接收数据，把每一行数据都转换为 CsvPerson 对象。

g. 用分隔符将文件中的每条数据分成各个属性，并使用名称指定文件中的属性顺序。

下面是 CsvPerson 的代码：

```
@Data
@AllArgsConstructor
@NoArgsConstructor
public class CsvPerson {
   private String name;
   private String gender;
   private Integer age;
}
```

4．在写入前对数据预处理

通过定义 ItemProcessor 在插入前对数据进行预处理：

```
@Bean
ItemProcessor<CsvPerson, Person> genderProcessor(){
   return item -> {
      Person person = new Person();
      BeanUtils.copyProperties(item, person);
      if (item.getGender().equals("M")){
         person.setGender("男");
      }else {
         person.setGender("女");
      }
      return person;
   };
}
```

在 ItemProcessor<CsvPerson, Person>中，ItemProcessor 为函数接口，CsvPerson 为输入类型，Person 为输出类型。本例将输入对象从 CsvPerson 类型转换成 Person 类型，并将性别的“M”和“F”转换为“男”和“女”。

Person 的代码如下。注意，Person 的定义其实是没有意义的，主要是为了帮助理解 ItemProcessor 的输入和输出。

```
@Data
@AllArgsConstructor
@NoArgsConstructor
public class Person {
   private Long id;
   private String name;
   private String gender;
   private Integer age;
}
```

5．配置数据写入

```
@Bean
JdbcBatchItemWriter<Person> itemWriter(DataSource dataSource){
```

```
    return new JdbcBatchItemWriterBuilder<Person>() //a
            .dataSource(dataSource) //b
            .sql("insert into person (name, gender, age) values
(:name, :gender, :age)").beanMapped() //c
            .build();
}
```

a. 同样适用 JdbcBatchItemWriterBuilder 来构建 JdbcBatchItemWriter。

b. 指定插入数据到指定的 DataSource 数据库。

c. 使用当前的 SQL 语句来插入数据，:name 从 Person 对象的 name 属性取值，其他不变。这里的 person 表是在前面章节新建的。

6. 配置 Step

```
@Bean
Step csvToMysqlStep(StepBuilderFactory stepBuilderFactory, //a
                ItemReader itemReader,
                ItemProcessor genderProcessor,
                ItemProcessor validatingItemProcessor,
                ItemWriter itemWriter){
    return stepBuilderFactory.get("csv导入到myql的step") //b
            .<CsvPerson, Person>chunk(5) //c
            .reader(itemReader) //d
            .processor(genderProcessor)//e
            .writer(itemWriter) //f
            .build();
}
```

a. 注入 StepBuilderFactory 的 Bean，由@EnableBatchProcessing 自动配置。

b. 通过 StepBuilderFactory 的 get()方法获得 StepBuilder，用来构建 Step；其中，get()方法中的字符为 Step 的名称。

c. 当从 ItemReader 中读取数据时，每读取 5 条数据就进行批写入。

d. 设置 Step 的 ItemReader。

e. 设置 Step 的 ItemProcessor。

f. 设置 Step 的 ItemWriter。

7. 配置 Job

```
@Bean
Job csvToMysqlJob(JobBuilderFactory jobBuilderFactory, //a
                Step csvToMysqlStep,
                MyJobListener myJobListener){
    return jobBuilderFactory.get("导入CSV到MySQL的任务") //b
            .start(csvToMysqlStep) //c
            .listener(myJobListener) //d
            .build();
}
```

a. 注入 JobBuilderFactory 的 Bean，同样由@EnableBatchProcessing 自动配置。

b. 通过 JobBuilderFactory 的 get()方法获得 JobBuilder，用来构建 Job；其中，get()方法中的字符为 Job 的名称。

c. 使用 start()方法执行 Step。

d. 使用 listener()方法为任务添加自定义的监听器。

通过任务监听器实现 JobExecutionListener 接口即可:

```
@Component
@Slf4j
public class MyJobListener implements JobExecutionListener {
   private long start;
   private long end;
   @Override
   public void beforeJob(JobExecution jobExecution) {
      start = System.currentTimeMillis();
   }

   @Override
   public void afterJob(JobExecution jobExecution) {
      end = System.currentTimeMillis();
      log.info("任务处理结束，耗时:" + (end - start) + "毫秒");
   }
}
```

8. 启动 Job

在默认情况下，Spring Boot 会自动执行 Spring Batch 任务。由于在配置中关闭了自动启动任务，所以需要使用 JobLaucher 来启动 Job，并使用 JobParameters 向 Job 传递参数。下面使用 CommandLineRunner 来启动 Job。

```
@Bean
CommandLineRunner jobClr(JobLauncher jobLauncher, //a
               Job job){//b
  return args -> {
    JobParameters jobParameters = new JobParametersBuilder() //c
         .addDate("time", new Date())
         .addString("input.file.name","people.csv") //d
         .toJobParameters();

    JobExecution jobExecution = jobLauncher.run(job, jobParameters); //e
  };
}
```

a. 注入 JobLauncher 的 Bean，同样由@EnableBatchProcessing 自动配置。

b. 注入前面配置的 Job 的 Bean。

c. 使用 JobParametersBuilder 构造 JobParameters。

d. 使用 addString()方法添加字符串类型所需的参数，此处设置 CSV 文件的路径，在 ItemReader 的 Bean 中读取。

e. JobLauncher 的 run()方法使用参数启动任务。

9. 检验结果

启动应用，控制台如图 10-4 所示。

```
o.s.b.c.l.support.SimpleJobLauncher      : Job: [SimpleJob: [name=导入csv到mysql的任务]]
o.s.batch.core.job.SimpleStepHandler     : Executing step: [csv导入到myql的step]
t.w.l.listener.MyJobListener             : 任务处理结束，耗时:228毫秒
o.s.b.c.l.support.SimpleJobLauncher      : Job: [SimpleJob: [name=导入csv到mysql的任务]]
com.zaxxer.hikari.HikariDataSource       : HikariPool-1 - Shutdown initiated...
com.zaxxer.hikari.HikariDataSource       : HikariPool-1 - Shutdown completed.
```

图 10-4

数据库 person 表如图 10-5 所示。

数据库 Spring Batch 配置表如图 10-6 所示。

修改	id	age	gender	name
编辑	91	20	男	aaa
编辑	92	21	女	bbb
编辑	93	22	男	ccc
编辑	94	23	女	ddd
编辑	95	24	男	eee
编辑	96	25	女	fff
编辑	97	26	男	ggg
编辑	98	27	女	hhh
编辑	99	28	男	iii
编辑	100	29	女	jjj

图 10-5

表	引擎?	校对?
BATCH_JOB_EXECUTION	InnoDB	utf8mb4_0900_ai_ci
BATCH_JOB_EXECUTION_CONTEXT	InnoDB	utf8mb4_0900_ai_ci
BATCH_JOB_EXECUTION_PARAMS	InnoDB	utf8mb4_0900_ai_ci
BATCH_JOB_EXECUTION_SEQ	InnoDB	utf8mb4_0900_ai_ci
BATCH_JOB_INSTANCE	InnoDB	utf8mb4_0900_ai_ci
BATCH_JOB_SEQ	InnoDB	utf8mb4_0900_ai_ci
BATCH_STEP_EXECUTION	InnoDB	utf8mb4_0900_ai_ci
BATCH_STEP_EXECUTION_CONTEXT	InnoDB	utf8mb4_0900_ai_ci
BATCH_STEP_EXECUTION_SEQ	InnoDB	utf8mb4_0900_ai_ci

图 10-6

10.3 小结

Spring Integration 和 Spring Batch 在日常开发中还是很常用的。Spring Integration 和 Spring Batch 涉及的内容特别多，本章只起到抛砖引玉的作用，如需更深入学习，请参考官方文档。

第 11 章 Spring Cloud与微服务

11.1 微服务基础

11.1.1 微服务和云原生应用

微服务是一些模块化、松耦合的服务，不同的服务关注不同的业务内容。我们可以独立设计、开发、测试并部署不同的微服务。使用微服务进行开发的核心是必须使用领域驱动设计进行指导。

专门为在云计算平台上运行而设计和架构的应用，称之为云原生应用。现在主流的应用运行平台有 Kubernetes、Cloud Foundry 和 Apache Mesos 等。

为了构建分布式微服务应用，云原生应用会实现被称之为“12 要素应用”的原则，更多内容可访问官网 https://12factor.net/zh_cn/ 获得。

（1）Codebase（基准代码）：一份基准代码，多份部署。

（2）Dependencies（依赖）：显式声明依赖关系。

（3）Config（配置）：在环境中存储配置。

（4）Backing Services（后端服务）：把后端服务当作附加资源。

（5）Build、Release 和 Run（构建、发布和运行）：严格分离构建和运行。

（6）Processes（进程）：以一个或多个无状态进程运行应用。

（7）Port Binding（端口转发）：通过端口绑定提供服务。

（8）Concurrency（并发）：通过进程模型进行扩展。

（9）Disposability（易处理）：快速启动和优雅终止可最大化健壮性。

（10）Dev/Prod Parity（开发环境和线上环境等价）：尽可能保持开发、预发布和线上环境相同。

（11）Logs（日志）：把日志当作事件流。

（12）Admin Processes（管理流程）：把后台管理任务当作一次性进程来运行。

11.1.2 领域驱动设计

除使用“12 要素应用”原则对技术架构进行指导外，还需使用“领域驱动设计”（DDD）对业务架构进行指导。

领域驱动设计主要分为战略设计和战术设计两部分。

（1）战略设计。

◎ 边界上下文（Bounded Context）：边界上下文是一种语义上的上下文边界。在这个边界内，领域模型的每一个组件都有相同的含义和行为。

◎ 通用语言（Ubiquitous Language）：在一个边界上下文内，团队的所有成员（包括技术人员和业务专家）使用相同的语言来描述软件模型，我们称这个语言为通用语言。通用语言既是口头表达语言，又是软件模型。

◎ 上下文映射（Context Mapping）：在一个大型的应用中会有多个边界上下文，而这些边界上下文之间不可避免的需要交互或数据共享。上下文映射指的是不同边界上下文之间的交互。

（2）战术设计。

◎ 实体（Entity）：对一个实际的事物进行抽象建模，每个实体都有唯一标识，用来和其他实体进行区分。实体是可变的，它会随着时间的推移而修改。

◎ 值对象（Value Object）：值对象等同于值，它是对一个不变的概念的整体进行建模。它没有唯一标识，是否相等需通过比较值对象的属性来确认。值对象不能被修改，只能被替换；它和值一样都是用来描述、量化或衡量实体的。

◎ 聚合（Aggregate）：聚合是由一个或多个实体组合而成的，其中，表示核心概念的实体叫作聚合根。在实际设计中，有“设计小聚合”的原则，一般情况下，一个聚合只包含一个实体类。聚合中不仅有各种属性值，还包含各种领域业务逻辑（真正的面向对象设计，而不是简单的数据模型）。这些业务逻辑只能存储在聚合内，不能泄露给调用的应用服务（Application Service）。

◎ 领域事件（Domain Event）：聚合之间的通信是通过发布领域事件来进行的，事件由聚合根发布。

◎ 领域服务（Domain Service）：有些领域业务逻辑不适合放在聚合中，这时就需要将这些业务逻辑放置在领域服务中。领域服务和聚合都关注业务逻辑，它们应放置在一起，且不能将业务逻辑泄露给调用的应用服务。

◎　库（Repository）：库是用来存储聚合的，每个聚合都有一个库，它们之间是一对一关系。库中对聚合的存储操作类似于集合类，和 Set 一样保证数据的唯一性。

11.2　Spring Cloud

在使用微服务开发系统时，这些微服务都是独立的系统，因而我们面临的是分布式系统开发。Spring Cloud 为分布式微服务开发提供了常见模式下的工具。这些常见模式将在后面进行讲解。

11.2.1　服务发现

服务发现是微服务架构的基本模式之一，分布式的微服务通过服务发现找到彼此并进行交互。Spring Cloud 支持的服务发现组件有：

◎　Spring Cloud Netflix Eureka；

◎　Spring Cloud Cluster；

◎　Spring Cloud Consul；

◎　Spring Cloud Zookeeper。

本节讲解 Spring Cloud Netflix Eureka。Spring Cloud Netflix 是 Spring Cloud 对 Netflix 开源软件的集成，它包含多个微服务常用组件。

1．服务发现服务端

（1）新建应用。

Group：top.wisely。

Artifact：discovery-server。

Dependencies：Eureka Server。

build.gradle 文件中的依赖如下：

```
dependencies {
    implementation
'org.springframework.cloud:spring-cloud-starter-netflix-eureka-server'
  //……
    }
```

（2）开启 Eureka Server：

```
@SpringBootApplication
@EnableEurekaServer
public class DiscoveryServerApplication {}
```

使用@EnableEurekaServer 注解开启对 Eureka Server 的支持。

（3）application.yml 外部配置：

```
server:
  port: 8761
eureka:
  client:
    register-with-eureka: false
    fetch-registry: false
```

在 Eureka Server 的依赖中包含了 Eureka Client，这是为了在构建高可用的 Eureka Server 集群时，Eureka Sever 彼此之间可互相发现。此处关闭了 Eureka 的客户端功能。

（4）启动 Eureka Server。

访问 http://localhost:8761/，可以看到 Eureka 的管理界面，如图 11-1 所示。Eureka 的 API 方法为/eureka/*。

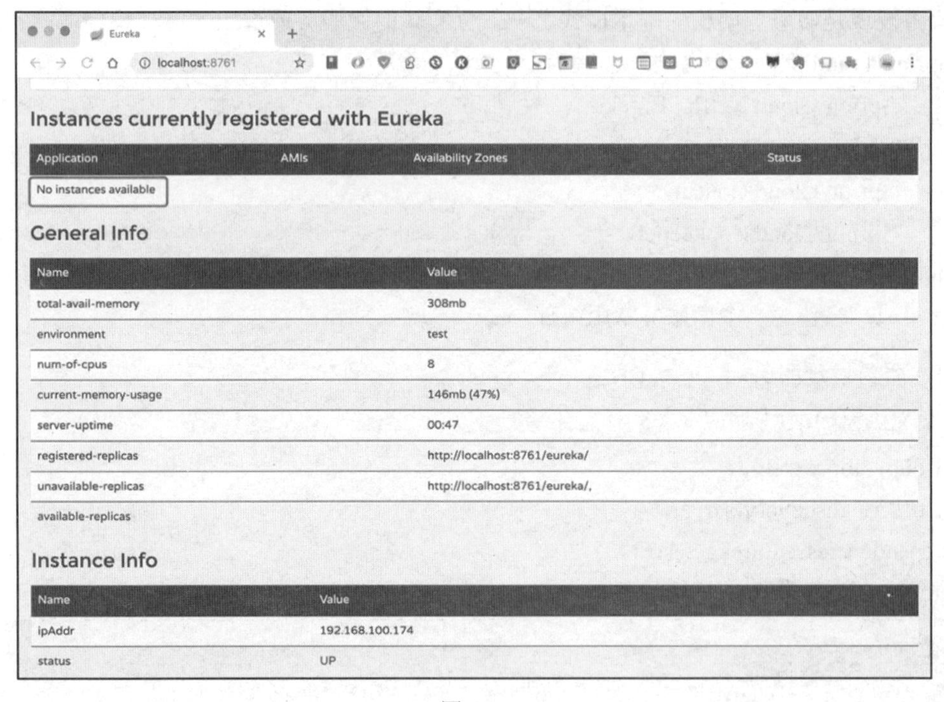

图 11-1

2. 服务发现客户端

（1）新建应用，信息如下。

Group：top.wisely。

Artifact：discovery-client。

Dependencies：Eureka Discovery Client、Spring Web Starter、Spring Boot Actuator 和 Lombok。

build.gradle 文件中的依赖如下：

```
dependencies {
    implementation
'org.springframework.cloud:spring-cloud-starter-netflix-eureka-client'
    implementation 'org.springframework.boot:spring-boot-starter-web'
  implementation 'org.springframework.boot:spring-boot-starter-actuator'
  compileOnly 'org.projectlombok:lombok'
    annotationProcessor 'org.projectlombok:lombok'
   //……
    }
```

（2）开启 Eureka Client：

```
@SpringBootApplication
@EnableEurekaClient
public class DiscoveryClientApplication {}
```

使用@EnableEurekaClient 注解开启对 Eureka Client 的支持。

（3）application.yml 外部配置：

```
spring:
  application:
    name: discovery-client #a
management:
  endpoints:
    web:
      exposure:
        include: shutdown
  endpoint:
    shutdown:
      enabled: true #b
server:
  port: 8080
eureka:
  client:
    service-url:
      default-zone: http://localhost:8761/eureka/ #c
```

a. 配置应用名称，作为注册到 Eureka Server 上的实例名。

b. 开启应用关闭端点。

c. 配置 Eureka Server 的路径。

（4）启动 Eureka Client。

这时 Eureka Server 管理界面会显示 DISCOVERY-CLIENT 微服务已注册，如图 11-2 所示。

（5）优雅关闭。

优雅关闭做了下面两件事情。

◎ 关闭应用。

◎ 从 Eureka Server 注销注册。

在 Postman 中，以 POST 方法访问 http://localhost:8080/actuator/shutdown，关闭注册在 Eureka Server 上的微服务，如图 11-3 所示。

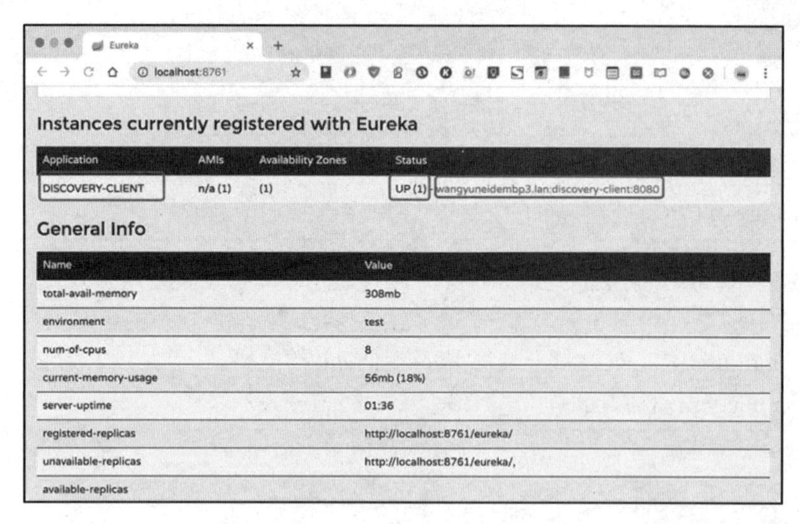

图 11-2

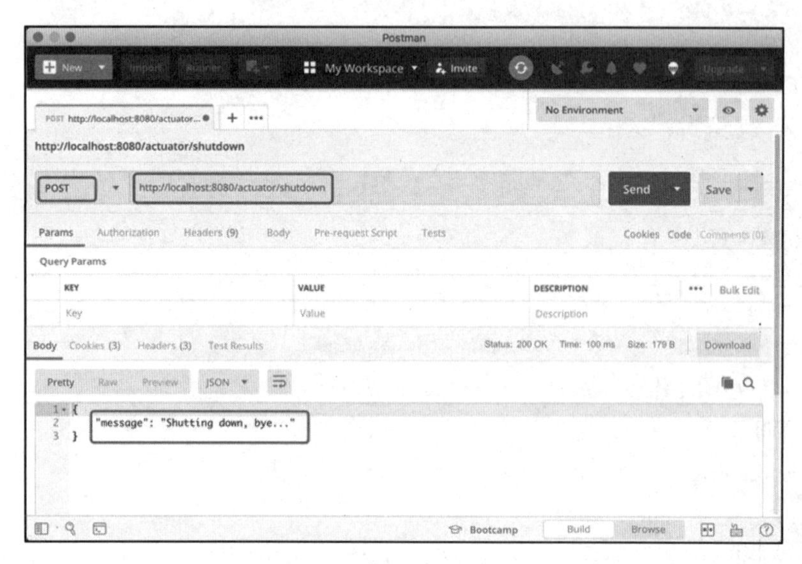

图 11-3

11.2.2 配置管理

使用分布式的配置管理（Spring Cloud Config）是微服务开发的重要模式之一，即所有应用的配置都由分布式的配置服务管理。Spring Boot 提供的配置管理组件有：

◎ Spring Cloud Config；

◎ Spring Cloud Netflix Archaius；

◎ Spring Cloud Consul；

◎ Spring Cloud Zookeeper。

本节以 Spring Cloud Config 作为示例。

1. 配置管理服务

（1）新建应用。信息如下。

Group：top.wisely。

Artifact：config-server。

Dependencies：Config Server、Eureka Discovery Client 和 Spring Boot Actuator。

build.gradle 文件中的依赖如下：

```
dependencies {
  implementation 'org.springframework.cloud:spring-cloud-config-server'
     implementation
'org.springframework.cloud:spring-cloud-starter-netflix-eureka-client'
  implementation 'org.springframework.boot:spring-boot-starter-actuator'
   //……
     }
```

（2）开启配置管理服务：

```
@SpringBootApplication
@EnableConfigServer //a
@EnableEurekaClient //b
public class ConfigServerApplication {}
```

a. 使用@EnableConfigServer 开启配置管理服务。

b. 使用@EnableEurekaClient 开启 Eureka 客户端。

（3）application.yml 配置：

```
spring:
  application:
    name: config-server
  profiles:
    active: native #1
  cloud:
    config:
      server:
        native:
          search-locations: classpath:/config/ #a
management:
  endpoints:
    web:
```

```
      exposure:
        include: shutdown
    endpoint:
      shutdown:
        enabled: true
server:
  port: 8888
eureka:
  client:
    service-url:
      default-zone: http://localhost:8761/eureka/ #b
```

a. Spring Cloud Config 的配置库支持文件系统、Git 等类型。本例使用本地文件作为演示，配置文件放置的目录为类路径下的 config 目录。

b. 将“配置管理”也配置为 Eureka Server 的客户端，这样“配置管理”的客户端可使用“服务发现”来找到并使用“配置管理”提供的配置。

（4）配置文件。

提供的配置依然使用 discovery-client 应用，在 src/main/resources/config 目录下新建文件：

```
discovery-client.yml
server:
  port: 9997
discovery-client-dev.yml
server:
  port: 9998
discovery-client-prod.yml
server:
  port: 9999
```

discovery-client 为应用名称，dev 和 prod 为 Profile，没有 Profile 的配置为默认配置。

（5）启动配置管理服务。

查看 Eureka Server 管理界面，可以看到配置管理服务已注册为服务，如图 11-4 所示。

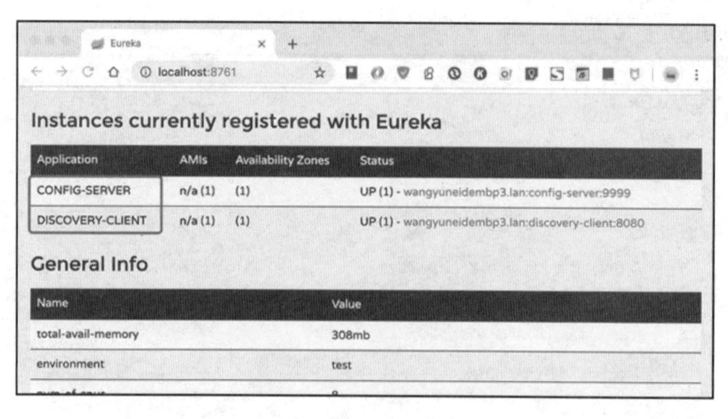

图 11-4

2. 配置客户端

继续使用 discovery-client 为配置客户端。

（1）添加 Config Client 依赖：

```
dependencies {
    //……
   implementation 'org.springframework.cloud:spring-cloud-starter-config'
      //……
}
```

（2）添加 bootstrap.yml：

```
spring:
  cloud:
    config:
      discovery:
        enabled: true
        service-id: config-server
```

因为使用了服务发现，所以在配置客户端时无须指定“配置管理”服务器地址，我们可通过服务名发现配置管理服务器。

这里可以看到我们使用的是 bootstrap.yml，而不是 application.yml。使用“配置管理”服务提供的配置时，使用 bootstrap.yml 进行配置，bootstrap.yml 的优先级高于 application.yml。

（3）使用远程默认配置启动。

首先用 shutdown 端点关闭 discovery-client，然后启动 discovery-client。端口配置将被配置管理服务的 discovery-client.yml 中的配置所覆盖，如图 11-5 所示。

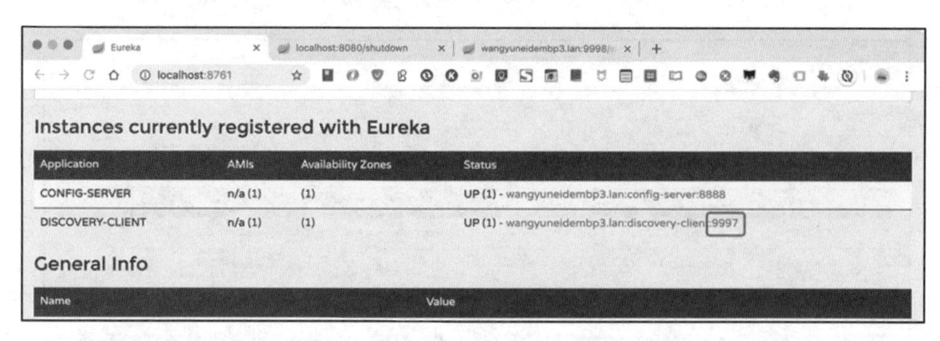

图 11-5

（4）以 dev Profile 启动。

同样优雅关闭 discovery-client（注意端口已变为 9997），编辑启动配置，如图 11-6 所示。

启动控制台，读取配置管理服务器的远程配置 discovery-client-dev.yml。此时端口配置将被远程配置覆盖，如图 11-7 所示。

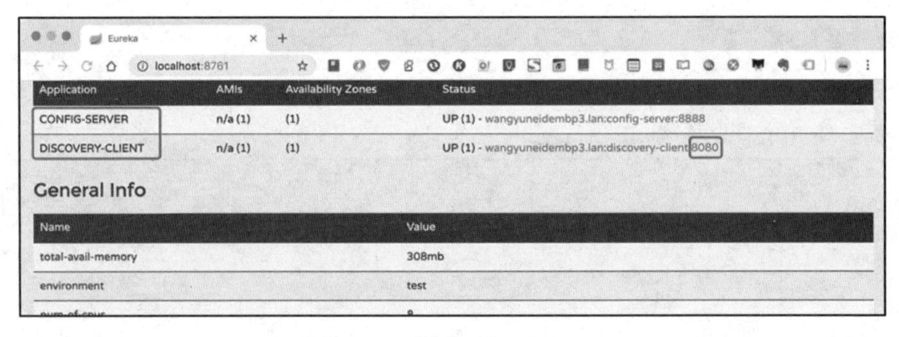

图 11-6

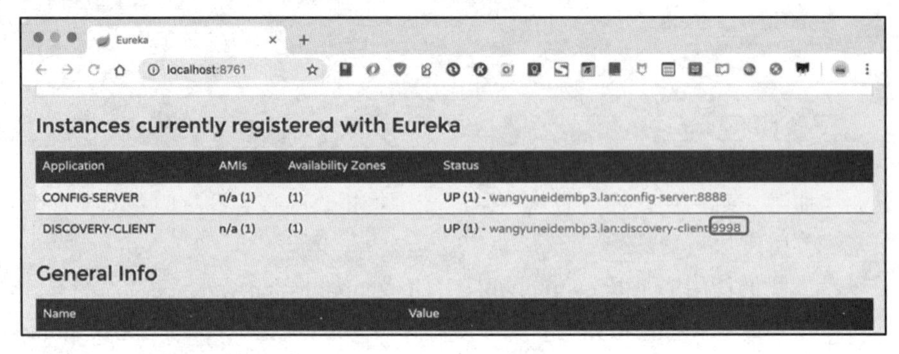

图 11-7

（5）以 prod Profile 启动。

同样优雅关闭应用（注意端口已变为 9998），并编辑启动配置，如图 11-8 所示。

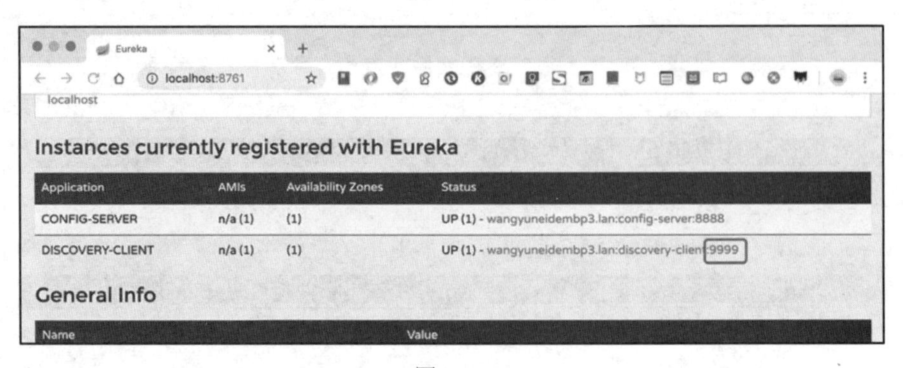

图 11-8

11.2.3 同步服务交互

分布式的微服务必然需要服务之间的交互，同步服务交互（Spring Cloud OpenFeign）是一种声明式 Web 服务客户端，我们只需使用 Spring MVC 注解即可调用远程服务。

1．服务提供者

依然使用 dicovery-client 提供 RESTful 服务。

（1）添加领域模型 Person：

```
@Data
@AllArgsConstructor
@NoArgsConstructor
public class Person {
   private Long id;
   private String name;
   private Integer age;
   private Integer serverPort; //存储当前服务器的端口
}
```

（2）添加演示控制器：

```
@RestController
@RequestMapping("/people")
public class PersonController {

   @Value("${server.port}")
   Integer port;

   @PostMapping
   public Person save(@RequestBody Person person){
      person.setServerPort(port);
      return person;
   }

   @GetMapping("/{id}")
   public Person getOne(@PathVariable Long id){
      return new Person(id, "wyf", 35 , port);
   }

   @GetMapping("/findByAge")
   public List<Person> findByAge(@RequestParam Integer age){
      return Arrays.asList(new Person(1l, "wyf", 35 , port),
                    new Person(2l, "foo", 35 , port));
   }

   @DeleteMapping("/{id}")
   public String delete(@PathVariable Long id){
      return "从端口为：" + port + "的服务删除";
   }

}
```

（3）启动 discovery-client。

我们分别使用远程默认配置、devProfile 和 prod Profile 启动 discovery-client，如图 11-9 所示。

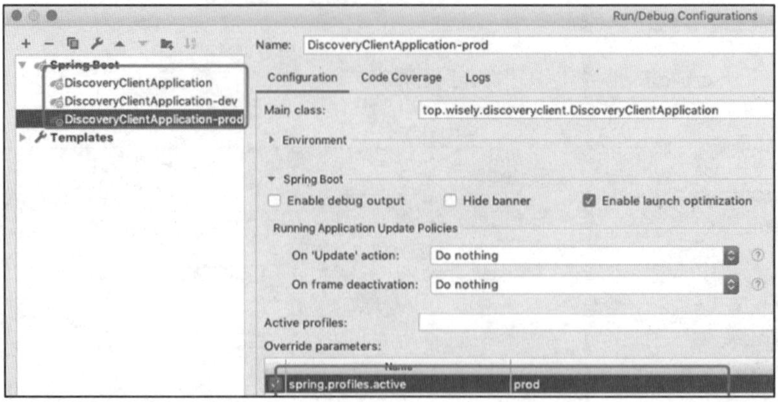

图 11-9

此时的 Eureka Server 如图 11-10 所示。

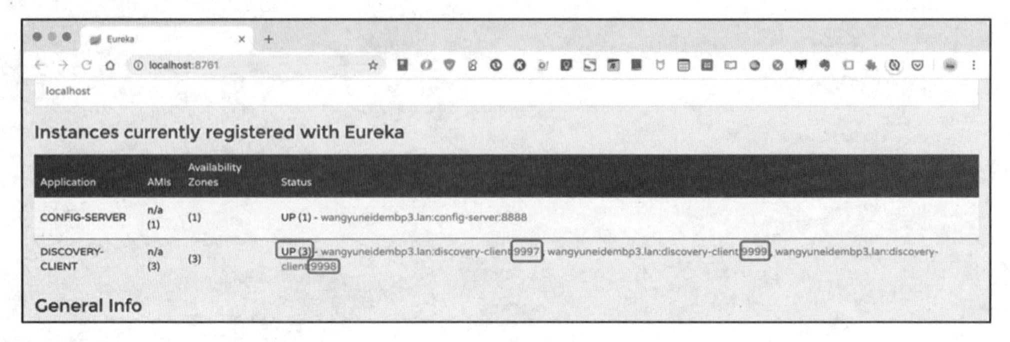

图 11-10

2．服务使用者

（1）新建应用。

Group：top.wisely。

Artifact：feign-client。

Dependencies：OpenFeign、Eureka Discovery Client、Spring Web Starter、Spring Boot Actuator 和 Lombok。

build.gradle 文件中的依赖如下：

```
dependencies {
    implementation 'org.springframework.cloud:spring-cloud-starter-openfeign'
    implementation
'org.springframework.cloud:spring-cloud-starter-netflix-eureka-client'
  implementation 'org.springframework.boot:spring-boot-starter-web'
    implementation 'org.springframework.boot:spring-boot-starter-actuator'
    compileOnly 'org.projectlombok:lombok'
```

```
    annotationProcessor 'org.projectlombok:lombok'
  //......
    }
```

（2）application.yml 配置：

```
spring:
  application:
    name: discovery-client
management:
  endpoints:
    web:
      exposure:
        include: shutdown
  endpoint:
    shutdown:
      enabled: true
server:
  port: 8082
eureka:
  client:
    service-url:
      default-zone: http://localhost:8761/eureka/
```

当然，也可以使用配置管理服务的配置，本书不做演示，它和 discovery-client 的用法保持一致。

（3）开启 Feign Client：

```
@SpringBootApplication
@EnableFeignClients
public class FeignClientApplication {}
```

使用@EnableFeignClients 开启对 Feign 客户端的支持。

（4）Feign 客户端声明：

```
@FeignClient("discovery-client")
public interface PersonClient {
    @PostMapping("/people")
    Person save(@RequestBody Person person);

    @GetMapping("/people/{id}")
    Person getOne(@PathVariable("id") Long id);

    @GetMapping("/people/findByAge")
    List<Person> findByAge(@RequestParam("age") Integer age);

    @DeleteMapping("/people/{id}")
    String delete(@PathVariable("id") Long id);
}
```

a. 使用@FeignClient 注解声明 Feign 客户端，使用服务名 discovery-client 指定要调用的微服务。

b. 使用 Spring MVC 的声明式注解映射远程微服务的方法；这里的@PathVariable 和@RequestParam 都需指定和服务端相同的参数名称。

Person 的定义：

```
@Data
@AllArgsConstructor
@NoArgsConstructor
public class Person {
   private Long id;
   private String name;
   private Integer age;
   private Integer serverPort;

   public Person(Long id, String name, Integer age) {
      this.id = id;
      this.name = name;
      this.age = age;
   }
}
```

（5）使用 Feign 客户端。

我们使用 CommandLineRunner 调用 Feign 客户端：

```
@Bean
CommandLineRunner personClientClr(PersonClient personClient){
  return args -> {
     System.out.println(personClient.save(new Person(1l, "wyf", 35)));
     System.out.println(personClient.getOne(1l));
     personClient.findByAge(35).forEach(System.out::println);
     System.out.println(personClient.delete(1l));
  };
}
```

启动 feign-client，如图 11-11 所示。

```
Person(id=1, name=wyf, age=35, serverPort=9998)
Person(id=1, name=wyf, age=35, serverPort=9999)
Person(id=1, name=wyf, age=35, serverPort=9997)
Person(id=2, name=foo, age=35, serverPort=9997)
从端口为：9998的服务删除
```

图 11-11

我们成功调用了 discovery-client 的服务，可以看到每次调用的都是不同的 discovery-client，这是因为 Eureka 和 Feign 都支持 Ribbon 客户端负载均衡技术，它默认按照 round-robin 算法调用即轮流调用。

11.2.4 异步服务交互

我们使用 Spring Cloud Stream 借助消息代理来实现微服务之间的异步交互。Spring Cloud Stream 目前支持 RabbitMQ 和 Kafka，RabbitMQ 和 Kafka 被称为 Binder。本节以 Kafka 作为示例进行演示。

Spring Cloud Stream 支持 Kafka 或 RabbitMQ 作为 Binder，本节以 Kafka 作为 Binder 进行演示。

本例演示两个微服务，一个叫 person-service，以 Person 为主体，也许需要使用 address-service 中的 Address 数据。在本例中，我们通过 Spring Cloud Stream 将 Address 数据同步到 person-service。同样，将 person-service 的 Person 信息同步到 address-service。这样两个微服务就都分享到了对方的数据，而不需要同步调用对方的接口了。

1. person-service

（1）新建应用。

Group：top.wisely。

Artifact：person-service。

Dependencies：Cloud Stream、Spring for Apache Kafka、Spring Web Starter、Spring Data MongoDB 和 Lombok。

build.gradle 文件中的依赖如下：

```
dependencies {
    implementation 'org.springframework.boot:spring-boot-starter-web'
    implementation 'org.springframework.cloud:spring-cloud-stream'
    implementation 'org.springframework.cloud:spring-cloud-starter-stream-kafka'
    implementation 'org.springframework.boot:spring-boot-starter-data-mongodb'
    compileOnly 'org.projectlombok:lombok'
    annotationProcessor 'org.projectlombok:lombok'
  //……
    }
```

（2）开启支持：

```
@SpringBootApplication
@EnableBinding({KafkaBindings.class})
public class PersonServiceApplication {}
```

使用@EnableBinding 开启对异步服务交互的支持，KafkaBindings 指定 Binding 信息。异步服务交互提供了三个类来指定绑定信息。

◎ Source：将数据输出到指定的 channel。

```
public interface Source {
  String OUTPUT = "output"; //输出的 channel 名称
  @Output(Source.OUTPUT)
```

```
  MessageChannel output(); //输出的 channel
}
```

◎ Sink：从 channel 中获得数据。

```
public interface Sink {
  String INPUT = "input"; //输入的 channel 名称
  @Input(Sink.INPUT)
  SubscribableChannel input(); //输入的 channel
}
```

Proccessor：从 input channel 中获取数据，对数据进行处理后将数据输出到 output channel。

```
public interface Processor extends Source, Sink {
}
```

本例我们使用自定义的 KafkaBindings：

```
public interface KafkaBindings {

    String ADDRESS_IN = "address-in"; // 输入 channel 名称
    String PERSON_OUT = "person-out"; //输出 channel 名称

    @Input(ADDRESS_IN)
    SubscribableChannel addressIn();

    @Output(PERSON_OUT)
    MessageChannel personOut();
}
```

（3）application.yml 配置：

```
spring:
  application:
    name: person-service

  data:
    mongodb:
      host: localhost
      port: 27017
      username: wisely
      password: zzzzzz
      database: first_db #a

  cloud:
    stream:
      kafka:
        binder: #b
          brokers: localhost
          defaultBrokerPort: 9092
      bindings: #c
        address-in: #d
```

```
        destination: addressTopic #e
        contentType: application/json
        group: addressInGroup #f
      person-out: #d
        destination: personTopic #e
        contentType: application/json
server:
  port: 8083
```

a. 配置连接 MongoDB。

b. 配置 Kafka Binder，前缀为 spring.cloud.stream.kafka.binder.*，通过指定 brokers 和端口连接 Kafka。

c. 配置 Bindings 信息，前缀为 spring.cloud.stream.bindings.*。

d. address-in 和 person-out 为 channel 名称。

e. destination 为 Topic 名称，默认会自动新建 Topic。

f. 若在当前微服务中启动了多个实例，则置 Comsumer Group 可以让这多个实例对数据进行负载处理。

（4）领域模型。

本微服务关注的 Person：

```
@Data
@AllArgsConstructor
@NoArgsConstructor
@Document(collection = "my-person")
public class Person {
    @Id
    private String id;
    private String name;
    private Integer age;
}
```

用来同步 address-service 的 Address：

```
@Data
@AllArgsConstructor
@NoArgsConstructor
@Document(collection = "my-address")
public class Address {
    @Id
    private Long id;
    private String province;
    private String city;
}
```

（5）数据访问 Repository：

```
public interface PersonRepository extends MongoRepository<Person, String> {
}
```

```
public interface AddressRepository extends MongoRepository<Address, Long> {
}
```

（6）事件发送。

此处将事件发送到 Kafka，让 address-service 去监听 Kafka 消费数据：

```
@Service
public class PersonService extends AbstractMongoEventListener { //a

   MessageChannel personOut;
         //b
   public PersonService(@Qualifier(KafkaBindings.PERSON_OUT) MessageChannel personOut)
{
      this.personOut = personOut;
   }

   @Override
   public void onAfterSave(AfterSaveEvent event) { //a
      if(event.getSource() instanceof Person){ //c
         super.onAfterSave(event);
         PersonSavedEvent personSavedEvent = new PersonSavedEvent(); //d
         BeanUtils.copyProperties(event.getSource(), personSavedEvent);
         Message message = MessageBuilder
               .withPayload(personSavedEvent)
               .build(); //e
         personOut.send(message); //f
      }
   }
}
```

a. 通过继承 AbstractMongoEventListener 的 onAfterSave()方法监听保存完成的事件，在 Person 被保存完成后，调用该方法。

b. 通过构造器注入输出的 channel，通过@Qualifier 指定 channel 名称。

c. 只有在保存 Person 时才会发送事件。

d. PersonSavedEvent 是我们向外部发送的真正事件。

e. 通过 MessageBuilder 构造消息。

f. 通过输出 channel 发送消息。

事件对象如下：

```
@Data
@AllArgsConstructor
@NoArgsConstructor
public class PersonSavedEvent {
   private String id;
   private String name;
   private Integer age;
}
```

（7）事件接收。

在此处监听 address-service 发送到 Kafka 的数据：

```
@Service
public class AddressService {

   AddressRepository addressRepository;

   public AddressService(AddressRepository addressRepository) {
      this.addressRepository = addressRepository;
   }

   @StreamListener(target = KafkaBindings.ADDRESS_IN) //a
   public void updateAddress(@Payload AddressSavedEvent event){
      Address address = new Address();
      BeanUtils.copyProperties(event, address);
      addressRepository.save(address); //b
   }
}
```

a. 通过@StreamListener 注解来监听消息，target 制定要监听的 channel。

b. 这里只是对数据进行了保存，当然，在实际应用中可能需要做一些复杂的业务处理。

事件对象如下：

```
@Data
@AllArgsConstructor
@NoArgsConstructor
public class AddressSavedEvent {
   private Long id;
   private String province;
   private String city;
}
```

（8）验证控制器。

person-service 负责对 Person 进行修改；它对 Address 没有所有权，所以只能查询：

```
@RestController
@RequestMapping("/people")
public class PersonController {
   PersonRepository personRepository;

   public PersonController(PersonRepository personRepository) {
      this.personRepository = personRepository;
   }

   @PostMapping
   public Person save(@RequestBody Person person){
      return personRepository.save(person);
   }
}
```

AddressController：

```
@RestController
@RequestMapping("/addresses")
public class AddressController {

   AddressRepository addressRepository;

   public AddressController(AddressRepository addressRepository) {
      this.addressRepository = addressRepository;
   }

   @GetMapping
   public List<Address> findAll(){
      return addressRepository.findAll();
   }
}
```

（9）正常启动应用。

2. address-service

address-service 和 person-service 刚好相反，它需要使用的 Person 数据是从 person-service 同步过来的。

（1）新建应用。

Group：top.wisely。

Artifact：person-service。

Dependencies：Cloud Stream、Spring for Apache Kafka、Spring Web Starter、Spring Data JPA、PostgreSQL Driver 和 Lombok。

build.gradle 文件中的依赖如下：

```
dependencies {
    implementation 'org.springframework.boot:spring-boot-starter-web'
    implementation 'org.springframework.cloud:spring-cloud-stream'
    implementation 'org.springframework.cloud:spring-cloud-starter-stream-kafka'
    implementation 'org.springframework.boot:spring-boot-starter-data-jpa'
    runtimeOnly 'org.postgresql:postgresql'
    compileOnly 'org.projectlombok:lombok'
    annotationProcessor 'org.projectlombok:lombok'
  //……
    }
```

（2）开启支持：

```
@SpringBootApplication
@EnableBinding(KafkaBindings.class)
@EnableAsync
public class AddressServiceApplication {}
```

```
public interface KafkaBindings {

   String PERSON_IN = "person-in"; //输入 channel 名称
   String ADDRESS_OUT = "address-out"; //输出 channel 名称

   @Input(PERSON_IN)
   SubscribableChannel personIn();

   @Output(ADDRESS_OUT)
   MessageChannel addressOut();
}
```

（3）applicaiton.yml 配置：

```
spring:
  application:
    name: address-service

  datasource:
    url: jdbc:postgresql://localhost:5432/first_db
    username: wisely
    password: zzzzzz
    driver-class-name: org.postgresql.Driver

  jpa:
    show-sql: true
    hibernate:
      ddl-auto: update

  cloud:
    stream:
      kafka:
        binder:
          brokers: localhost
          defaultBrokerPort: 9092
      bindings:
        person-in:
          destination: personTopic
          contentTypem : application/json
          group: personInGroup
        address-out:
          destination: addressTopic
          contentType: application/json
server:
  port: 8084
```

（4）领域模型：

```
@Data
@AllArgsConstructor
@NoArgsConstructor
```

```
@Entity
public class Address {
   @Id
   @GeneratedValue
   private Long id;
   private String province;
   private String city;

   @DomainEvents //发布领域事件
   Collection<Object> domainEvents(){
      List<Object> events= new ArrayList<Object>();
      events.add(new AddressSavedEvent(this.id, this.province, this.city));
      return events;
   }

   @AfterDomainEventPublication
   void callbackMethod() {
      domainEvents().clear();
   }
}
```

用来存储 person-service 发布的事件数据：

```
@Data
@AllArgsConstructor
@NoArgsConstructor
@Entity
@Table(name = "another_person")
public class Person {
   @Id
   private String id;
   private String name;
   private Integer age;
}
```

（5）数据访问 Repository：

```
public interface AddressRepository extends JpaRepository<Address, Long> {
}
public interface PersonRepository extends JpaRepository<Person, String> {
}
```

（6）事件发送：

```
@Service
public class AddressService {
   MessageChannel addressOut;

   public AddressService(@Qualifier(KafkaBindings.ADDRESS_OUT) MessageChannel
addressOut) {
      this.addressOut = addressOut;
   }
```

```
    @Async
    @TransactionalEventListener //监听数据库事件
    public void handleAddressSavedEvent(AddressSavedEvent addressSavedEvent){
        Message message = MessageBuilder
                .withPayload(addressSavedEvent)
                .build();
        addressOut.send(message);
    }
}
```

AddressSavedEvent 和 person-service 一致。

（7）事件接收：

```
    PersonRepository personRepository;

    public PersonService(PersonRepository personRepository) {
        this.personRepository = personRepository;
    }

    @StreamListener(target = KafkaBindings.PERSON_IN)
    public void updateAddress(@Payload PersonSavedEvent event){
        Person person = new Person();
        BeanUtils.copyProperties(event, person);
        personRepository.save(person);
    }

}
```

PersonSavedEvent 与 person-service 一致。

（8）验证控制器。

同样，address-service 负责对 Address 数据进行修改，对 Person 数据则只能查询。

```
@RestController
@RequestMapping("/addresses")
public class AddressController {

    AddressRepository addressRepository;

    public AddressController(AddressRepository addressRepository) {
        this.addressRepository = addressRepository;
    }

    @PostMapping
    public Address save(@RequestBody Address address){
        return addressRepository.save(address);
    }
}
```

在本微服务查询 Person 数据：

```
@RestController
@RequestMapping("/people")
public class PersonController {
   PersonRepository personRepository;

   public PersonController(PersonRepository personRepository) {
      this.personRepository = personRepository;
   }

   @GetMapping
   public List<Person> findAll(){
      return personRepository.findAll();
   }
}
```

（9）正常启动应用。

3．验证结果

在 person-service 中保存 Person 数据，在 address-service 中可正确查询 Person 数据，如图 11-12。

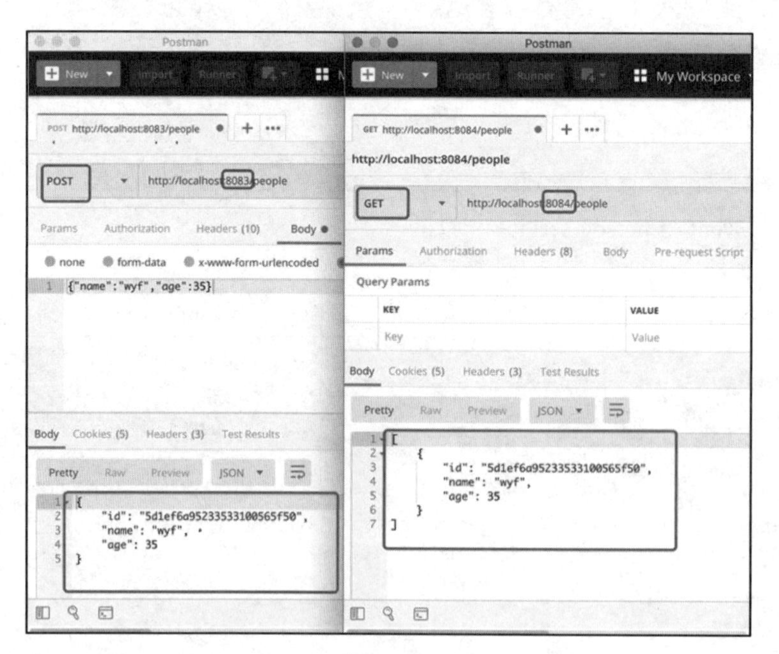

图 11-12

同样，在 address-service 中写入 Address 数据，在 person-service 中也可正常查询，如图 11-13 所示。

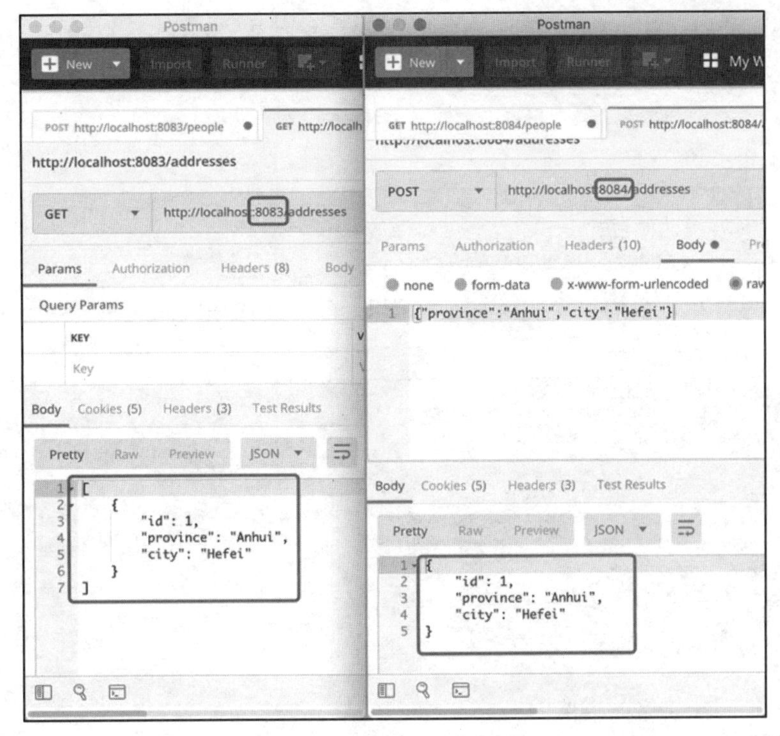

图 11-13

11.2.5　响应式异步交互

我们还可以使用 Spring Cloud Stream Reactive 进行响应式的异步交互开发。本例演示 Spring Cloud Stream Reactive 使用响应式的 Flux 的示例。

1．reactive-stream-producer

（1）新建应用。

Group：top.wisely。

Artifact：reactive-stream-producer。

Dependencies：Reactive Cloud Stream、Spring for Apache Kafka、Spring Reactive Web 和 Lombok。

build.gradle 文件中的依赖如下：

```
dependencies {
    implementation 'org.springframework.boot:spring-boot-starter-webflux'
    implementation 'org.springframework.cloud:spring-cloud-stream-binder-kafka'
    implementation 'org.springframework.cloud:spring-cloud-stream-reactive'
    implementation 'org.springframework.kafka:spring-kafka'
```

```
    compileOnly 'org.projectlombok:lombok'
    annotationProcessor 'org.projectlombok:lombok'
  //……
    }
```

（2）开启配置：

```
@SpringBootApplication
@EnableBinding(Source.class)
public class ReactiveStreamProducerApplication {}
```

本例演示非常简单，我们只需使用 Source 类即可，它的输出 channel 名称为 output。

（3）application.yml 配置：

```
spring:
  application:
    name: reactive-stream-producer

  cloud:
    stream:
      kafka:
        binder:
          brokers: localhost
          defaultBrokerPort: 9092
      bindings:
        output: # 输出 channel 的名称
          destination: msgTopic # Topic 名称
          contentType: application/json
server:
  port: 8085
```

（4）事件。

在这里定义我们要发送的事件：

```
@Data
@AllArgsConstructor
@NoArgsConstructor
public class MessageEvent {
    private String msg;
    private Date time;
}
```

（5）发送事件：

```
@Service
public class ProducerService {

    @StreamEmitter //a
    public void send(@Output(Source.OUTPUT)FluxSender output){ //b
        output.send(Flux.interval(Duration.ofSeconds(5))
                .map(i -> new MessageEvent(UUID.randomUUID().toString(),new
```

```
Date())));  //c
    }
}
```

a. 把@StreamEmitter 注解在方法上，让方法具备发送消息到输出 channel 的能力。

b. 把@Output 注解在指定输出的 channel 上，使用 FluxSender 向指定 channel 发送数据。

c. 向 channel 每隔 5 秒钟发送一次事件。

2. reactive-stream-consumer

（1）新建应用。

Group：top.wisely。

Artifact：reactive-stream-consumer。

Dependencies：Reactive Cloud Stream、Spring for Apache Kafka、Spring Reactive Web 和 Lombok。

build.gradle 文件中的依赖如下：

```
dependencies {
    implementation 'org.springframework.boot:spring-boot-starter-webflux'
    implementation 'org.springframework.cloud:spring-cloud-stream-binder-kafka'
    implementation 'org.springframework.cloud:spring-cloud-stream-reactive'
    implementation 'org.springframework.kafka:spring-kafka'
    compileOnly 'org.projectlombok:lombok'
    annotationProcessor 'org.projectlombok:lombok'
  //……
    }
```

（2）开启支持：

```
@SpringBootApplication
@EnableBinding(Sink.class)
public class ReactiveStreamConsumerApplication {}
```

使用 Sink 从名为 input 的 channel 中获取数据。

（3）application.yml：

```
spring:
  application:
    name: reactive-stream-consumer

  cloud:
    stream:
      kafka:
        binder:
          brokers: localhost
          defaultBrokerPort: 9092
      bindings:
        input: # 输入 channel 的名称
```

```
      destination: msgTopic
      contentType: application/json
      group: msgInGroup
server:
 port: 8086
```

（4）事件：

```
@Data
@AllArgsConstructor
@NoArgsConstructor
public class MessageEvent {
   private String msg;
   private Date time;
}
```

（5）接收事件：

```
@Service
public class ConsumerService {

   @StreamListener(target = Sink.INPUT)
   public void consume(Flux<MessageEvent> messageEventFlux){
      messageEventFlux.subscribe(System.out::println);
   }
}
```

这里的处理很简单，将提供者的消息打印出来即可。

3. 验证

首先启动 reactive-stream-consumer，然后启动 reactive-stream-producer。reactive-stream-consumer 的控制台如图 11-14 所示。

```
MessageEvent(msg=2fba0019-ed0c-48c8-8b58-58f8617f6fdb, time=Fri Jul 05 20:00:34 CST 2019)
MessageEvent(msg=0a75ec8f-6099-44a0-8e86-7852c0d48ccc, time=Fri Jul 05 20:00:39 CST 2019)
MessageEvent(msg=41d5c683-a454-46cf-87be-448e6f8b3c6d, time=Fri Jul 05 20:00:44 CST 2019)
MessageEvent(msg=f224cf78-8b40-45b2-9979-f6bb3e9303d7, time=Fri Jul 05 20:00:49 CST 2019)
MessageEvent(msg=f65282d0-fd37-4afc-941a-492924ceb1c8, time=Fri Jul 05 20:00:54 CST 2019)
MessageEvent(msg=9d5e3f52-fcc2-43f9-be6d-a8ea45de4118, time=Fri Jul 05 20:00:59 CST 2019)
MessageEvent(msg=4e350a85-7636-469b-af40-9b33cc9dab6f, time=Fri Jul 05 20:01:04 CST 2019)
MessageEvent(msg=ab05dfad-56c3-40da-9ae2-adb608ceeea5, time=Fri Jul 05 20:01:09 CST 2019)
```

图 11-14

11.2.6 应用网关：Spring Cloud Gateway

API Gateway 模式是常用的微服务模式之一，通过实现一个网关服务作为外部客户端访问内部分布式微服务的唯一入口。Spring 支持的 Gateway 模式的组件有：

- ◎ Spring Cloud Gateway；
- ◎ Spring Cloud Netflix Zuul。

本例采用 Spring Cloud Gateway 进行演示。

（1）新建应用。

Group：top.wisely。

Artifact：reactive-stream-consumer。

Dependencies：Gateway、Config Client、Eureka Discovery Client 和 Spring Boot Actuator。

build.gradle 文件中的依赖如下：

```
dependencies {
    implementation 'org.springframework.cloud:spring-cloud-starter-config'
    implementation 'org.springframework.boot:spring-boot-starter-actuator'
    implementation 'org.springframework.cloud:spring-cloud-starter-gateway'
    implementation
'org.springframework.cloud:spring-cloud-starter-netflix-eureka-client'
  //……
    }
```

（2）在“配置管理”服务进行配置。

在 config-server 的 src/main/resources/config 目录下，添加 gateway.yml：

```
server:
 port: 81

spring:
  cloud:
    gateway:
      discovery:
        locator:
          enabled: true # a
          lower-case-service-id: true # b
```

a. Gateway 通过 Discovery Server 找到在其上注册的微服务，自动将它们通过在 Gateway 地址在附加/serviceId/**路径来访问这些微服务。

b. 默认的服务名为大写，通过此属性修改为小写。

（3）开启 Eureka 客户端。

```
@SpringBootApplication
@EnableEurekaClient
public class GatewayApplication {}
```

（4）application.yml：

```
spring:
 application:
   name: gateway

management:
 endpoints:
```

```
    web:
      exposure:
        include: shutdown
  endpoint:
    shutdown:
      enabled: true

eureka:
  client:
    service-url:
      default-zone: http://localhost:8761/eureka/
```

（5）bootstrap.yml：

```
spring:
  cloud:
    config:
      discovery:
        enabled: true
        service-id: config-server
```

（6）验证：

按顺序启动 discovery-server、config-server、discovery-client 和 Gateway。访问 http://localhost:81/ discovery-client/people/1，实际访问的地址为 http://localhost:9997/people/1，如图 11-15 所示。

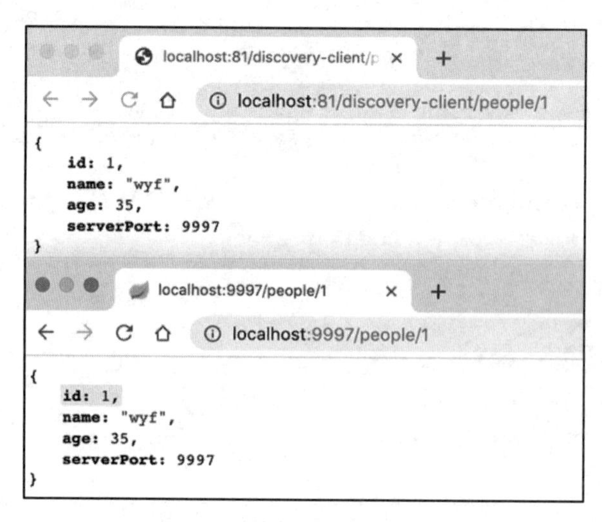

图 11-15

（7）修改默认。

如果访问的路径为 discovery-id，则可以对 config-server 中的配置文件做如下修改：

```
server:
  port: 81

spring:
  cloud:
    gateway:
      discovery:
        locator:
          enabled: true
          lower-case-service-id: true
      routes:
      - id: discoveryClient # a
        uri: lb://discovery-client #b
        predicates:
        - Path=/dis/** #c
        filters:
        - RewritePath=/dis/(?&lt;segment&gt;.*), /$\{segment} #d
```

a. 当前路由的 id。

b. lb 是负载均衡的缩写，discovery-client 为服务名。

c. 将微服务的路径映射为/dis/**。

d. 因为访问的路径为/dis/people/形式，所以路径要用正则表达式做匹配。

此时访问的路径被修改成了 http://localhost:81/dis/people/1。

11.2.7　认证授权

第 7 章对 OAuth2 进行了演示。在微服务环境下，Spring Cloud 提供了 Spring Cloud Security OAuth2，但遗憾的是，目前 Spring Cloud Security OAuth2 和 Spring Cloud Gateway 并不兼容。但是这并不影响我们使用 Gateway 向下游微服务转发 Token。

本节的示例结合 Authorization Server、Resource Server、Gateway 以及客户端来演示微服务下的认证授权。

1. OAuth 2.0 Authorization Server

下面使用第 7 章的 auth-server 针对微服务进行一定的改造。

（1）添加 Spring Cloud 相关依赖：

```
ext {
  set('springCloudVersion', "Hoxton.BUILD-SNAPSHOT") //版本可能会有变化
}

dependencies {
  implementation
'org.springframework.cloud:spring-cloud-starter-netflix-eureka-client'
  implementation 'org.springframework.boot:spring-boot-starter-actuator'
  implementation 'org.springframework.cloud:spring-cloud-starter-config'
```

```
      //……
  }
  testImplementation 'org.springframework.security:spring-security-test'
}

dependencyManagement {
  imports {
    mavenBom
"org.springframework.cloud:spring-cloud-dependencies:${springCloudVersion}"
  }
}
```

（2）配置管理。

在 config-server 的 config 目录下新建 auth-server.yml，将原 auth-server 的配置放入其中：

```
spring:
  datasource:
    url: jdbc:mysql://localhost:3306/first_db?useSSL=false
    username: root
    password: zzzzzz
    driver-class-name: com.mysql.cj.jdbc.Driver
    initialization-mode: always
    continue-on-error: true
  jpa:
    show-sql: true
    hibernate:
      ddl-auto: update
  security:
    oauth2:
      resourceserver:
        jwt:
          public-key-location: classpath:public.txt

security:
  oauth2:
    authorization:
      jwt:
        key-store: classpath:keystore.jks
        key-store-password: pass1234
        key-alias: wisely
        key-password: pass1234
      tokenKeyAccess: permitAll()
      checkTokenAccess: isAuthenticated()
      realm: wisely

server:
  port: 8087
```

把 application.yml 修改为如下内容：

```yaml
spring:
  application:
    name: auth-server

management:
  endpoints:
    web:
      exposure:
        include: shutdown
  endpoint:
    shutdown:
      enabled: true

eureka:
  client:
    service-url:
      default-zone: http://localhost:8761/eureka/
```

新增 bootstrap.yml：

```yaml
spring:
  cloud:
    config:
      discovery:
        enabled: true
        service-id: config-server
```

（4）开启 Eureka 客户端：

```java
@SpringBootApplication
@EnableEurekaClient
public class AuthServerApplication {}
```

2．OAuth 2.0 Resource Server

用第 7 章中的 resource-server 针对微服务进行一定的改造。

（1）添加 Spring Cloud 依赖。

和 auth-server 一样，添加 Spring Cloud 相关依赖。

（2）配置管理。

在 config-server 的 config 目录下新建 resource-server.yml，将原 resource-server 的配置放入其中：

```yaml
spring:
  security:
    oauth2:
      resourceserver:
        jwt:
          public-key-location: classpath:public.txt
```

```
server:
  port: 8088
```

同样，把 application.yml 修改为如下内容：

```
spring:
  application:
    name: resource-server

management:
  endpoints:
    web:
      exposure:
        include: shutdown
  endpoint:
    shutdown:
      enabled: true

eureka:
  client:
    service-url:
      default-zone: http://localhost:8761/eureka/
```

新增 bootstrap.yml：

```
spring:
  cloud:
    config:
      discovery:
        enabled: true
        service-id: config-server
```

（3）开启 Eureka 客户端。

```
@SpringBootApplication
@EnableEurekaClient
public class ResourceServerApplication {}
```

3. Gateway

使用前面的 Gateway，对其进行修改。

配置新的微服务代理路径。

在 config-server 下的 gateway.yml 中添加 auth-server 和 resource-server 代理配置：

```
spring:
  cloud:
    gateway:
      discovery:
        locator:
          enabled: true
          lower-case-service-id: true
      routes:
```

```
  - id: uaa
    uri: lb://auth-server
    predicates:
    - Path=/uaa/**
    filters:
    - RewritePath=/uaa/(?&lt;segment&gt;.*), /$\{segment}
  - id: resource1
    uri: lb://resource-server
    predicates:
    - Path=/res/**
    filters:
    - RewritePath=/res/(?&lt;segment&gt;.*), /$\{segment}
```

4. 客户端调用

（1）获取 Access Token。

通过访问 http://localhost:81/uaa/oauth/token 获取 Token，此时端口已经修改为 Gateway 的端口，并通过 uaa 转向到 auth-server 微服务，如图 11-16 所示。

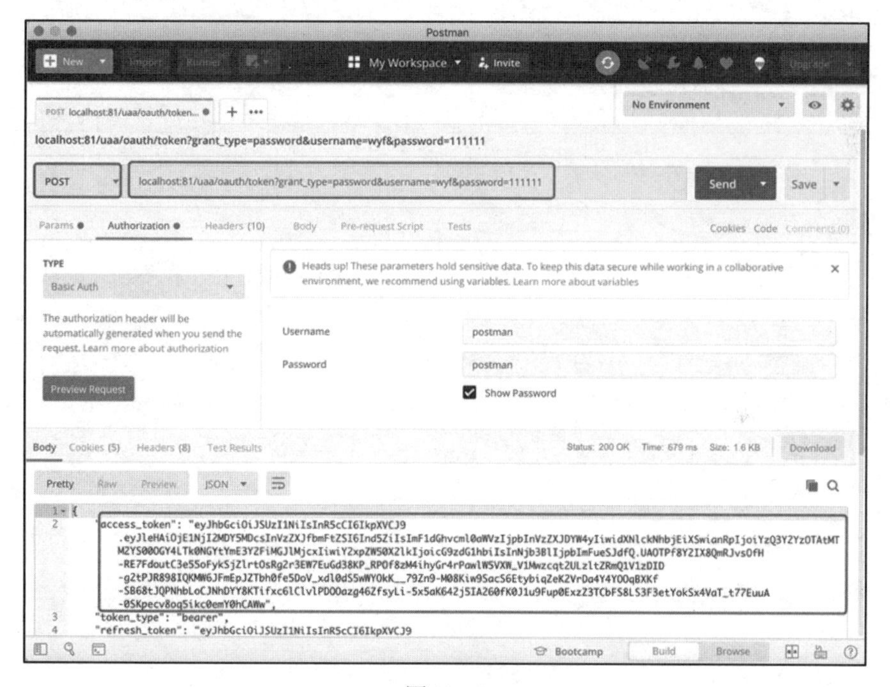

图 11-16

（2）访问 Resouce Server，如图 11-17 所示。

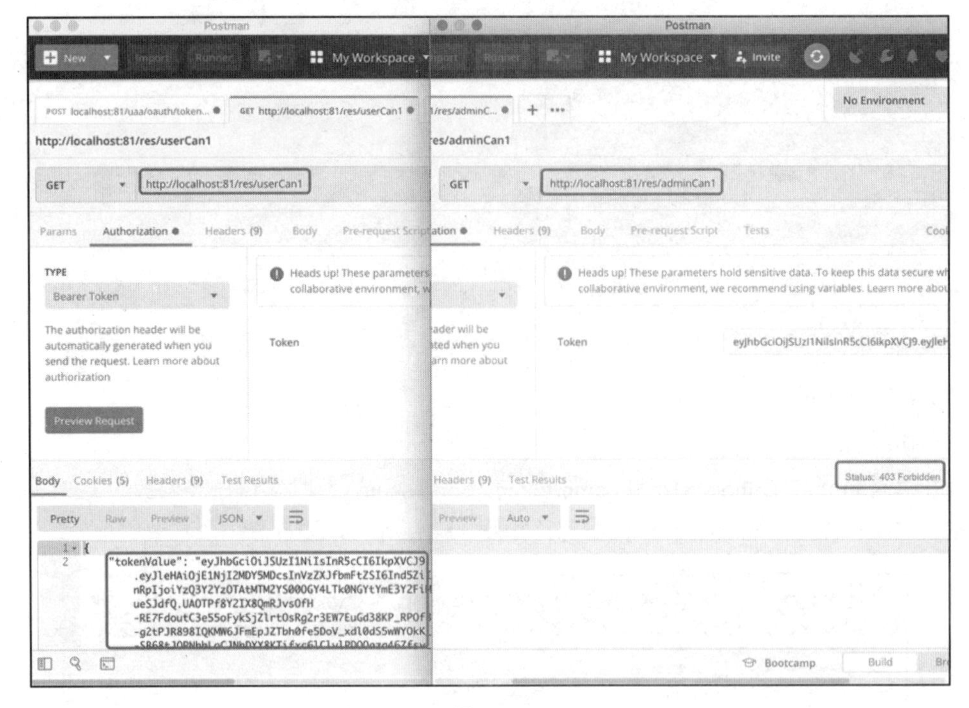

图 11-17

5. 微服务互调

当微服务之间互相调用时，需要将 Access Token 传递到被调用的微服务完成认证授权。

（1）为 resource-server 添加方法，用于远程调用：

```
@RestController
public class SecurityController {

        //……

   @GetMapping("/remote1")
   @PreAuthorize("hasAuthority('userCan1')")
   public String remote1(){
      return "验证调用 remote1";
   }

   @GetMapping("/remote2")
   @PreAuthorize("hasAuthority('adminCan1')")
   public String remote2(){
      return "验证调用 remote2";
   }

}
```

（2）新建调用微服务。

下面新建一个微服务来调用 resource-server 的服务，这个新的微服务也是一个 Resource Server。

Group：top.wisely。

Artifact：feign-service。

Dependencies：Eureka Discovery Client、Spring Boot Acturator、Config Client、OAuth2 Resource Server、Spring Security 和 OpenFeign。

build.gradle 文件中的依赖如下：

```
dependencies {
  implementation 'org.springframework.boot:spring-boot-starter-actuator'
  implementation
'org.springframework.boot:spring-boot-starter-oauth2-resource-server'
  implementation 'org.springframework.boot:spring-boot-starter-security'
  implementation 'org.springframework.cloud:spring-cloud-starter-config'
  implementation
'org.springframework.cloud:spring-cloud-starter-netflix-eureka-client'
  implementation 'org.springframework.cloud:spring-cloud-starter-openfeign'
   //……
}
```

（3）外部配置。

在 config-server 的 config 目录下新增 feign-service.yml：

```
spring:
  security:
    oauth2:
      resourceserver:
        jwt:
          public-key-location: classpath:public.txt

server:
  port: 8089
```

application.yml 中的配置如下：

```
spring:
  application:
    name: feign-service

management:
  endpoints:
    web:
      exposure:
        include: shutdown
  endpoint:
    shutdown:
```

```
    enabled: true

eureka:
  client:
    service-url:
      default-zone: http://localhost:8761/eureka/
```

bootstrap.yml：

```
spring:
  cloud:
    config:
      discovery:
        enabled: true
        service-id: config-server
```

将 resource-server 的 resources 目录下的 public.txt 文件拷贝到 feign-service 的 resources 目录下。

（4）开启支持：

```
@SpringBootApplication
@EnableEurekaClient
@EnableFeignClients
public class FeignServiceApplication {}
```

（5）Security 配置：

```
@EnableGlobalMethodSecurity(prePostEnabled = true)
@EnableWebSecurity
public class WebSecurityConfig extends WebSecurityConfigurerAdapter {
   @Override
   protected void configure(HttpSecurity http) throws Exception {
      http.authorizeRequests()
             .anyRequest().authenticated()
             .and()
             .oauth2ResourceServer().jwt().jwtAuthenticationConverter(jwt -> {
         Collection<SimpleGrantedAuthority> authorities = ((Collection<String>)
jwt.getClaims().get("authorities")).stream()
               .map(SimpleGrantedAuthority::new)
               .collect(Collectors.toSet());
         return new JwtAuthenticationToken(jwt, authorities);
      });
   }

   @Bean
   RequestInterceptor bearerHeaderAuthRequestInterceptor(){
      return requestTemplate -> {
         ServletRequestAttributes attributes = (ServletRequestAttributes)
RequestContextHolder.getRequestAttributes();
         HttpServletRequest request = attributes.getRequest();
         requestTemplate.header("Authorization", request.getHeader("Authorization"));
```

```
        };
    }

}
```

这里的配置和 resource-server 一致，不同的是，我们自定义了 RequestInterceptor，用于将请求头部信息的 Token 传递给 resource-server。这个 RequestInterceptor 的 Bean 是微服务互相调用实现的核心。

Spring Cloud Security OAuth2 还提供了 OAuth2FeignRequestInterceptor 来实现相同的功能，本例不引用 Spring Cloud Security OAuth2 的依赖。

（6）Feign 客户端：

```
@FeignClient("resource-server")
public interface ResourceClient {

    @GetMapping("/remote1")
    public String invokeRemote1();

    @GetMapping("/remote2")
    public String invokeRemote2();

}
```

（7）控制器：

```
@RestController
public class FeignController {
    ResourceClient client;

    public FeignController(ResourceClient client) {
        this.client = client;
    }

    @GetMapping("/remote1")
    @PreAuthorize("isAuthenticated()")
    public String remote1(){
        return client.invokeRemote1();
    }

    @GetMapping("/remote2")
    @PreAuthorize("isAuthenticated()")
    public String remote2(){
        return client.invokeRemote2();
    }

}
```

（8）验证。

此时访问 feign-service 远程调用 resource-server 的控制器路径。

使用 Access Token 分别访问 http://localhost:81/fgn/remote1 和 http://localhost:81/fgn/remote2，如图 11-18 所示。

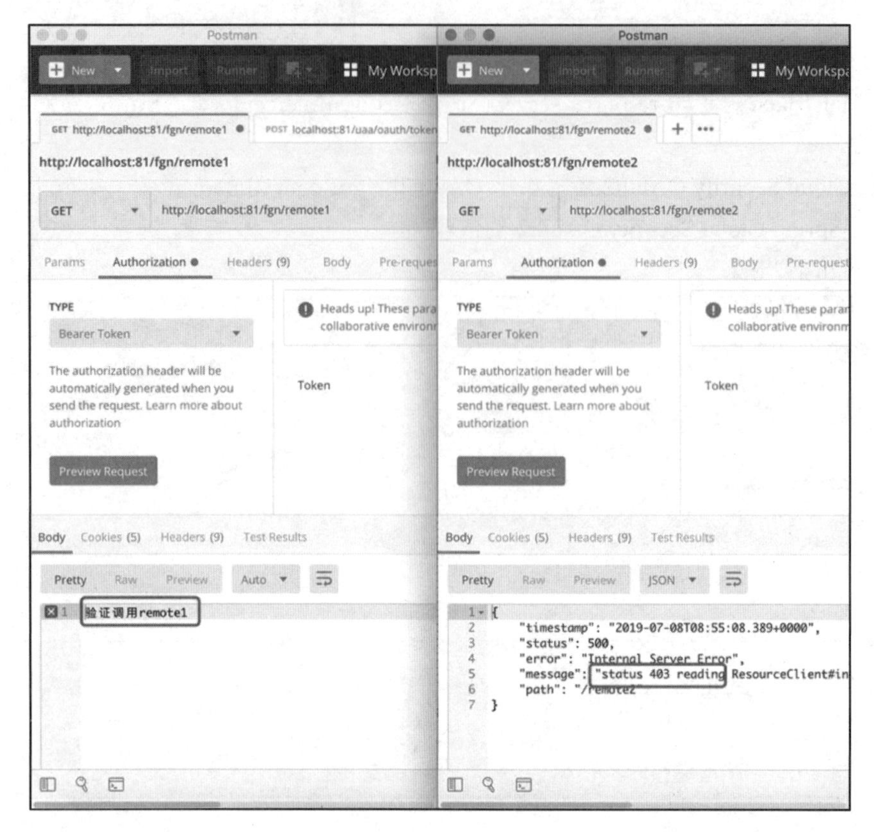

图 11-18

11.3 小结

本章介绍了使用 Spring Cloud 进行微服务开发的常用组件，但是这些组件在生产环境中往往不是自己搭建的，而是借助云平台的特性实现的，所以本章的很多介绍并没有深入。下一章我们将一起学习应用最广泛的云平台 Kubernetes。

第 12 章 Kubernetes与微服务

在使用 Spring Cloud 开发微服务时，需要通过编码和部署解决分布式微服务所遇到的常见问题，包括服务发现、负载均衡和 API 网关，等等。但在应用部署时，很难实现微服务的自动水平扩展。这时就需要 Kubernetes 这样的平台帮助我们解决应用部署及分布式微服务所面临的常见问题。

12.1 Kubernetes

Kubernetes 是一个开源平台系统，它可用来自动化部署、扩展和管理容器化（如 Docker）的应用。本章将介绍 Kubernetes 的基础知识及周边相关知识，涉及的内容较多，且每一项知识都值得读者进行专门的深入学习。本章的主要目的是将这些技术和 Spring Boot 的生产部署运维结合起来。

12.1.1 安装

1. 安装客户端

Kubernetes 的客户端名称是 kubectl，具体安装方法请参考官网。

输入下面的命令，确认客户端安装完成，如图 12-1 所示。

```
$ kubectl version
```

```
(c) 2017 Microsoft Corporation。保留所有权利。

C:\Windows\system32>kubectl version
Client Version: version.Info{Major:"1", Minor:"15", GitVersion:"v1.15.0", GitCommit:"e8462b5b5dc2584fdcd18e6bcfe9f1e4d970a529", GitTreeState:"clean", BuildDate:"2019-06-19T16:40:16Z", GoVersion:"go1.12.5", Compiler:"gc", Platform:"windows/amd64"}
Unable to connect to the server: dial tcp [::1]:6445: connectex: No connection could be made because the target machine actively refused it.

C:\Windows\system32>
```

图 12-1

2. 安装 minikube

Docker Desktop 内置了单节点的 Kubernetes 集群。开启 Kubernetes 功能可参考云栖社区的博文《Docker 社区版中 Kubernetes 开发》https://yq.aliyun.com/articles/508460。本节将基于常用的 minikube 进行演示。

（1）安装 VirtualBox。

minikube 默认使用的虚拟化技术为 VirtualBox，前面在使用 Docker Desktop 时已经安装过 VirtualBox。读者可以去官网下载最新的版本。

（2）安装 minikube，如图 12-2 所示，（具体参见 WebJar 官网）。

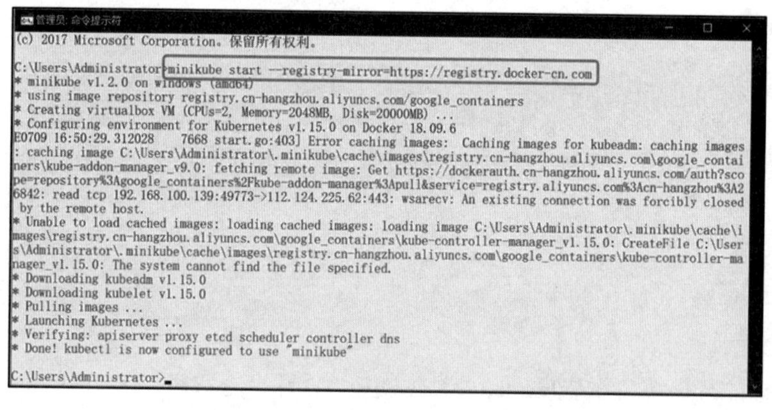

图 12-2

在开发测试时通常通过 minikube 来使用 Kubernetes，下载和安装可参考云栖社区的博文《Minikube-Kubernetes 本地实验环境》。

启动 minikube，此时关闭 Docker Desktop，以免产生影响。

```
$ minikube start --registry-mirror=https://*.mirror.aliyuncs.com
```

--registry-mirror 使用个人的阿里云“容器镜像服务控制台”（阿里云控制台→弹性计算→容器镜像服务）的“镜像加速器”的“加速器地址”，可自行申请。

在 Windows 下请关闭 Hyper-V，若需要使用 Hyper-V 作为虚拟化技术，请参考上面的安装文档。

（3）有客户端查询集群信息，如图 12-3 所示。

```
$ kubectl cluster-info
```

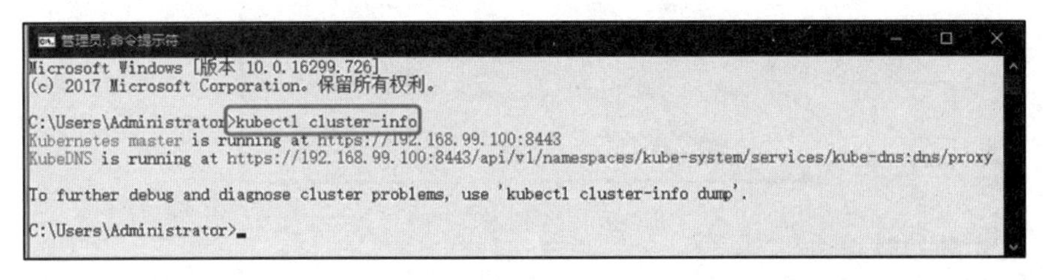

图 12-3

（4）启动 Dashboard，如图 12-4 所示。

```
$ minikube dashboard
```

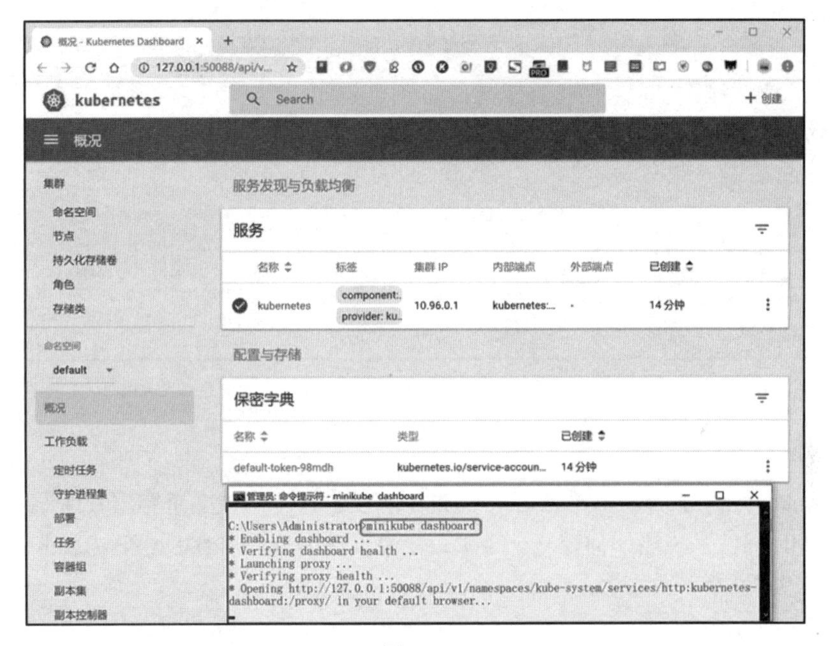

图 12-4

12.1.2　Kubernetes 基础知识

1. 节点 （Node）

Node 即为 Kubernetes 的节点；minikube 为单节点的集群，只有一个 Node。

（1）通过客户端命令查看节点，如图 12-5 所示。

kubectl get nodes

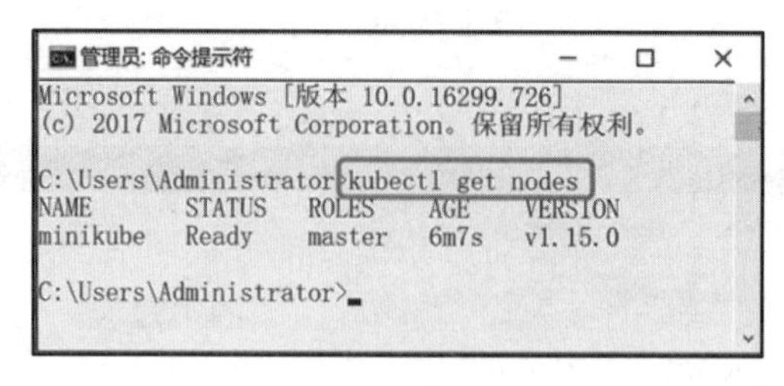

图 12-5

（2）通过 Dashboard 查看节点，如图 12-6 所示。

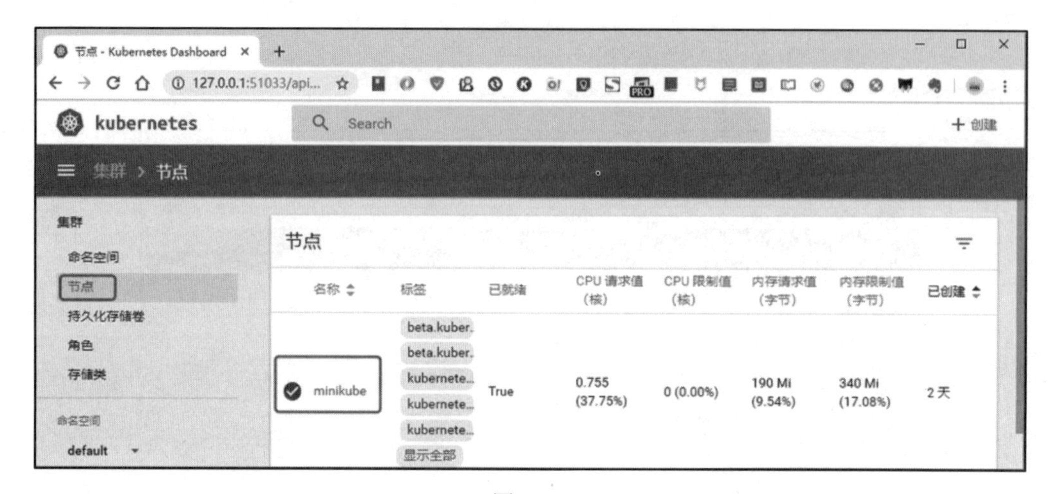

图 12-6

2. 容器组（Pod）

Pod 是 Kubernetes 中的最小单元，一个 Pod 可以包含一个或多个容器（Docker）。我们可以用一个 YAML 文件来描述如何建立一个 Pod。可以使用 Intellij IDEA 安装 Kubernetes Plugin，进行 YAML 的编辑。

（1）tomcat-pod.yml：

```
apiVersion: v1
kind: Pod # 类型为 Pod
metadata:
  name: tomcat
  labels:
    app: mytomcat # label 为元数据，用于给其他组件匹配使用
spec:
```

```
containers:
- name: tomcat
  image: tomcat:8 # Tomcat Docker 镜像
  resources:
    limits:
      memory: "128Mi" # 限制容器内存
      cpu: "500m" # 限制容器 CPU
  ports:
    - containerPort: 8080
```

（2）部署 Pod。

使用 kubectl 部署 Pod。

```
$ kubectl create -f tomcat-pod.yaml
```

这时会下载 Tomcat 的 Docker 镜像，需要一段时间才能完成。

（3）查看 Pod。

在客户端查看 Pod，如图 12-7 所示。

```
$ kubectl get pods
```

```
wangyuneideMBP3:basic wangyunfei$ kubectl get pods
NAME      READY    STATUS     RESTARTS    AGE
tomcat    1/1      Running    0           6m11s
```

图 12-7

通过 Dashboard 查看 Pod，如图 12-8 所示。

图 12-8

（4）查看 Pod 详情。

在客户端查看 Pod 详情，如图 12-9 所示。

```
$ kubectl describe pods/tomcat
```

```
wangyuneideMBP3:basic wangyunfei$ kubectl describe pods/tomcat
Name:         tomcat
Namespace:    default
Priority:     0
Node:         minikube/10.0.2.15
Start Time:   Thu, 11 Jul 2019 21:34:13 +0800
Labels:       app=mytomcat
Annotations:  <none>
Status:       Running
IP:           172.17.0.5
Containers:
  tomcat:
    Container ID:   docker://9e52505d48806a0175dc40040a2a003771d6447556c5dea12a9e468b91840f31
    Image:          tomcat:8
    Image ID:       docker-pullable://tomcat@sha256:2540479366f71fa58609145c07bbff8a24cfaa7aa772094e3a3145e7147325a1
    Port:           8080/TCP
    Host Port:      0/TCP
    State:          Running
      Started:      Thu, 11 Jul 2019 21:34:15 +0800
    Ready:          True
    Restart Count:  0
    Limits:
      cpu:     500m
      memory:  128Mi
    Requests:
      cpu:        500m
      memory:     128Mi
    Environment:  <none>
    Mounts:
      /var/run/secrets/kubernetes.io/serviceaccount from default-token-tdnm2 (ro)
Conditions:
  Type              Status
  Initialized       True
  Ready             True
  ContainersReady   True
  PodScheduled      True
Volumes:
  default-token-tdnm2:
    Type:        Secret (a volume populated by a Secret)
    SecretName:  default-token-tdnm2
    Optional:    false
QoS Class:       Guaranteed
Node-Selectors:  <none>
Tolerations:     node.kubernetes.io/not-ready:NoExecute for 300s
                 node.kubernetes.io/unreachable:NoExecute for 300s
Events:
  Type    Reason     Age   From               Message
  ----    ------     ----  ----               -------
  Normal  Scheduled  42s   default-scheduler  Successfully assigned default/tomcat to minikube
  Normal  Pulled     40s   kubelet, minikube  Container image "tomcat:8" already present on machine
  Normal  Created    40s   kubelet, minikube  Created container tomcat
  Normal  Started    40s   kubelet, minikube  Started container tomcat
```

图 12-9

单击 Pod 列表中的“tomcat”查看 Pod 的详情，如图 12-10 所示。

相应的资源在 Dashboard 中都有相关信息展示，后面将减少 Dashboard 的相关截图。

（5）登录容器，如图 12-11 所示。

```
$ kubectl exec -it tomcat -- /bin/bash
```

图 12-10

```
wangyuneideMBP3:~ wangyunfei$ kubectl exec -it tomcat -- /bin/bash
root@tomcat:/usr/local/tomcat# ls
BUILDING.txt  CONTRIBUTING.md  LICENSE  NOTICE  README.md  RELEASE-NOTES
root@tomcat:/usr/local/tomcat# pwd
/usr/local/tomcat
root@tomcat:/usr/local/tomcat# 
```

图 12-11

3. 副本集（Replica Set）

Replica Set 可用来控制 Pod 副本的运行数量，同样使用 YAML 文件来部署 Replica Set。

（1）tomcat-rs.yaml：

```yaml
apiVersion: apps/v1
kind: ReplicaSet  #类型为 ReplicaSet
metadata:
  name: tomcat-rs
  labels:
    app: mytomcat
spec:
  replicas: 3
  selector:
    matchLabels:
      app: mytomcat #匹配 Pod 的元数据，将此副本集应用到该 Pod
  template:
    metadata:
      labels:
        app: mytomcat
    spec:
      containers:
```

```
    - name: tomcat
      image: tomcat:8
```

（2）部署 Replica Set：

```
$ kubectl create -f tomcat-rs.yaml
```

（3）查看 Replica Set 和 Pod，如图 12-12 所示。

```
$ kubectl get rs
$ kubectl get pods
```

```
wangyuneideMBP3:basic wangyunfei$ kubectl get rs
NAME        DESIRED   CURRENT   READY   AGE
tomcat-rs   3         3         3       5s
wangyuneideMBP3:basic wangyunfei$ kubectl get pods
NAME              READY   STATUS    RESTARTS   AGE
tomcat            1/1     Running   0          17s
tomcat-rs-54tbz   1/1     Running   0          13s
tomcat-rs-ts4pk   1/1     Running   0          13s
```

图 12-12

4. 服务（Service）

Service 将一组 Pod 组织在一起，外界不仅可以通过 Service 名来访问，还可以实现 Pod 的负载均衡。这里类似于第 11.2.1 节的“服务发现”功能。

tomcat-service.yaml。

```
apiVersion: v1
kind: Service # 类型为 Service
metadata:
  name: tomcat-service
  labels:
    app: mytomcat
spec:
  type: NodePort #Service 类型
  ports:
    - port: 30000
      nodePort: 30000 # 节点端口号
      targetPort: 8080 # 目标端口
  selector:
    app: mytomcat # 匹配 Pod
```

Service 有以下 3 种类型。

◎ NodePort：通过 Node 的静态端口暴露服务。

◎ LoadBalancer：使用云平台提供商的 Load Balancer（如阿里云会提供一个公网 IP 地址）。

◎ ClusterIP：只能在 Kubernetes 内部可以访问的服务。

（1）部署 Service。

```
$ kubectl create -f tomcat-service.yaml
```

（2）查看并描述服务，如图 12-13 所示。

```
$ kubectl get services
$ kubectl describe services/tomcat-service
```

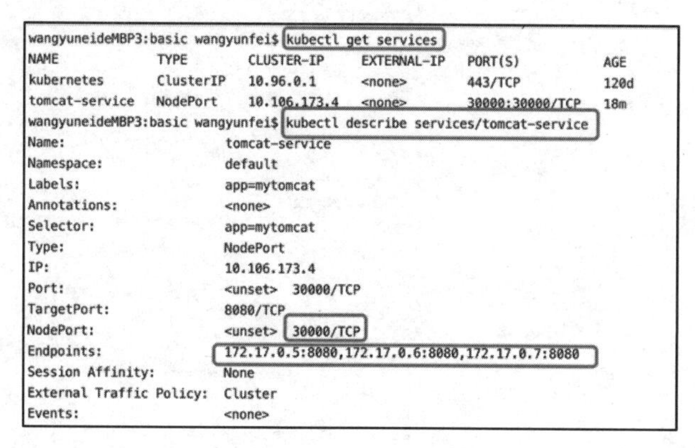

图 12-13

（3）访问服务。

这里使用的是 NodePort 访问服务，即 minikube 使用的是 VirtualBox 的 IP 地址，可以通过下面命令获得 Node 的 IP 地址。

```
$ minikube ip
```

或者使用下面的命令直接打开服务，显示如图 12-14 所示。

```
$ minikube service tomcat-service
```

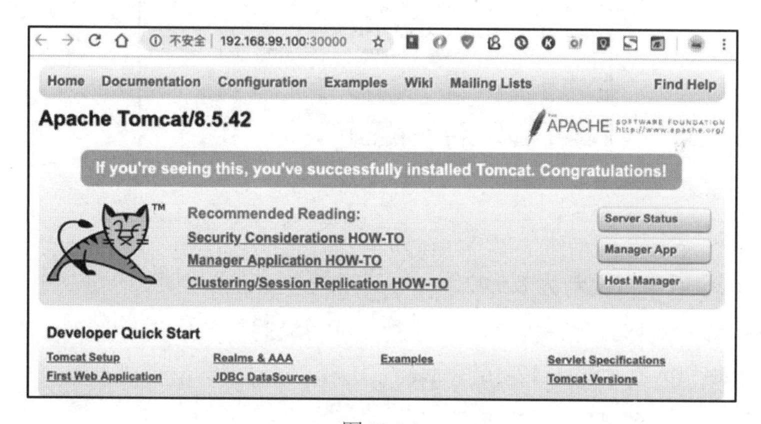

图 12-14

5. 部署（Deployment）

Deployment 可管理 Replica Set 和 Pod 的部署，它具备更新 Replica Set 和回滚部署的能力。

（1）tomcat-deployment.yaml。

```
apiVersion: apps/v1
kind: Deployment #
metadata:
  name: tomcat-deployment
  labels:
    app: mytomcat
spec:
  replicas: 4
  selector:
    matchLabels:
      app: mytomcat
  template:
    metadata:
      labels:
        app: mytomcat
    spec:
      containers:
        - name: tomcat
          image: tomcat:8
```

（2）部署 Deployment。

```
$ kubectl create -f tomcat-deployment.yaml
```

（3）查看 Deployment，如图 12-15 所示。

```
$ kubectl get deployments
$ kubect get pods
```

```
wangyuneideMBP3:basic wangyunfei$ kubectl get deployments
NAME                READY   UP-TO-DATE   AVAILABLE   AGE
tomcat-deployment   4/4     4            4           11s
wangyuneideMBP3:basic wangyunfei$ kubectl get pods
NAME               READY   STATUS    RESTARTS   AGE
tomcat             1/1     Running   0          2m19s
tomcat-rs-54tbz    1/1     Running   0          2m15s
tomcat-rs-bf7g7    1/1     Running   0          15s
tomcat-rs-ts4pk    1/1     Running   0          2m15s
```

图 12-15

（4）更新部署。

将 Tomcat 镜像版本号更新为 8.5.42。

```
$ kubectl set image deployment/tomcat-deployment tomcat=tomcat:9
```

执行下面的命令，显示如图 12-16 所示。

```
$ minikube service tomcat-service
```

若页面没有更新，请多刷新几次或清理浏览器缓存。

（5）更新 Dashboard，如图 12-17 和图 12-18 所示。

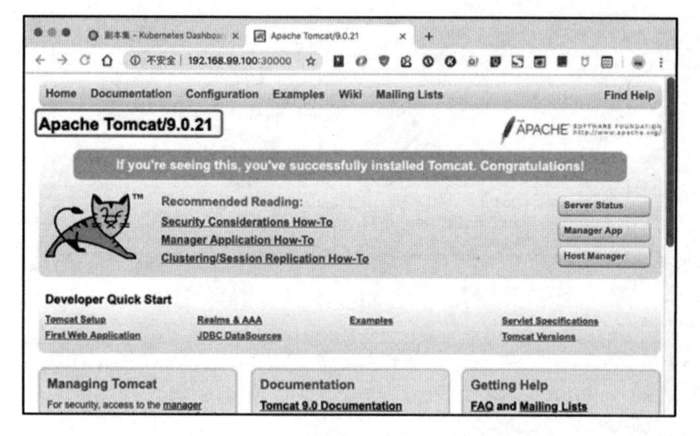

图 12-16

修改 部署

```
1  {
2    "kind": "Deployment",
3    "apiVersion": "extensions/v1beta1",
4    "metadata": {
5      "name": "tomcat-deployment",
6      "namespace": "default",
7      "selfLink": "/apis/extensions/v1beta1/namespaces/default/d
8      "uid": "b441b19a-358e-41b1-bbf9-1772b8980c15",
9      "resourceVersion": "79589",
10     "generation": 2,
11     "creationTimestamp": "2019-07-11T13:38:15Z",
12     "labels": {
13       "app": "mytomcat"
14     },
15     "annotations": {
16       "deployment.kubernetes.io/revision": "2"
17     }
18   },
19   "spec": {
20     "replicas": 5,
21     "selector": {
22       "matchLabels": {
23         "app": "mytomcat"
24       }
```

取消　复制　更新

图 12-18

更新后 Pod 的副本个数为 5 个，如图 12-19 所示。

```
wangyuneideMBP3:basic wangyunfei$ kubectl get pods
NAME                                READY   STATUS    RESTARTS   AGE
tomcat                              1/1     Running   0          17m
tomcat-deployment-677c87b74d-2sgkd  1/1     Running   0          11m
tomcat-deployment-677c87b74d-5glxt  1/1     Running   0          11m
tomcat-deployment-677c87b74d-gc9wd  1/1     Running   0          11m
tomcat-deployment-677c87b74d-l4c5w  1/1     Running   0          11m
tomcat-deployment-677c87b74d-s9qmf  1/1     Running   0          2m21s
```

图 12-19

（6）回滚。

查看回滚历史，显示如图 12-20 所示。

```
$ kubectl rollout history deployments/tomcat-deployment
```

```
wangyuneideMBP3:basic wangyunfei$ kubectl rollout history deployments/tomcat-deployment
deployment.extensions/tomcat-deployment
REVISION  CHANGE-CAUSE
1         <none>
2         <none>
```

图 12-20

回滚到第一个版本，如图 12-21 所示。

```
$ kubectl rollout undo deployment.v1.apps/tomcat-deployment --to-revision=1
```

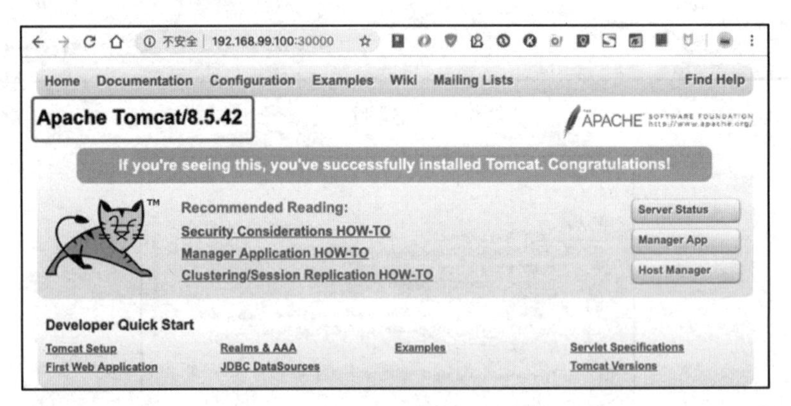

图 12-21

6. 访问权（Ingress）

访问权用来定义外部访问集群内部服务的规则，这里类似于“Gateway”的功能。

（1）在 minikube 下，开启 NGINX Ingress Controller。

```
$ minikube addons enable ingress
```

此步操作需下载两个镜像，请耐心等待。使用下面的命令可查看 NGINX Ingress Controller

的 Pod 是否已经安装完成，如图 12-22 所示。

```
$ kubectl get pods -n kube-system
```

```
wangyuneideMBP3:basic wangyunfei$ kubectl get pods -n kube-system
NAME                                        READY   STATUS    RESTARTS   AGE
coredns-6967fb4995-8tk4t                    1/1     Running   6          2d18h
coredns-6967fb4995-ldslz                    1/1     Running   6          2d18h
default-http-backend-66664b9769-z8ntk       1/1     Running   2          12h
etcd-minikube                               1/1     Running   3          2d18h
kube-addon-manager-minikube                 1/1     Running   3          2d18h
kube-apiserver-minikube                     1/1     Running   3          2d18h
kube-controller-manager-minikube            1/1     Running   3          2d18h
kube-proxy-gl889                            1/1     Running   3          2d18h
kube-scheduler-minikube                     1/1     Running   3          2d18h
kubernetes-dashboard-95564f4f-b6brw         1/1     Running   5          2d18h
nginx-ingress-controller-7b465d9cf8-sw5gr   1/1     Running   0          12h
storage-provisioner                         1/1     Running   7          120d
```

图 12-22

-n kube-system 查看的是 namespace 为 kube-system 下的 Pod，前面所有操作的 namespace 都是 default。

若 nginx-ingress-controller 镜像提示拉取失败，则可先使用下面的命令登录 minikube。

```
$ minikube ssh
```

使用下面的命令手动拉取镜像。

```
$ docker pull
quay.io/kubernetes-ingress-controller/nginx-ingress-controller:0.23.0
```

镜像版本可能会有变化。

（2）编写 Ingress 文件。

ingress.yaml：

```
apiVersion: networking.k8s.io/v1beta1
kind: Ingress
metadata:
  name: my-ingress
  annotations:
    nginx.ingress.kubernetes.io/rewrite-target: /$1
    nginx.ingress.kubernetes.io/ssl-redirect: "false"
spec:
  rules:
    - http:
        paths:
          - path: /tomcat(.*) #映射路径
            backend:
              serviceName: tomcat-service # 服务名
              servicePort: 30000 # 服务端口号
```

（3）创建 Ingress。

```
$ kubectl create -f ingress.yaml
```

（4）查看 Ingress，如图 12-23 所示。

```
$ kubectl get ingress
$ kubectl describe ingress/my-ingress
```

```
wangyuneideMBP3:basic wangyunfei$ kubectl get ingress
NAME         HOSTS   ADDRESS     PORTS   AGE
my-ingress   *       10.0.2.15   80      13m
wangyuneideMBP3:basic wangyunfei$ kubectl describe ingress/my-ingress
Name:             my-ingress
Namespace:        default
Address:          10.0.2.15
Default backend:  default-http-backend:80 (172.17.0.7:8080)
Rules:
  Host  Path  Backends
  ----  ----  --------
  *
        /tomcat(.*)   tomcat-service:30000 (172.17.0.2:8080)
Annotations:
  nginx.ingress.kubernetes.io/rewrite-target:  /$1
  nginx.ingress.kubernetes.io/ssl-redirect:    false
Events:
  Type    Reason  Age                  From                      Message
  ----    ------  ----                 ----                      -------
  Normal  CREATE  13m                  nginx-ingress-controller  Ingress default/my-ingress
  Normal  UPDATE  2m12s (x5 over 12m)  nginx-ingress-controller  Ingress default/my-ingress
```

图 12-23

（5）通过 Ingress 访问。

通过 http://192.168.99.100/tomcat 访问 Tomcat，如图 12-24 所示。

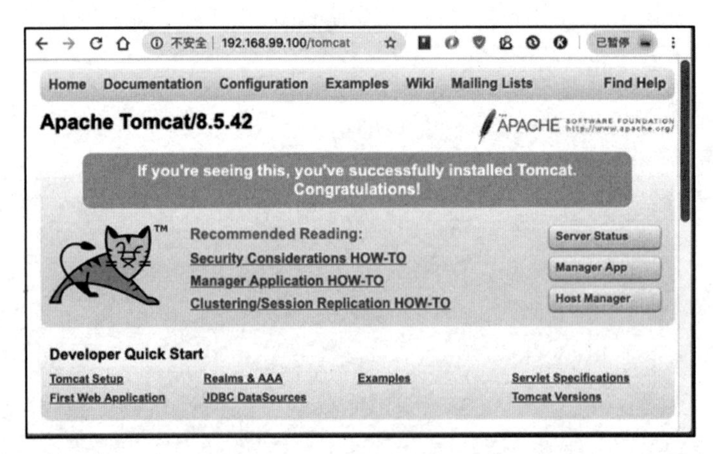

图 12-24

7. 数据存储

Kubernetes 可以通过 Storage Class 来定义存储，并使用 Persistent Volume Claim 来动态索取存储。Kubernetes 支持的 Storage Class 可查看官网，本节使用 minkube 默认的 Storage Class：standard。

（1）查看存储类，如图 12-25 所示。

```
$ kubectl get storageclass
$ kubectl describe storageclass/standard
```

```
wangyuneideMBP3:basic wangyunfei$ kubectl get storageclass
NAME                PROVISIONER                AGE
standard (default)  k8s.io/minikube-hostpath   121d
wangyuneideMBP3:basic wangyunfei$ kubectl describe storageclass/standard
Name:                  standard
IsDefaultClass:        Yes
Annotations:           storageclass.beta.kubernetes.io/is-default-class=true
Provisioner:           k8s.io/minikube-hostpath
Parameters:            <none>
AllowVolumeExpansion:  <unset>
MountOptions:          <none>
ReclaimPolicy:         Delete
VolumeBindingMode:     Immediate
Events:                <none>
```

图 12-25

（2）定义 Persistent Volume Claim。

pvc.yml：向 Storage Class 声明的存储 standard 索取 500MB 的存储。

```
apiVersion: v1
kind: PersistentVolumeClaim
metadata:
  name: my-pvc
spec:
  accessModes:
    - ReadWriteMany #访问类型
  storageClassName: standard # Storage Class
  resources:
    requests:
      storage: 500Mi # 索取 500MB
```

访问类型有 3 种。

◎　ReadWriteOnce：可以被一个 node 挂载读写。

◎　ReadOnlyMany：可以被多个 node 挂载只读。

◎　ReadWriteMany：可以被多个 node 挂载读写。

通过下面的命令创建 Persistent Volume Claim。

```
$ kubectl create -f pvc.yml
```

查看 Persistent Volume Claim，如图 12-26 所示。

```
$ kubectl get pvc
$ kubectl describe pvc/my-pvc
```

此时 Kubernetes 创建了一个 Persistence Volume，它代表存储本身。通过下面的命令可查看新建的 Persistence Volume，如图 12-27 所示。

```
$ kubectl get pv
$ kubectl describe pv/pvc-4cde457a-ad37-45ab-89ca-034145e70f29
```

```
wangyuneideMBP3:basic wangyunfei$ kubectl get pvc
NAME     STATUS   VOLUME                                     CAPACITY   ACCESS MODES   STORAGECLASS   AGE
my-pvc   Bound    pvc-4cde457a-ad37-45ab-89ca-034145e70f29   500Mi      RWX            standard       54s
wangyuneideMBP3:basic wangyunfei$ kubectl describe pvc/my-pvc
Name:          my-pvc
Namespace:     default
StorageClass:  standard
Status:        Bound
Volume:        pvc-4cde457a-ad37-45ab-89ca-034145e70f29
Labels:        <none>
Annotations:   control-plane.alpha.kubernetes.io/leader:
                 {"holderIdentity":"ce03801a-a44d-11e9-b001-080027fbc0f7","leaseDurationSeconds":15,"acquireTime":
               pv.kubernetes.io/bind-completed: yes
               pv.kubernetes.io/bound-by-controller: yes
Finalizers:    [kubernetes.io/pvc-protection]
Capacity:      500Mi
Access Modes:  RWX
VolumeMode:    Filesystem
Mounted By:    <none>
Events:
  Type    Reason                 Age   From                                                                    Message
  ----    ------                 ----  ----                                                                    -------
  Normal  Provisioning           105s  k8s.io/minikube-hostpath ce03801a-a44d-11e9-b001-080027fbc0f7  External pro
  Normal  ProvisioningSucceeded  105s  k8s.io/minikube-hostpath ce03801a-a44d-11e9-b001-080027fbc0f7  Successfully
```

图 12-26

```
wangyuneideMBP3:basic wangyunfei$ kubectl get pv
NAME                                       CAPACITY   ACCESS MODES   RECLAIM POLICY   STATUS   CLAIM            STORAGECLASS   REASON   AGE
pvc-4cde457a-ad37-45ab-89ca-034145e70f29   500Mi      RWX            Delete           Bound    default/my-pvc   standard                8m52s
wangyuneideMBP3:basic wangyunfei$ kubectl describe pv/pvc-4cde457a-ad37-45ab-89ca-034145e70f29
Name:            pvc-4cde457a-ad37-45ab-89ca-034145e70f29
Labels:          <none>
Annotations:     hostPathProvisionerIdentity: ce037fbd-a44d-11e9-b001-080027fbc0f7
                 pv.kubernetes.io/provisioned-by: k8s.io/minikube-hostpath
Finalizers:      [kubernetes.io/pv-protection]
StorageClass:    standard
Status:          Bound
Claim:           default/my-pvc
Reclaim Policy:  Delete
Access Modes:    RWX
VolumeMode:      Filesystem
Capacity:        500Mi
Node Affinity:   <none>
Message:
Source:
    Type:          HostPath (bare host directory volume)
    Path:          /tmp/hostpath-provisioner/pvc-4cde457a-ad37-45ab-89ca-034145e70f29
    HostPathType:
Events:            <none>
```

图 12-27

（3）使用 Persistent Volume Claim。

use-pvc-deployment.yaml：

```
apiVersion: apps/v1
kind: Deployment
metadata:
 name: use-pvc-deployment
spec:
 replicas: 2
 selector:
   matchLabels:
     app: another-tomcat # 匹配下面的定义的 Pod
 template:
   metadata:
     labels:
```

```
      app: another-tomcat
  spec:
    volumes:
      - name: demo # 通过 persistentVolumeClaim 定义 Volume，名称为 demo
        persistentVolumeClaim:
          claimName: my-pvc
    containers:
      - name: another-tomcat
        image: tomcat:9
        ports:
          - containerPort: 8080
        volumeMounts: # 挂在 volume
          - mountPath: "/files" #挂载在容器中的/files 目录下
            name: demo
```

部署。

```
$ kubectl create -f use-pvc-deployment.yaml
```

查看新部署的 Pods，如图 12-28 所示。

```
$ kubectl get pods -l app=another-tomcat
```

```
wangyuneideMBP3:basic wangyunfei$ kubectl get pods -l app=another-tomcat
NAME                                 READY   STATUS    RESTARTS   AGE
use-pvc-deployment-54c7d45784-fhz7v  1/1     Running   0          95s
use-pvc-deployment-54c7d45784-j2fdg  1/1     Running   0          95s
```

图 12-28

单击 Dashboard 上的“容器组”菜单，分别打开两个容器组。在每个容器组详情页面上单击“运行命令”选项，如图 12-29 所示。

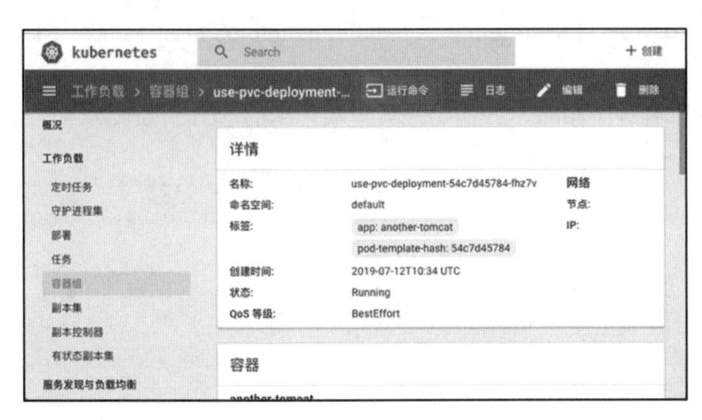

图 12-29

在一个容器的/files 目录下新建文件，同时在另外一个容器中能看见这个文件，如图 12-30 所示。

图 12-30

12.1.3 Helm

Helm 是 Kubernetes 包管理器，我们可以使用 Helm 来安装应用。一个 Helm 包叫作 Chart，存储和发布 Chart 的库叫作 Repository。

（1）安装 Helm 客户端。

macOS：使用 homebrew。

```
$ brew install kubernetes-helm
```

Windows：使用 Chocolatey（安装请参考 Chocolatey 官网）。

```
$ choco install kubernetes-helm
```

更详细的信息请参考：https://github.com/helm/helm/blob/master/docs/install.md 。

（2）初始化。

安装 Helm 服务端组件 tiller，并初始化客户端使用的 Repository。

```
$ helm init --upgrade -i
registry.cn-hangzhou.aliyuncs.com/google_containers/tiller:v2.14.2
--stable-repo-url  https://kubernetes.oss-cn-hangzhou.aliyuncs.com/charts/

$ helm repo add incubator
https://kubernetes.oss-cn-hangzhou.aliyuncs.com/charts-incubator/

$ helm repo update
```

（3）查看和检索包，如图 12-31 所示。

```
$ helm ls # 查看集群已安装 Chart，目前没有安装
$ helm search postgres # 搜索 Repository 的包
```

```
wangyuneideMBP3:helm wangyunfei$ helm search postgres
NAME                    CHART VERSION   APP VERSION     DESCRIPTION
stable/postgresql       0.9.1                           Object-relational database management system (ORDBMS) wit...
incubator/patroni       0.6.1           1.3-p4          Highly available elephant herd: HA PostgreSQL cluster.
stable/gcloud-sqlproxy  0.2.3                           Google Cloud SQL Proxy
wangyuneideMBP3:helm wangyunfei$ helm search mysql
NAME                            CHART VERSION   APP VERSION     DESCRIPTION
incubator/mysqlha               0.1.1                           MySQL cluster with a single master and zero or more slave...
stable/mysql                    0.3.5                           Fast, reliable, scalable, and easy to use open-source rel...
stable/percona                  0.3.0                           free, fully compatible, enhanced, open source drop-in rep...
stable/percona-xtradb-cluster   0.0.2           5.7.19          free, fully compatible, enhanced, open source drop-in rep...
stable/gcloud-sqlproxy          0.2.3                           Google Cloud SQL Proxy
stable/mariadb                  2.1.6           10.1.31         Fast, reliable, scalable, and easy to use open-source rel...
```

图 12-31

（4）安装包。

通过下面的命令安装 PostgreSQL。

```
$ helm install stable/postgresql
```

一般来说，需要对默认的属性进行配置。可以在命令执行目录，配置一个 values.yaml 文件来覆盖配置。

Helm 的库的源码地址参见 GitHub。PostgreSQL 的具体配置文档可参考：https://github.com/helm/charts/tree/master/stable/postgresql。

values.yaml：

```
postgresqlUsername: wisely # 数据库管理员账号
postgresqlPassword: zzzzzz # 数据库管理员密码
postgresqlDatabase: first_db # 数据库
persistence:
  storageClass: standard # 数据使用 minikube 的 standard 存储类
```

安装 PostgreSQL，如图 12-32 所示。

```
$ helm install --name my-postgres -f values.yaml stable/postgresql
```

当前包的默认服务类型为 ClusterIP，首先转发 Service 端口到本机，然后使用 Navicat 连接 Kubernetes 中的 PostgreSQL，如图 12-33 所示。

```
$ kubectl port-forward svc/my-postgres-postgresql 5432:5432
```

svc 为 service 的缩写，上面的 pod 也可缩写为 po，deployment 可缩写为 deploy。

```
wangyuneideMBP3:postgresql wangyunfei$ helm install --name my-postgres -f values.yaml stable/postgresql
NAME:   my-postgres
LAST DEPLOYED: Fri Jul 12 21:26:18 2019
NAMESPACE: default
STATUS: DEPLOYED

RESOURCES:
==> v1/Pod(related)
NAME                       READY  STATUS    RESTARTS  AGE
my-postgres-postgresql-0   0/1    Init:0/1  0         0s

==> v1/Secret
NAME                    TYPE    DATA  AGE
my-postgres-postgresql  Opaque  1     0s

==> v1/Service
NAME                             TYPE       CLUSTER-IP    EXTERNAL-IP  PORT(S)   AGE
my-postgres-postgresql           ClusterIP  10.102.65.8   <none>       5432/TCP  0s
my-postgres-postgresql-headless  ClusterIP  None          <none>       5432/TCP  0s

==> v1beta2/StatefulSet
NAME                    READY  AGE
my-postgres-postgresql  0/1    0s


NOTES:
** Please be patient while the chart is being deployed **

PostgreSQL can be accessed via port 5432 on the following DNS name from within your cluster:

    my-postgres-postgresql.default.svc.cluster.local - Read/Write connection
To get the password for "wisely" run:

    export POSTGRES_PASSWORD=$(kubectl get secret --namespace default my-postgres-postgresql -o jsonpath="{.data.pos

To connect to your database run the following command:

    kubectl run my-postgres-postgresql-client --rm --tty -i --restart='Never' --namespace default --image docker.io/



To connect to your database from outside the cluster execute the following commands:

    kubectl port-forward --namespace default svc/my-postgres-postgresql 5432:5432 &
    PGPASSWORD="$POSTGRES_PASSWORD" psql --host 127.0.0.1 -U wisely -d first_db -p 5432
```

图 12-32

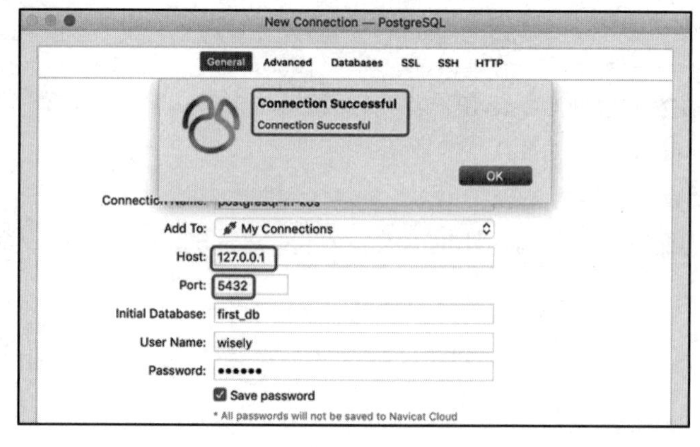

图 12-33

（4）卸载包。

```
$ helm delete my-postgres --purge
```

12.1.4 DevOps

1. DevOps

DevOps 是一种将软件开发（dev）和运维部署（ops）结合起来的过程、方法和系统的统称。使用 DevOps 后，开发和部署运维将成为一个完整的流程整体。

下面创建一个完整的研发环境。

◎ Docker Resgistry：用于存放 Docker 镜像的仓库，这里使用阿里云的镜像服务。

◎ 代码服务器：用来作为源码版本控制，这里使用 GitHub。

◎ 部署：使用 Helm 对应用进行部署。

◎ Jenkins：开源持续集成交付平台，这里使用 Helm 安装 Jenkins。

在整个 DevOps 的开发过程中最重要的是 Jenkins，通过定义 Jenkins 的流程，（pipeline）将开发和部署的各个环节串联起来。本章演示的 DevOps 流程如下。

（1）向 GitHub 提交代码。

（2）触发 Jenkins 流程。

（3）从流程中拉取 GitHub 代码。

（4）在流程中，把 Spring Boot 程序编译成 Jar 包。

（5）在流程中，将 Jar 包编译到 docker 镜像，并推送到阿里云镜像服务。

（6）在流程中，使用 Helm 拉取阿里云中的应用镜像部署应用。

2. 修改 minikube 配置

（1）创建高配置的 minikube。

```
$ minikube stop # 停止当前minikube
$ minikube delete # 删除当前minikube
$ minikube start --registry-mirror=https://*.mirror.aliyuncs.com --cpus 8
--memory 8192 --disk-size 150g
```

（2）配置 Helm。

```
$ helm init --upgrade -i
registry.cn-hangzhou.aliyuncs.com/google_containers/tiller:v2.14.2
--stable-repo-url  https://kubernetes.oss-cn-hangzhou.aliyuncs.com/charts/

$ helm repo add incubator
https://kubernetes.oss-cn-hangzhou.aliyuncs.com/charts-incubator/

$ helm repo add aliyun-stable
https://aliacs-app-catalog.oss-cn-hangzhou.aliyuncs.com/charts/ # 添加阿里
# 云Helm库

$ helm repo update
```

12.1.5 安装 Jenkins

在线 Helm 库的 Jenkins Chart 版本较低，我们可以直接下载使用 GitHub 中的 Jenkins Chart 源码进行部署。其位置在 stable/jenkins 下，可拷贝 Jenkins Chart 的目录到你需要的位置，如图 12-34 所示。

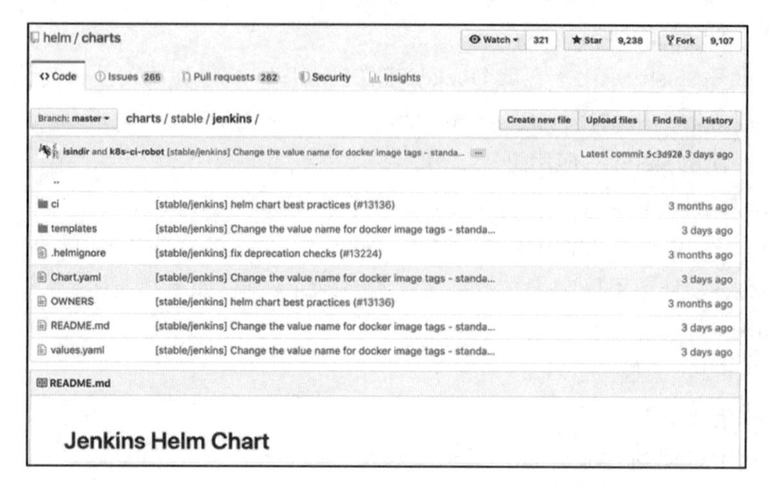

图 12-34

定制安装的 values.yaml：

```
master:
  serviceType: NodePort
  installPlugins:
    - kubernetes:1.17.2
    - workflow-job:2.33
    - workflow-aggregator:2.6
    - credentials-binding:1.19
    - git:3.10.1
    - blueocean:1.17.0
```

安装 Jenkins 和上述插件。最核心的插件是 kubernetes 插件，它可以在 Kubernetes 集群里动态运行 Jenkins 代理。在每一个代理启动时，kubernetes 插件都会为其创建 Kubernetes Pod，使用定义的 Docker 镜像运行，在编译结束后停止运行。我们会在后面定义 Jenkins 流程的 Jenkinsfile 文件中演示。

执行下面的命令安装 Jenkins。

```
$ helm install --name my-jenkins -f values.yaml jenkins/
```

上面命令中的 jenkins/为 Jenkins Chart 的源码目录，命令执行在源码的上级目录。执行结果如下。

通过

```
$ kubectl get --namespace default -o jsonpath="{.spec.ports[0].nodePort}"
services my-jenkins
```

获得当前端口号为 30705。

Jenkins 的访问地址为 http://192.168.99.101:30705/login，登录的账号为 wisely，密码为 zzzzzz。登录后如图 12-35 所示。

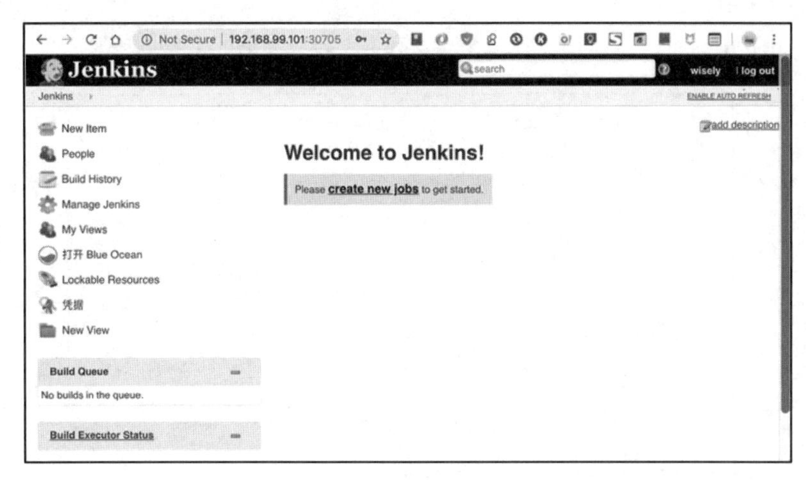

图 12-35

最后，为安装时自动创建的用户 my-jenkins 绑定集群管理角色。

```
$ kubectl create clusterrolebinding my-jenkins-cluster-rule
--clusterrole=cluster-admin --serviceaccount=default:my-jenkins
```

12.1.6　微服务示例

1. 新建应用

Group：top.wisely。

Artifact：demo-service。

Dependencies：Spring Web Starter 和 Spring Boot Actuator。

build.gradle 文件中的依赖如下：

```
dependencies {
    implementation 'org.springframework.boot:spring-boot-starter-actuator'
    implementation 'org.springframework.boot:spring-boot-starter-web'
  //……
}
```

新建一个 HelloController 类作为演示：

```
@RestController
public class HelloController {
   @GetMapping("/")
   public String hello(){
      return "Hello Spring Boot in k8s";
   }
}
```

2. 将代码发布到 GitHub

（1）在 GitHub 页面新建应用，名称为 demo-service，如图 12-36 所示。

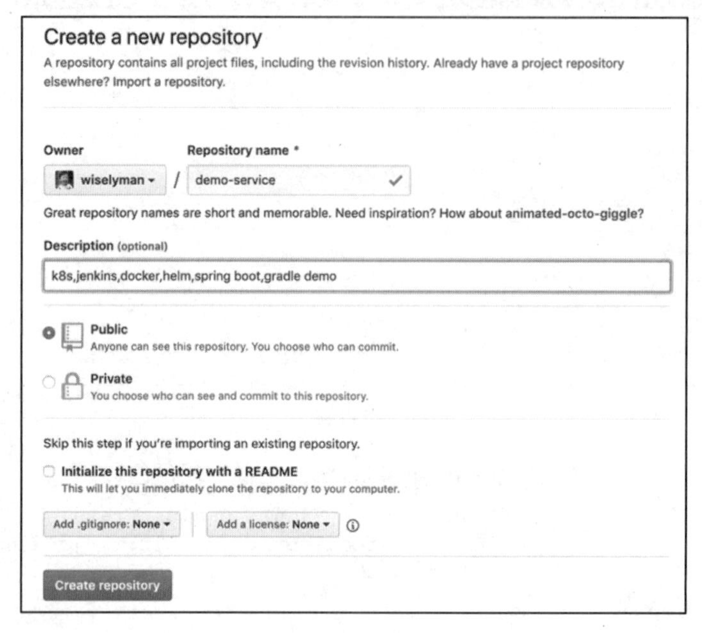

图 12-36

（2）通过 git 命令将源码推送到 GitHub，如图 12-37 所示。

```
$ cd demo-service # 进入新建的源码目录
$ git init # 初始化 Git
$ git add . # 添加所有的文件到 Git
$ git commit -m "first commit" # 提交代码到本地 Git
$ git remote add origin https://github.com/wiselyman/demo-service.git
# 添加远程 Git
$ git push -u origin master # 推送代码到远程 GitHub，此时可能需要输入 Github 账号
# 和密码
```

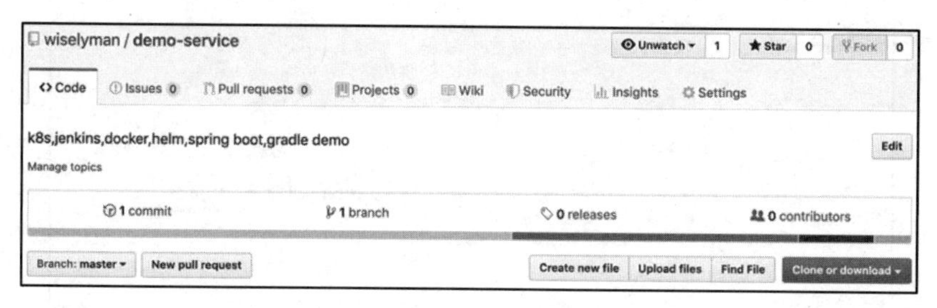

图 12-37

12.1.7 镜像仓库和 Dockerfile

1. 配置阿里云镜像仓库

在阿里云“容器镜像服务”控制台的“默认实例”下的“镜像仓库”页面，单击“创建镜像仓库”，如图 12-38 所示。

创建镜像仓库
1 仓库信息
2 代码源
地域 华东1（杭州）
* 命名空间 wiselyman
* 仓库名称 demo-service
长度为2-64个字符，可使用小写英文字母、数字，可使用分隔符"_"、"-"、"."（分隔符不能在首位或末位）
仓库类型 公开 私有
* 摘要 k8s,jenkins,docker,helm,spring boot,gradle demo
长度最长100个字符
描述信息
支持Markdown格式
下一步 取消

图 12-38

单击“下一步”按钮后，选择“本地仓库”选项，使用命令行推送镜像到仓库，如图 12-39 所示。

图 12-39

创建成功后，仓库列表会显示刚才创建的 demo-service，如图 12-40 所示。

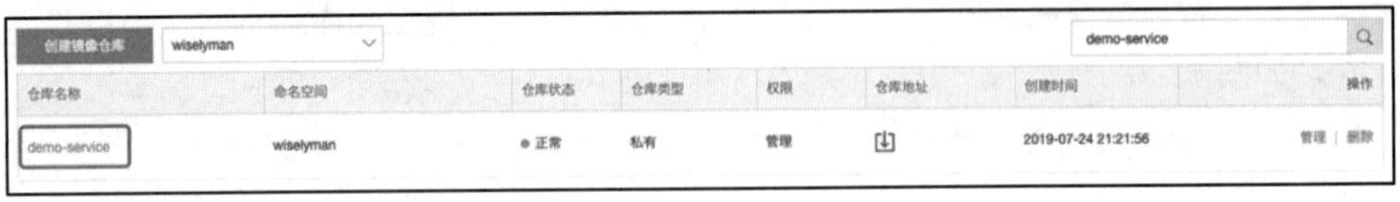

图 12-40

单击 demo-service，连接可查看仓库的连接登录方式，如图 12-41 所示。

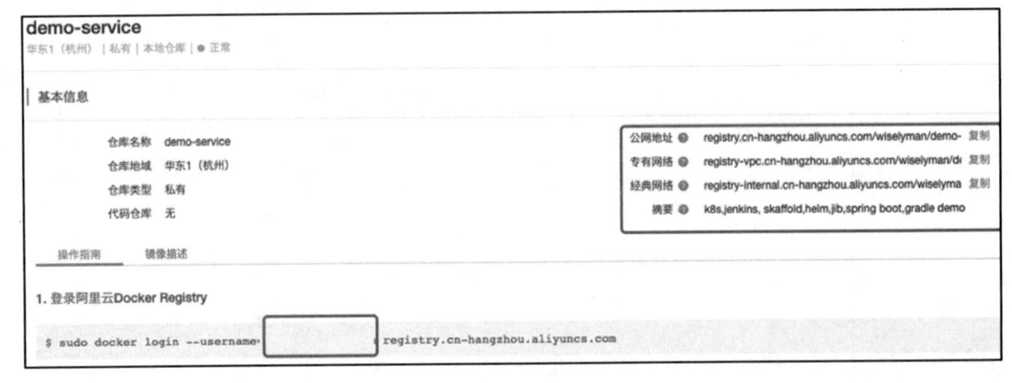

图 12-41

2. 编写 Dockerfile

在应用目录下添加 Dockerfile 文件，内容如下：

```
FROM java:8
RUN mkdir /app
ADD build/libs/demo-service-*.jar /app/app.jar # 编译后的 jar 包位于 build/libs 目录下
ADD runboot.sh /app/ # 添加启动的 shell 脚本 root.sh
RUN bash -c 'touch /app/app.jar'
WORKDIR /app
```

```
RUN chmod a+x runboot.sh
EXPOSE 8080
CMD /app/runboot.sh # 运行 runboot.sh
RUN echo "Asia/Shanghai" > /etc/timezone; # 调整时区
```

启动的 shell 脚本 runboot.sh 的内容如下：

```
java -Djava.security.egd=file:/dev/./urandom -jar /app/app.jar
```

3. 设置 Docker 编译环境

虽然不需要 Dockerfile 自己编译镜像，但是需要一个 Docker 运行环境来编译镜像，而 minikube 就包含一个 Docker 运行环境。

使用命令：

```
$ minikube docker-env
```

macOS 输出：

```
export DOCKER_TLS_VERIFY="1"
export DOCKER_HOST="tcp://192.168.99.101:2376"
export DOCKER_CERT_PATH="/Users/wangyunfei/.minikube/certs"
# Run this command to configure your shell:
# eval $(minikube docker-env)
```

Windows 输出：

```
SET DOCKER_TLS_VERIFY=1
SET DOCKER_HOST=tcp://192.168.99.101:2376
SET DOCKER_CERT_PATH=C:\Users\Administrator\.minikube\certs
REM Run this command to configure your shell:
REM @FOR /f "tokens=*" %i IN ('minikube docker-env') DO @%i
```

按照提示，在 macOS 下执行下面的命令：

```
$ eval $(minikube docker-env)
$ echo $DOCKER_HOST # 查看是否设置成功
```

在 Windows 下执行下面的命令：

```
@FOR /f "tokens=*" %i IN ('minikube docker-env') DO @%i
echo $DOCKER_HOST
```

此时 Docker 的编译环境已经设置完毕。

4. 编译和推送镜像

在当前应用目录下，可执行下面命令编译程序、编译镜像，以及推送镜像到仓库。

（1）登录阿里云镜像服务：

```
$ docker login --username=******@**.com registry.cn-hangzhou.aliyuncs.com
```

登录时会提示输入阿里云镜像仓库的密码：

```
wangyuneideMBP3:demo-service wangyunfei$ docker login --username=■ ■  ■■  com registry.cn-hangzhou.aliyuncs.com
Password:
Login Succeeded
```

（2）编译打包应用：

```
$ ./gradlew bootJar
```

编译生成 app.jar 到 build\libs 目录下，如图 12-42 所示。

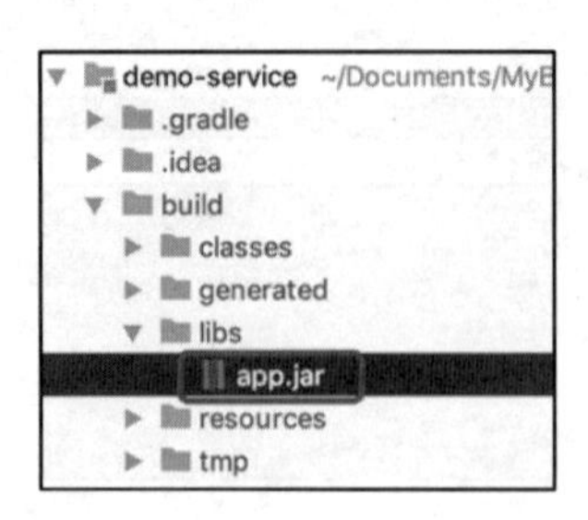

图 12-42

（3）编译镜像，如图 12-43 所示。

```
$ docker build -t
registry.cn-hangzhou.aliyuncs.com/wiselyman/demo-service:0.0.1-SNAPSHOT .
```

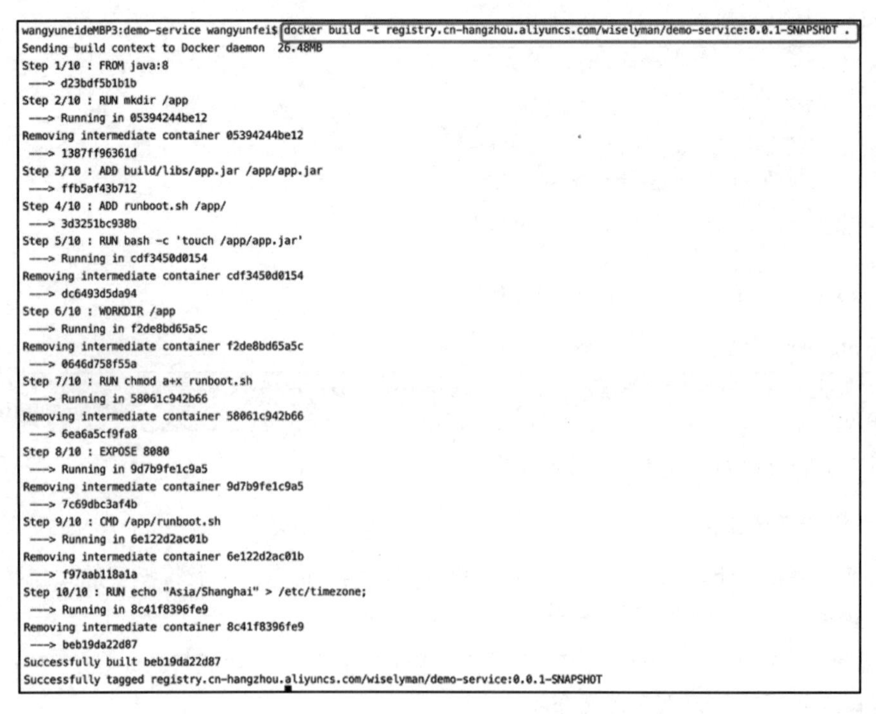

图 12-43

推送到阿里云镜像服务：

```
$ docker push
registry.cn-hangzhou.aliyuncs.com/wiselyman/demo-service:0.0.1-SNAPSHOT
```

控制台输出结果如图 12-44 所示。

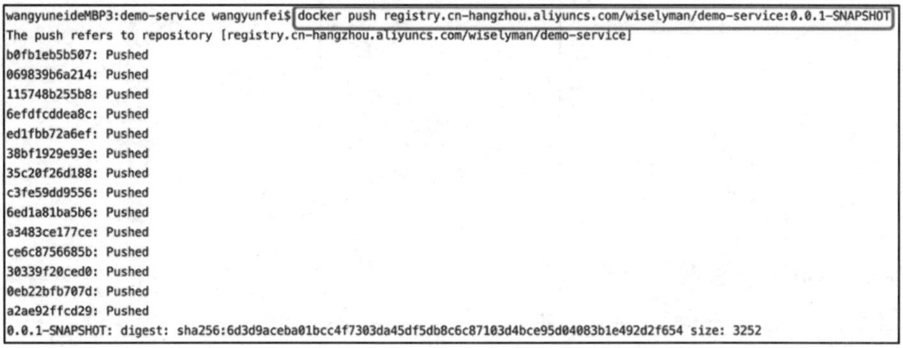

```
wangyuneideMBP3:demo-service wangyunfei$ docker push registry.cn-hangzhou.aliyuncs.com/wiselyman/demo-service:0.0.1-SNAPSHOT
The push refers to repository [registry.cn-hangzhou.aliyuncs.com/wiselyman/demo-service]
b0fb1eb5b507: Pushed
069839b6a214: Pushed
115748b255b8: Pushed
6efdfcddea8c: Pushed
ed1fbb72a6ef: Pushed
38bf1929e93e: Pushed
35c20f26d188: Pushed
c3fe59dd9556: Pushed
6ed1a81ba5b6: Pushed
a3483ce177ce: Pushed
ce6c8756685b: Pushed
30339f20ced0: Pushed
0eb22bfb707d: Pushed
a2ae92ffcd29: Pushed
0.0.1-SNAPSHOT: digest: sha256:6d3d9aceba01bcc4f7303da45df5db8c6c87103d4bce95d04083b1e492d2f654 size: 3252
```

图 12-44

在 demo-service 仓库下的“镜像版本”中可查看刚才推送的 Docker 镜像，如图 12-45 所示。

阿里云　搜索　费用　工单　备案　企业　支持与服务

demo-service

华东1（杭州）｜私有｜本地仓库｜● 正常

基本信息
仓库授权
触发器
镜像版本
镜像同步

镜像版本

版本	镜像ID	状态	Digest	镜像大小	最后更新时间
0.0.1-SNAPSHOT	beb19da22d87...	● 正常	6d3d9aceba01bcc4f7303da45df5db8c6c87103d4bce95d04083b1e492d2f654	267.876 MB	2019-07-29 10:38:57

图 12-45

注意，这里执行命令的目的是为了验证结果和理解原理，在开发中是不需要手动执行命令的。最后将本节的代码提交并推送到 GitHub。

12.1.8　使用 Helm 打包应用

1. 生成 Helm Chart

手动使用 Kubernetes 的 Deployment、Service 等编写 YAML 来部署还是很烦琐的，下面使用 Helm 命令来生成 Chart。Chart 是由 Kubernetes 的基本部署文件组成的。

进入应用目录，执行下面的命令：

```
$ cd demo-service
$ helm create demo-service # 此 demo-service 为 Chart 名称
```

这时在 demo-service 应用目录下生成了名为 demo-service 的 Chart 目录，如图 12-46 所示。

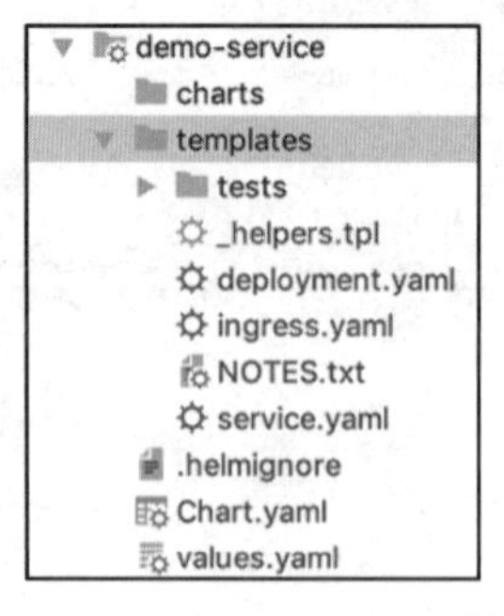

图 12-46

2. Helm Chart 文件解释

从图 13-45 可以看到，Chart 的主要功能都在 templates 目录下，这些文件的作用如下。

◎ templates/deployment.yaml：Kubernetes 的 Deployment YAML 文件。

◎ templates/service.yaml：Kubernetes 的 Service YAML 文件。

◎ templates/ingress.yaml: Kubernetes 的 Ingress YAML 文件。

◎ templates/_helpers.tpl：可以在模板中使用的助手函数。

◎ templates/NOTES.txt：安装 Helm Chart 时显示的说明文档。

◎ Chart.yaml：Helm Chart 的元数据信息。

◎ values.yaml：用在模板中的变量。

在 templates 下的 Kubernetes 模板文件，如 service.yaml：

```
apiVersion: v1
kind: Service
metadata:
  name: {{ include "demo-service.fullname" . }}
  labels:
{{ include "demo-service.labels" . | indent 4 }}
spec:
  type: {{ .Values.service.type }}
  ports:
    - port: {{ .Values.service.port }}
      targetPort: http
      protocol: TCP
      name: http
  selector:
    app.kubernetes.io/name: {{ include "demo-service.name" . }}
    app.kubernetes.io/instance: {{ .Release.Name }}
```

通过{{ .Values.service.type }}形式定义模板。在 values.yaml 中定义变量，替代模板中的值。

```
service.type: NodePort
```

当默认的模板文件不符合需求时，我们可以像对常规的 Kubernetes 的部署文件一样编辑它们，并配置成模板，最后使用 values.yaml 中的变量进行定制。

3. 定制 Chart

（1）添加阿里云镜像仓库的 Kubernetes Secret（保密字典），执行下面的命令，指定阿里云镜像库的账号和密码，secret 名称为 regsecret。

```
$ kubectl create secret docker-registry regsecret
--docker-server=registry.cn-hangzhou.aliyuncs.com
--docker-username=******@**.com --docker-password=******
--docker-email=******@**.com
```

（2）修改 templates/service.yaml：

```
apiVersion: v1
kind: Service
metadata:
  name: {{ include "demo-service.fullname" . }}
  labels:
{{ include "demo-service.labels" . | indent 4 }}
spec:
  type: {{ .Values.service.type }}
  ports:
    - port: {{ .Values.service.port }}
      targetPort: 8080 # 容器端口号为 8080
      protocol: TCP
      name: {{ .Values.service.name }} # 服务名由外部定制
  selector:
    app.kubernetes.io/name: {{ include "demo-service.name" . }}
    app.kubernetes.io/instance: {{ .Release.Name }}
```

（3）定制 deployment.yaml：

```
apiVersion: apps/v1
kind: Deployment
metadata:
  name: {{ include "demo-service.fullname" . }}
  labels:
{{ include "demo-service.labels" . | indent 4 }}
spec:
# ……
      containers:
        - name: {{ .Chart.Name }}
          image: "{{ .Values.image.repository }}:{{ .Values.image.tag }}"
          imagePullPolicy: {{ .Values.image.pullPolicy }}
          ports:
            - name: http
              containerPort: 8080 # 容器端口号
```

```
          protocol: TCP
        livenessProbe:
          httpGet:
            path: /actuator/health # 使用 Spring Boot Actuator 提供的健康端点进行活跃探# 测
            port: 8080
          initialDelaySeconds: 130
          timeoutSeconds: 10
          failureThreshold: 10
        readinessProbe:
          httpGet:
            path: /actuator/health # 使用 Spring Boot Actuator 提供的健康端点进行就绪探# 测
            port: 8080
          initialDelaySeconds: 60 # 不要设置太小事件进行检测，这时应用可能在启动中
          timeoutSeconds: 5
#……
```

（4）定制 values.yaml，下面未列出的部分没有修改，保持默认即可：

```
replicaCount: 2 # 运行实例个数

image:
  repository: registry.cn-hangzhou.aliyuncs.com/wiselyman/demo-service # 镜像仓库# 地址
  tag: 0.0.1-SNAPSHOT # 镜像版本
  pullPolicy: IfNotPresent

imagePullSecrets: [name: regsecret] # 阿里云镜像仓库访问授权
nameOverride: ""
fullnameOverride: ""

service:
  name: demo-service # 服务名
  type: NodePort # 服务类型为 NodePort
  port: 8080
#……
```

（5）使用 Helm 部署应用。在应用目录 demo-service 下执行下面的命令进行部署：

```
$ helm install --name demo-service demo-service/
```

控制台输出结果如图 12-47 所示。

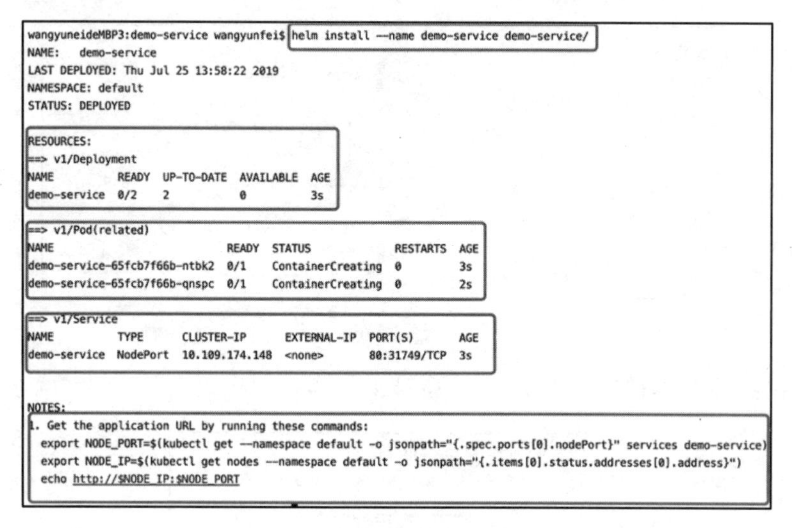

图 12-47

使用下面的命令查看服务端口号：

```
$ kubectl get --namespace default -o jsonpath="{.spec.ports[0].nodePort}"
services demo-service
```

当前环境随机分配的端口号为 32740；访问 http://192.168.99.101:32740/，可以看到部署的应用，如图 12-48 所示。

图 12-48

验证成功后删除应用：

```
$ helm del --purge demo-service
```

同样，这里只是进行验证演示，在生产中我们不会直接调用 Helm 命令进行部署。最后将 Helm Chart 的代码提交并推送至 GitHub。

12.1.9　Jenkins 流程

现在使用 Jenkins 的流程（pipeline）将这个部署的过程串起来，流程使用 Jenkinsfile 来表示。

1. 添加“凭据”

在 Jenkins 的页面单击左侧菜单的“凭据”选项，分别添加 GitHub 的凭据和阿里云镜像仓库的凭据，如图 12-49 所示。“凭据”即 credentials，指的是账号信息（虽然我们建的代码仓库是公开的，但在生产环境下，代码仓库一般都是私有的，需要账号和密码）。

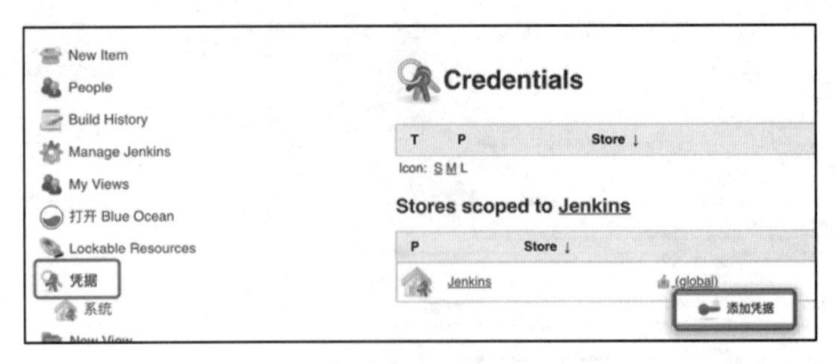

图 12-49

添加 GitHub 的凭据信息，如图 12-50 所示。

Kind: Username with password
Scope: 全局 (Jenkins, nodes, items, all child items, etc)
Username: ?126.com
Password: ....
ID: github

图 12-50

以同样的方式将阿里云镜像仓库的账号信息添加到凭据中，此时凭据列表如图 12-51 所示。

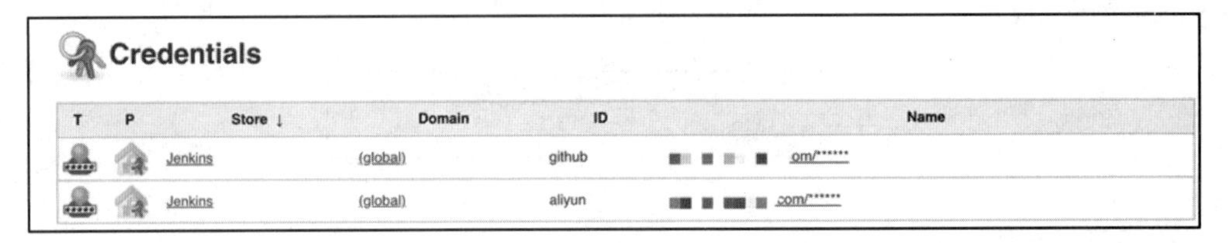

Credentials

T	P	Store ↓	Domain	ID	Name
		Jenkins	(global)	github	om/******
		Jenkins	(global)	aliyun	.com/******

图 12-51

2. 编写 Jenkinsfile

Jenkins 使用 Jenkinsfile 来定义自动部署流程；在应用目录下新增 Jenkinsfile，内容如下：

```
podTemplate(label: 'demo-service',serviceAccount: 'my-jenkins',containers: [ //a
  containerTemplate(name: 'gradle', image: 'gradle:5.5.1-jdk8', command: 'cat',
ttyEnabled: true), //b
  containerTemplate(name: 'docker', image: 'docker', command: 'cat', ttyEnabled: true),
//c
  containerTemplate(name: 'helm', image: 'lachlanevenson/k8s-helm:latest', command:
'cat', ttyEnabled: true) //d
],
volumes: [
  hostPathVolume(mountPath: '/home/gradle/.gradle', hostPath: '/tmp/jenkins/.gradle'),
  hostPathVolume(mountPath: '/var/run/docker.sock', hostPath: '/var/run/docker.sock')
```

```
]) {
  node('demo-service') {
      stage('Build Source Code to Jar file') { //e
          git url: 'https://github.com/wiselyman/demo-service.git', credentialsId:
'github', branch: 'master' //f
          container('gradle') {
            sh "gradle bootJar"  //g
            env.version =  sh(returnStdout: true, script: 'gradle properties -q | grep
"version:" | awk \'{print $2}\'').trim() //h
          }
      }

      stage('Build Docker Image and Push to Docker Registry') { //i
          container('docker'){
              docker.withRegistry("http://registry.cn-hangzhou.aliyuncs.com",
"aliyun"){
                docker.build("wiselyman/demo-service:${env.version}").push(env.version)
              } //j
          }
      }

      stage('Deploy to Kubernetes Cluster'){ //k
          container('helm'){
            sh "helm upgrade --install --force --set image.tag=${env.version}
demo-service  demo-service/" //l
      }
    }
  }
}
```

a. 定义 podTemplate，此处的 my-jenkins 为 Jenkins 用户，在前面已为它授权。

b. 编译应用的 Gradle 容器模板。

c. 编译 Docker 的 docker 容器模板。

d. 部署应用的 Helm 容器模板。

e. 流程第一阶段：编译应用为 jar 包。

f. 从 GitHub 拉取应用源码。

g. 编译打包应用。

h. 获取当前应用的版本号，并放置在环境变量中。

i. 流程第二阶段：编译应用 Docker 镜像并推送至阿里云镜像仓库。

j. 编译 Docker 镜像并推送至阿里云；其中凭据为 aliyun，版本从环境变量中获取。

k. 流程第三阶段：使用 Helm 部署应用。

l. 使用 helm upgrade --install 命令即可部署或更新应用。

编写完成后，提交代码并推送至 GitHub。

3. 创建任务

在 Jenkins 首页单击“create new job”选项，首先输入任务名称 demo-service，然后选择任务类型为“流水线”，最后单击“Pipeline”选项，如图 12-52 所示。

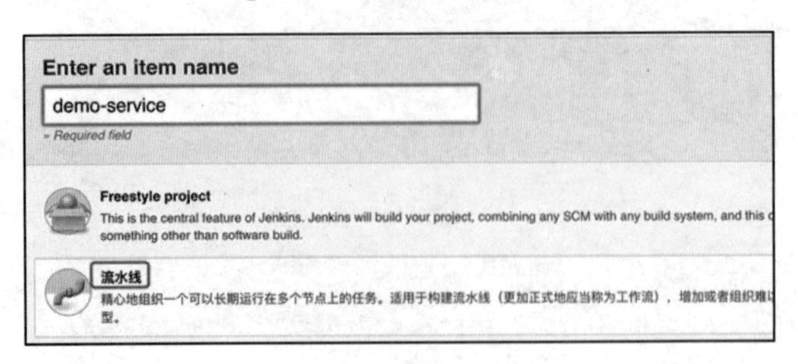

图 12-52

单击“OK”按钮后进入下一页。选择“Pipeline”选项，从 GitHub 文件中获取流程描述文件 Jenkinsfile，然后单击“Save”按钮，如图 12-53 所示。

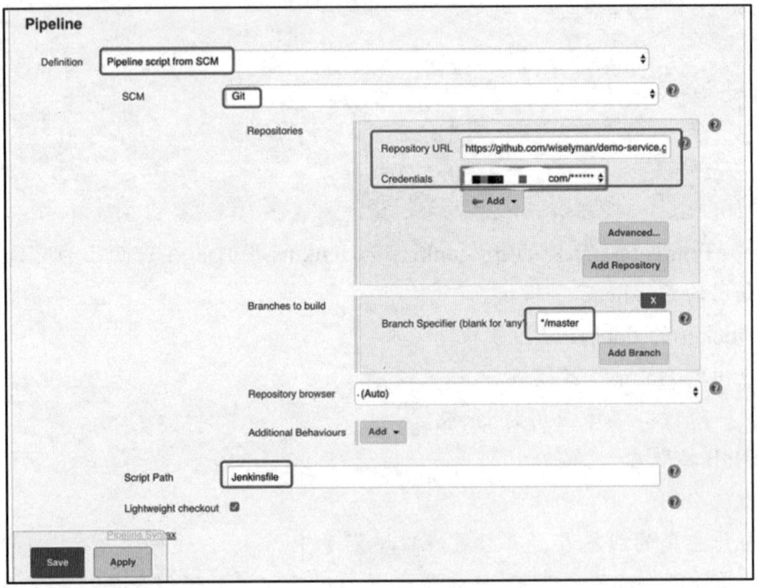

图 12-53

此时任务列表会显示刚才新建的任务，如图 12-54 所示。

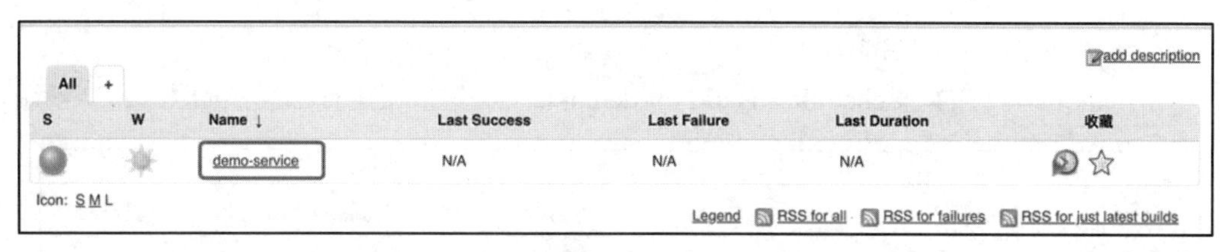

add description

All　+

S	W	Name ↓	Last Success	Last Failure	Last Duration	收藏
		demo-service	N/A	N/A	N/A	

Icon: S M L

Legend　RSS for all　RSS for failures　RSS for just latest builds

图 12-54

4. 配置任务运行节点

在 Jenkins 首页单击“Manage Jenkins”选项，选择“Manage Nodes”选项，显示如图 12-55 所示。

S	Name ↓	Architecture	Clock Difference	Free Disk Space	Free Swap Space	Free Temp Space	Response Time
	master	Linux (amd64)	In sync	116.67 GB	0 B	116.67 GB	0ms
	Data obtained	43 min	43 min	43 min	43 min	43 min	43 min

Refresh status

图 12-55

单击齿轮处的“Configure”按钮，在设置页面将“0”调整为“1”，并单击“Save”按钮，如图 12-56 所示。

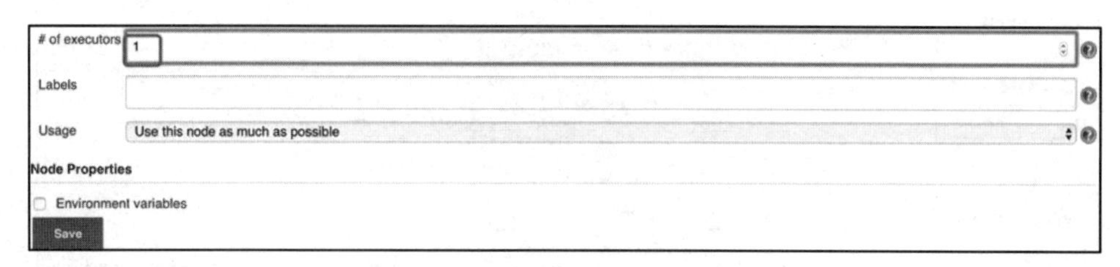

图 12-56

5. 部署应用

编译应用代码 HelloController：

```
@RestController
public class HelloController {
   @GetMapping("/")
   public String hello(){
      return "Hello Spring Boot in k8s using Jenkins";
   }
}
```

在 build.gradle 中修改应用的版本：

```
version = '0.0.2-SNAPSHOT'
```

提交代码并推送至 GitHub。

单击 Jenkins 首页的“Schedule a Build for demo-service”按钮进行部署，如图 12-57 所示。

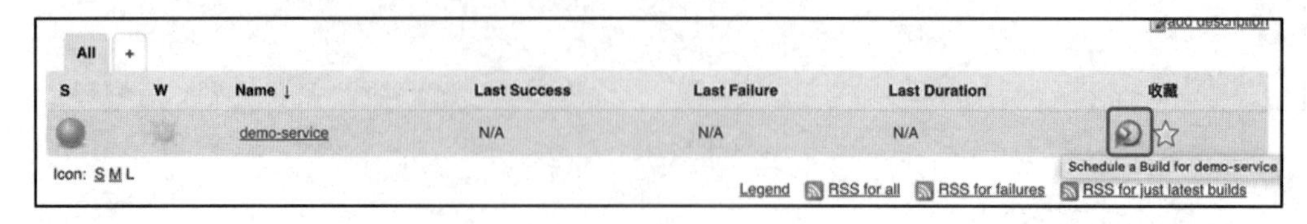

图 12-57

单击“demo-service”选项，左下角出现当前正在执行的任务，如图 12-58 所示。

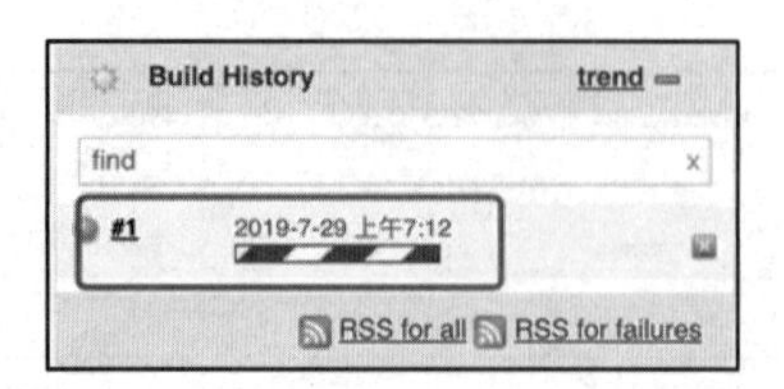

图 12-58

单击当前任务后选择“Console Output”选项，可以查看当前的进度详情，如图 12-59 所示。

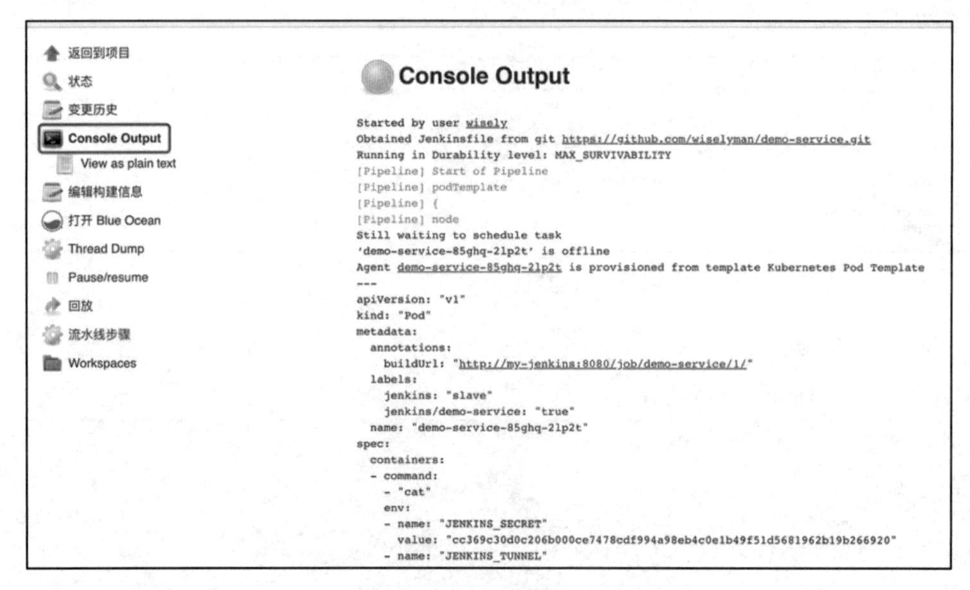

图 12-59

还可以通过安装的 Blue Ocean 插件，查看当前的任务进度。单击页面上的“打开 Blue Ocean”选项，可以更直观地观察任务进度，如图 12-60 所示。

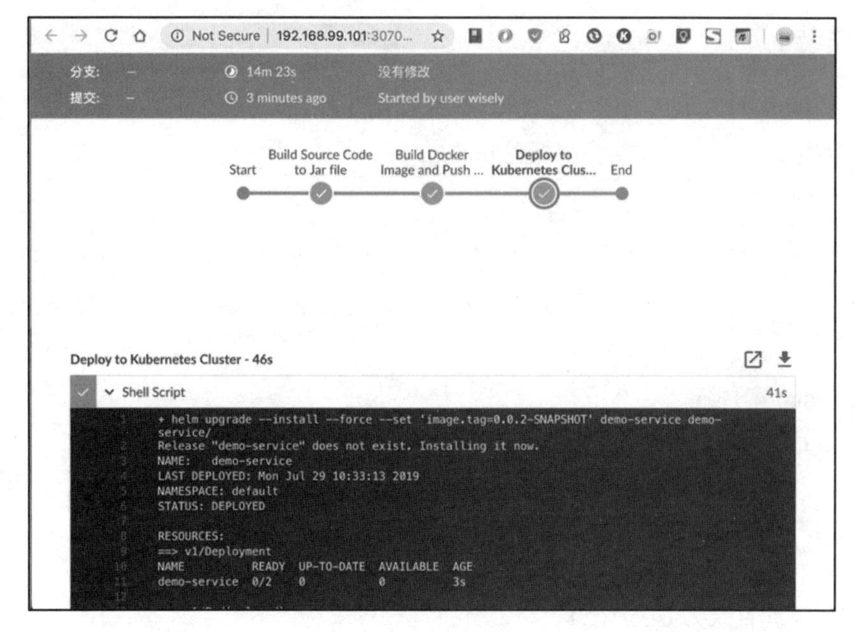

图 12-60

通过下面命令获得端口号：

```
$ kubectl get --namespace default -o jsonpath="{.spec.ports[0].nodePort}"
services demo-service
```

本例的端口号为 31260，访问 http://192.168.99.101:31260，如图 12-61 所示。

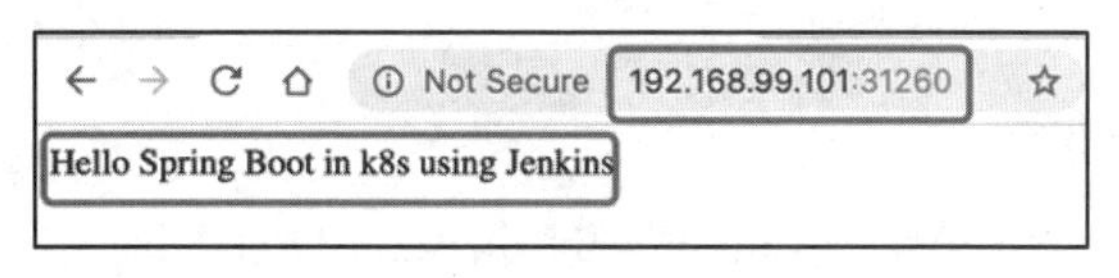

图 12-61

12.2 Service Mesh 和 Istio

Service Mesh，中文译为“服务网格”，它可以作为分布式微服务的基础设施（Spring Cloud 中的很多组件都是为了解决这些问题）。在使用服务网格后，无须通过编码、安装即可获得分布式微服务常用的组件。

使用 Istio 部署微服务，Istio 会向应用容器组（Pod，可以有多个容器）添加一个代理容器（Envoy）。通过代理容器，Istio 提供了统一的方式来集成微服务、管理微服务之间的网络流量，并对常用的微服务架构模式提供了支持。

12.2.1 安装 Istio

1. minikube 准备

为了安装 Istio，下面建立新的高配置 minikube：

```
$ minikube stop
$ minikube delete
$ minikube start --registry-mirror=https://*.mirror.aliyuncs.com --cpus 4
--memory 16384
$ minikube tunnel # 为 Istio 提供 Load Balancer，此句命令为阻碍操作，后续操作可打# 开
新的控制台
```

2. 安装 Istio 客户端

（1）下载 Istio 最新发行版。

访问 https://github.com/istio/istio/releases，下载 Istio 最新发行版，支持 Linux、macOS 和 Windows，请选择合适的版本下载。

（2）解压压缩包。

◎ 解压缩的目录为 istio-1.2.4。

◎ install/kubernetes/helm 为 Istio Chart 的源码。

◎ bin/istioctl 为 Istio 客户端。

（3）Istio 客户端。

◎ Windows：在“环境变量”中将 istioctl 路径添加到 Path 中。

◎ macOS/Linux：在 istio-1.2.4 目录下，执行下面的命令。

```
$ export PATH=$PWD/bin:$PATH
```

3. 使用 Helm 安装 Istio

（1）添加 Helm tiller 的 Service account，在 istio-1.2.4 目录下执行下面的命令：

```
$ kubectl apply -f install/kubernetes/helm/helm-service-account.yaml
```

（2）使用 tiller 账户安装 Tiller：

```
$ helm init --service-account tiller --upgrade -i
registry.cn-hangzhou.aliyuncs.com/google_containers/tiller:v2.14.2
```

（3）安装 istio-init chart：

```
$ helm install install/kubernetes/helm/istio-init --name istio-init --namespace
istio-system
```

（4）安装 istio chart：

```
$ helm install install/kubernetes/helm/istio --name istio --namespace
istio-system --set grafana.enabled=true --set tracing.enabled=true --set
kiali.enabled=true --set kiali.createDemoSecret=true
```

（5）确认安装成功。

安装时间较长，请耐心等待，可用下面的命令查看安装情况，如图 12-62 所示。

```
$ kubectl get pods -n istio-system #确保所有容器组的状态都是 Running
```

```
wangyuneideMBP3:istio-1.2.4 wangyunfei$ kubectl get pods -n istio-system
NAME                                      READY   STATUS      RESTARTS   AGE
grafana-6575997f54-qqd92                  1/1     Running     0          20m
istio-citadel-7fff5797f-94gml             1/1     Running     0          20m
istio-galley-74d4d7b4db-6v9h9             1/1     Running     0          20m
istio-ingressgateway-686854b899-zfpp8     1/1     Running     0          20m
istio-init-crd-10-d7tkf                   0/1     Completed   0          22m
istio-init-crd-11-vpctw                   0/1     Completed   0          22m
istio-init-crd-12-vfp4g                   0/1     Completed   0          22m
istio-pilot-7fdcbd6f55-5kst7              2/2     Running     0          20m
istio-policy-79f647bb6-7qwx6              2/2     Running     6          20m
istio-sidecar-injector-578bfd76d7-dlb6w   1/1     Running     0          20m
istio-telemetry-cb4486d94-l7xbr           2/2     Running     7          20m
istio-tracing-555cf644d-xpdxm             1/1     Running     0          20m
kiali-6cd6f9dfb5-95hxn                    1/1     Running     0          20m
prometheus-7d7b9f7844-gls5g               1/1     Running     0          20m
```

图 12-62

使用客户端命令查看版本，如图 12-63 所示。

```
$ istioctl version
```

```
wangyuneideMBP3:istio-1.2.4 wangyunfei$ istioctl version
client version: 1.2.4
citadel version: 1.2.4
galley version: 1.2.4
ingressgateway version: 94746ccd404a8e056483dd02e4e478097b950da6-dirty
pilot version: 1.2.4
policy version: 1.2.4
sidecar-injector version: 1.2.4
telemetry version: 1.2.4
```

图 12-63

（6）为 default 命名空间开启自动 sidecar 代理注入：

```
$ kubectl label namespace default istio-injection=enabled
```

12.2.2 微服务示例

1. 新建应用

Group：top.wisely。

Artifact：demo-istio-service。

Dependencies：Spring Web Starter 和 Spring Boot Actuator。

build.gradle 文件中的依赖如下：

```
dependencies {
    implementation 'org.springframework.boot:spring-boot-starter-actuator'
    implementation 'org.springframework.boot:spring-boot-starter-web'
  //……
}
```

新建一个 HelloController 类作为演示：

```
@RestController
public class HelloController {
   @GetMapping("/")
   public String hello(){
      return "Hello Spring Boot with Istio";
   }
}
```

2. Docker 镜像

（1）配置阿里云镜像仓库。

在阿里云“容器镜像服务”控制台的“默认实例”下的“镜像仓库”页面，单击“创建镜像仓库”，新建名为 demo-istio-service 的镜像仓库。同样，代码源选择“本地仓库”，可以通过命令行推送镜像到镜像仓库。

（2）编写 Dockerfile。

在应用根目录下添加 Dockerfile：

```
FROM java:8
RUN mkdir /app
ADD build/libs/demo-istio-service-*.jar /app/app.jar
ADD runboot.sh /app/
RUN bash -c 'touch /app/app.jar'
WORKDIR /app
RUN chmod a+x runboot.sh
EXPOSE 8080
CMD /app/runboot.sh
RUN echo "Asia/Shanghai" > /etc/timezone;
```

runboot.sh 中的内容如下：

```
$ java -Djava.security.egd=file:/dev/./urandom -jar /app/app.jar
```

（3）编译推送镜像到阿里云镜像仓库。

设置 Docker 编译环境。

登录阿里云镜像仓库，在应用根目录下执行下面的命令：

```
$ docker login --username=******@**.com registry.cn-hangzhou.aliyuncs.com
$ ./gradlew bootJar
$ docker build -t
registry.cn-hangzhou.aliyuncs.com/wiselyman/demo-istio-service:0.0.1-SNAPSHO
T .
$ docker push
registry.cn-hangzhou.aliyuncs.com/wiselyman/demo-istio-service:0.0.1-SNAPSHO
T
```

（4）向 Kubernetes 添加阿里云镜像仓库账号的保密字典：

```
$ kubectl create secret docker-registry regsecret
--docker-server=registry.cn-hangzhou.aliyuncs.com
--docker-username=******@**.com --docker-password=******
--docker-email=******@**.com
```

3. 部署微服务

下面编写 YAML 文件来部署微服务。将下面四项内容都放置在根目录下的 all.yml 中，每项内容之间用---隔开。

（1）Service：

```
apiVersion: v1
kind: Service
metadata:
  name: demo-istio-service
  labels:
    app: demo-istio-service
spec:
  ports:
    - name: http
      port: 8080
  selector:
    app: demo-istio-service
```

（2）Deployment：

```
---
apiVersion: apps/v1
kind: Deployment
metadata:
  name: demo-istio-service
spec:
  replicas: 1
  selector:
    matchLabels:
      app: demo-istio-service
  template:
    metadata:
      labels:
        app: demo-istio-service
        version: v1
    spec:
      imagePullSecrets:
        - name: regsecret
      containers:
        - name: demo-istio-service
          image:
registry.cn-hangzhou.aliyuncs.com/wiselyman/demo-istio-service:0.0.1-SNAPSHOT
```

```
        imagePullPolicy: IfNotPresent
        ports:
          - containerPort: 8080
        livenessProbe:
          httpGet:
            path: /actuator/health
            port: 8080
          initialDelaySeconds: 130
          timeoutSeconds: 10
          failureThreshold: 10
        readinessProbe:
          httpGet:
            path: /actuator/health
            port: 8080
          initialDelaySeconds: 60
          timeoutSeconds: 5
```

（3）Gateway：

```
---
apiVersion: networking.istio.io/v1alpha3
kind: Gateway
metadata:
  name: my-gateway
spec:
  selector:
    istio: ingressgateway
  servers:
    - port:
        name: http
        number: 80
        protocol: HTTP
      hosts:
        - '*'
```

（4）Virtual Service：

```
---
apiVersion: networking.istio.io/v1alpha3
kind: VirtualService
metadata:
  name: demo-istio-service
spec:
  hosts:
    - "*"
  gateways:
    - my-gateway
  http:
  - match:
      - uri:
          prefix: /demo-istio
```

```
  rewrite:
    uri: /
  route:
    - destination:
        host: demo-istio-service
        port:
          number: 8080
```

（5）部署微服务。

在应用根目录执行下面的命令：

```
$ kubectl apply -f all.yml
```

当然，也可以使用 Jenkins 自动化部署流程。在上面的流程中，是通过 lachlanevenson/k8s-helm 容器使用 Helm 命令部署的，也可以通过 lachlanevenson/k8s-kubectl 容器使用上面的命令进行部署。

（6）访问微服务。

通过下面的命令获取当前应用的访问地址，如图 12-64 所示。

```
$ kubectl get svc istio-ingressgateway -n istio-system
```

```
wangyuneideMBP3:demo-istio-service wangyunfei$ kubectl get svc istio-ingressgateway -n istio-system
NAME                   TYPE           CLUSTER-IP      EXTERNAL-IP     PORT(S)
                                                                                                 AGE
istio-ingressgateway   LoadBalancer   10.108.185.95   10.108.185.95   15020:30393/TCP,80:31380/TCP,443:31390/TCP,3140
0:31400/TCP,15029:30017/TCP,15030:31814/TCP,15031:32195/TCP,15032:31008/TCP,15443:30650/TCP    22m
```

图 12-64

在浏览器中输入 http://10.108.185.95/demo-istio，可访问微服务的演示控制器，如图 12-65 所示。

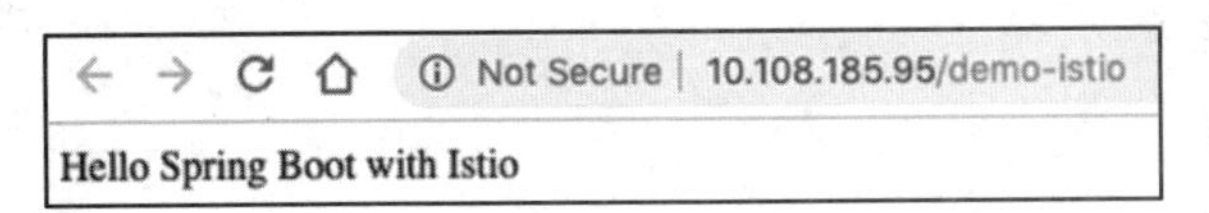

图 12-65

4. 使用 Kiali 监控

（1）转发端口：

```
$ kubectl port-forward \
    $(kubectl get pod -n istio-system -l app=kiali \
    -o jsonpath='{.items[0].metadata.name}') \
    -n istio-system 20001
```

（2）访问 Kiali（http://localhost:20001），账号和密码均为 admin，如图 12-66 和图 12-67 所示。

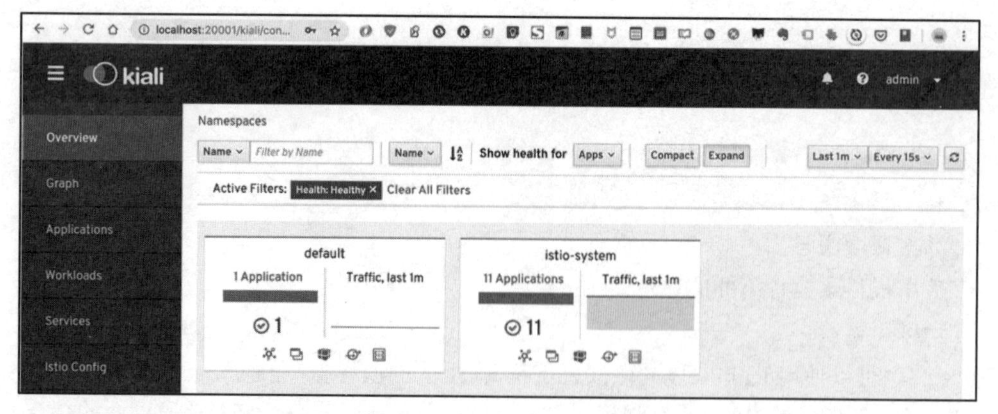

图 12-66

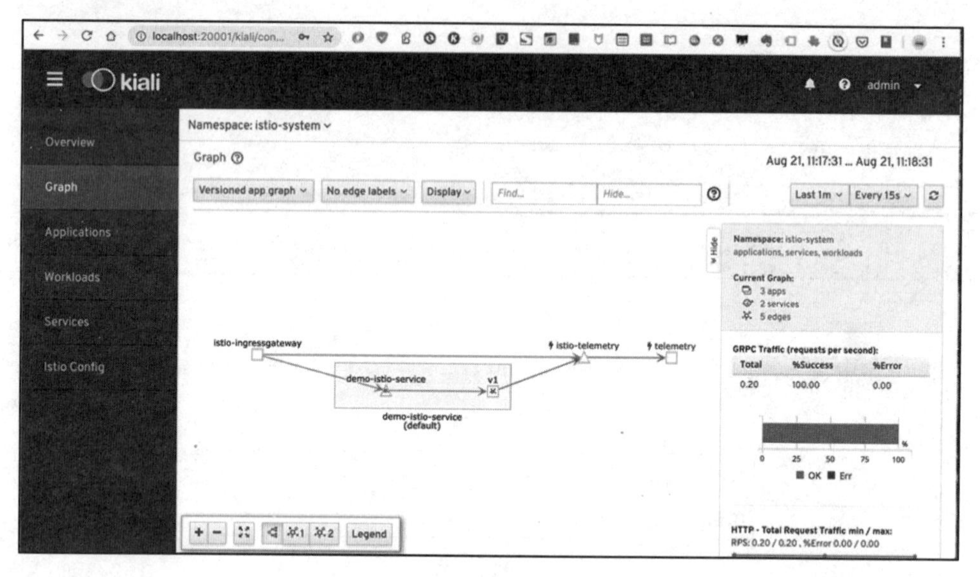

图 12-67

5. 使用 Grafana 监控指标

（1）转发 Grafana 端口：

```
$ kubectl -n istio-system port-forward \
   $(kubectl -n istio-system get pod -l app=grafana \
   -o jsonpath='{.items[0].metadata.name}') 3000
```

（2）访问 Grafana：http://localhost:3000，单击“Home”→“istio”选项，如图 12-68 所示。单击“Istio Service Dashboard”选项，如图 12-69 所示。

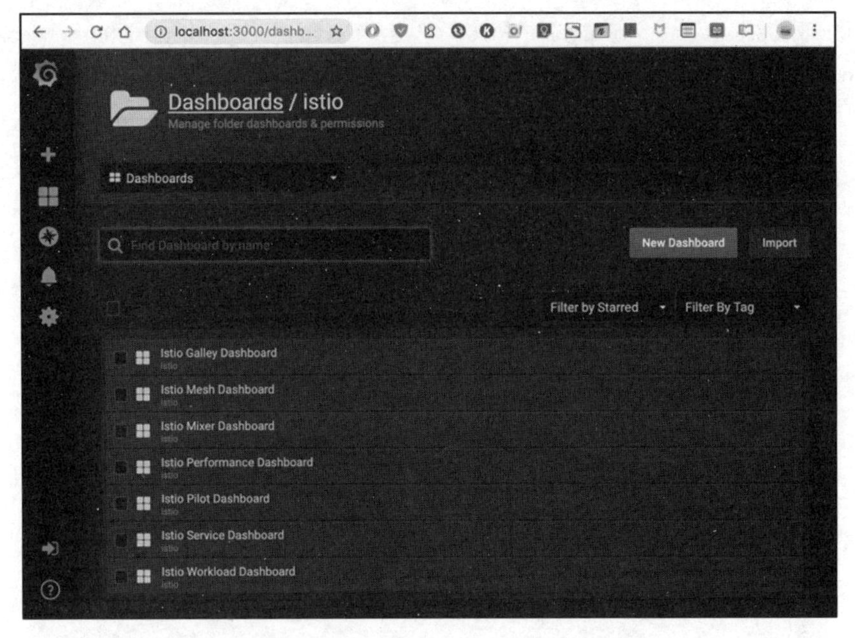

图 12-68

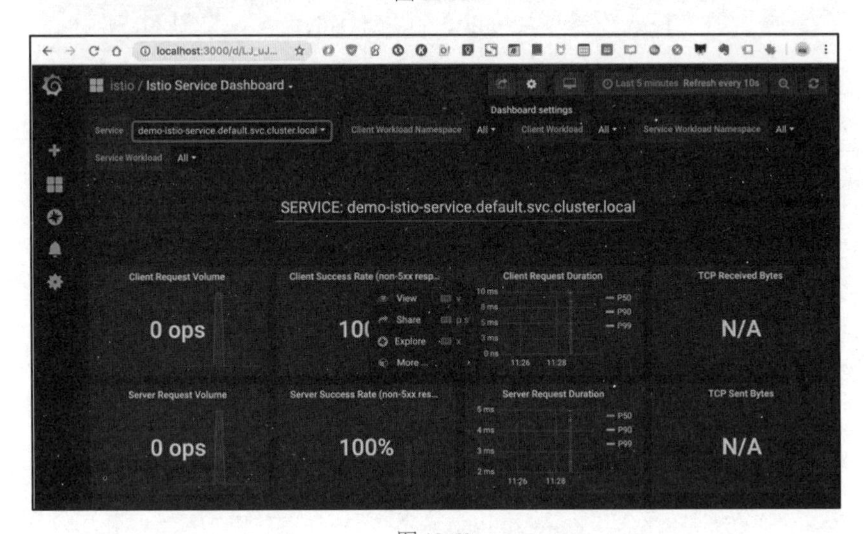

图 12-69

6. 使用 Jaeger 进行链路追踪

（1）转发 Jaeger 端口：

```
$ kubectl port-forward -n istio-system \
   $(kubectl get pod -n istio-system -l app=jaeger \
   -o jsonpath='{.items[0].metadata.name}') 16686
```

（2）访问 Jaeger：http://localhost:16686，单击“Service”→“demo-istio-service.default”→“Find Traces”选项，如图 12-70 所示。

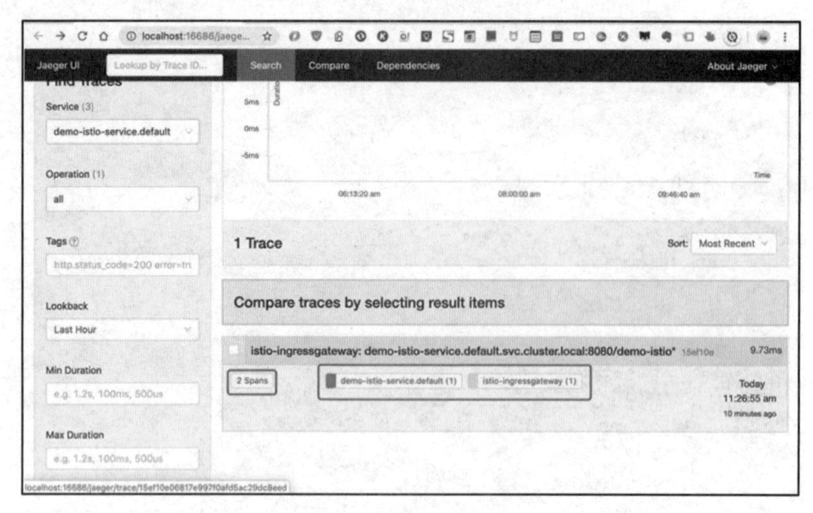

图 12-70

单击图 13-69 中的“Trace”选项，可以看到更详细的信息，如图 12-71 所示。

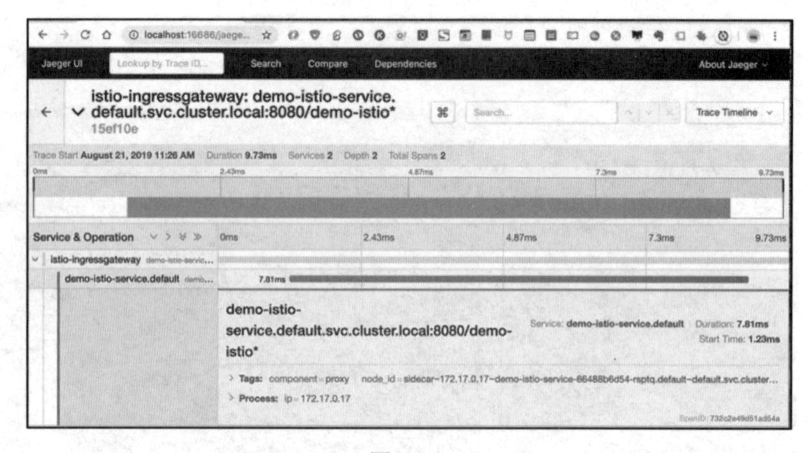

图 12-71

12.3 小结

本章学习了 Kubernetes 和 Jenkins 的基础知识，通过 Jenkins 的自动化流程，将 Spring Boot 应用开发和部署运维串联起来。这样，开发和部署运维就成为了有机的整体，而不是割裂的彼此，从而可以极大地提高开发效率。